S0-ABY-199

Animal Nutrition and Feeding

James R. Gillespie

DELMAR PUBLISHERS INC. ®

DELMAR STAFF

Developmental Editor: Marjorie A. Bruce
Production Editor: Cynthia Haller

For information, address Delmar Publishers, Inc.
2 Computer Drive West, Box 15-015
Albany, NY 12212-9985

Interior design by Rebecca Evans, ips Publishing, Inc. Composition by ips Publishing, Inc., Laura Welch, director.
Cover Photo: Grant Heilman Photography

COPYRIGHT © 1987
BY DELMAR PUBLISHERS INC.

All rights reserved. No part of this work covered by the copyright hereon may be reproduced or used in any form or by any means—graphic, electronic, or mechanical, including photocopying, recording, taping, or information storage and retrieval systems—without written permission of the publisher.

Printed in the United States of America
Published simultaneously in Canada
by Nelson Canada,
a division of International Thomson Limited

10 9 8 7 6 5 4 3 2 1

Library of Congress Cataloging-in-Publication Data

Gillespie, James R.
    Animal nutrition and feeding.

    Includes index.
    1. Animal nutrition. I. Title.
SF95.G55 1987   636.08'52   86-8982
ISBN 0-8273-2354-9
ISBN 0-8273-2355-7 (instructor's guide)

CONTENTS

Index of Tables
Preface

1
INTRODUCTION TO ANIMAL NUTRITION

The agricultural industry,   1
Livestock and poultry feeding,   1
Occupations that relate to livestock,   7
Functions served by animals,   8

2
DIGESTION IN ANIMALS

Digestive systems,   14
Nonruminant (monogastric) digestive
   system,   14
Ruminant digestive system,   17
Avian digestive system,   21
Absorption of nutrients,   21
Metabolism,   22
Nutrient transport,   22

3
ENERGY NUTRIENTS

Energy terminology,   26
Energy nutrients,   27
Carbohydrates,   27
Lipids (fats and oils),   30
Sources of energy—concentrates,   32
Sources of energy—forages,   34
Sources of energy—byproducts,   34
Sources of energy—fats,   35
Sources of energy—molasses,   35
Functions of energy nutrients,   35
Effects of deficiency of energy,   36
Energy requirements in the ration,   37

4
PROTEIN

Proteins,   39
Amino Acids,   40
Limiting Amino Acid,   43
Functions of proteins,   44
Protein deficiency,   44
Protein as energy source,   45
Unavailable feed protein,   45
Protein solubility,   46
Biological value of protein,   46
Plant protein supplements,   46
Fat extraction methods,   47
Soybeans as a protein source,   47
Urea,   47
Urea Toxicity,   48
Nitrates/nitrites,   48
Other nonprotein nitrogen
   sources,   49
Byproduct feeds as protein
   sources,   49
Grass and legume forages as protein
   sources,   50
Grain as a protein source,   50

5
MINERALS

Minerals defined,   53
Functions of minerals,   53
Deficiency symptoms,   54
Sources of minerals,   54
Minerals in the ration,   54
Calcium,   55
Phosphorus,   58

iii

Calcium/phosphorus ratio, 59
Salt (Sodium and Chlorine), 59
Potassium, 60
Magnesium, 61
Sulphur, 62
Iron, 63
Manganese, 63
Copper, 64
Zinc, 65
Molybdenum, 66
Selenium, 66
Cobalt, 67
Iodine, 68
Fluorine, 69
Chromium, 70
Silicon, Nickel, Strontium, Tin,
   Vanadium, 70
Aluminum, 70
Boron, 70
Toxic minerals, 71

6

VITAMINS, FEED ADDITIVES, AND WATER

Vitamins, 75
Vitamin A, 76
Vitamin D, 79
Vitamin E, 81
Vitamin K, 82
B Complex Vitamins, 83
Vitamin C (Ascorbic Acid), 89
Feed additives, 90
Water, 93

7

CLASSIFICATION AND USE OF FEEDS

Classification of feeds, 100
International feed nomenclature, 100
International feed classes, 101
International feed number, 101

Forms of feed, 101
Roughages, 103
Concentrates, 103
Uses of feed in an animal's body, 104
Underutilized sources of feed, 108

8

FEED QUALITY AND FEED ANALYSIS

Feed quality, 115
Harvesting hay to maintain
   quality, 115
Production of quality silage, 115
Collecting feed samples for
   analysis, 121
Proximate analysis of feedstuffs, 123
Limitations of proximate analysis, 125
Feed composition basis, 126
Forage evaluation by Van Soest
   method, 127
Digestion trials, 128
Determining net energy values of
   feed, 128
Other measures of ration value, 129
Feeding trials, 130
Feed regulations, 130

9

RATIONS: SELECTING, BALANCING, MIXING

Formulating rations, 133
Economics of ration selection, 134
General principles for balancing
   rations, 135
Steps in balancing a ration, 136
Use of the Pearson Square to balance
   a ration, 137
Use of algebraic equations to balance
   rations, 139

Using fixed ingredients when
formulating diets, 141
Using computers to formulate
rations, 142
Preparation of feeds, 143
Mechanical processing, 143
Heat treatment processing of
grains, 146
Feed mixing, 146
Feed storage, 148

10

ENVIRONMENT AND NUTRITION

Effective ambient temperature, 153
Feed intake, 155
Effect of temperature on forage
intake, 155
Water and the environment, 155
Effect of environment on nutritional
efficiency, 157
Adjusting nutrition in relation to
environment, 158

11

FEEDING BEEF CATTLE

General guidelines for balancing
beef rations, 165
Pasture for beef breeding herd, 170
Creep-feeding calves, 171
Growth implants, 172
Backgrounding calves, 173
Feeding replacement heifers, 175
Feeding bulls, 176
Finishing cattle, 178
Urea in finishing rations, 181

12

FEEDING DAIRY CATTLE

Nutrient requirements, 185
Methods of feeding dairy cows, 186
Guidelines for feeding lactating dairy
cows, 190
Guidelines for feeding dry cows, 192
Feeding herd replacements, 192
Balancing rations for dairy cattle, 195
Feeding and reproduction, 201

13

FEEDING SWINE

Nutrient requirements, 204
Selecting feeds for swine, 206
Roughages, 210
Minerals, 210
Vitamins, 211
Additives, 211
Feeding the breeding herd, 211
Feeding baby pigs, 214
Feeding market pigs, 215

14

FEEDING SHEEP

Nutritional needs, 219
Feeding the breeding flock, 221
Feeding lambs, 225
Feeding replacement ewes, 228

15

FEEDING GOATS

Nutrient requirements, 231
Herbage and browse utilization, 233
Dairy goats, 233
Angora goats, 236

16

FEEDING HORSES

Nutrient requirements—energy, 239
Nutrient requirements—protein, 240
Nutrient requirements—
    minerals, 241
Nutrient requirements—vitamins, 241
Nutrient requirements—water, 242
Roughage, 242
Grain, 248
Protein supplements, 249
Pelleting rations, 250
Antibiotics, 251
Feeding horses—general, 251
Feeding breeding horses, 253
Feeding growing horses, 255

17

FEEDING POULTRY

Nutrient requirements of poultry, 261
Feeds for poultry, 265
Feed preparation, 266
Preparing rations, 266
Feeding management, 267
Grit, 267
Feeding replacement pullets, 267
Feeding laying hens, 269
Feeding broilers, 270
Feeding turkeys, 273
Feeding ducks and geese, 275

18

MYCOTOXINS

Mycotoxins—definition, 278
General effects of mycotoxins, 278
Incidence of mycotoxicoses in the
    United States, 280
Ochratoxin, 280

Pasture grass mycotoxins, 280
Fusarium toxins, 282
Slobber, 284
Aflatoxin, 284
Detecting mycotoxins in feed, 286
Mycotoxin residues, 286
Using feeds containing
    mycotoxins, 287

BIBLIOGRAPHY, 293

GLOSSARY, 302

PROBLEM SECTION, 311

APPENDIX

Table 1.   Nutrient Requirements for Growing-Finishing Steer Calves and Yearlings (Daily Nutrients per Animal),   316

Table 2.   Nutrient Requirements for Growing-Finishing Heifer Calves and Yearlings (Daily Nutrients per Animal),   318

Table 3.   Nutrient Requirements for Beef Cattle Breeding Herd (Daily Nutrients per Animal),   320

Table 4.   Nutrient Requirements for Growing-Finishing Steer Calves and Yearlings (Nutrient Concentration in Diet Dry Matter),   322

Table 5.   Nutrient Requirements for Growing-Finishing Heifer Calves and Yearlings (Nutrient Concentration in Diet Dry Matter),   324

Table 6.   Nutrient Requirements for Beef Cattle Breeding Herd (Nutrient Concentration in Diet Dry Matter),   326

Table 7.   Daily Nutrient Requirements of Sheep (100% Dry Matter Basis),   328

Table 8.   Nutrient Content of Diets for Sheep (Nutrient Concentration in Diet Dry Matter),   330

Table 9.   Daily Nutrient Requirements of Goats,   332

Table 10.   Nutrient Requirements of Growing-Finishing Swine Fed *Ad Libitum* (Percent or Amount per Kilogram of Diet),   334

Table 11.   Daily Nutrient Requirements of Growing-Finishing Swine Fed *Ad Libitum,*   335

Table 12.   Nutrient Requirements of Breeding Swine: Percent or Amount per Kilogram of Diet,   336

Table 13.   Daily Nutrient Requirements of Breeding Swine,   337

Table 14.   Nutrient Requirements of Horses (Daily Nutrients per Horse), Ponies, 200 kg (440 lb) Mature Weight,   338

Table 15.   Nutrient Requirements of Horses (Daily Nutrients per Horse), 400 kg (880 lb) Mature Weight,   339

Table 16.   Nutrient Requirements of Horses (Daily Nutrients per Horse), 500 kg (1,100 lb) Mature Weight,   340

Table 17.   Nutrient Requirements of Horses (Daily Nutrients per Horse), 600 kg (1,320 lb) Mature Weight,   341

Table 18. Nutrient Concentration in Diets for Horses and Ponies Expressed on 100 percent Dry Matter Basis, **342**

Table 19. Nutrient Concentration in Diets for Horses and Ponies Expressed on 90 percent Dry Matter Basis, **343**

Table 20. Daily Nutrient Requirements per Female (Single-comb White Leghorns and Similar Breeds), **344**

Table 21. Daily Nutrient Requirements per Chicken (Chickens of Broiler Strains), **346**

Table 22. Protein and Amino Acid Requirements of Egg-Type and Meat-Type Chickens, **348**

Table 23. Protein and Amino Acid Requirements of Broilers, **349**

Table 24. Vitamin, Linoleic Acid, and Mineral Requirements of Chickens (In Percentage or Amount per Kilogram or Pound of Feed—As-fed Moisture Basis), **350**

Table 25. Protein and Amino Acid Requirements of Turkeys, **351**

Table 26. Vitamin, Linoleic Acid, and Mineral Requirements of Turkeys (In Percentage or Amount per Kilogram or Pound of Feed—As-Fed Moisture Basis), **352**

Table 27. Nutrient Requirements of Ducks (In Percentage or Amount per Kilogram or Pound of Feed—As-fed Moisture Basis), **353**

Table 28. Nutrient Requirements of Pheasants and Quail (In Percentage or Amount per Kilogram or Pound of Feed—As-fed Moisture Basis), **354**

Table 29. Nutrient Requirements of Geese (In Percentage or Amount per Kilogram or Pound of Feed—As-fed Moisture Basis), **356**

Table 30. Feed Composition, **358**

Table 31. Amino Acid Composition of Common Feed, **370**

Table 32. Conversion Factors, **374**

Table 33. Abbreviations used in Table 32, **377**

Table 34. Daily Nutrient Requirements of Dairy Cattle (Metric System), **378**

Table 35. Daily Nutrient Requirements of Lactating and Pregnant Cows (Metric System), **382**

Table 36. Recommended Nutrient Content of Rations for Dairy Cattle (Metric System), **383**

Table 37. Maximum Dry Matter Intake Guidelines, **384**

Table 38.   Daily Nutrient Requirements of Dairy Cattle (Pounds),   385

Table 39.   Daily Nutrient Requirements of Lactating and Pregnant Cows (Pounds),   389

Table 40.   Recommended Nutrient Content of Rations for Dairy Cattle (Pounds),   390

Table 41.   Feed Composition—Fat and Fatty Acids,   391

Table 42.   Feed Composition—Mineral Supplements,   392

UNIFORM STATE FEED BILL,   397
INDEX,   412

INDEX OF TABLES

Table 1-1.   Kinds and quantities of feed consumed by livestock and poultry, feeding years 1965-66 and 1979-80. (Measured in feed units <corn equivalents>.)   2

Table 1-2.   Wet raw manure produced by different animals.   11

Table 2-1.   Summary of digestion.   23

Table 2-2.   Capacities of digestive system of selected species (ranges indicate different ages, breeds, sizes).   25

Table 4-1.   Essential and nonessential amino acids for swine and poultry.   43

Table 5-1.   Maximum tolerable levels of dietary minerals for domestic animals.   56

Table 6-1.   Conversion of beta-carotene to vitamin A for different animal species.   78

Table 6-2.   Recommended limits of concentration of some potentially toxic substances in drinking water for livestock.   95

Table 7-1.   Feed names using applicable descriptors.   101

Table 7-2.   Stage of maturity terms for plants.   102

Table 7-3.   Feed classes.   104

Table 7-4.   Normal temperatures of farm animals.   107

Table 11-1.   Rations for wintering dry, pregnant beef cows (wt. 1,000 to 1,100 lb; 453 to 499 kg).   166

Table 11-2.   Wintering rations for bred heifers (wt. 800-900 lb; 363-408 kg).   167

Table 11-3.   Lactating rations for cows in drylot (wt. 1,000 lb; 499 kg).   168

Table 11-4.   First-calf heifers—lactation rations, drylot.   169

Table 11-5.   Mixtures for creep feeding calves—self-fed (100 lb. (45 kg) mix).   173

Table 11-6.   Rations for growing replacement heifers (450-500 lb; 204-227 kg, expected gain 1 to 1.25 lb (0.45-0.56 kg) daily).   175

Table 11-7.   Feed mixtures for weaned bull calves (weaning to about 700 lb (318 kg)).   176

Table 11-8.   Feed mixtures for bulls over 700 lb (318 kg).   177

Table 11-9. Beef cattle rations for cattle on full feed. Steer calves—initial weight 400-450 pounds (181-204 kg). **180**

Table 14-1. Rations that may be self-fed to ewes. **223**

Table 15-1. Rations and concentrate mixtures for lactating dairy goats. **235**

Table 16-1. Stallion (breeding season)—sample concentrate mixes. **254**

Table 16-2. Pregnant mares—sample concentrate mixes. **254**

Table 16-3. Lactating mares—sample concentrate mixes. **255**

Table 16-4. Foals, creep feeding—sample concentrate mixes. **255**

Table 16-5. Foals, six months to one year—sample concentrate mixes. **256**

Table 16-6. Yearlings and two-year-olds—sample concentrate mixes. **256**

Table 16-7. Mature working horses—sample concentrate mixes. **257**

Table 17-1. Replacement pullets, egg laying breeds—example feed formulations. **268**

Table 17-2. Suggested lighting program for replacement pullets—egg laying breeds. **269**

Table 17-3. Suggested feeder and waterer space for replacement layers. **269**

Table 17-4. Suggested feeder and waterer space for laying chickens. **270**

Table 17-5. Example mash formulation for laying hens. **271**

Table 17-6. Suggested feeder and waterer space for broilers. **272**

Table 17-7. Suggested feed formulations for broilers. **272**

Table 18-1. Mycotoxicoses commonly found in the United States. **279**

PREFACE

CONTENT AND ORGANIZATION

ANIMAL NUTRITION AND FEEDING is a comprehensive text designed to present the principles of animal nutrition, types of feeds and feeding guidelines. It is recommended for animal nutrition courses and for animal science programs where nutrition and feeding is a segment of the program. The text addresses the major environmental and human health issues facing those producing animals and animal products for human consumption.

The information presented in the text is based on the latest data available from research being conducted at universities, agricultural research stations, the USDA, and the National Research Council.

Following an introduction comparing the livestock industry to the other areas of agricultural production, and stressing the importance of the industry to the economy, Chapter 2 reviews the digestive systems and the digestive process for nonruminants, ruminants and avians. A thorough explanation is provided of how nutrients are absorbed and utilized by various species of animals. Chapters 3-8 provide an in-depth examination of feed nutrients, feed classification, use, quality and analysis. Several methods of analyzing feeds are presented, including proximate analysis, the Kjeldahl process, and the Van Soest method. Both digestion trials and feeding trials are explained in detail. The importance of the information gained from these trials is stressed as it applies to determining practical rations for farm use. The section concludes with a discussion of regulations relating to feeds, including the Uniform State Feed Bill (given in the Appendix), published by the Association of the American Feed Control Officials.

Chapter 9, on selecting, balancing and mixing rations, describes the general principles of ration formulation and the economics of providing rations meeting animal nutritional needs. Several methods of balancing rations are presented (with examples), including the Pearson Square, use of algebraic equations, and the use of computers. The chapter concludes with a detailed explanation of the preparation of feeds by mechanical processing and heat treatment. Feed storage is also thoroughly described.

Chapter 10 discusses the influence of environmental factors on nutrition. Environmental stress alters the efficiency of energy use by animals. Adjustments in feed formulation may be required to counteract the effects of air temperature, air movement, humidity, precipitation, drought and heat radiation.

Chapters 11-17 present information for feeding specific species of livestock including beef cattle, dairy cattle, swine, sheep, goats, horses and poultry. Each chapter covers general guidelines for rations for the particular species and for special populations within the species (such as feeding calves and feeding bulls, or feeding laying hens and feeding broilers, and so on). Also described are the nutrient requirements of the species and recommendations for feeding.

Chapter 18 discusses the effects of mycotoxins when ingested by animals from feeds. The primary effect of mycotoxins in animals is lowering of productivity. The chapter discusses the sources and effects of the following mycotoxins: aflatoxin, ochratoxin, pasture grass toxins, fusarium toxins and trochothecenes.

A bibliography provides additional sources of information for livestock nutrition and feeding. Animal nutrition terms are defined in the glossary.

The Problem Section contains additional problems to be solved by students, including proximate analysis problems and ration balancing problems. The Appendix consists of 42 tables containing current feed composition and nutrient requirement data from the National Research Council.

FEATURES OF THE TEXT

- Objectives are provided at the beginning of each chapter to focus attention on the information and skills to be learned.
- Numerous tables are presented within the chapters to summarize ration information and recommendations.
- Each chapter concludes with a summary to highlight the important concepts and recommendations.
- Feeding recommendations are provided by species and for specific subpopulations within the species.
- The latest feed and ration data is presented from a variety of sources, including universities, agricultural research stations, the USDA, and the National Research Council.
- Ration formulation using computers is discussed with recommendations for the use of electronic spreadsheets with templates.
- Environmental considerations and their effects on animal feeding and energy utilization are described.
- A separate chapter discusses mycotoxins, their harmful effect on livestock, and guidelines for minimizing the presence of mycotoxins in feeds.
- Each chapter contains numerous review questions, and problems where appropriate, to test student comprehension of the chapter content.
- The text discusses feed additives and the potential for carryover in animal products for human consumption.
- The Appendix contains 42 tables specifying nutrient requirements for livestock and feed composition.
- The Uniform State Feed Bill is presented in its entirety after the Appendix.

The Instructor's Guide contains an outline of each chapter, answers to the review questions at the end of each chapter, solutions to problems included in the text, comprehensive test questions covering the text content, answers to the test questions, solutions to the problems contained in the Appendix of the text, and transparency masters for instructor use.

ABOUT THE AUTHOR

James R. Gillespie, the author of the text, has extensive training and experience in the field of livestock production and agriculture education. He received a BS and MS degree in Agricultural Education from Iowa State University and also holds a degree as Educational Specialist, School Administration, from Western Illinois University. Mr. Gillespie has taught vocational agriculture courses at the high school and adult education levels. In addition to other agriculture-related positions, Mr. Gillespie has also been self-employed in farming. He is currently employed by the Illinois State Board of Education, Department of Adult, Vocational and Technical Education, Program Approval and Evaluation Section as Regional Vocation Administrator, Region II. Mr. Gillespie is a member of the Illinois Council of Local Administrators, the Illinois Vocational Association, the Honor Society of Phi Kappa Phi, and Phi Delta Kappa.

ACKNOWLEDGEMENTS

The author wishes to express appreciation to the following individuals for their critiques of the manuscript in various stages. Their recommendations provided valuable guidance in refining the text content.

For evaluating the table of contents and early drafts of portions of the manuscript:

Bruce Moos, Linn Benton Community College, Albany, OR 97321
Zachary Estes, Western Oklahoma State College, Altus, OK 73521
Sherrill Baumgartner, Harcum Junior College, Bryn Mawr, PA 19010
Glenn Lyons, Western Iowa Technical Community College, Sioux City, IA 51102
Charles St. Jean, Columbus Technical Institute, Columbus, OH 43216
Ronald P. Boulton, Hennepin Technical Centers, Brooklyn Park, MN 55455
Terry N. Teeple, Fort Steilacoom Community College, Tacoma, WA 98498
Alan Snedegar, Hawkeye Institute of Technology, Waterloo, IA 50704
D.W. Scheid, Madison Area Technical College, Madison, WI 53703
Arthur Brieske, Western Wisconsin Technical Institute, Lacrosse, WI 54601
Park C. Romney, Eastern Arizona College, Thatcher, AZ 85552
Albert Adams, North Iowa Community College, Mason City, IA 50401
Dale Diedrich, District One Technical Institute, Eau Claire, WI 54701

For providing detailed revisions of the entire manuscript:

Glenn Wehner
North East Missouri State University
Kirksville, MO 63501

Dwight Hurley
Vernon Regional Junior College
Vernon, TX 76384

Robert Ritter
University of Minnesota Technical
   College
Waseca, MN 56093

Thomas C. Jenkins
Agricultural Technical Institute
The Ohio State University, Wooster,
   OH 44691

William L. Vandegrift
The University of Georgia
Athens, GA 30602

Edward Leal
Modesto Junior College
Modesto, CA 95350

1

INTRODUCTION TO ANIMAL NUTRITION

OBJECTIVES

After completing this chapter you will be able to:

- Describe the importance of livestock feeding in today's agriculture.
- List facts relating to the kinds and quantities of feeds fed to livestock and poultry.
- List occupations in which a knowledge of animal nutrition is needed.
- Describe the uses of animals.

THE AGRICULTURAL INDUSTRY

Agriculture is the largest industry in the United States. The total assets of agriculture exceeds one trillion dollars, which is equal to approximately 70 percent of the capital assets of all manufacturing corporations in the country.

The production of livestock is an important part of the total agricultural industry. About one-half of farm cash receipts come from the sale of livestock and livestock products, with the rest coming from the sale of crops. More than $106.3 billion is spent yearly for major farm production expenses. Of this total, approximately $18.5 billion is spent for purchased feed and $10.5 billion for purchased livestock, representing approximately 27.3 percent of the total amount spent for major farm production expenses.

LIVESTOCK AND POULTRY FEEDING

Feed Costs

The cost of feed represents from one-half to more than three-fourths of the total cost of raising livestock. Careful attention to animal nutrition can help reduce feed costs and thus increase the potential profit from livestock. However, the lowest-cost rations may not be the most profitable. Feeding efficiency must be considered when selecting the feeds to include in a ration.

It is estimated that 65 to 80 percent of the total cost of production in a swine enterprise is for feed. Poultry feed costs range from 55 percent (layers) to 65 percent (broilers and turkeys) of the total cost. Dairy feed costs range from 50 to 60 percent of the total cost of milk production. Feed costs for a cattle finishing operation are approximately 70 percent of the total cost of feeding cattle. Feed for feeder lambs represents about 50 percent of the total cost of production.

Research and Feeding

The work of research scientists in agriculture has resulted in greatly improved methods of feeding livestock. Growing knowledge of animal nutrition has brought tremendous progress in feed technology. Genetic improvement in livestock and poultry has also contributed to more efficient feeding. Bulk formulation, mixing, transporting, and distribution of feeds have improved because of knowledge gained through research.

Labor

The amount of labor needed to produce livestock and poultry has declined. Reduced labor requirements, along with improved technology, have resulted in the development of large livestock and poultry feeding operations in the United States.

Disease Control

Improved feed technology has also resulted in better control and eradication of disease. It is possible to safely administer drugs through feed because of the development of precision feed metering and mixing equipment.

Two Groups of Feeds

The two general groups of feeds used in livestock rations are concentrates and roughages. *Concentrates* are feeds that are low in fiber and high in total digestible nutrients. Concentrates include grains such as corn, oats, barley, wheat, and rye. Protein feeds such as soybean meal, cottonseed meal, linseed meal, meat and bone meal, fish meal, and tankage are also concentrates.

Concentrates provide approximately 40 percent of the feed used, harvested roughages provide 19 percent, and pasture and rangeland provide 41 percent. About one-half of the concentrates fed are processed as commercial formula feeds. Table 1-1 shows the kinds and quantities of feed consumed by livestock and poultry.

Table 1-1. Kinds and quantities of feed consumed by livestock and poultry, feeding years 1965-66 and 1979-80. (Measured in feed units <corn equivalents>.)

Feed Materials	1965-66 feeding year (1,000 tons)	Percent of total	1979-80 feeding year (1,000 tons)	Percent of total
Grains:				
Corn	81,540	16	127,234	23
Other feedgrains	32,090	7	24,438	5
Wheat and rye	3,584	1	2,673	1
Protein feeds	31,725	6	44,248	8
Byproduct feeds	9,072	2	8,048	2
Other	2,407	1	2,974	1
Total concentrates.	160,418	33	209,615	40
Hay	49,403	10	63,000	12
Other harvested roughages	26,300	5	37,800	7
Pasture	249,144	52	211,254	41
Total roughage	324,847	67	312,054	60
Total, all feeds	485,265	100	521,669	100

Source: *USDA, Fact Book of U.S. Agriculture,* Miscellaneous Publication Number 1063, November 1981.

The use of corn as a feed has increased in recent years compared to other feedgrains, wheat, and rye, Figure 1-1. The use of all concentrates has increased, while the use of roughages for livestock feed has decreased. However, the current trend in adult ruminant (cattle and sheep) feeding is toward the increased use of pasture and roughages.

Roughages are feeds that are high in fiber and low in total digestible nutrients. Roughages include hay, pasture, and silage, and may be grasses or legumes, Figure 1-2.

Important Feed Components

Six feed components are important when balancing rations for livestock: carbohydrates, lipids, protein, minerals, vitamins, and water. It is necessary to match the nutrients provided by the feeds to the requirements of the animal to have a properly balanced ration. Properly balanced rations are necessary to maximize growth, production, and economic return. Tables of nutrient requirements and feed composition are provided in the Appendix of this text.

Figure 1-1. Corn is an important feed grain for livestock feeding. It may also be harvested as silage to provide roughage in the ration.

Figure 1-2. Red clover is a legume which may be used for hay or pasture to provide roughage in the ration.

Digestive Systems

The digestive system of the animal species being fed must be considered when balancing rations. *Ruminant* animals have a multi-compartmented stomach and are able to utilize large quantities of roughage in their diet. *Nonruminants* have single-compartment stomachs and require higher levels of concentrates in their diet, Figures 1-3 and 1-4. Cattle and sheep are examples of ruminants, swine and poultry are examples of nonruminants. Horses are nonruminants but, because of a greatly enlarged cecum in their digestive system, they are able to utilize large quantities of roughage in their diet, Figure 1-5.

Energy

Carbohydrates, lipids, and excess protein provide energy for animals. A minimum level of energy is required for maintenance of body functions. Additional energy is needed for growth, reproduction, gain, production of milk and fiber, and work. Feed energy may be measured by total digestible nutrients, net energy, digestible energy, and metabolizable energy. Efficient production depends on animal energy requirements being met through properly balanced rations.

Protein

The need for protein varies with the age, sex, body weight, and productivity of the animal. Protein needs increase with increased productivity. Rations need to be properly balanced to meet the protein requirements of the animal.

Nonruminant animals have a requirement for amino acids, which are the component parts of protein. There are ten essential amino acids needed in swine diets and fourteen essential amino acids needed in poultry diets. If the ration meets the amino acid requirements of the nonruminant animal, it is adequate regardless of its protein level.

Ruminant animals are able to synthesize amino acids in the rumen and thus have a requirement for protein or nitrogen in the ration. Synthetic proteins which supply nitrogen, such as urea, may be used by ruminant animals to meet their amino acid requirements.

Figure 1-3. Swine are nonruminants and require large amounts of concentrates in their ration.

Figure 1-4. Poultry utilize mainly concentrates in their diet.

Figure 1-5. Horses can utilize roughage in their diet because they have an enlarged cecum in their digestive system.

Minerals

All animals require minerals in the diet; the major minerals are calcium, phosphorus, and salt (sodium chloride). Calcium and phosphorus must be included in the diet in the proper ratio to each other. Other minerals that are needed in small quantities are called *trace minerals*. These are often added to the ration through the use of trace-mineralized salt.

Vitamins

Ruminants are able to synthesize most of the vitamins they require, with the exception of vitamin A, which needs to be added to the ration. Nonruminants generally need vitamins A, D, E, K, B_{12}, riboflavin, pantothenic acid, niacin, and choline added to the ration, so vitamin premixes are often used to add vitamins to the ration, Figure 1-6.

Water

Water is essential because it is necessary for many body processes. Water requirements vary with the kind and amount of feed being consumed, environmental temperature, and the activity level of the animal. While part of the water requirement is contained in the feed being consumed, it is important to have a fresh, clean supply of water always available for all livestock.

Feed Additives

Feed additives are not nutrients in the generally accepted sense of the word. They are substances added to the ration to promote growth, increase the rate of gain, improve feed conversion, or reduce the level of disease. Feed additives are used in small amounts and are added to the ration through the use of premixes, Figure 1-7.

There is some controversy over the subtherapeutic use of antimicrobials in animal feeds. The continued use of antimicrobials in livestock feeding results in the development of microorganisms that are resistant to the antimicrobial being used. There is some indication that this may result in adverse effects on human health when the animal products are consumed by humans. For

Figure 1-6. Vitamins are added to the ration by using vitamin premixes.

Figure 1-7. A premix is used to add feed additives to the ration.

example, approximately 25 percent of Salmonella bacteria are now resistant to antibiotics such as sulfanilamides, tetracycline, and streptomycin, which are generally used at subtherapeutic levels in livestock feeding. A 1980 report from the National Academy of Sciences on subtherapeutic use of microbials concluded that a human health hazard was neither proven nor disproven from the use of antibiotics in animal feeding. However, since then there has been some evidence that outbreaks of Salmonella in humans might be linked to the consumption of beef from animals fed subtherapeutic levels of antibiotics.

The Food and Drug Administration since 1977 has proposed a ban on the use of penicillin and tetracycline in animal feeding because these two antibiotics are widely used to treat human health problems. Antibiotics not generally used in human health care could be substituted in animal feeding. Penicillin and tetracycline are not currently banned for livestock feeding in the United States. In 1971 and 1973, some European countries began limiting the use of selected antibiotics in livestock feeding.

More study is needed in this aspect of livestock feeding. Regulations are subject to change; therefore, livestock feeders need to be aware of current regulations regarding the use of feed additives and to follow these regulations carefully.

OCCUPATIONS THAT RELATE TO LIVESTOCK

Approximately 20 percent of the labor force in the United States works in an agriculturally related occupation. About 3.7 million workers are directly engaged in the production of agricultural products on the farm. Another 15 to 16 million workers are required to store, transport, process, and sell farm products. Approximately 3 million workers supply the seeds, fertilizers, and other supplies needed by farmers.

Many of these workers are in occupations related to livestock production. For example, about 360,000 workers are employed in the meat and poultry industry, and the dairy industry employs about 177,500 workers.

Occupations that relate more closely to animal nutrition include:

animal-science teacher
beef farmer
beef farmhand
cattle feeder
cattle rancher
dairy barnman
dairy choreman
dairy farmer
dairy farm worker
dairy herdsman
dairy management specialist
dairy nutrition specialist
dairy scientist
extension service livestock specialist
feedlot supervisor
feedlot maintenance worker
feed salesman
horse farmhand
horse stable attendant
horse trainer
livestock farmhand
livestock farmer
manager of retail feed and supply store
nutrition researcher
nutrition physiologist
poultry farmer
poultry farmhand
poultry nutrition researcher
stock ranch supervisor
swine farmer
swine farmhand
sheep farmer
sheep farmhand
veterinarian
veterinarian's assistant
vocational agriculture teacher

FUNCTIONS SERVED BY ANIMALS

Animals serve many functions in society. They provide food, clothing, and byproducts for human use. Some animals are important for recreation. In some societies, animals still provide much of the power used in agriculture. Animals contribute to conservation of natural resources and help to stabilize the farm economy. They concentrate bulky feeds into more easily transported forms.

Conversion of Feed into Food

Livestock convert feed grains and roughages into food for human consumption. In recent years, there has been some controversy over the use of feed grains as livestock feed. In the face of world food shortages, it has been suggested that this is not the most efficient use of limited resources.

Plants use about 1 percent of the solar energy reaching the earth's surface. They convert about 5 percent of this captured energy to a form suitable for human food. There is a loss of energy when plants are fed to animals subsequently used for human food. It is estimated that an acre of grain can support five times as many humans if the grain is used directly for human food rather than first being fed to animals and then the animals or animal products used for human food. This disparity in potential for human food production from an acre of ground is sometimes cited to support the view that using livestock for human food is not an efficient use of limited resources.

There is a significant difference in human diets in different parts of the world, which may be partly attributable to differences in the amounts of grain produced. People in developed nations get about 37 percent of their food calories from grain cereals, whereas people in underdeveloped and developing nations get about 62 percent of their food calories from cereal grains.

Nonruminant animals (swine and poultry) are fed large amounts of grain wherever they are raised because of their relative inability to effectively utilize roughages in their diet. Ruminants are fed largely on roughages in nations with limited grain supplies. In the United States and some other nations of the world where grain is plentiful and relatively low in price, ruminants are fed rations high in grains during the finishing period. High energy rations fed to beef cattle during the finishing period contain approximately 70 percent concentrates, including cereal grains, oilseed meals, and animal protein feeds. However, during the entire lifetime of a beef animal, approximately 80 percent of the total feed ingested comes from roughages such as pasture, range vegetation, hay, fodder, straw, and silage. These are feeds which generally cannot be used directly by humans for food. Animals fed these types of feeds add to the total human food supply and increase the efficiency of energy use in the food system.

Fish meal, meat and bone meal, milling and fermentation byproducts, and tankage make up about 30 percent of the feed fed to swine and poultry in the United States. These feeds are not used for human consumption. Roughages such as hay, silage, and grain-milling byproducts make up about 75 percent of the total energy intake of dairy and beef cattle in the United States. Again, these are feeds not used directly for human consumption.

Rangeland not suitable for cereal grain production produces forage plants that are consumed by cattle and sheep. Thus these forages, unsuitable for direct human use as food, are transformed into human food by ruminants. Approximately 50 percent of the land area of the continental United States consists of native and natural grasslands and forest lands, which produce vegetation that may be grazed by livestock. Crop residues may also be converted into human food by being fed to ruminants.

Animals provide protein, energy, vitamins, and minerals in the human diet. Of these, protein is the most important component from animal sources. Proteins for human use can be supplied from vegetable sources. However, it requires a careful use of a variety of vegetable sources of protein to provide all of the essential amino acids needed by humans. It is much easier to provide a balance of needed amino acids by including meat in the human diet.

The most important livestock sources of protein and energy in the world are swine, beef, poultry, and sheep, Figure 1-8. There are differences among these animals in the efficiency with which they convert feed into food for human use. Poultry for egg production are the most efficient producers of protein for human use, followed by dairy cattle for milk production, poultry for broiler production, turkeys, swine, beef, and sheep. The ranking for energy efficiency is: poultry for egg production, dairy cattle for milk production, swine, broilers, turkeys, beef, and sheep, Figure 1-9.

Clothing

Livestock provide fiber and skins for the production of clothing. The use of synthetic fibers for clothing production has reduced the demand for fiber from animals. However, because most synthetic fibers are oil-based and the price of oil has risen dramatically in recent years, animal fiber will continue to be an important resource in human society.

Leather is used for shoes, belts, gloves, and clothing as well as for other products used by humans. From 5 to 10 percent of the market value of animals comes from the sale of hides. Leather has some characteristics that make it superior to synthetics for the production of clothing. It can allow air to pass through, is more durable, and is warmer than clothing made from synthetics.

Byproducts

Byproducts are any products from the animal carcass other than meat. This includes blood, bone, skin, fat, intestine, brain, stomach, and

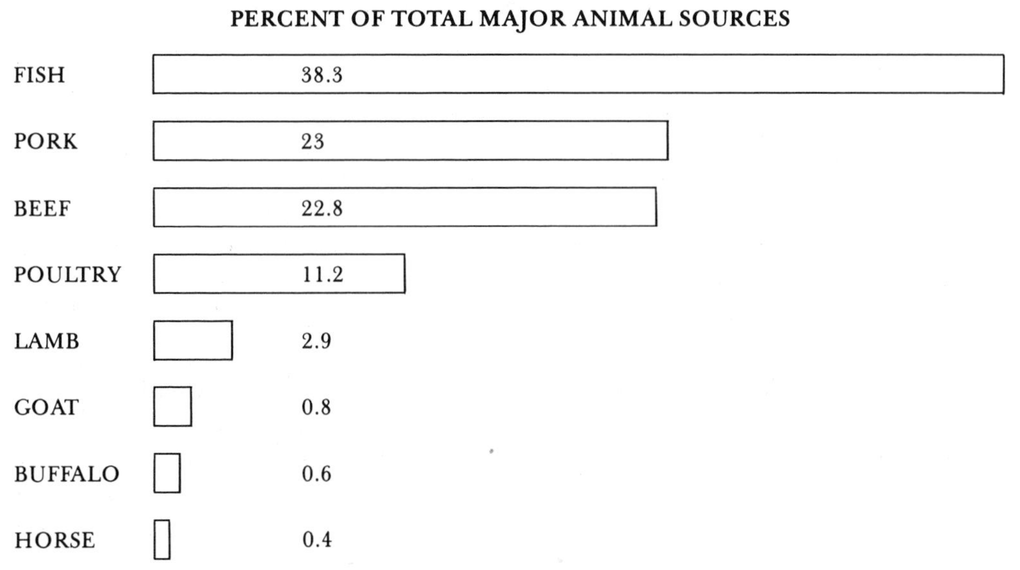

PERCENT OF TOTAL MAJOR ANIMAL SOURCES

FISH	38.3
PORK	23
BEEF	22.8
POULTRY	11.2
LAMB	2.9
GOAT	0.8
BUFFALO	0.6
HORSE	0.4

Source: Vocational Agriculture Service, *Animals in World Agriculture*, VAS 1058, University of Illinois, 1980.

Figure 1-8. The eight major animal sources of the world's food.

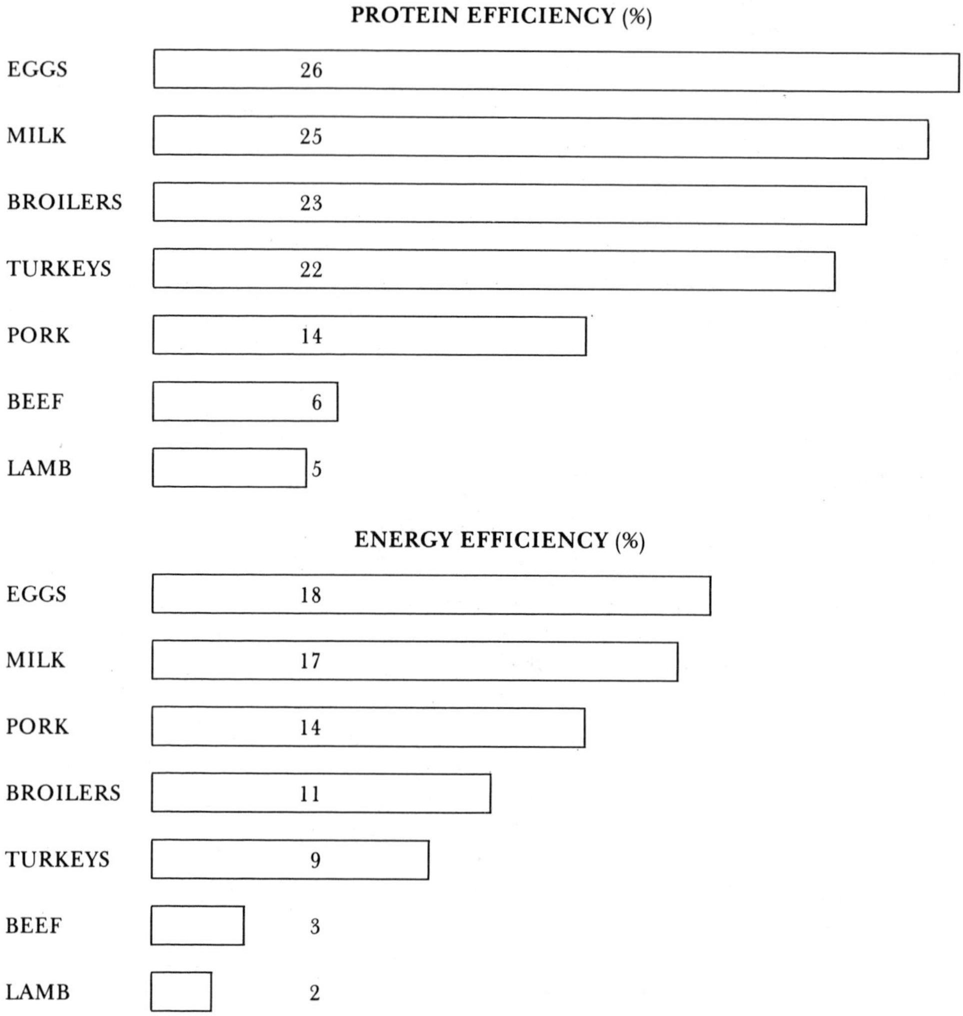

PROTEIN EFFICIENCY (%)

EGGS	26
MILK	25
BROILERS	23
TURKEYS	22
PORK	14
BEEF	6
LAMB	5

ENERGY EFFICIENCY (%)

EGGS	18
MILK	17
PORK	14
BROILERS	11
TURKEYS	9
BEEF	3
LAMB	2

Sources: Vocational Agriculture Service, *Animals in World Agriculture,* VAS 1058, University of Illinois, 1980. National Academy of Sciences, *Plant and Animal Products in the U.S. Food System,* Washington, D.C., 1978.

Figure 1-9. Protein and energy conversion efficiency for major farm animals.

various glands. Many useful products are made from byproducts of animal carcasses.

Hooves and horns are used in the production of glue and gelatin. Inedible fats are used in the production of cosmetics, waxes, soap, lubricants, and printing ink. Extracts from the endocrine glands include insulin, cortisone, and thromboplastin. *Insulin* is used in the treatment of diabetes. *Cortisone* is used for the treatment of rheumatoid arthritis and *thromboplastin* is a coagulant used in surgery.

Animal byproducts are also used in the manufacture of perfumes, fertilizers, candles, lanolin, and glycerine as well as many other products—animal byproducts make valuable contributions to society.

Recreation and Power

In the early days of agriculture, animals such as horses and oxen were used mainly for power and transportation. Today horses are used primarily for recreation in the United States. Horseback riding and racing are both major recreational activities using horses. Also, livestock shows and fairs provide recreation for many people.

In many parts of the world, animals are still a major source of power in agriculture. In Asia, Africa, and Latin America where income and farm productivity are low, the use of mechanical power on the farm has not seen the growth that has occurred in the United States and other major industrial countries. The increasing costs of fuel and tractors will tend to slow the conversion from animal power to machine power in these areas.

Conservation

Livestock produce large quantities of manure, Table 1-2. This is used for fertilizer by approximately 40 percent of the farmers in the world. In Europe and the United States, there has been a trend toward disposing of animal manure in lagoons and landfills. However, the rising costs of chemical fertilizers may produce a reversal of this trend.

The use of manure as fertilizer produces some long-term benefits that are not apparent with chemical fertilizers. Manure adds organic matter to the soil, which improves soil tilth and helps break up hardpan soils. The increase in crop yields is slower with manure than with chemical fertilizers, but a combination of the two can produce short-term yield increases with long-term benefits from the addition of the organic matter to the soil.

Animal manure can also be used as a fuel source. In many parts of the world, dried animal manure is burned as a fuel for cooking and to heat homes. Approximately 8 to 12 percent of the world's population depends on dried manure for a fuel source.

Animal manures can be used as a raw material in methane gas digesters. The use of methane gas converters has increased worldwide as a result of the energy crisis brought about by higher oil prices. Fuel for the electricity, cooking, and heating needs of an average U.S. farm could be supplied by the manure from about 40 cows. Some large farms and feedlots in the United States have

Table 1-2. Wet raw manure produced by different animals.

	Live Weight		Daily Production						Yearly Production		Moisture Content
	lb	kg	lb	kg	gal	litre	ft³	m³	ton	metric tonne	%
Beef	500	227	45.0	20.4	4.5	17.03	0.7	0.02	8.2	7.44	80-90
Beef	1,000	454	60.0	27.2	7.5	28.39	1.0	0.03	11.0	9.98	80-90
Pig	35	16	2.3	1.04	0.27	1.02	0.038	0.001	0.42	0.38	90
Pig-Growing	100	45	7.5	3.4	1.1	4.16	0.13	0.004	1.2	1.09	75-80
Sow-Gestation (limit fed)	275	125	8.9	4.04	1.1	4.16	0.15	0.004	1.6	1.45	90
Boar (limit fed)	350	159	11.0	4.99	1.4	5.3	0.19	0.005	2.0	1.8	90
Sheep	100	45	4.0	1.8	0.60	2.27	0.1	0.003	0.73	0.66	70-75
Laying hen	4	1.8	0.21	0.095	0.025	0.09	0.0035	0.0001	0.038	0.034	55-75
Broiler	4	1.8	0.28	0.127	0.032	0.12	0.004	0.0001	0.051	0.046	55-75
Turkey	15	6.8	0.55	0.249	0.07	0.26	0.009	0.0002	0.1	0.09	68
Horse	1,000	454	45.0	20.4	11.0	41.64	1.5	0.04	8.2	7.44	65-79
Dairy cow	1,000	454	83.5	37.9	10.2	38.6	1.4	0.04	15.2	13.8	87
Dairy cow	1,400	635	115.0	52.2	13.7	51.9	1.8	0.05	21	19.1	87

Source: Gillespie, James R., *Modern Livestock and Poultry Production*, 2nd ed, Delmar Publishers, Albany, NY, 1983

built bio-gas plants to utilize the animal manure produced.

Higher feed costs have led to research on the possibility of using animal manures as a supplement in feeds, but this recycling is still in the experimental stage.

Stabilization of Farm Economy

The production of livestock helps make better use of the land, labor, and capital inputs to agriculture. Farm income is increased by livestock production and risk is spread over more enterprises, making the farmer less dependent on one or two sources of income. Labor and income are spread more evenly throughout the year.

Concentrate Bulky Feeds

In some farming regions, the major crops are forages, which are bulky to transport to market. Raising livestock in these areas helps to convert these bulky feeds to a form more easily and cheaply transported to market.

SUMMARY

Livestock are an important part of agriculture in the United States. The sale of livestock and livestock products accounts for almost half the total farm income each year in the United States. From one-half to three-fourths of the total cost of raising livestock is feed cost in spite of research that has resulted in greatly improved methods of feeding livestock.

Feeds are generally classified as concentrates or roughages. Concentrates consist mainly of the feed grains and protein supplements, while roughages are the hay, pasture, and silage used for livestock feed. Ruminants can utilize large quantities of roughage in their diets while nonruminants generally utilize more concentrates. The horse, which is a nonruminant, utilizes roughage very effectively in its diet.

Six important components of livestock diets are carbohydrates, lipids, protein, minerals, vitamins, and water. Energy is provided by carbohydrates, lipids, and excess protein. Amino acids are provided by protein feeds. Calcium, phosphorus and salt are the major minerals required in livestock diets. Ruminants can generally synthesize needed vitamins with the exception of vitamin A, but nonruminants need vitamin supplements added to the diet. A clean, fresh supply of water is essential for all livestock.

Feed additives are used in small amounts in the diet to promote growth, production, rate of gain, or to reduce disease levels. Care must be taken to follow all current regulations in the use of feed additives.

There are many occupations available to people who have a knowledge of livestock nutrition.

Animals convert many plant products not directly usable as human food into a form that may be used by humans for food. They thus increase the total human food supply and increase the efficiency of energy use in the food system.

Animals perform many other functions of value to humans.

REVIEW

1. Describe the importance of livestock production in the total agricultural industry in the United States.

2. The cost of feed is what percent of the total cost of raising each kind of livestock?

3. Name, define, and give examples of the two kinds of feeds generally used in livestock feeding.

4. What has been the recent trend in the relative amounts of concentrates, roughages, and commercial feeds fed to livestock in the United States?

5. Name the six components of feed that are important when balancing rations for livestock.

6. Why are ruminant animals able to use more roughage in their diet than non-ruminant animals?

7. Which feed components provide energy for animals?

8. How many essential amino acids are required for swine and how many for poultry?

9. Name the major minerals needed in livestock rations.

10. Which vitamins usually need to be supplied to nonruminants?

11. Name some factors that affect the water intake of animals.

12. Why are feed additives used in livestock rations?

13. Why is there some concern over the use of antibiotics in livestock feeding?

14. What percent of the labor force in the United States works in agriculturally related occupations?

15. How many of these workers are involved directly in the production of agricultural products on the farm?

16. List 10 occupations in which a knowledge of animal nutrition is important for job success.

17. List the major functions of animals in human society.

18. What are the eight major animal sources of food in the world and approximately what percent of the total does each source supply?

19. Compare the relative efficiencies of the major farm animals in converting feed to protein and energy for human consumption.

20. Why does the livestock industry add to the human food base rather than decrease the food base?

21. Describe how animals are important in providing clothing for human use.

22. What are some important byproducts of the livestock industry?

23. What are some recreational activities that animals provide?

24. Discuss the use of animal power in the world today.

25. How does animal manure aid in the conservation of natural resources?

26. How do livestock help stabilize the farm economy?

27. Of what value are livestock in areas where forages are the major crop?

2

DIGESTION IN ANIMALS

OBJECTIVES

After completing this chapter you will be able to:

- Describe the nonruminant (monogastric), ruminant, and avian digestive systems.
- Describe the process of digestion in animals.
- Describe the absorption of nutrients in animals.

DIGESTIVE SYSTEMS

Digestion is a process that breaks feed down into simple substances that can be absorbed by the body. This process involves mechanical, chemical, and enzymic action. The simple compounds that result from digestion are absorbed into the bloodstream and lymph system.

The *digestive tract* (also referred to as the *gastrointestinal tract* or the *alimentary tract*) is the passage through the body that begins with the mouth and ends with the anus (vent in poultry). After feed is consumed, it is subjected to the various digestive processes as it moves through this passage.

There are three major kinds of digestive systems in animals: nonruminant (monogastric), ruminant, and avian. Swine and horses have nonruminant systems. Cattle, sheep, and goats have ruminant systems. The avian system is that of poultry. While both swine and horses are considered nonruminant animals, there are major differences in their digestive systems. These differences will be discussed in the nonruminant section of this unit.

NONRUMINANT (MONOGASTRIC) DIGESTIVE SYSTEM

The digestive system of *nonruminant* animals (those having a single-compartment stomach) includes the mouth, teeth, tongue, salivary glands, esophagus, stomach, small intestine, liver, pancreas, cecum, large intestine, rectum, and anus. Figure 2-1 shows the digestive system of swine and Figure 2-2 shows the digestive system of the horse.

Mouth

The mouth contains the teeth, tongue, and salivary glands. Chewing action in the mouth breaks, cuts, and tears up the feed, which is then mixed with saliva produced by three paired salivary glands located under the lower jaw and under the ears. Saliva contains water, mucin, bicarbonate salts, and enzymes (in some species). Horse saliva does not contain any enzymes; however, in swine, saliva contains the enzymes salivary amylase and salivary maltase.

Enzymes are organic catalysts that cause and/or speed up the digestive action but remain unchanged in the process. Amylase

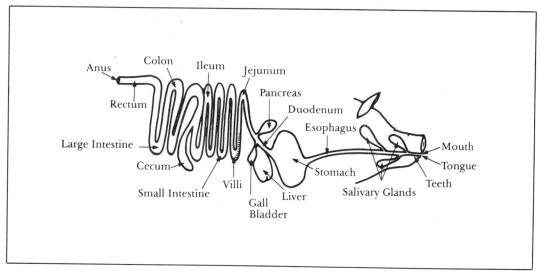

Figure 2-1. Digestive system of a pig.

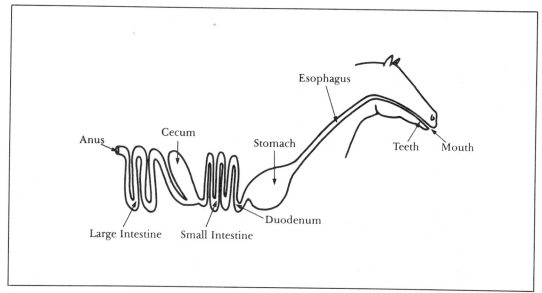

Figure 2-2. Digestive system of a horse.

changes some starch to maltose or malt sugar and maltase changes maltose to glucose. The saliva is slightly alkaline in reaction. Amylase acts only in a slightly alkaline solution. A weak acid solution will stop the enzyme action.

Saliva stimulates the taste nerves, water in the saliva moistens the feed for chewing and swallowing, and mucin lubricates the feed for swallowing. Bicarbonate salts act as a buffer to regulate the pH of the stomach.

The tongue aids in gathering feed into the mouth, mixing it, and directing the feed to the throat for swallowing. The chewed (*masticated*) feed enters the esophagus.

Esophagus

The *esophagus* is the tubelike passage which leads from the mouth to the stomach. A series of muscle contractions (*peristaltic waves*) and pressure differences carries the feed down the esophagus to the stomach. A valve called the *cardia*, located at the end of the esophagus, prevents the feed in the stomach from coming back into the esophagus.

Stomach

The *stomach*, which is a pear shaped, muscular organ, receives the feed, where it is further broken down by muscles in the stomach wall and mixed with digestive juices. Gastric juice is secreted by glands in the wall of the stomach. It begins to flow as soon as the masticated feed enters the stomach. Gastric juice contains about 0.2 to 0.5 percent hydrochloric acid. When the gastric juice is mixed with the feed it stops the action of the enzyme amylase.

Gastric juice also contains the enzymes pepsin, rennin, and gastric lipase. As soon as the amylase action is stopped these enzymes begin to act on the feed. Protein in the feed is broken down by pepsin into polypeptides. The casein of milk is curdled by the rennin. This prevents the milk from passing through the system in an undigested form. In young animals, gastric lipase acts on butterfat to split it into glycerol and fatty acids. Gastric lipase has little effect on the digestive process in adult animals.

The wall of the stomach is lined with muscle, which churns and squeezes the feed. This forces the liquid portion on into the small intestine. The gastric juices continue to act on the solid portion still in the stomach.

The stomach of the horse has less muscular activity than that found in other species, which causes feed to tend to accumulate in layers leading to an increased tendency toward digestive disorders. The stomach of the horse is relatively small, compared to other species, in relation to the size of the animal. It is, therefore, necessary to feed the horse smaller amounts of feed at one time but provide more frequent feedings.

Small Intestine

The first section of the small intestine is the duodenum where secretions from the pancreas, liver, and intestinal walls occur and active digestion takes place. Secretion from the liver is called *bile* and is stored in the gallbladder from where it is secreted into the duodenum. The horse does not have a gallbladder; therefore, bile is secreted continuously from the liver into the duodenum.

The middle section of the small intestine is the *jejunum* and the last section is the *ileum*. Nutrient absorption occurs in these two sections of the small intestine. Absorption of nutrients is discussed later in this unit.

The partially digested feed in the stomach is called *chyme*. Chyme is an acid, semifluid, grey, pulpy mass. The chyme moves from the stomach to the small intestine where it is mixed with three digestive juices: pancreatic juice, bile, and intestinal juice.

Pancreatic juice is secreted by the pancreas, a small gland located between the folds of the small intestine. This secretion is emptied into the small intestine through the pancreatic duct, which is attached to the duodenum. Pancreatic juice contains the enzymes: trypsin, chymotrypsin, carboxypeptidase, pancreatic amylase, pancreatic lipase, and others.

Proteins that have been broken down by pepsin are digested further by trypsin, chymotrypsin, and carboxypeptidase. These enzymes convert polypeptides to oligopeptides, dipeptides, and amino acids. Polypeptides, oligopeptides, and dipeptides are combinations of amino acids that have not yet been broken down into simple amino acids and are sometimes referred to as intermediate protein breakdown products.

Starch in the feed is changed into maltose by pancreatic amylase. Pancreatic amylase does more of the digesting of starch than salivary amylase, is found in larger quantities, and has longer to act than does the salivary amylase.

Pancreatic lipase breaks down the fats in the feed into fatty acids, glycerol, and monoglycerides.

The liver produces a yellowish-green, alkaline, bitter liquid called *bile*. Bile is stored in the gallbladder and is emptied into the small intestine in the duodenum region. The horse does not have a gallbladder; therefore, the bile is secreted directly into the small intestine from the liver.

Bile does not contain any enzymes but does help emulsify fats (separate the fat globules) and acts as a solvent for fatty acids. This aids in their digestion. Emulsifiying fats increases the surface area upon which the enzyme lipase can act.

Intestinal juice is produced by glands in the wall of the small intestine. The enzymes aminopeptidase, dipeptidase, sucrase, maltase, and lactase are found in the intestinal juice. Aminopeptidase breaks down protein to peptides and amino acids. Dipeptidase breaks down peptides to amino acids. Sucrase, maltase, and lactase break starches and sugars down into the simple sugars, glucose, fructose, and galactose.

Large Intestine

In swine, the first part of the large intestine is the cecum, which has little or no function. The middle and biggest part of the large intestine is the colon. The rectum is the last part and terminates in the anus.

The *cecum* or blind gut is located at the beginning of the large intestine. In most non-ruminant animals this organ has little function. However, it serves an important function in the horse. In the horse, the large intestine makes up approximately 60 percent of the total digestive tract. The large intestine of the horse is divided into the cecum, large colon, small colon, and rectum, Figure 2-2.

The horse can use fairly large amounts of roughage in its ration because of the presence of bacteria in the cecum and colon. These bacteria digest hemicellulose and cellulose, ferment the carbohydrates to volatile fatty acids, and synthesize B-vitamins and amino acids. Some volatile fatty acids are absorbed from the cecum in the horse. It is presently believed that vitamins and amino acids synthesized in the large intestine of the horse are of limited value in horse nutrition. Because the large intestine of the horse usually contains substantial quantities of ingested material, impaction occurs easily.

In all species, undigested, unabsorbed, and undigestable material passes from the small intestine to the large intestine. The main function of the large intestine is to absorb water from the material passing through. In the horse, the small colon is the site of most of the water resorption. Enzyme activity continues in the large intestine resulting in continued breakdown of fiber and the synthesis of some protein. Some water soluble vitamins and vitamin K are also synthesized in the large intestine. Mucus is added to the material to facilitate passage.

Material that is not absorbed or digested is called *feces*. Muscles in the wall of the large intestine move the feces to the rectum where it is eliminated from the body through the anus. The *anus* is the external opening at the end of the digestive tract.

RUMINANT DIGESTIVE SYSTEM

Mouth

The saliva of ruminants does not contain enzymes to help digest the starches. It does contain buffers (sodium bicarbonate) which neutralize the fatty acids produced in the rumen. Therefore, the rumen contents are maintained at approximately a pH of 6-6.5. This helps promote microbial growth in the rumen.

Mature cows produce about 12 gallons of saliva per day while sheep produce about 2 gallons per day. Saliva contains nitrogen, phosphorus, and sodium utilized by rumen microorganisms.

Stomach

Ruminants such as cattle and sheep can digest large amounts of roughage because their digestive system is different from nonruminant animals. The major difference is the stomach of the ruminant, which has four compartments: the rumen, reticulum, omasum, and abomasum, Figure 2-3.

The *rumen*, or paunch, is the first compartment of the stomach. The *reticulum*, or honeycomb, is the second part. There is no clear partition between these two parts of the ruminant stomach. The cardia (lower end of the esophagus) is common to both compartments. No enzymes are secreted in these two parts of the ruminant stomach. Together, they make up about 85 percent of the stomach of the mature ruminant.

The third compartment is the *omasum*, or manyplies. It is about 8 percent of the stomach of the adult ruminant. The omasum contains strong muscles in its walls. The *abomasum*, or true stomach, is the fourth part of the ruminant stomach, and makes up about 7 percent of the stomach.

Ruminants eat rapidly, swallowing much of their feed without chewing. The solid portion of the feed goes into the rumen. The liquid part also goes into the rumen but passes quickly to the reticulum, then through the omasum and on into the abomasum.

Two muscular folds called the *esophageal groove* form a passageway from the cardia to the omasum. When closed, this passageway directs material from the esophagus directly to the omasum and when it is open, the material goes into the rumen and the reticulum. Its major function appears to be to allow milk ingested by a nursing animal to bypass fermentation in the rumen. This groove is almost nonfunctional in the adult ruminant.

Rumination

After the ruminant has filled the rumen with feed it lies down to chew its cud. This process is called *rumination*. Cattle will spend about

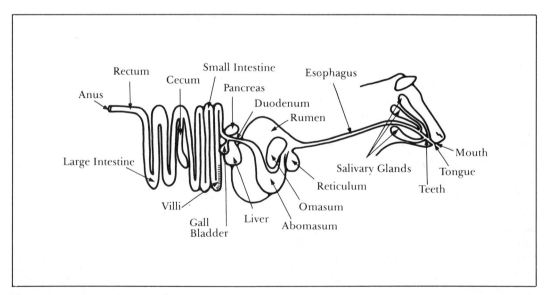

Figure 2-3. Digestive system of a cow.

five to seven hours per day in ruminating. This time is broken up into six to eight rumination periods.

Regurgitation is the process of forcing the feed back into the mouth for chewing. This is done by a series of muscular contractions and pressure in the rumen and reticulum. The animal breathes in with a closed glottis. This causes a drop in pressure in the thorax and esophagus. The pressure in the rumen is now greater, forcing the cud into the esophagus where it is carried to the mouth with the help of muscular action. More saliva is mixed with the feed as it is chewed and it is then re-swallowed. If it has been thoroughly chewed, it will then pass into the reticulum and on into the omasum.

Microorganisms in the Rumen

The rumen and the reticulum contain millions of microorganisms called bacteria and protozoa. Rumen bacteria are single-celled members of the plant kingdom. Rumen protozoa are single-celled members of the animal kingdom. Together, these tiny organisms feed on the fibrous material in the rumen. They digest cellulose and complex starch, synthesize protein from nonprotein nitrogen, and synthesize vitamins.

The three general types of rumen bacteria important in digestion are *Streptococci, Lactobacilli*, and a group of cellulolytic bacteria. The easily available carbohydrates (starches and sugars) are digested and fermented by the *Streptococci* and *Lactobacilli* bacteria. About 50-65 percent of the starch is digested in the rumen. Almost all of the soluble sugar is completely digested in the rumen. Various volatile fatty acids are produced as a result of this action. These are mainly acetic, propionic, and butyric acids. They are the primary energy source in ruminants. About 50-60 percent of the carbohydrates in the ration are converted to these volatile fatty acids and absorbed from the rumen. A ration high in concentrates and young, tender forage will result in a high population of *Streptococci* and *Lactobacilli* in the rumen. Less acetic acid (may be reduced to 40 percent of total) and more propionic acid (may be increased to 40 percent of total) is produced with high concentrate rations or when finely ground forage is fed.

In the cellulolytic group are bacteria with the scientific names of *Bacteriodes succinogenes* and *Ruminococcus flavefaciens*. These bacteria digest and ferment 30-50 percent of the cellulose and hemicellulose part of the feed. High roughage rations result in a high population of this group of bacteria with acetic acid (60-70 percent of total) being the main volatile fatty acid produced. Smaller amounts of propionic acid (15-20 percent of total) and butyric acid (5-15 percent of total) are produced.

Protozoa store readily available carbohydrates (polysaccharides), produce protozoal protein, and ferment cellulose material. The protozoa are the last microorganisms to develop in the rumen of the young calf.

Protein in the rumen is converted to ammonia, organic acids, and amino acids. Some of the microbes in the rumen use ammonia for growth and the production of microbial protein. Nonprotein nitrogen (NPN) sources such as urea and ammonium salts provide the ammonia the microbes need. The NPN is converted into amino acids, used by the microbes to form microbial protein. These microbes are then digested in the abomasum and small intestine which supplies the body with 65-80 percent of the necessary amino acids. As most amino acids can be synthesized in the rumen, it is not necessary to supply large quantities of amino acids in the ration.

Microorganisms in the rumen can synthesize the B vitamins and vitamin K needed by the ruminant. There is usually enough vitamin synthesis in the rumen to meet the needs of the animal.

Functions of the Rumen

There is a continual flow of feed materials into and out of the rumen (paunch). It acts like a large fermentation vat and accounts for about 50-85 percent of the total utilization of the digestible dry matter in the ration. The saliva, which is mixed with the feed, helps to control the pH of the rumen. Shifts in the predominate kind of microorganisms that develop in the rumen depend on the kind of ration being fed.

There is a constant mixing of the feed in the rumen. This action helps to break down the feed material into smaller particle sizes, which exposes more surface area to bacterial action. The smaller-sized material can also pass more quickly through the rumen. Feed material stays in the rumen and reticulum area from about two hours to several days. The kind of feed fed influences the amount of time it takes for feed to pass on into the rest of the digestive system. Concentrates pass more quickly through the rumen, while roughages take longer to digest.

Papillae are fingerlike projections lining the interior wall of the rumen. These increase the surface area and therefore increase the absorption ability of the rumen wall. Some absorption of the end products from the microbial action also occurs through the lining of the rumen.

Bacterial action in the rumen produces large quantities (30-50 quarts per hour) of gas. This is mainly carbon dioxide and methane. This gas must be removed or the animal will bloat. Usually the pressure from the gas causes belching (eructation) which eliminates the gas from the rumen. A small amount of this gas is absorbed by the bloodstream and then eliminated through the lungs.

Functions of the Reticulum (honeycomb)

There is no clear separation between the rumen and the reticulum. The two parts of the ruminant digestive system are often considered one unit. The reticulum contains the same bacteria and protozoa as does the rumen. No enzymes are secreted in the reticulum.

The reticulum wall is lined with intersecting ridges that form honeycomblike projections. Hardware such as nails and wire that are ingested are trapped in this area and generally do not move on through the digestive system. These items tend to collect in the reticulum and may puncture the lining causing "hardware disease." Sometimes a magnet is used in treating the animal; the magnet is placed in the throat of the animal. The magnet passes through to the reticulum where it attracts the bits of metal and holds them so they will not injure the animal.

Food is moved back and forth between the rumen and the reticulum by regular contractions originating in the reticulum. These contractions occur two to three times per minute. This movement separates the small feed particles from the larger ones. The coarser feed stays in the rumen for further digestion, while the smaller particles pass on into the omasum.

Functions of the Omasum. The omasum (manyplies) grinds and squeezes the feeds. There is little or no digestive action in the omasum. The grinding and squeezing help to break up the feed and squeeze some of the liquid out of it. The material leaving the omasum is 60-70 percent drier than the material entering the omasum.

Functions of the Abomasum. The abomasum is often called the *true stomach* of the ruminant. Digestion proceeds in the abomasum in much the same manner as it does in the monogastric digestive system.

Digestive juices are added to the feed and it is moistened. These digestive juices contain enzymes which digest the protein. There is little or no digestion of fat, cellulose, or starch in the abomasum. The pH of the material in the abomasum is 3.5 to 4.0, which results from the higher acidity of the digestive juices added to the feed in this area.

Enough moisture is added to the feed in the abomasum to make it highly fluid as it passes on into the small intestine. Further digestion and absorption occurs in the small intestine. Material not absorbed is then passed on to the large intestine and eventually excreted as feces.

AVIAN DIGESTIVE SYSTEM

The avian digestive system is quite different from the nonruminant and ruminant systems, Figure 2-4. Poultry have no teeth so the feed is not chewed. The tongue helps push the feed into the gullet. The feed passes down the gullet to the crop, where it is softened by saliva and secretions from the crop wall. The crop acts as a storage area for the feed.

From the crop the feed moves to the glandular stomach (proventriculus) where it remains for two to three minutes. Digestive juices are secreted by the glandular stomach and mixed with the feed. Hydrochloric acid and the enzyme pepsin are the principal gastric juices secreted in the glandular stomach. Pepsin starts the breakdown of proteins to amino acids.

The feed next moves to the gizzard (ventriculus), which is heavily muscled. The gizzard is lined with a thick, horny membranelike material called the *epithelium*. Muscular action in the gizzard breaks down the feed into smaller particles. There is a further mixing of proventricular digestive juices with the feed in the gizzard.

Grit (such as crushed granite or oyster shells) is sometimes fed to poultry to help the gizzard break down the feed. These are not added to a mixed ration, but are fed free choice. The feeding of grit may be especially helpful for poultry that are fed whole grains.

After being broken up in the gizzard, the feed moves into the small intestine. Digestive juices which break down the nutrients into simple forms are secreted in the small intestine. Most of the digestion of the feed takes place here. Muscle contraction in the small intestine moves the material through to the ceca and large intestine.

There are two blind pouches called *ceca* where the small intestine joins the large intestine. The ceca are about seven inches long. While these contain some bacteria, they have little effect on the digestion of fiber. They are usually filled with soft, undigested feed.

Material not digested and absorbed into the bloodstream moves on into the large intestine and to the cloaca. There is little digestion or absorbtion of feed material in the large intestine.

The *cloaca* is an enlarged part where the large intestine, urinary, and genital systems meet. The end of the digestive system is the vent. Feces from the large intestine are passed out of the body through the vent. Eggs from the oviduct also pass out through the vent.

The liver and pancreas are accessory organs to the digestive system. Food does not pass through these organs; however, they secrete substances which aid in the digestion of feed in much the same manner as in the monogastric system.

ABSORPTION OF NUTRIENTS

Absorption is the process of taking nutrients from the digested feed into the blood and lymph systems. The nutrients are then distributed to the cells and tissues of the body where they are utilized in the metabolism of the animal.

In nonruminants most absorption takes place from the small intestine, with a lesser

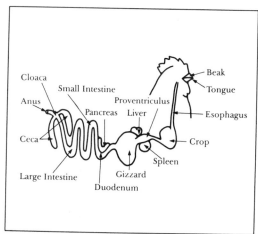

Figure 2-4. Digestive system of a chicken.

amount being absorbed from the large intestine. In ruminants there is some absorption of nutrients through the wall of the rumen.

Villi are small cone-shaped projections found in the wall of the small intestine. Each villus contains a network of blood capillaries through which nutrients enter the bloodstream. Protein is converted to amino acids in digestion. Starches and sugars are converted to glucose, fructose, and galactose when digested. Crude fiber is converted to short-chained fatty acids or glucose by digestion. These nutrients pass into the blood capillaries by osmosis through the semipermeable membranes of the digestive tract, then through the liver and into the bloodstream for circulation through the body. The two methods of absorption are *diffusion* and *active transport*. Diffusion is the movement of molecules from an area of high concentration to one of low concentration. Electrolytes and fructose move by diffusion. Active transport is the movement of molecules from one area to another, requiring the expenditure of energy. Amino acids and glucose move by active transport.

When fats are digested they are converted into fatty acids and glycerol. This material is called *chyle*. The chyle is absorbed by the lacteal (lymphatic vessel) in the villus and carried by the lymphatic system. It passes through the thoracic duct in the neck into the circulatory system.

Dissolved minerals and water are absorbed into the bloodstream through the villi. In the large intestine, nutrients are absorbed directly into the bloodstream through capillaries in the wall of the intestine.

METABOLISM

Metabolism is the sum of the chemical and physical changes continually occurring in living organisms and cells utilizing nutrients after they are absorbed from the digestive system. Both the formation and repair of body tissues (anabolism) and the breakdown of body tissue (catabolism) into simpler substances and waste products with the oxidation of nutrients to provide energy are included in metabolic processes.

NUTRIENT TRANSPORT

Nutrients, in the water soluble form, are primarily carried by the blood in the animal's body from where they are absorbed to where they are utilized. Animal cells are surrounded by lymph. The nutrients pass out of the capillaries by diffusion into the lymph. They then enter the cell by osmosis to provide the materials for metabolism.

Nutrients serve several purposes in the body. They are first used for maintenance to keep the animal alive. Oxidation provides heat for maintaining body temperature and energy for movement. Nutrients that are not used for basic maintenance may be used for other purposes. These include (1) growth and fattening; (2) fetal development; (3) production of milk and eggs; (4) wool and mohair production; and (5) work. These functions of nutrients are discussed in greater detail in other units.

SUMMARY

Digestion is breaking feed down into simple substances that can be absorbed by the body. Monogastric (single stomach) digestive systems are found in swine and horses. Horses, however, have an enlarged cecum that permits them to utilize more roughage than swine can use. Ruminant (four-compartment stomach) digestive systems are found in cattle, sheep, and goats. Poultry have an avian digestive system.

Digestion occurs when feeds are broken up mechanically and acted upon by enzymes and other digestive juices in the digestive system. Ruminants have large populations of bacteria in the rumen, which act upon the feeds and synthesize amino acids and vitamins needed by the animal. Because of the bacterial action and the larger capacity of the ruminant system, these animals can utilize more roughage in their rations than can animals with monogastric or avian digestive systems.

Poultry have no teeth; therefore, the feed is broken up mechanically in the gizzard. The overall capacity of the avian digestive system is less than that of the monogastric or ruminant systems. Poultry utilize little roughage in their ration.

Most absorption of nutrients after digestion takes place in the small intestine, although some absorption occurs in the rumen. There is a small amount of absorption from the large intestine. Material that is undigested and unabsorbed passes through the large intestine and is excreted as feces.

Metabolism refers to the chemical and physical changes occurring after the feed nutrients are absorbed into the bloodstream. Animals use nutrients for maintenance, energy, growth, reproduction, production, and work. Tables 2-1 and 2-2 summarize the functions of the various components of the digestive systems of animals and the digestive capacities for selected species.

Table 2-1. Summary of digestion.

Location/ Source	Digestive Juice	Enzyme/ Secretion	Action/Function	Comment
Mouth (Salivary glands)	Saliva	Salivary Amylase	Acts on starch / change to maltose.	Of little importance. None in ruminants.
		Salivary Maltase	Acts on maltose / change to glucose.	Saliva adds moisture to feed. Small amount in poultry.
Rumen & Reticulum			Microorganisms act on:	Synthesize essential amino acids, B complex vitamins, vitamin K.
			protein/nonprotein nitrogen to form essential amino acids.	
			starch/sucrose/cellulose to form volatile fatty acids (mainly acetic, proprionic, butyric), methane, carbon dioxide, and heat.	
			fat to form fatty acids and glycerol.	
			glycerol to form propionic acid.	
Omasum			Grinds and squeezes feed/ removes some liquid.	Little digestive action in the omasum.
Stomach/ Abomasum in ruminants/ proventriculus in avian. (wall of stomach)	Gastric juice	Hydrochloric acid	Stops action of salivary amylase.	
		Pepsin	Acts on protein / change to proteoses, polypeptides, and peptides.	
		Rennin	Acts on milk / curdles the casein.	
		Gastric lipase	Acts on fat / forms fatty acids and gylcerol.	

continued

Table 2-1. (continued)

Location/ Source	Digestive Juice	Enzyme/ Secretion	Action/Function	Comment
Gizzard in avian			Grinds and mixes feed.	Digestive juices continue to act on feed.
Small intestine (Pancreas)	Pancreatic juice	Trypsin and Chymotrypsin	Acts on proteins, proteoses, polypeptides, and peptides / produces proteoses, peptones, peptides, and amino acids.	
		Pancreatic amylase	Acts on starch / change to maltose.	Small amounts in ruminants
		Pancreatic lipase	Acts on fat / forms glycerol, fatty acids, and monoglycerides.	
		Carboxy- peptidase	Acts on peptides / forms peptides and amino acids.	
(Liver)	Bile		Acts on fats / forms glycerol and soap.	
(Intestinal wall)	Intestinal juice	Intestinal peptidase (formerly called erepsin)	Acts on remaining proteins, proteoses, peptones, and peptides / produces amino acids.	
		Maltase	Acts on maltose / changes to glucose.	Small amounts in ruminants.
		Sucrase	Acts on sucrose / changes to glucose and fructose.	Small amounts in ruminants.
		Lactase	Acts on lactose / changes to glucose, fructose, and galactose.	Large amounts in young mammals.
		Nucleotidase	Acts on nucleoproteins / forms nucleotides, nucleosides, purines, pyrimidines, phosphoric acid.	
Cecum in horse			Bacterial action digests roughage.	
Large intestine		Cellulase	Acts on cellulose / forms volatile fatty acids.	Mostly in the horse.
			Some digestion continues as material moves from the small intestine to the large intestine.	

Table 2-2. Capacities of digestive system of selected species (ranges indicate different ages, breeds, sizes).

Organ / Species	Swine		Horse		Cattle		Sheep / Goat	
	(qts)	(liters)	(qts)	(liters)	(qts)	(liters)	(qts)	(liters)
Rumen					80-192	75.7-181.6	25	23.6
Reticulum					4-12	3.8-11.4	2	1.9
Omasum					8-20	7.6-18.9	1	0.9
Abomasum					8-24	7.6-22.7	4	3.8
Stomach in nonruminants	8	7.57	8-19	7.6-18				
Small intestine	10	9.5	27-67	25.5-63.4	65-69	61.5-65.3	10	9.5
Cecum	1-1.5	0.95-1.4	14-35	13.2-33.1	10	9.5	1	0.9
Large Intestine	9-11	8.5-10.4	41-100	38.8-94.6	25-40	23.6-37.8	5-6	4.7-5.7
Total	28-30.5	26.5-28.87	90-221	85.1-209.1	200-367	189.3-347.2	48-49	45.2-46.2

REVIEW

1. Define digestion and digestive system.
2. Name the three major kinds of digestive systems and give examples of animals with each type.
3. Name the parts of the monogastric digestive system and briefly describe the function of each.
4. Define and give examples of enzymes.
5. Describe the action of enzymes in the digestive process.
6. Define chyme.
7. Describe the function of the liver.
8. How does the digestive system of the horse differ from that of swine?
9. Name the four major compartments of the stomach of a ruminant.
10. Describe the function of each compartment.
11. Define rumination.
12. Describe regurgitation in the ruminant and tell how it relates to the digestive process.
13. Name the major microorganisms found in the rumen and describe their function.
14. Name the parts of the avian digestive system and describe the function of each part.
15. Describe how absorption of nutrients occurs.
16. Define and briefly describe metabolism.

3

ENERGY NUTRIENTS

OBJECTIVES

After completing this chapter you will be able to:

- Define terms associated with energy.
- Describe the energy nutrients.
- List sources of energy nutrients.
- Describe the functions of energy nutrients.
- Describe symptoms of deficiencies of energy in the ration.

ENERGY TERMINOLOGY

The following terms are used to express measurements of energy:

- *Calorie* (cal) measures the amount of heat energy required to raise the temperature of one gram of water from $14.5°C$ to $15.5°C$. A calorie is defined as containing 4.1860 international joules.
- *Kilocalorie* (kcal) is 1,000 calories.
- *Megacalorie* (Mcal) is 1,000,000 calories.

The following terms are associated with descriptions of energy in feeds:

- *Gross Energy* (GE) is the total amount of heat released when a substance is completely oxidized in a bomb calorimeter under 25 to 30 atmospheres of oxygen. The gross energy is usually expressed as kilocalories/kilogram or kilocalories/pound of feed. Digestible, metabolizable, or net energy values of a feed cannot exceed the gross energy value of that feed.

- *Digestible Energy* (DE) is the gross energy of the feed consumed minus the gross energy excreted in the feces.
- *Metabolizable Energy* (ME) is the gross energy of the feed consumed minus the energy in the feces, urine, and gaseous products of digestion. In nonruminants and poultry the amount of gas produced by digestion is usually small; therefore, ME values are not corrected for gaseous products of digestion.
- *Heat Increment* (HI) is that portion of the metabolizable energy which is used for digestion or metabolism of absorbed nutrients into body tissue. In cold weather this heat may help keep the animal warm. During hot weather the heat increment may reduce production by increasing the animal's body temperature above the neutral thermal zone.
- *Net Energy* (NE) is the metabolizable energy minus the heat increment. It is used for growth, maintenance, production, work, fetal development, and heat production.

The utilization of dietary energy in the animal's body is illustrated in Figure 3-1.

ENERGY NUTRIENTS

Carbohydrates and lipids (fats and oils) are the major sources of energy in livestock rations. Some energy is also derived from the protein in the ration. Of these sources, carbohydrates are the most important because they are readily available, easily digested in greatest quantities in most feeds, and generally lower in cost. Fats and oils are the second most important source of energy for livestock. However, during warm weather, it is difficult to store feeds that are high in fat content because they tend to become rancid. This makes the feed unpalatable and animals are reluctant to eat it. In some cases, rancid feed may cause digestive disturbances, making the animal sick. Proteins are seldom fed for their energy content because of the higher cost of this source.

CARBOHYDRATES

Carbohydrates are organic compounds made of carbon (C), hydrogen (H), and oxygen (O). The percentage of each by molecular weight is: carbon (40 percent), hydrogen (7 percent), and oxygen (53 percent). The hydrogen and oxygen in the carbohydrate compound are in the same proportion as in water (H_2O). Growing plants produce carbohydrates by photosynthesis ($6CO_2 + 6H_2O$ + energy from the sun $\rightarrow C_6H_{12}O_6$ (glucose) $+ 6O_2$). Carbohydrate compounds found in plants include starch, sugars, hemicellulose, cellulose, pectins, gums, and lignin. The simplest of these compounds are the sugars. Sugars are the most easily digested, while cellulose and lignin are more difficult to digest.

Carbohydrates in feed occur as compound substances, that is, two or more molecules are combined to make complex forms. For example, lactose, which is a compound sugar, is made up of glucose and galactose.

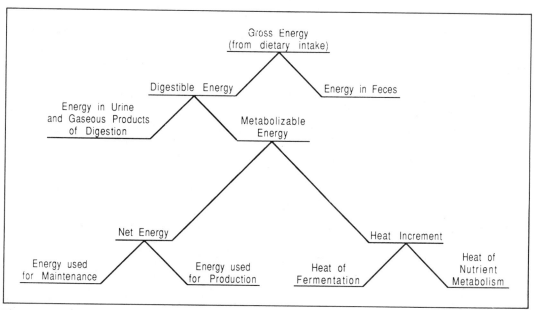

Figure 3-1. Utilization of dietary energy by the animal.

During digestion, carbohydrates in the feed are changed to simpler forms. The glucose supply in the bloodstream of animals is maintained at a level of 0.05 to 0.1 percent concentration. Because glucose can be used by all animals, it is the most important of the sugars found in the body.

A *monosaccharide* is a simple sugar that is not decomposable by hydrolysis. Hexoses are simple sugars containing six carbon atoms while pentoses are sugars containing five carbon atoms. The most common of the hexoses are glucose (blood sugar), fructose (found in ripe fruit), and galactose (found in milk). The pentoses are rarely found in a free form but do occur in complex carbohydrates. Figure 3-2 shows the molecular structure of the three monosaccharides glucose, fructose, and galactose.

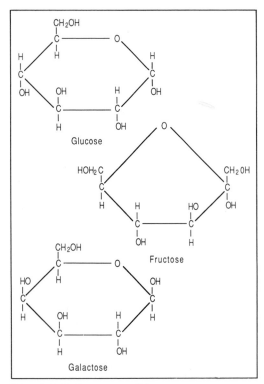

Figure 3-2. Structures of three common monosaccharides (hexoses).

When simple hexose sugars are combined, compound sugars are formed. *Disaccharides* contain two sugar molecules. The most important of these in animal nutrition are cellobiose, maltose (malt sugar), and lactose (milk sugar). Figure 3-3 shows the molecular structure of these three common disaccharides.

When simple sugars are combined and the water split out, complex carbohydrates called *polysaccharides* (many sugars) are formed. Complex sugars, starch, hemicellulose, and cellulose are examples of polysaccharides.

Polysaccharides contain more than two sugar molecules. Starch contains many glucose molecules with an alpha linkage and cellulose contains many glucose molecules with a beta linkage. Figure 3-4 shows the molecular structure of starch and cellulose.

Hemicellulose and gums are a mixture of hexose and pentose molecules. Pectins contain a mixture of hexose and pentose molecules combined with salts of complex acids.

About 75 percent of all the dry matter in plants is carbohydrates. The more easily digested forms of carbohydrates are generally found stored in the seeds, roots, and tubers of the plant. The fiber cells of the plant contain hemicellulose and cellulose which are harder to digest. As a result of digestion, some of the hemicellulose and cellulose are converted to glucose. Because digestion of hemicellulose and cellulose requires more energy, they are less efficient sources of energy for the animal.

Carbohydrates are generally divided into two groups called *fiber* and *nitrogen-free extract* (NFE). Fiber, which contains hemicellulose, cellulose, and lignin, is harder to digest than NFE. Hemicellulose is easier to digest than cellulose, while lignin is almost completely indigestible. Lignin has a different proportion of carbon, hydrogen, and oxygen than the other polysaccharides and thus is not considered a true carbohydrate. The presence of lignin in a feed may reduce the digestibility of other nutrients contained in the feed. The NFE group includes sugar, starch, some

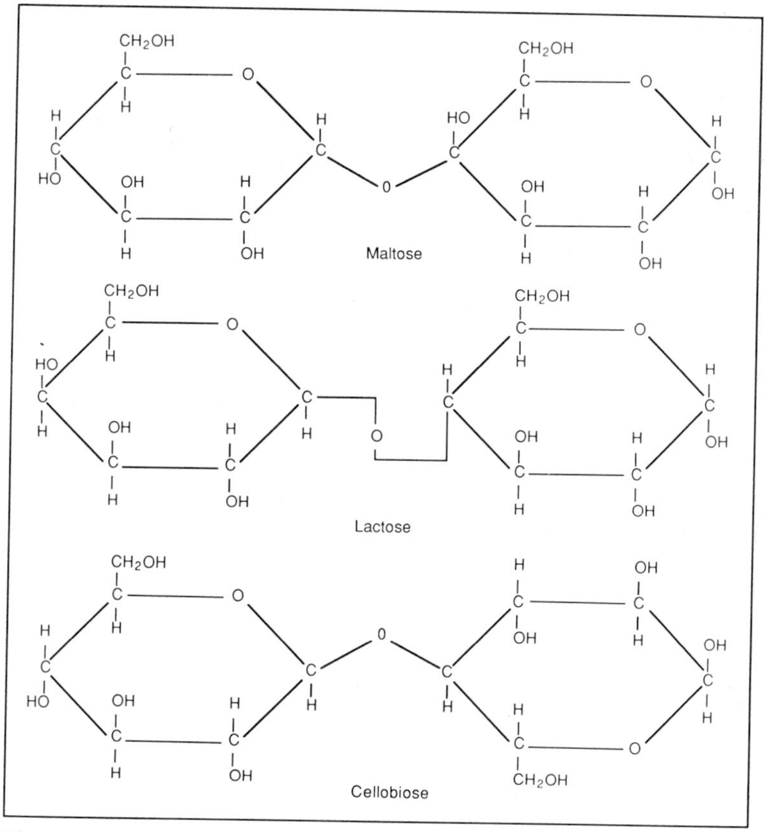

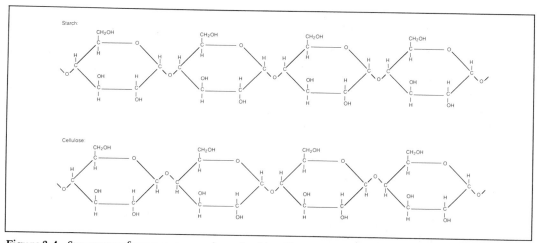

Figure 3-3. Structures of three common disaccharides.

Figure 3-4. Structures of two common polysaccharides. Shown are several molecules which make up part of the long chain polymers which make up these two polysaccharides. Both starch and cellulose are made up of glucose molecules which are bonded together, but the molecules are arranged differently in the two.

hemicelluloses, and the more soluble parts of the celluloses and pentosans. Pentosans are pentose-based polysaccharides.

Starch is made up of many molecules of glucose. Plants store energy in the form of starch in grain. Grains have a high feeding value because the starch is easily digested.

The ability of animals to use fibrous sources of carbohydrates is related to the presence or absence of bacteria in the digestive system. Ruminants, which have a high bacteria population in the rumen, make use of energy from the fiber portion of the ration. Bacterial action breaks the fiber down into volatile fatty acids absorbed through the rumen wall. Roughages in the ration of ruminants can provide much of the maintenance energy needed by the animal. Animals with monogastric and avian digestive systems have less ability to utilize energy from fiber. The young of all species require more easily digested energy sources than do more mature animals. To prevent excessive weight gains, it is recommended that the level of fiber in the ration be increased for mature breeding animals. When animals are being raised for breeding purposes, higher levels of fiber can be used in the ration to keep them from getting too fat.

LIPIDS (FATS AND OILS)

Chemically, lipids are made up of carbon, hydrogen, and oxygen. The percentage of each by molecular weight is: carbon (77 percent), hydrogen (12 percent), and oxygen (11 percent). Because there is more carbon and hydrogen and less oxygen in the molecule, lipids (fats and oils) supply approximately 2.25 times as much energy as an equal weight of carbohydrates. A typical gram of carbohydrate yields approximately 4.2 kcal of gross energy when completely oxidized while a gram of a typical fat yields 9.45 kcal.

Fats are solids and oils are liquids at body temperature. In animal nutrition, both are generally referred to as fats.

Lipids are classified as simple, compound, and derived. Simple lipids are true fat and waxes. True fats are formed when fatty acids are combined with glycerol, Figure 3-5. Waxes are esters of fatty acids with alcohols other than glycerol.

Compound lipids are esters, which contain groups in addition to an alcohol and fatty acid. The phospholipids lecithin, cephalin, and sphingomyelin contain phosphoric acid and nitrogen. Glycolipids do not contain phosphoric acid but do contain carbohydrates and nitrogen. Lipoproteins are lipids that are bound to protein in blood and other body tissues.

Substances derived from simple or compound lipids by hydrolysis are called *derived lipids*. Fatty acids and sterols are derived lipids.

Depending on the source of the fat, fatty acids are either saturated or unsaturated. The structure of fatty acids is chains of carbon atoms, which are from 2 to 24 or more carbon

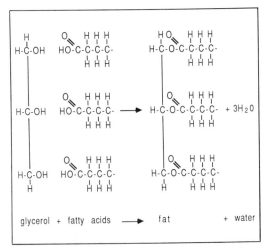

Figure 3-5. Synthesis of a fat. When three molecules of water are removed, three molecules of fatty acid bond to one molecule of glycerol producing a fat. Only a portion of the carbon chain is shown in this example. The carbon chain may range from 2 to 24 or more carbon atoms in length. Note that the fatty acids have a carboxyl group (COOH) at one end. The carboxyl group functions as a unit making the compound an acid.

atoms in length and have a carboxyl group on the end. A saturated fatty acid has two hydrogen atoms attached to each carbon atom with the carbon atom at the end having three hydrogen atoms attached. Figure 3-6 shows examples of the structural formulas of saturated and unsaturated fatty acids. An unsaturated fatty acid molecule contains one or more pairs of double-bond carbon atoms in the chain from which the hydrogen atom has been removed. This allows the molecule to combine with oxygen or certain other chemical elements.

Oxidation or hydrolysis will cause unsaturated fatty acids to become rancid. This causes bad flavors and/or odors in the feed and reduces the value of the feed because it is less palatable and livestock do not like to eat unpalatable feed. Storing feeds with high fat content in a cool place, adding antioxidants, or hydrogenation (adding hydrogen) will slow the development of rancidity in the feed.

The unsaturated fatty acids linoleic, linolenic, and arachidonic are called *essential fatty acids* (EFA) because they are considered to be necessary for normal health. Rations fed on the farm normally contain enough of these fatty acids to meet the needs of animals. Digestion separates the fatty acid from the glycerol molecule. The fatty acid is absorbed into the lacteals and then the bloodstream, where it moves to various parts of the body and is combined with glycerol to form fat. Fat deposits in the muscle tissue are called *marbling*. Fat stored in adipose (fatty) tissue contains reserve energy, which the animal could use to help sustain life for a while if its feed supply were cut off.

Fats are soluble in ether and some other organic solvents. Ether is used in feed analysis to extract the fat from the feed; therefore, the substances dissolved are called *fats* or *ether extract*.

Fats are used to raise the energy level of the diet and/or improve the flavor, texture, and palatability of the feed. Added fat will reduce

the dustiness of the feed. Show animals are sometimes fed rations high in fat to improve the glossiness of the hair coat.

Rations for adult ruminant animals should contain no more than 3-5 percent fat and 15-20 percent fat for nonruminants. Because fats carry fat-soluble vitamins, some fat in the ration is desirable. Too much fat in the ration will reduce feed intake and increase the chances of scouring (diarrhea). Young animals receiving a high level of milk or milk replacer in the diet will be getting 25-40 percent fat in their diet on a dry matter basis, a desirable level of fat intake for animals of this age.

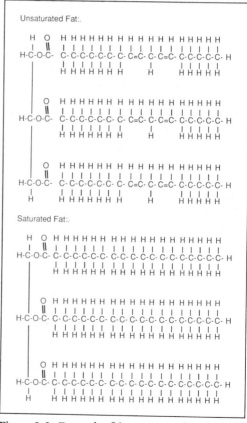

Figure 3-6. Example of the structure of unsaturated and saturated fat. The fatty acids of an unsaturated fat have fewer hydrogen atoms and contain double bonds between some of the carbon atoms.

A high level of unsaturated fat in swine diets can result in soft pork. This is especially a problem in large garbage feeding hog operations or when a high level of raw soybeans or raw peanuts are included in the diet. Soft pork lowers the carcass quality and value. A high intake of saturated fat does not result in soft pork.

SOURCES OF ENERGY— CONCENTRATES

A major source of energy nutrients is the grains and grain byproducts. These feeds are called *energy concentrates* or *basal feeds* when their crude protein content is less than 18 percent in the air-dry state.

Shelled Corn

One of the highest energy feeds available for use in livestock rations is shelled corn. It is the most widely grown and used feed grain crop. Corn produces more pounds of total digestible nutrients (TDN) per acre than any other feed grain. It is an economical and superior source of energy for livestock. Consideration must be given to amount to feed, frequency of feeding, and combinations with other feeds to get the most efficient use of this high energy feed without causing digestive problems.

Corn and Cob Meal

A mix of corn and cob meal contains about 10 percent less energy than shelled corn, because of the fiber content in the cob. Corn and cob meal is a good energy source for ruminants, horses and mules, and may be used in the rations of mature hogs. It should not be fed to growing-fattening hogs because they do not have the ability to digest and use much of the cob content. The cob portion may irritate the digestive tract of growing-fattening hogs, which can lead to intestinal infection. When used in rations for ruminants and horses and mules, other roughages must be reduced to allow for the higher fiber content of the corn and cob meal. The fiber content of corn and cob meal is 9 percent compared to only 2 percent in shelled corn. When corn and cob meal is used in dairy rations, this higher fiber content helps keep dairy cows on feed and maintain fat test in the milk.

Ground Snapped Corn

Ground snapped corn is made up of the grain, cob, and shucks and is considerably higher in fiber and lower in TDN than shelled corn. It is comparable to oats as an energy feed. Ground snapped corn is seldom used because of the high labor requirement in harvesting. It may be used in much the same manner as corn and cob meal in the ration.

Corn Starch and Corn Oil

Corn starch and corn oil are derived from corn grain and are fairly pure forms of starch and oil. These energy sources are usually not used in regular livestock feeding, but are sometimes used in the purified diets of experimental animals.

Oats

Oats have about 85 percent of the energy of shelled corn. They are higher in crude protein than shelled corn and add fiber and bulk to the ration. When fed to ruminants, oats help to maintain the rumen function. Oats are not a good fattening feed but are used extensively in rations for horses, young growing stock, show stock, and breeding animals. Oats are generally rolled, crimped, or ground for feeding.

Oat Groats

Oat groats consists of the kernel of the oat grain. While oat groats is very high in feeding value, it is generally too expensive to use in livestock rations. Sometimes oat groats is used in special diets such as early weaning rations for pigs.

Barley

Barley is almost equal to corn in energy value but lies between corn and oats in fiber content. It is used in the ration in a manner similar to oats. Barley may replace up to 50 percent of the corn in rations for fattening animals. The grain content of the ration may be decreased by 10 percent if barley replaces all of the corn. To improve palatability, barley is usually steam rolled (flaked), crimped, or coarsely ground for inclusion in the ration. Barley is sometimes cooked to improve its palatability when used in rations for beef show animals. When using barley in dairy rations, make the adjustment in the ration gradually.

Wheat

While wheat is high in both energy and protein it is generally not used in livestock rations because of its higher price relative to other feeds. When it is favorably priced in relation to other feeds, it makes a good source of energy in the ration. Wheat is similar to corn in composition and feeding value. If used in livestock rations, it is generally included at low levels in a mix with other grains because it is rapidly digested and may cause digestive disturbances. Wheat is generally coarse ground or cracked for livestock feeding.

Wheat Bran

Wheat bran is the seed coat of wheat, removed in the manufacture of wheat flour. The main use of wheat bran in livestock rations is to provide bulk, a mild laxative, and as a source of phosphorus. Wheat bran is also a source of energy and protein in the ration. Because of its relative high price, bulk, and laxative effect, the use of wheat bran is usually limited to not more than 10 percent of the ration. It is seldom used in rations for feedlot steers or growing-fattening hogs. Wheat bran is sometimes used in rations for horses, dairy cows, brood sows, and beef show animals.

Wheat Middlings and Shorts

Wheat middlings and shorts are byproducts of the manufacture of wheat flour. They are lower in fiber and higher in energy than bran and are used mainly as a source of energy in livestock rations. Because the fiber content is quite variable (2-9 percent), the energy value also varies considerably depending on the source and the kind of process used in manufacture. Wheat middlings and shorts are often used in swine rations, but seldom used in rations for other classes of livestock.

Grain Sorghum

There are many varieties of grain sorghum including milo, kafir, and various hybrids. Most grain sorghum is grown in the western part of the United States in the semi-arid regions where corn does not grow well. Grain sorghum is similar to shelled corn in composition and may replace up to 100 percent of the corn in a feedlot ration. It is generally rolled or ground when included in livestock rations.

Rye

Rye is usually used for bread for human consumption and has limited use as a livestock feed. It is not as valuable as corn, wheat, or grain sorghum for livestock rations. Rye is sometimes contaminated with ergot (a fungus), which makes it less palatable for livestock. Ergot contaminated rye can be toxic to livestock, especially swine, if fed at high levels. The use of rye in livestock rations should be limited to no more than one-third of the ration. A high level of rye feeding in dairy rations results in a hard, unsatisfactory butter from the milk. When rye is fed to livestock it should be coarsely ground or rolled to increase palatability.

Hominy Feed

Hominy feed has a little more protein and fat and is higher in fiber than corn. It is about equal to corn in feeding value and can be substituted on an equal basis for corn in the ration. When hominy feed is used as a major energy source in swine rations, it tends to produce soft pork.

Rice Bran

Rice bran is the seed coat and germ removed when rice is polished for human consumption. It is similar to wheat bran in feeding value for livestock. Rice bran is used mainly in dairy rations although it may be fed to other classes of livestock. Because of its high fat and fiber content and low palatability, it should be limited to not more than one-third of the ration.

SOURCES OF ENERGY—FORAGES

Forages (roughages) can supply some of the energy needs in the livestock ration, although they are not as concentrated a source of energy as the grains. The value of forages for livestock feed is highly dependent on the time of harvesting. As forage plants mature, the crude fiber content (cellulose and lignin) increases, which lowers the digestibility of the feed. When forages are harvested as silage, more of the nutritional value of the plant is preserved.

Corn Silage

Corn silage, which contains almost 50 percent grain on a dry matter basis, is an excellent energy source for certain classes of livestock, such as dairy cattle. Sorghum and small grain silages are lower in energy content than corn silage.

Straws

Straws, such as oat, barley, and wheat, are low in energy value and are not used as a major source of energy in the ration. Straw may be used if additional fiber is needed in the ration.

Corn Stover

Corn stover, when it is properly supplemented, may be used in rations for heifers and dry milk and beef cows, although it is generally low in nutritional value.

Pastures

Properly managed pastures can be a good source of nutrients for livestock. Rotate and fertilize pastures to get the best yield and nutritional value. The quality of the pasture must be closely watched and the ration supplemented with good quality stored forages when necessary.

SOURCES OF ENERGY—BYPRODUCTS

Several byproduct feeds may be used as energy sources in livestock rations. Some common byproduct feeds are dried citrus pulp, dried beet pulp, potato meal, dried sweet potatoes or sweet potato meal, dried bakery products, and dried whey.

Dried Citrus Pulp

Dried citrus pulp is a byproduct of the citrus fruit industry and is made up of the remaining pulp after the juice is removed and sometimes cull fruit that cannot be used for juice. Generally the pulp is dried for shipping but it can be fed wet if the feeder is located near a processing plant. Dried citrus pulp is used mainly in dairy cattle rations but can be fed to beef cattle. It is seldom used in rations for other classes of livestock. While dried citrus pulp is high in fiber content, it is considered an energy feed. It is usually limited to not more than 20-25 percent of the ration.

Dried Beet Pulp

Dried beet pulp is a byproduct of the sugar beet industry. It is usually used in dairy cattle rations and occasionally in rations for horses, beef cattle, and sheep. The major uses of dried beet pulp in rations is to add bulk, make the ration more palatable, and as a mild laxative. It is only secondarily an energy source in the ration. Dried beet pulp should not replace more than 20 percent of the grain in the ration.

Potato Meal

Potato meal may be used as a substitute for grain in dairy and beef rations. When it is cooked it can be substituted for grain in swine rations.

Sweet Potatoes

Dried sweet potatoes or sweet potato meal is rarely used in farm feeding of livestock but is used in feeding experiments. Sweet potatoes produce a high yield of carbohydrates but are expensive to feed. When sweet potatoes are limited to not more than 50 percent of the ration, their feeding value for cattle and sheep is about equal to corn. Generally swine do not like sweet potatoes and do not do well when fed rations containing this feed.

Dried Bakery Product

Dried bakery products (a mix of stale bakery products such as bread, cake, cookies, etc.) can be a good substitute for corn in cattle and swine rations if its use is limited to not more than 20 percent of the ration. It is similar to corn in composition but does have a high salt content, and contains 12-16 percent fat.

Dried Whey

Dried whey is a byproduct of the cheese manufacturing industry and consists of the milk remaining after the casein and fat have been removed. It is very high in milk sugar and is considered to be an energy feed, although it is also relatively high in protein. Dried whey is generally used in poultry rations and in early weaning rations for other classes of livestock.

SOURCES OF ENERGY—FATS

Feed grade animal fat is a byproduct of packing plants, poultry processing plants, and animal rendering plants. Animal fat is an economical source of energy used in the manufacturing of commercial mixed feeds. Commercial feed mixes will contain 1-7 percent animal fat, depending on the type of feed. Animal fat in the feed reduces the dustiness of the feed, improves its color, texture, and palatability, reduces wear on feed mixing equipment, and improves pelleting characteristics. Feeds containing animal fat are often treated with antioxidants to prevent the feed from becoming rancid in storage. Beef and dairy cattle rations may contain up to 5 percent animal fat, while swine rations may have up to 20 percent animal fat.

SOURCES OF ENERGY— MOLASSES

A variety of molasses feeds are available to provide energy in livestock rations. The common types of molasses are cane or blackstrap, beet, citrus, and wood. All of the molasses are concentrated water solutions of sugars, hemi-celluloses, and minerals and are similar in feeding value per pound of dry matter. Cane or blackstrap molasses is the most commonly used of the various types of molasses available. Molasses is used in rations for cattle, sheep, and horses but is seldom used in swine rations because it may cause scouring. It improves palatability, aids rumen microbial activity, reduces the dustiness of the ration, and serves as a binder when feeds are pelleted. Molasses is usually limited to not more than 10-15 percent of the ration.

FUNCTIONS OF ENERGY NUTRIENTS

Energy nutrients are needed for the maintenance of life in the animal. The work of the vital functions of the animal is called its *basal metabolism.* The basal metabolism of an animal is defined as the heat production of the animal while it is at rest and not digesting food. Vital life processes such as the beating of the heart, maintenance of blood pressure, transmission of nerve impulses, breathing, and the work of other internal organs depends on a supply of energy nutrients.

Energy is also used in the secretion of milk and the production of eggs and wool. Energy nutrients are oxidized in the muscles to keep them in a state of tension. An animal requires more energy standing than lying down and even more energy to move about or do work.

Energy supplies the fuel which maintains the body temperature of the animal. There are several sources of heat to maintain body temperature. These include the work of the vital organs, utilization of nutrients, normal activity of the animal, work done by the animal, and shivering. The lower critical temperature for a given animal is that point at which the body must increase the oxidation of nutrients to increase body temperature. The animal begins to shiver when it reaches critical temperature. Lower critical temperature will vary depending on the species of animal, how much hair or wool covering it has, how fat it is, the feeding level, how much activity the animal is doing, and the movement and humidity level of the air. When animals are on full feed they seldom reach lower critical temperature unless the weather is extremely cold.

It is only after all the maintenance needs of the animal are met that energy nutrients can be used for growth or production. Some energy is lost through the feces, urine, and gases produced in the body.

Fattening livestock requires a large amount of energy nutrients. Energy not used for other needs is deposited as fat within the body tissues. The deposition of fat in the tissues makes the meat tender, juicy, and gives it a better flavor. Fat is the highest cost gain in animals.

The energy requirements of lactating animals is almost twice as high as for those not producing milk. This energy must also be in the form of net energy. A shortage of energy in the ration will limit milk production although the animal will use body fat to some extent for milk production if the ration is deficient in energy.

Energy is used for the development of the fetus in pregnant animals. It is important that rations during the gestation period have sufficient energy to maintain the animal in a healthy condition without its getting too fat. Horses, for example, need about a 12 percent increase in energy in the ration during the last 90 days of pregnancy. The amount of energy provided in the diet must be increased during this time because the horse reduces its voluntary intake of feed as the fetus increases in size.

When feeding horses, the amount of energy needed for no body weight change and the normal activities of the nonworking horse is called the *maintenance requirement.* Energy above the needs for maintenance must be provided in the ration of the working horse. A number of factors affect the amount of energy needed, including the intensity and duration of work, the condition and training of the horse, the ability and weight of the rider or driver, the degree of fatigue of the horse, and the environmental conditions under which the horse is performing work.

EFFECTS OF DEFICIENCY OF ENERGY

A deficiency of energy in the ration causes a number of problems for all classes of livestock. Typical of these are:

1. Slower growth in the young.
2. A delay in the onset of puberty.
3. A decrease in milk yield in lactating females.

4. A shortened lactation period.
5. A loss in body weight.
6. Several kinds of reproductive problems including reduced fertility and delayed estrus.
7. In sheep, a reduction in wool quantity and quality.
8. A higher mortality rate.
9. A lowered resistance to disease and parasites.
10. Weakness, generally poor condition, and an unthrifty appearance.
11. Hypoglycemia.
12. A loss of subcutaneous fat.
13. A reduction in levels of blood glucose, calcium, and sodium.

ENERGY REQUIREMENTS IN THE RATION

The requirements for energy in the ration vary with specific classes of livestock. These requirements are discussed in the chapters which deal with feeding each class of livestock.

SUMMARY

Fuel is supplied to the body by the energy nutrients. Energy is used for digestion, absorption of nutrients, breathing, heart action, movement of muscles, tissue building, production of milk, eggs, wool, and mohair, reproduction, waste formation and excretion, and to supply heat to maintain body temperature. Some feed energy is lost through the feces, urine, and gases produced in the body. Excess energy not used to sustain life is stored as body fat.

The major sources of energy in livestock rations are carbohydrates, fats, and oils. A small amount of energy comes from the protein in the ration, but carbohydrates are the most important source of energy in the ration. Feed grains, especially corn, provide most of the carbohydrates in livestock rations. Some energy also comes from forages and by-products.

A deficiency of energy in the ration causes many problems, including slow growth, poor reproduction, and a generally unthrifty condition.

REVIEW

1. Why are energy nutrients needed in livestock rations?
2. Define basal metabolism.
3. What are some of the vital life processes that require energy nutrients?
4. List other functions of energy nutrients in addition to the vital life processes.
5. List the sources of heat to maintain the body temperature of the animal.
6. What is meant by the term *critical temperature* and how is it important in livestock management?
7. What is meant by "net energy"?
8. Is all the energy in the ration used by the animal? Explain.
9. Discuss the energy needs of animals for milk production, pregnancy, and work.
10. Which three nutrients are the major sources of energy in livestock rations?
11. Which of these is the most important and why?

12. Describe carbohydrates.
13. Which carbohydrates are the most easily digested and which are the hardest to digest?
14. What are hexoses?
15. Name the most common hexoses in the animal body.
16. How are compound sugars formed?
17. Name the most important compound sugars in the animal body.
18. What parts of the plant store the most easily digested carbohydrates?
19. Discuss the digestion of fiber.
20. Compare the amount of energy supplied by fats and oils as compared to carbohydrates.
21. Why do fats have more energy value than carbohydrates?
22. Describe saturated and unsaturated fatty acids.
23. Name three essential fatty acids.
24. Why are fats added to the rations of livestock?
25. List and describe three major concentrate sources of energy grown or used in your area.
26. Discuss the use of forages as an energy source in livestock rations.
27. Name and describe two byproduct feeds used in livestock rations.
28. What is a common source of fats for livestock rations?
29. Discuss the use of molasses as an energy source in livestock rations.
30. List some common effects of a deficiency of energy in livestock rations.

4

PROTEIN

OBJECTIVES

After completing this chapter you will be able to:

- Describe protein.
- List sources of protein.
- Describe the functions of protein.
- Describe the symptoms of protein deficiency in the animal.
- Discuss nonprotein nitrogen sources.

PROTEINS

Proteins are long and complex organic compounds that are formed when amino acids are combined with each other into polymers. Chemically, proteins contain carbon, hydrogen, oxygen, nitrogen, sulphur, and sometimes phosphorus and iron. By molecular weight, the percentage of each is: carbon (53 percent), hydrogen (7 percent), oxygen (23 percent), and nitrogen (16 percent). When present, sulphur, phosphorus, and iron each make up less than 1 percent of the molecular weight of the protein molecule. Protein molecules are large, with molecular weights ranging from about 5000 to several million. The internal organs and soft structures of the body are composed primarily of proteins. Proteins have a wide variation of shape, solubility, biological function, chemical composition, and physical properties.

All proteins in the food chain originate from plants that combine simple sugars and carbohydrates (produced by photosynthesis) with nitrogen and small amounts of phosphorus, sulphur, and iron, to form amino acids. The digestion of plant protein by the animal provides the amino acids used in the body to form animal protein. Each protein molecule is made up of hundreds of amino acids. Different kinds of protein result from different proportions of amino acids in the protein molecule.

The plant cell nucleus and protoplasm is mainly protein. Most of the protein in plants is found in the leaves, petioles, and seeds.

In addition to proteins, plants contain simpler nitrogenous compounds called *nonprotein nitrogen* (NPN). Some NPNs are simple forms of amino acids or combinations of amino acids, which are not as complex as proteins. Young growing plants contain simple nitrogenous compounds called *amides*. Only small amounts of amides are found in seeds and mature plants but immature plants may have over 30 percent of their nitrogen in this form.

Crude protein refers to all the nitrogenous compounds found in a feed. This includes protein and nonprotein nitrogen such as urea and ammonium salts. The nitrogen content of protein averages 16 percent. The amount of crude protein in the feed is found by determining the nitrogen content and multiplying by 6.25 (100 divided by 16 = 6.25). The feed sample is treated chemically, causing it to release ammonia. The ammonia is then titrated to determine the percent of nitrogen it contains. Sometimes the crude protein content is determined by first subtracting the NPN content such as nitrates, nitrite, ammonia, and urea before multiplying by the 6.25 factor. A commercial feed tag shows the guaranteed minimum crude protein content of the feed as a percent of the weight.

Not all crude protein in a feed is digestible. Feeds sometimes contain substances such as hair, hooves, and feathers, which contain nitrogen but are not digestible. About 60 percent of a roughage ration and 75-85 percent of a high-concentrate ration is digestible, depending on the species of animal. *Digestible protein* is a ration's approximate amount of protein available for use by the animal. It is the difference between the protein content of the feed and what is found in the feces.

Because ruminants can utilize both protein and NPN through microbial action in the rumen, the use of the crude protein value is valid and realistic when balancing ruminant rations. Both nonruminant and ruminant animals can utilize the simple nitrogenous compounds (which consist of amino acids and combinations of amino acids) in the same manner as protein. However, nonruminant animals cannot use the amides as a substitute for the essential amino acids.

AMINO ACIDS

Amino acids are organic acids containing one or more alpha-amino groups (NH_2) that form the building blocks of proteins. The alpha-amino group combines with a carboxyl group ($-COOH$) to form the amino acid. The NH_2 group and the COOH group are both attached to the same carbon atom in all amino acids in proteins. The R in the structural formula shown in Figure 4-1 represents side chains, which may be very simple or quite complex. For example, glycine has a single hydrogen atom attached as a side chain. Tryptophan is a complex amino acid that contains two carbon ring structures attached as a side chain. The number of different amino acids in a protein varies from a low of 3 or 4 to a high of 14 or 15. The average protein contains 100 or more amino acids, which are linked by peptide bonds. Peptide bonds are formed by condensation reactions between the COOH groups and the NH_2 groups, producing polypeptide chains, Figure 4-2.

Although commonly referred to as amino acids, proline and hydroxyproline are imino acids formed from NH and COOH groups.

There are 20 to 22 amino acids commonly found in proteins. Animal tissues contain 20 or more of these amino acids. Amino acids are classified by the number of acidic and basic groups found in the molecule. An excess of carboxyl groups results in an acidic amino acid, while an excess of basic groups results in a basic amino acid. Amino acids containing one carboxyl and one basic group are classified as neutral amino acids. Figure 4-3 shows the structural formulas for a number of the common amino acids.

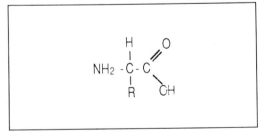

Figure 4-1. General structure of amino acids. The R represents side chains which may be very simple or quite complex.

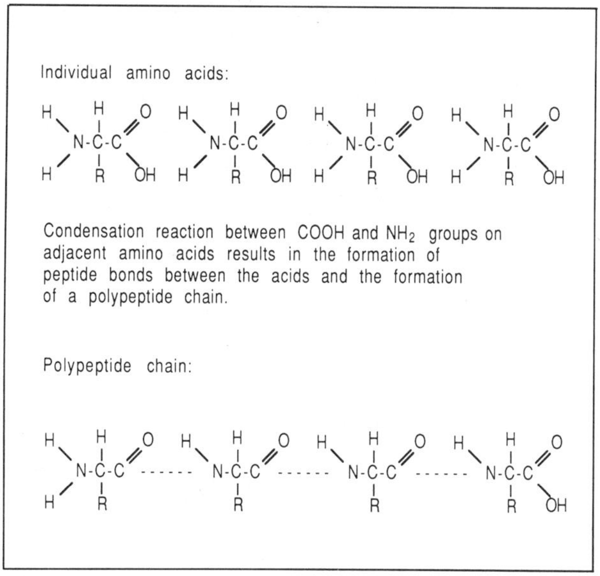

Figure 4-2. Formation of a polypeptide chain.

Essential amino acids are those which must be provided in the ration of nonruminant animals because the animals cannot synthesize them fast enough to meet their needs. Ruminant animals can generally synthesize the essential amino acids by rumen microbial action at a rate sufficient to meet their needs. Nonessential amino acids are needed by animals but are synthesized in the body from other amino acids and, therefore, do not have to be provided in the ration for either nonruminant or ruminant animals. There are 10 essential and 13 nonessential amino acids for swine, while 14 amino acids are considered essential for poultry, Table 4-1.

The quality of a protein is related to its amino acid content. High quality proteins have a good balance of the essential amino acids. Poor quality proteins are deficient in amount or balance of the essential amino acids.

When feeding nonruminant animals, the amino acid content of the protein is of greater importance than the percent of protein present in the feed. Swine and poultry do not have a protein requirement per se; rather, they have a requirement for specific individual amino acids and nonessential nitrogen. If the amino acid levels in the diet are adequate, the percent of protein it contains has little meaning. For example, swine may grow better on a 12 percent protein ration that contains good quality proteins than on a 16 percent ration made up of poor quality proteins. Swine, horses, and poultry all need a variety of protein sources in the ration to provide the balance of essential amino acids needed.

It has long been believed that cattle, sheep, and goats can meet their amino acid needs from plant proteins and NPN sources such as urea, because they can synthesize the essential amino acids in the rumen. The rumen microbes digest the available protein, synthesize the essential amino acids, and then pass through the digestive tract where they are digested and utilized by the ruminant to provide the needed amino acids.

Recent research indicates that good quality proteins may be reduced in quality by fermentation and microbial synthesis in the rumen.

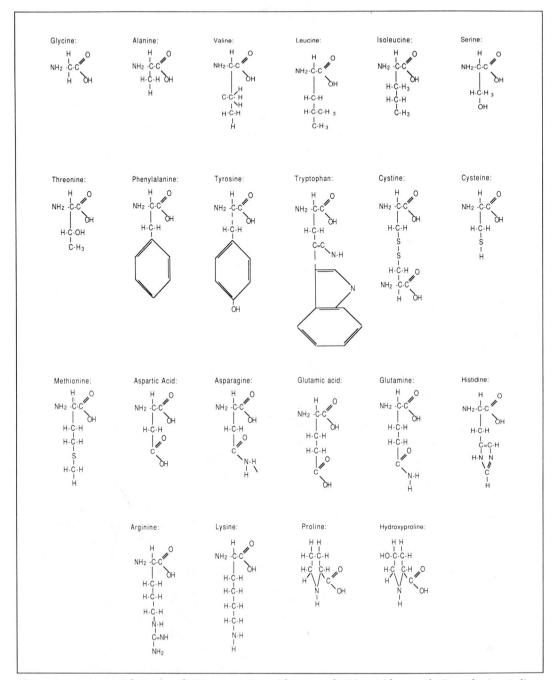

Figure 4-3. Structural formulas of common amino acids. Neutral amino acids are: Glycine, Alanine, Valine, Leucine, Isoleucine, Serine, Threonine, Phenylaline, Tyrosine, Tryptophan, Cystine, Cysteine, and Methionine. Acidic amino acids are: Aspartic acid, Asparagine, Glutamic acid, and Glutamine. Basic amino acids are: Histidine, Arginine, and Lysine. Proline and Hydroxyproline are imino acids. A hexagon in the structural formula represents a six-carbon ring.

Table 4-1. Essential and nonessential amino acids for swine and poultry.

Essential	Nonessential
Arginine	Alanine
Histidine	Aspartic acid
Isoleucine	Citrulline
Leucine	Cysteine
Lysine	Cystine
Methionine*	Glutamic acid
Phenylalanine**	Glycine
Threonine	Hydroxyglutamic acid
Tryptophan	Hydroxyproline
Valine	Norleucine
	Proline
Additional essential for poultry:	Serine
	Tyrosine
Alanine	
Aspartic acid	
Glycine	
Serine	

*Part can be replaced with Cystine.
**Part can be replaced with Tyrosine.

This may cause a deficiency in one or more of the essential amino acids. Quality protein can be protected in the rumen by encapsulating to prevent exposure to the fermentation in the rumen. Treating the protein with heat or chemicals (such as tannic acid and formaldehyde) before feeding reduces the tendency toward ruminal degradation. When the high quality protein is protected, the microbes are more likely to use NPN sources in fermentation. More research is needed in this phase of animal nutrition.

The concept of metabolizable protein is currently being studied as a method of describing amino acids available to the animal after rumen digestion. It is believed that metabolizable protein will more accurately describe the needs of the ruminant animal for maintenance, growth, and milk production.

LIMITING AMINO ACID

Limiting amino acid is the term for the essential amino acid that is present in the lowest amount in the feed (when expressed as a percent of its requirement). Essential amino acids are required in the ration in definite proportions. The amount required varies with the function of the amino acid but is always the same for the same function in a given species of animal. Even if other essential amino acids are present in surplus, they can only be used to the extent that the one in shortest supply is available. For example, in swine rations of shelled corn, lysine is the limiting amino acid because it is deficient in corn.

Amino acids may supplement each other when two different protein feeds have different amounts of a limiting amino acid. When the two protein sources are mixed in the ration and one has an excess of the limiting amino acid that is also deficient in the other protein source, the animal will draw amino acid from both sources to overcome the limiting effect. Because of this supplementary effect, it is recommended that more than one source of protein be supplied in the diets of nonruminants.

FUNCTIONS OF PROTEINS

Proteins are necessary for all plant and animal life because they are an essential part of all living tissue. The muscles, cartilage, ligaments, nerves, brain, blood cells, internal organs, skin, hair, wool, feathers, hooves, horns, and bones of animals all contain protein. Secretions in the body including enzymes, hormones, mucin, and milk all require specific amino acids. Protein is required for fetal development during the gestation period.

While protein is needed for maintenance, finishing, work, and wool production, the greatest need is for growth, reproduction, and lactation. There is no other nutrient that can replace protein in the diet. Depending on the species, the minimum level of protein needed in the ration ranges from 8 to 21 percent. More protein is required for young animals growing rapidly than for more mature animals. Protein requirements are also higher during the gestation and lactation periods than at other times.

PROTEIN DEFICIENCY

A shortage of protein in the diet results in a variety of symptoms, depending on the species and the degree of deficiency. Generally, depressed performance and higher production costs result from dietary deficiencies of protein.

Dairy Cattle

A limited amount of protein is stored in the blood, liver, and muscles of cattle. When the diet is deficient in protein, this reserve is quickly used up and signs of deficiency appear.

A chronic deficiency of protein in the diet depresses the appetite, resulting in reduced feed intake. This causes a combined deficiency of both protein and energy.

When a protein deficiency continues over a long period of time, there is a reduction in general health and vigor as well as a reduction in milk production. During pregnancy, a protein deficiency results in a slower growth rate of the fetus and the production of poor quality colostrum. A protein deficiency in the diets of young calves results in slow growth.

During lactation the solids-not-fat content of the milk is reduced. Cows will lose more weight than normal during early lactation and do not gain weight normally during the later stages of lactation. High producing cows are especially susceptible to a reduction in body condition during lactation.

Protein deficiency may result in lowered immunity and reduced hormone secretions. This can cause a higher level of infectious and metabolic diseases than normal in cattle.

Beef Cattle

The major symptom of a protein deficient diet in beef cattle is appetite depression, which results in a reduced energy intake. This can cause delayed or irregular estrus in breeding females. A loss of weight, slow growth, and reduced milk production will also result from a protein deficiency in the diet.

Sheep

Protein deficiency in the diet of sheep also depresses appetite, reducing feed intake and resulting in an energy shortage. Poor growth rate, poor muscular development, and loss of weight will result. The efficiency of feed utilization is reduced when the diet is deficient in protein. Wool production is lower and reproductive problems also appear. When the protein deficiency is drastic, severe digestive disturbances, anemia, and edema will result.

Goats

When the dietary crude protein level drops below 6 percent, appetite is depressed, so feed intake is reduced. This results in a combined shortage of protein and energy, which lowers feed utilization and reduces rumen function. An extended deficiency of protein will slow down fetal development, resulting in lower birth weights and slower kid growth. Milk production is lower during the gestation period when the diet is deficient in protein over an extended period of time.

Horses

The primary indication of protein deficiency in horses is a depressed appetite. Mature horses will lose weight and young horses will grow more slowly and not develop properly. Reduced fertility in mares and lower milk production during gestation also are caused by protein deficiency.

Swine

Protein deficiency in swine slows down growth and makes the pigs unthrifty, leaving them with an increased susceptibility to bacterial infection. Other signs of protein deficiency include anemia, gross edema, and increased fat concentration in the liver.

Poultry

Borderline protein deficiency in the diet of poultry results in poor growth and feathering, reduced egg size, and lower egg production. The hatchability of eggs is not affected by protein deficiency. There is also a tendency for increased fat deposit in the carcass and liver. A deficiency of lysine in the diet will cause a lack of melanin pigment in black or reddish-colored feathers.

When the protein deficiency is more severe, feed intake will stop. This results in a cessation of egg production in four or five days, body weight loss, the resorption of the ova, constipation, and death.

Deficiencies in the amino acids leucine, isoleucine, and phenylalanine result in deformity of the tongue.

Too wide a ratio of calorie-to-protein intake in the diet reduces growth rate and egg production. Too much feed is consumed, which results in excessive carcass and liver fat deposits and a poor conversion of feed into meat or eggs.

PROTEIN AS ENERGY SOURCE

When more protein is included in the diet than is needed for body functions, the excess may be used as an energy source. The nitrogen is split off the protein molecule and excreted in the urine. The remaining material is used for energy needs or stored as body fat. Because protein feeds are generally more expensive than energy feeds, this is not an economical source of energy. Rations should be properly balanced to be economical. Protein that is not usable because of a limiting amino acid could be used as an energy source even though the overall diet does not contain an excess of protein. However, this is not desirable from the standpoint of proper growth and development of the animal; every effort should be made to have the correct balance of amino acids in the diet.

UNAVAILABLE FEED PROTEIN

When feeds are improperly handled they sometimes become overheated in storage. Low moisture silage or haylage and hay that is stored when it is too wet are often damaged by overheating. When feed overheats, it becomes charred, browned, or caramelized. The protein content is denatured and the nitrogen compounds are not as readily available.

When calculating rations, adjustments for heat damaged feed need to be made when there is visible evidence of such damage. Feed that has moderate heat damage should be calculated to contain only 80 percent of the crude protein shown in feed composition tables. When a forage is dark brown or blackened, it may contain less than 50 percent of the feed composition table value of crude protein in usable form. Additional protein must be added to the ration to compensate for heat damaged feeds.

Feed may be analyzed in a laboratory to determine its nutrient content. If the acid detergent fiber component contains nitrogen, this is an indication that there will be some unavailable protein in the feed.

PROTEIN SOLUBILITY

Some feeds contain a high level of water soluble crude protein. For example, direct cut grass silages, sorghum silages, and high moisture corn may contain over 60 percent water soluble crude protein. This protein is poorly utilized when all silage rations are fed.

Water soluble crude protein is quickly attacked in the rumen by bacteria enzymes and degraded to simpler compounds and ammonia. When the released ammonia is greater than the capacity of the bacteria to make use of it, there is an immediate loss of part of the dietary nitrogen by excretion in the urine.

Byproduct concentrate feeds often contain a high level of soluble crude protein, which makes them a poor choice in the ration when feeding silages. Other concentrates such as corn grain are a better choice because they contain less soluble crude protein. Caution must be exercised, however, because the solubility of grain protein is increased when fermentation occurs in moist storage. On the other hand, care must be taken not to underestimate the protein needs in the ration because some of the crude protein is unavailable or highly soluble.

BIOLOGICAL VALUE OF PROTEIN

In feed, the percent of digestible protein that is retained by the animal for use is a measure of the biological value of the protein. Comparing the excreted protein in the feces and urine with the protein intake in the diet gives a measure of the digestible protein value of a feed.

A protein feed with a good balance of the essential amino acids will show less excreted protein, thus having a high biological value— a good quality protein. A protein feed that is deficient in one or more of the essential amino acids has a low biological value (larger quantities of protein are excreted) and is referred to as a low quality protein. In general, animal protein feeds have a higher biological value than plant proteins because they contain a better balance of amino acids.

PLANT PROTEIN SUPPLEMENTS

Feeds high in amino acids are called *protein supplements*. A protein supplement generally contains over 20 percent crude protein. Many commercial protein supplements are generally produced for particular classes of livestock. These supplements are usually blends of animal and plant protein sources and may contain minerals, vitamins, and antibiotics. Protein supplements made for ruminants often have urea added.

The protein supplements are divided into two general groups based on their origin and method of processing. These two categories are *animal origin* and *plant origin*.

Protein supplements of animal origin are considered to be of higher quality because they contain a better balance of the essential amino acids. These supplements include meat meal, fish meal, condensed fish solubles, dried whey, casein, dried milk albumin, and dried skim milk. Meat packing and rendering plants are a source of meat-meal supplements. Surplus milk and milk products provide sources of milk-based supplements.

Not all of the animal origin protein supplements are of high quality. For example, feather meal (from poultry processing plants) contains 85 percent crude protein, most of which is not easily digested.

The major source of plant protein supplements is oil seed byproducts. The most commonly used is soybean meal, generally the most economical of the plant protein supplements, containing about 45 percent (normal range 43-48 percent) protein on a dry matter basis. Most commercial supplements contain a large amount of soybean meal.

Cottonseed and linseed meal are also important plant protein sources. Other oil seed sources include rapeseed meal, sunflower meal, safflower meal, peanut meal, and coconut meal. The oil seed meals vary in composition and feeding value depending on the source, the amount of hull and/or seed coat left in the meal, and the method of extracting the oil. Dehydrated alfalfa meal, while not one of the oil seed byproducts, is also an important source of plant protein.

FAT EXTRACTION METHODS

Over the years, three methods of extracting fat from oil seeds have been used: hydraulic, expeller, and solvent.

The original method, widely used in southern United States for many years, is the hydraulic press. This is a mechanical method of extraction, leaving a lot of the oil in the meal. It is not a continuous operation and is rarely used today.

The expeller method was developed for use in the soybean industry and is also a mechanical process. It uses a screw press and is a continuous operation. This method is still used widely in the cotton belt but has been generally replaced with the solvent method in the soybean industry.

The solvent method is a chemical process using hexane as the solvent. This method was originally developed in Germany and its use began in the United States in the late 1940s. Solvent extraction removes almost all the fat from the meal. In some cases it is necessary to put back some fat in the meal to maintain the desired fat level for feeding purposes. New fat-extraction plants usually use the solvent process. Practically all soybean meal is currently being produced by the solvent process.

SOYBEANS AS A PROTEIN SOURCE

Unprocessed soybeans can be used in grain mixes for ruminants if the price is low enough.

Soybeans are a little lower in protein content (37-38 percent) than soybean meal and contain about 17-18 percent fat. The use of soybeans in the ration should be limited to not more than 20 percent of the total grain mix. To prevent diarrhea or throwing the animals off-feed, gradually adjust the amount of soybeans in the ration over a period of time.

Soybeans contain the enzyme urease, which releases ammonia from urea; therefore raw soybeans should not be fed in a beef diet already containing urea. This is not a problem when feeding soybean meal because the heat used in its processing eliminates the urease. While it is not necessary to roast or extrude soybeans for ruminant feeding, this treatment will increase the palatability and stability of the feed. Soybeans used in swine diets should be roasted or extruded for maximum protein utilization and maximum animal performance.

Raw soybeans also contain an antitrypsin factor that prevents the action of the enzyme trypsin in nonruminants such as swine and poultry. Cooking soybeans destroys this factor.

Whole cooked soybeans can be used in swine diets and may result in about a 4-8 percent increase in feed efficiency, attributable to the higher fat content. Rate of gain is usually not improved and more research is needed to learn the effect on carcass quality. Special equipment is required to uniformly heat raw soybeans to destroy the antitrypsin, likely making raw soybeans a more expensive protein source than soybean meal.

UREA

Urea, a nonprotein nitrogen compound (N_2H_4CO) in crystalline form, is white, odorless and contains 45 percent nitrogen. It has a protein equivalent of 281 percent (45 percent $N \times 6.25$). Urea is manufactured by combining atmospheric nitrogen with ammonia and carbon dioxide. It contains carbon, oxygen, nitrogen, and hydrogen.

Urea is the most common of the nonprotein nitrogen sources used in ruminant rations and use should be limited to not more than one-third of the total protein in the ration. In beef cattle rations, urea can be used to provide all the supplemental protein needed for animals over 600 pounds (272 kg).

Urea is generally not used in nonruminant animal rations, but can be used in ruminant rations because the rumen microbes have the ability to utilize the nitrogen in the urea, forming amino acids needed by the bacteria. The enzyme urease, which is secreted by rumen organisms, releases ammonia from the urea. The ammonia reacts with organic acids during fermentation in the rumen to form ammonium salts of organic acids. The rumen microbes then utilize the ammonium salts to form microbial body proteins, which can be used by the animal.

When urea is included in the ration, it is necessary to assure a ready supply of easily fermented energy sources. This supplies the energy needed by the rumen microbes to form protein from the urea. Corn, grain sorghum, and molasses make good energy sources in a urea ration because the rate of release of carbon-skeletons (usually fatty acids) comes close to matching the rate of release of ammonia from the urea. While silages can be fed as energy sources in urea rations, the rate of release of the carbon-skeletons is much slower and, therefore, silages are not as good as grains for energy sources.

Urea is not palatable and must be thoroughly mixed in the ration to be acceptable to the animal. Mixing urea with molasses increases its palatability. Add urea to the ration gradually over a seven to ten day period when it has not been previously fed. This period of adjustment is necessary to allow the rumen bacteria to adjust to this source of nitrogen. If urea is added abruptly to the ration, a reduction in feed intake, production, or sudden death may result. The use of urea should be limited in the early stages of lactation for dairy cattle because they need higher quality protein during this

time. Do not top-dress urea on feed because it has a bitter taste and feed intake will be reduced.

UREA TOXICITY

Urea is a normal byproduct of protein metabolism in animals and is not toxic, because of its slow and controlled release. However, the ammonia produced by microbe activity in the rumen may be toxic if more is released than can be completely utilized by the microbes. It is absorbed into the bloodstream through the wall of the rumen and may cause death. This is more likely to happen when there is not enough of the readily fermentable carbohydrates in the ration and/or the level of urea in the ration exceeds one-third of the total protein nitrogen.

Two mechanisms operate to keep ammonia below a toxic level in the blood. One is the conversion of ammonia to microbial protein by microbes in the rumen. The other occurs in the liver where ammonia is combined with carbon dioxide to form the less toxic urea. The urea is released into the bloodstream and is ultimately excreted, mainly through the urine in mammals. In poultry it is converted to uric acid prior to excretion.

NITRATES/NITRITES

Plants sometimes take in more nitrogen than can be metabolized to form amino acids and proteins. Nitrate accumulation can be as high as 5 percent of the dry weight of the plant. Plants with a high accumulation of nitrates may have a toxic effect on livestock. Causes of high nitrate accumulation include excess levels of nitrate in the soil, dry conditions followed by adequate soil moisture, an imbalance in the soil fertility program, or other growing conditions that slow down plant growth.

Annual grasses are more likely to accumulate excess nitrates than are legumes or perennial grasses. The leaves and stems of plants are more likely to accumulate excess nitrates than the seeds; therefore, cereal grains generally do not cause a nitrate problem in livestock feeding.

Symptoms of the toxic effects of nitrate poisoning include poor appetite, slower gains, lower egg and milk production, watering of the eyes, rough hair coat, and lower conception rate. Pinkeye and foot rot are secondary infections that may more readily occur if the animal is weakened by nitrate toxicity. Acute toxic effects include noninfectious abortions, stillborn calves, drastic drop in milk production, rapid pulse, frequent urination, diarrhea, trembling, staggering gait, rapid breathing, prostration, convulsions, coma, and sudden death. Nitrate poisoning is treated with intravenously administered methylene blue in a 5 percent glucose solution or a 1.8 percent sodium sulfate solution. Begin treatment immediately upon observing the symptoms of poisoning.

Feeds with a high nitrate content are less palatable; therefore, the lower production and gain may be because of lower feed intake. Reproduction in swine does not appear to be affected by a long-term consumption of nitrates.

Nitrate poisoning can be prevented by maintaining a proper soil fertility program based on soil analysis. Feeding ruminants at least 3,000 I.U. of vitamin A per 100 pounds body weight per day will reduce the effects of nitrate poisoning. Forages suspected of having a high nitrate content should be mixed with forages lower in nitrate content, to reduce the nitrate intake. When ruminants are on complete roughage rations, feed higher energy sources such as molasses and make sure the diet is adequately supplemented with minerals such as copper, iron, manganese, and magnesium. The nitrate level of drinking water provided to livestock should be checked, as water is sometimes a source of nitrate poisoning.

Nitrates in the feed become toxic when they are reduced to nitrites in the blood. Nitrites oxidize the ferrous iron of the hemoglobin to the ferric iron of methemoglobin, which cannot transport or release oxygen to the tissues. Monogastric animals are more tolerant of higher nitrate levels in the feed than are ruminants because there is less reduction of nitrate to nitrite by nonruminants. Horses are more susceptible to nitrite toxicity than other nonruminant animals, because of the greater similarity of their lower digestive tract to that of ruminants.

OTHER NONPROTEIN NITROGEN SOURCES

There are other nonprotein nitrogen sources that may be used in rations for ruminants. These include biuret, ammoniated molasses, ammoniated beet pulp, ammoniated cottonseed meal, ammoniated citrus pulp, ammoniated rice hulls, diammonium phosphate, monoammonium phosphate, and ammonium polyphosphate. Ammonia in anhydrous (without water) or water form can be added to corn silage or other feeds to provide an NPN source. Adding ammonia to silage improves its preservation because of increased fermentation. The digestibility of straw and other low quality roughages is improved by the addition of ammonia. Biuret is similar to urea but releases ammonia more slowly in the rumen and is, therefore, less toxic. It is, however, more expensive than urea and not widely available. These NPN sources are acted upon by rumen microbes in a similar manner as urea, providing amino acids, but none of these sources of NPN are as widely used as urea in the ration.

BYPRODUCT FEEDS AS PROTEIN SOURCES

Brewer's grain, a byproduct of the brewing industry, is high in protein and a fair source of energy. Because it contains approximately 80 percent water, large amounts are needed to provide any substantial dry matter intake.

Wheat bran averages 14 percent protein and improves the palatability of the ration. It provides bulk and fiber in the grain mix and has a laxative effect.

Malt sprouts are high in protein but are sometimes bitter and are usually mixed with other feeds to improve palatability.

Beet pulp, cereal grain screenings (milling industry byproduct), and whey will add some protein to the ration but are usually not fed primarily for their protein content. Beet pulp improves the palatability of a ration and adds digestible fiber and bulk. Screenings can be an economical buy but the protein content is quite variable and screenings are often unpalatable and difficult to digest. Whey is economical in dried or liquid form but is relatively low in protein.

GRASS AND LEGUME FORAGES AS PROTEIN SOURCES

Legume forages contain more protein than grass forages but both are lower in protein than the oil meals. The crude protein content of forages is the best overall indicator of their feeding value in livestock rations. Forages harvested in the early stages of growth have a higher protein level than more mature forages, which have a higher fiber level, though less protein and energy.

GRAIN AS A PROTEIN SOURCE

Grains vary widely in protein level and are not fed primarily for their protein content but as a source of energy. However, the protein content of grains must be taken into account when formulating rations. Wheat, oats, rye, and milo all have higher protein contents than shelled corn. Corn, which is the most widely fed grain in livestock rations, is deficient in lysine. While a high lysine corn has been developed by plant geneticists, it has not gained wide acceptance among corn growers because of generally lower yields, so it is available on only a limited basis.

Deficiencies of amino acids in the grains are important in formulating rations for nonruminant animals, but generally are not considered in rations for ruminants. Adding protein supplements overcomes deficiencies in essential amino acids for nonruminant animals.

SUMMARY

Proteins are complex organic compounds formed from amino acids. Different combinations of amino acids make different kinds of proteins. Plants also contain simpler nitrogenous compounds called *nonprotein nitrogen.* Most of the protein in plants is found in the leaves, petioles, and seeds.

Crude protein refers to all the nitrogenous compounds in feed and is found by multiplying the nitrogen content of the plant by 6.25. Not all crude protein is digestible.

Amino acids are organic compounds, which form the building blocks of proteins. There are 10 essential amino acids that must be supplied in the rations of swine and 14 for poultry. Ruminant animals can synthesize the essential amino acids through action of the microbes in the rumen.

High quality proteins contain a good balance of the essential amino acids, while low quality proteins are deficient in one or more of the essential amino acids. A *limiting amino acid* refers to an essential amino acid that is deficient in a feed. The animal can utilize other amino acids in the feed only to the extent that the limiting amino acid is present in the ration.

Proteins are an essential part of all living tissue. No other nutrient can substitute for protein in the diet. Young animals require more protein in the diet than more mature animals. More protein is also required for gestation and lactation than at other times.

The major symptom of protein deficiency in the diet is the suppression of appetite with a resulting lowered feed intake and an accompanying shortage of energy. Poor fetal growth, reduced resistance to infectious diseases, lower milk production, lower wool production, anemia, and edema are also symptoms of protein deficiency in the diet.

Protein not used for other purposes may be converted to energy or fat in the animal's body. Because protein is higher in price than other sources of energy, it is not economical to feed excess protein.

Feed damaged by overheating may contain unavailable protein. Some feeds contain a high level of soluble crude protein, which may be rapidly lost in the animal's body.

Protein supplements generally contain over 20 percent crude protein and have a high level of amino acids. The two general types of protein supplements are animal and plant. Soybean meal is the most commonly used protein supplement in livestock feeding.

Urea is a synthetic nitrogen compound that provides a nonprotein nitrogen source in the rations of ruminant animals. The rumen microbes have the ability to utilize urea as a nitrogen source, converting it into microbial proteins and amino acids that can be utilized by the animal. Care must be taken to limit the amount of urea in the ration in order to prevent urea toxicity.

Excess nitrate in the feed can cause toxicity, especially in ruminants. Lower gains, digestive and reproductive problems, and death can result from nitrate toxicity. Reducing the nitrate level in the feed is the best way to prevent this problem. Affected animals may be treated with a methylene blue solution.

Byproduct feeds, forages, and grains all contain some protein. Generally these must be supplemented by protein supplements to achieve the desired balance of amino acids in the ration.

REVIEW

1. What are proteins?
2. In what part of the plant is most of the protein stored?
3. What are nonprotein nitrogen compounds?
4. What is crude protein?
5. How is the crude protein content of a feed determined?
6. What is digestible protein?
7. Compare the utilization of protein and nonprotein nitrogen by ruminant and nonruminant animals.
8. What are amino acids?
9. What is the difference between essential and nonessential amino acids?
10. What is meant by the quality of a protein?
11. Discuss protein quality as it relates to formulating rations for ruminant and nonruminant animals.
12. What is meant by a limiting amino acid and how does this relate to ration formulation?

13. Name the functions of protein in the animal's body.

14. At what stages of the animal's life are the protein requirements the greatest?

15. List some signs of protein deficiency in the animal's diet.

16. What is the relationship between protein deficiency and energy nutrition?

17. Discuss the use of protein as an energy source in the diet.

18. What may cause protein in the feed to be unavailable?

19. What are some indications that there might be unavailable protein in the feed?

20. What adjustments may need to be made when formulating a ration to compensate for unavailable protein in the feed?

21. What feeds may have a high level of soluble protein?

22. Why may a high level of soluble protein in the feed be a problem when formulating rations?

23. Discuss the biological value of protein.

24. What are protein supplements?

25. What are the two categories of protein supplements?

26. What is the most commonly used protein supplement and why is it the most common?

27. What are some other common protein supplements?

28. Describe the three methods of extracting fat from oil meals.

29. Discuss the use of soybeans as a protein supplement.

30. What is urea and what is it made from?

31. Why can urea be used in ruminant rations but not in rations for nonruminant animals?

32. Why is a good supply of readily fermentable energy sources necessary when using urea in the ration?

33. Why should the adjustment to urea feeding be made gradually?

34. Discuss the problem of urea toxicity.

35. In addition to urea, list some other nonprotein nitrogen sources.

36. List and discuss some byproduct feeds that may be used as protein sources.

37. Discuss grass and legume forages as protein sources.

38. Discuss grains as a protein source.

39. Discuss the problem of nitrate poisoning and how it can be prevented.

5

MINERALS

OBJECTIVES

After completing this chapter you will be able to:

- Describe minerals used in animal nutrition.
- List sources of minerals for animal feeding.
- Describe the functions of minerals in animal nutrition.
- Describe the deficiency symptoms caused by a lack of minerals in the ration.
- Discuss the requirements for minerals in the ration.

MINERALS DEFINED

When an organic material, such as a feed, is burned, the ash that is left is the mineral content of the material. Minerals are inorganic substances needed in very small amounts by animals. It is known that at least 18 mineral elements are needed by various animal species.

These minerals are divided into two groups based on the quantities needed by animals. Those needed in larger amounts are referred to as *major* or *macro* minerals and those needed in relatively small amounts are called *trace* or *micro* minerals. The major or macro minerals include sodium, chlorine, calcium, phosphorus, potassium, magnesium, and sulphur. Trace or micro minerals include chromium, cobalt, copper, fluorine, iodine, iron, manganese, molybdenum, selenium, silicon, and zinc.

FUNCTIONS OF MINERALS

Although the mineral content of animal bodies is small, ranging from 2 to 5 percent depending on the species, minerals perform many essential functions. Minerals provide material for the growth of bones, teeth, and soft tissues. They regulate many of the vital chemical processes in the body, aid in muscular stimulation and activity, reproduction, digestion of feed, repair of body tissue, formation of new tissue, and the release of body heat for energy. Minerals are an essential part of the blood, body fluids, and some secretions in the body. They help regulate the acid-base balance to maintain body fluids at a pH of about 7.0. They are also essential for the utilization of some vitamins in the body.

Bones contain about 25 percent minerals made up mainly of calcium (36 percent), phosphorus (17 percent), and magnesium (0.8 percent). The presence of minerals in the bones gives rigidity and strength to the skeletal structure of the body.

Some minerals are found in hair, hoofs, horns, blood cells, and other soft tissues of the body. Minerals, and enzymes, hormones, and vitamins that contain minerals, help regulate metabolic cycles and molecular concentration of body fluids, making them physiologically compatible with body tissues. Minerals also affect nerve irritability.

Milk contains all the significant minerals except iron, being 5.8 percent mineral on a dry matter basis. Minerals are an essential part of the ration for lactating animals.

Calcium and phosphorus are essential for the production of eggs. Calcium is one of the major components of the eggshell and phosphorus is necessary for the transport of calcium for eggshell formation.

DEFICIENCY SYMPTOMS

Some geographic areas produce feeds deficient in minerals because of mineral deficiencies in the soil. The most common deficiencies are iodine, manganese, cobalt, copper, magnesium, iron, and zinc.

A mineral deficiency may also result because of poor utilization by the animal. An interrelationship exists between the presence or absence of some minerals and the utilization of other minerals. Excessive amounts of some minerals in the diet can inhibit utilization of other minerals. The chemical form of the mineral and the presence of binding or chelating substances (such as oxalates, phytates, and fats) can affect the absorption and utilization of minerals. These interrelationships are described later in this chapter in the discussion of specific minerals.

A lack of iron in the blood reduces its ability to carry oxygen to the body cells. An iron, copper, or cobalt deficiency in young pigs causes anemia. When calcium and phosphorus are deficient, the bones and teeth do not form properly.

Mineral deficiencies may result in physical problems such as anemia, hypocalcemia (milk fever), parakeratosis, goiter, or death. Mineral deficiencies also result in inefficient feed utilization, poor gains, and lowered production of meat, milk, eggs, and wool.

SOURCES OF MINERALS

Commercial feeds such as protein supplements or mineral premixes usually contain both major and trace minerals. Feed tags show the guaranteed minimum and maximum percent of calcium, minimum percent of phosphorus, and minimum and maximum percent of salt in the feed. Major minerals are guaranteed as individual elements except sodium and chloride, which are guaranteed as the compound salt. Trace minerals are guaranteed as a minimum percent of the weight of the feed in the bag.

Calcium and phosphorus are usually supplied in commercial feeds by the addition of monocalcium phosphate, dicalcium phosphate, ground limestone, steamed bone meal, or calcium carbonate. Mineralized salt and mineral blocks are often used to provide additional minerals in the ration.

MINERALS IN THE RATION

Minerals are usually added to the ration either by feeding them free choice or including them in the mixed ration. Sometimes a combination of both these methods is used. Care must be taken not to have an imbalance or an excess of minerals in the ration, Table 5-1.

Trace mineral needs must be carefully monitored if there is a deficiency of one or more of them in the feeds grown or used in the area. The feeding of poor quality roughage or high- or all-concentrate rations may result in deficiencies in one or more of the essential trace minerals.

Salt, calcium, and phosphorus are the minerals most likely to be needed in the rations of farm animals. About 70 percent of the mineral content of an animal's body is calcium and phosphorus.

Trace-mineralized salt, calcium, and phosphorus are usually included in mixed feeds in rations. Trace-mineralized salt is generally used because it adds the necessary trace minerals, costs very little, and is not harmful to animals. Other minerals are not generally added to the mix unless special circumstances indicate a need for them. Trace-mineralized salt is included at 0.25 to 0.50 percent of the total ration, and calcium and phosphorus are added as needed to balance the ration. Minerals may also be fed free choice, in addition to those mixed in the ration.

When livestock are on pasture or are not being fed a concentrate feed, then minerals must be supplied free choice. This may be done with a multicompartment mineral feeder, with different minerals or mineral mixes being supplied in several compartments. However, research has shown that animals do not do an adequate job of balancing their mineral needs when given a free choice of several minerals. Unfortunately, the equipment for feeding minerals in compartments is expensive, and it takes time to fill each compartment of the mineral feeder.

Another method is to feed a single-mineral mix containing the necessary minerals in a single-compartment mineral feeder. When animals are being fed grain as a major part of the diet, a practical mix that may be used is 1 part trace-mineralized salt; one to two parts defluorinated phosphate, dicalcium phosphate, or steamed bone meal; and 1 to 2 parts ground limestone or oystershell flour. When the animal's diet is mainly pasture, hay, or silage, a mix that may be used is 1 part trace-mineralized salt and 1 to 3 parts of defluorinated phosphate, dicalcium phosphate, or steamed bone meal. Forages are higher in calcium than grains, so additional calcium (above what is contained in the phosphorus

source) is not needed when livestock are being fed mainly forages.

CALCIUM

Major Functions

The bones and teeth of an animal contain 99 percent of the calcium found in its body. Calcium is also an important part of milk and eggs, so it is essential in the diets of lactating animals and laying hens. Calcium is also important for proper nerve and muscle functioning and for maintaining the acid-base balance of the body fluids.

Deficiency Symptoms

A deficiency of calcium will result in abnormal bone growth and weak bones in all classes of livestock. Milk fever may occur when cattle are deficient in calcium shortly before or after calving.

Young animals that are deficient in calcium may develop *rickets*. The normal amount of calcium is not deposited in the growing bones. The joints become enlarged and the bones become weak, soft, and deformed and are easily broken. Young cattle are stiff, with swollen joints, bent knees, and arching backs. In young pigs the characteristic symptom of rickets is stiffness of the legs, with an unthrifty appearance and poor gain.

Older animals with a calcium deficiency may develop *osteomalacia*, also known as *osteoporosis* or *stiffs*. The calcium in the spongy part of the bones is withdrawn first to meet the needs of the animal. If the deficiency continues, the calcium is withdrawn from the shafts and other structural parts of the bone. The bones become porous and weak and will break easily. Often, injury to the joints causes the animal to become lame or stiff.

Calcium deficiency in pregnant females may manifest itself as a paralysis of the hindquarters. Calcium is withdrawn from the bones to meet the needs of the developing young. This is more likely to occur with brood sows

Table 5-1. Maximum tolerable levels of dietary minerals for domestic animals.

Element	Species Cattle	Sheep	Swine	Poultry	Horse
Aluminum[1], ppm	1,000	1,000	(200)	200	(200)
Antimony, ppm	—	—	—	—	—
Arsenic, ppm					
Inorganic	50	50	50	50	(50)
Organic	100	100	100	100	(100)
Barium[1], ppm	(20)	(20)	(20)	(20)	(20)
Bismuth, ppm	(400)	(400)	(400)	(400)	(400)
Boron, ppm	150	(150)	(150)	(150)	(150)
Bromine, ppm	200	(200)	200	2,500	(200)
Cadmium[2], ppm	0.5	0.5	0.5	0.5	0.5
Calcium[3], %	2	2	1	Laying hen 4.0 Other 1.2	2
Chromium, ppm					
Chloride	(1,000)	(1,000)	(1,000)	1,000	(1,000)
Oxide	(3,000)	(3,000)	(3,000)	3,000	(3,000)
Cobalt, ppm	10	10	10	10	(10)
Copper, ppm	100	25	250	300	800
Fluorine[4], ppm	Young 40 Mature dairy 40 Mature beef 50 Finishing 100	Breeding 60 Finishing 150	150	Turkey 150 Chicken 200	(40)
Iodine, ppm	50[5]	50	400	300	5
Iron, ppm	1,000	500	3,000	1,000	(500)
Lead[4], ppm	30	30	30	30	30
Magnesium, %	0.5	0.5	(0.3)	(0.3)	(0.3)
Manganese, ppm	1,000	1,000	400	·2,000	(400)
Mercury[4], ppm	2	2	2	2	(2)
Molybdenum ppm	10	10	20	100	(5)
Nickel, ppm	50	(50)	(100)	(300)	(50)
Phosphorus[3], %	1	0.6	1.5	Laying hen 0.8 Other 1.0	1
Potassium, %	3	3	(2)	(2)	
Selenium, ppm	(2)	(2)	2	2	(3)
Silicon[1], %	(0.2)	0.2	—	—	—
Silver, ppm	—	—	(100)	100	—
Sodium Chloride, %	Lactating 4 Nonlactating 9	9 —	8 —	2 —	(3) —

continued

Table 5-1. (continued)

Element	Species Cattle	Sheep	Swine	Poultry	Horse
Strontium, ppm	2,000	(2,000)	3,000	Laying hen 30,000 Other 3,000	(2,000)
Sulphur, %	(0.4)	(0.4)	—	—	—
Tin, ppm	—	—	—	—	—
Titanium[6], ppm	—	—	—	—	—
Tungsten, ppm	(20)	(20)	(20)	20	(20)
Uranium, ppm	—	—	—	—	—
Vanadium, ppm	50	50	(10)	10	(10)
Zinc, ppm	500	300	1,000	1,000	(500)

[1]As soluble salts of high bioavailability. Higher levels of less-soluble forms found in natural substances can be tolerated.

[2]Levels based on human food residue consideration.

[3]Ratio of calcium to phosphorus is important.

[4]As sodium fluoride or fluorides of similar toxicity. Fluoride in certain phosphate sources may be less toxic. Morphological lesions in cattle teeth may be seen when dietary for the young exceeds 20 ppm, but a relationship between the lesions caused by fluoride levels below the maximum tolerable levels and animal performance has not been established.

[5]May result in undesirably high iodine levels in milk.

[6]No evidence of oral toxicity has been found.

[7]Continuous long-term feeding of minerals at the maximum tolerable levels may cause adverse effects. The listed levels were derived from toxicity data on the designated species. The levels in parenthesis were derived by interspecific extrapolation. Dashes indicate that data were insufficient to set a maximum tolerable level.

Source: National Research Council, *Mineral Tolerance of Domestic Animals,* National Academy of Sciences, Washington, D.C., 1980.

than with ruminants, because the grain rations usually fed brood sows are deficient in calcium unless properly supplemented.

Interrelationships

The ratio of calcium to phosphorus, magnesium, manganese, and zinc must be correct for the proper utilization of these minerals. The calcium:phosphorus (Ca:P) ratio should be 1:1 to 2:1. An excessive amount of calcium in the ration will result in poor utilization of these other minerals. An excess of phosphorus in the form PO_4 decreases the absorption of calcium in the body. An excess of magnesium decreases the absorption of calcium, replaces calcium in the bone with mag-

nesium, and causes an increase in the excretion of calcium from the body. The vitamin D level is also critical because a deficiency of vitamin D in the ration prevents the proper utilization of calcium.

Calcium toxicity may occur with an excess of calcium in the diet. This may cause a decrease in the absorption of other minerals, the calcification of soft body tissues, and the formation of kidney stones.

Sources

Grains, grain byproducts, straw, dried mature grasses, and protein supplements from plant sources contain the least amount of calcium, ranging from 0.01 to 0.15 percent. Grass forages generally contain from 0.35 to 0.75 percent calcium, while legume forages and animal-origin protein supplements are the highest in calcium content, ranging from 1 to 2 percent on a dry weight basis. Plants grown on well-fertilized soils are higher in calcium content than those grown on poorer soils. Other good feed sources of calcium include fish meal, milk, skim milk, citrus pulp, and citrus molasses. Rations that are high in grain need a higher level of calcium supplementation, while legume forage rations need little or no added calcium.

Calcium supplements are derived from two basic groups of materials, calcium carbonates and calcium phosphates. Calcium carbonate ($CaCO_3$) materials, containing 35 to 40 percent calcium, have little or no phosphorus content. Typical materials in this group include ground limestone, oystershell flour, and marble dust, all about equal in value as sources of calcium in the ration; therefore, the cheapest available locally can be used. The major difference in selling prices is the cost of transporting materials to the local market.

Calcium phosphate materials contain about 30 percent calcium and 14 to 20 percent phosphorus. Typical materials in this group include steamed bone meal, defluorinated rock phosphates, and dicalcium phosphates. These materials are also about equal in value as sources of calcium and phosphorus. However, they are usually more expensive than supplements containing only calcium, so should be used only when a phosphorus supplement is needed in addition to the calcium. Any of the calcium supplements may be fed free choice in a mineral mix or added to the ration at the proper level.

PHOSPHORUS

Major Functions

Phosphorus is a vital element in many body functions and is involved in interrelationships with a number of other major and trace minerals. Approximately 80 percent of the phosphorus in the body is found in the bones and teeth; therefore, it is a key element in their proper growth and development. Phosphorus affects appetite, milk and egg production, reproduction, conversion of carotene to vitamin A, and utilization of vitamin D. Phosphorus is a component of protein in the soft tissues, and is involved in other metabolic processes.

Deficiency Symptoms

A deficiency of phosphorus may result in rickets or osteomalacia (see CALCIUM for a discussion of these). Phosphorus deficiency may also result in poor appetite, slow gains, lower milk and egg production, reproductive problems, poor utilization of vitamin D, deficiency of vitamin A, and a generally unthrifty appearance. While it is not a specific symptom of phosphorus deficiency, the deliberate eating of soil and chewing on other nonfeed objects may indicate a phosphorus deficiency in the diet. This condition is referred to as *pica*.

Interrelationships

Excessive amounts of calcium and magnesium in the diet reduces phosphorus absorption. In ruminants, an excess of phosphorus may cause urinary calculi. As mentioned above, the Ca:P ratio in the diet should be 1:1 to 2:1.

Sources

Feeds that are good sources of phosphorus include wheat bran, wheat middlings, cottonseed meal, linseed meal, meat scraps, tankage, fish meal, and dried skim milk. Legume and grass pastures grown on fertile soils are fairly good sources of phosphorus. Corn and sorghum silages and fodders are a little lower in phosphorus than the legume and grass pastures. Corn stover and small grain straw is very low in phosphorus, and mature, weathered grass forage is generally low in phosphorus.

Cereal grains and grain byproducts are fairly high in phosphorus, but much of it is not readily available to nonruminant animals because it is in the form of phytin, an organic phosphorus compound. An adequate vitamin D level in the diet improves the assimilation of phytin phosphorus. Cattle and sheep can make better use of phytin phosphorus than can swine and poultry. Cattle will utilize about 60 percent and swine about 50 percent of the phosphorus from plant sources. Poultry and swine should be supplied with phosphorus from inorganic sources.

Mineral supplements may be used to provide phosphorus in the ration. The type of supplement to use depends on the amounts of calcium and phosphorus in the ration from other sources and the desired calcium/phosphorus ratio. Supplement sources of phosphorus include steamed bone meal, defluorinated phosphates, dicalcium phosphate, monocalcium phosphate, ammonium phosphates, phosphoric acids, and soft phosphate with colloidal clay. Raw, unprocessed rock phosphate must have the excess fluorine removed by heating before use as a mineral supplement. Fertilizer grade superphosphate should not be used as a phosphorus supplement because of the excess fluorine content.

Phosphorus supplements may be included in a mineral mix fed free choice and/or added to the ration at the proper level.

CALCIUM/PHOSPHORUS RATIO

The ratio of calcium to phosphorus in the diet affects the utilization of both minerals. The optimum ratio varies with the specie, type of feed being fed, and the vitamin D level in the ration. The calcium/phosphorus ratio is not as critical when adequate vitamin D is found in the ration or is available from the animal's exposure to sunlight.

Ratios in the range of 1:1 to 1:2 are generally adequate for nonruminants such as swine and horses. A wider range (1:1 to 7:1)

appears to be satisfactory for ruminants; however, ratios below 1:1 are not adequate for any species of livestock. Specifically, it is recommended that dry cows and heifers have a ratio of 1.3:1; lactating animals a ratio of 1.5:1; and swine a ratio of 1.2:1.

SALT (SODIUM AND CHLORINE)

Functions

Salt (NaCl) contains the mineral elements sodium and chlorine, which are needed by all classes of animals. Cattle, sheep, and horses usually require more salt supplementation than swine or poultry because of the higher levels of forages in their diets. Most grains and forages produced on nonirrigated soils are low in sodium and chlorine content.

Sodium and chlorine are important for maintaining osmotic pressure in the body cells. The assimilation of nutrients and the removal of waste from the cells depend on the maintenance of the proper osmotic pressure. Sodium is the major mineral responsible for maintaining a neutral pH level in the body tissues. Chlorine is essential for the formation of hydrochloric acid in the digestive juices. Both minerals affect muscle and nerve activity.

The requirements for salt in the diet vary with the specie, type of feed fed, activity of the animal, air temperature, water salinity, and production. Milk is especially rich in salt; therefore lactating animals have a higher salt requirement. Heavy sweating causes a rapid loss of salt from the body, increasing the salt requirement of the animal.

Deficiency

Animals that are temporarily deprived of salt may develop an abnormal appetite for dirt, manure, or urine. Symptoms of salt deficiency are slow to develop because the salt in the body is recycled when the intake is low. While there are no specific symptoms of salt deficiency, some general indications include an unthrifty appearance, slow growth, rough hair coat, and poor performance.

Sodium deficiency may result in corneal lesions in the eye. Reproduction is also affected, with males becoming infertile and females having a delayed sexual maturity.

Ration Requirements

Salt may be mixed in the ration at a level of 0.25 to 0.50 percent of the ration and/or fed free choice. During the lactation period, include salt at about 1 percent of the ration for cattle, sheep, and horses. It is generally recommended that a ration for swine should include about 0.50 percent salt. Including salt in the ration helps to ensure that all animals get an adequate amount and improves ration palatability.

Sources

Supplemental salt may be added in block form, as loose salt, or in the mineral mix. The price and availability of the various sources are the main considerations when deciding which to use. It is a common practice to use salt as the carrier for trace minerals because of the improved palatability of the mix. Salt fed to cattle on pasture often has organic iodine added to prevent foot rot and magnesium oxide added to help prevent grass tetany.

Plain block salt contains only salt, yellow block salt has sulphur added, and red, brown, or purple block salt has trace minerals added. Block salt is easy to use, does not need to be protected from wet weather, stimulates salivation, and there is little danger of the animal overeating salt. However, some animals may not get enough salt or may develop sore tongues if it is provided only in the block form.

Loose salt may be plain or have trace minerals added. It is easy for the animal to consume, but needs protection from rain or snow. Another disadvantage is that animals starved for salt may eat too much of the loose form. When salt is fed loose, make sure there is an adequate water supply to prevent toxic reactions from overeating salt.

Salt may also be included as 20 to 50 percent of the overall mineral mix in either plain or trace-mineralized form. Including salt as part of the mineral mix improves the palatability of the mix, helping to ensure that less palatable minerals will be eaten by the animal. The mix needs to be protected from wet weather by a mineral feeder, and an adequate supply of water must be available.

Toxicity

Toxicity rarely develops from the consumption of salt unless animals that overeat salt are restricted in their access to water. Swine are especially susceptible to salt toxicity under these circumstances, with death sometimes occurring within a few hours of the overconsumption. When there is no restriction on water intake, the upper limit for salt intake is high, because the excess salt is excreted in the urine. Levels as high as 15 percent of the ration have been given to cattle on pasture to limit grain intake, with no toxicity resulting.

Nonruminants in general are subject to salt toxicity when dietary levels are above 8 percent. Symptoms include a staggering gait, blindness, nervous disorder, and hypertension.

Animals that have been salt starved may overeat if given unlimited access to salt and overeating may cause digestive upset or death. Under these circumstances, hand feed salt daily, gradually increasing the amount of salt being fed until some is being left in the feeder each day. After that, the salt may be fed free choice.

POTASSIUM

Functions

Potassium affects the osmotic pressure and acid-base balance of the body fluids, muscle activity, and the digestion of carbohydrates.

While it is essential for life, most rations contain enough potassium and, therefore, it seldom needs to be added to the ration.

Sources

Forages are especially high in potassium, containing 3 to 4 percent on a dry weight basis in the early growth stage. Grains and concentrates contain from 0.3 to 0.7 percent potassium. Animals generally need less than 1 percent potassium in the ration on a dry weight basis. An optimum level of potassium in the ration of growing-finishing steers is 0.6 to 0.8 percent of the ration, while baby pigs need a level of about 0.26 percent in their diet. Dairy cattle need approximately 0.8 percent potassium on a dry weight basis. While cattle and sheep on high-grain concentrate diets may need some additional potassium, no general recommendations have been developed for supplementing these rations with potassium. Potassium chloride may be used as a source of potassium in the ration; however, high-grain rations for swine generally meet their needs for potassium.

Deficiency Symptoms

The symptoms of potassium deficiency in the diet are not specific. Some indications of possible deficiency include poor appetite, lower feed efficiency, slow growth, emaciation, stiffness, diarrhea, and decreased milk production in lactating animals.

Excess in Diet

An excess of potassium in the diet may result in poor assimilation of calcium and magnesium. The resulting magnesium deficiency leads to poor retention of potassium and, hence, a potassium deficiency. A high potassium intake will result in increased urine output as the animal tries to excrete the excess amount. A toxic level of potassium in the diet will result in diarrhea, tremors, and heart failure.

Soils high in potassium produce forage crops that have a lower magnesium and calcium content. If this forage is the main diet of ruminants, there is an increased danger of grass tetany because of the magnesium deficiency.

MAGNESIUM

Functions

Magnesium plays a role in activating several enzyme systems in the body, in the proper maintenance of the nervous system, in carbohydrate digestion, and in the utilization of phosphorus, zinc, and nitrates. It is a constituent of bone and is necessary for normal skeletal development. While magnesium is essential for life, it is present in only small amounts in the body, ranging from 0.02 to 0.05 percent of the total body weight of the animal.

Deficiency Symptoms

A low level of magnesium in the ration may result in a decreased utilization of phosphorus. Magnesium deficiency can cause vasodilation (dilation and relaxation of the blood vessels).

An acute magnesium deficiency may result in grass tetany (sometimes called *grass staggers* or *wheat poisoning*). Older, lactating animals are generally affected, but younger animals may develop grass tetany. Levels of magnesium below 0.001 percent in the blood serum are associated with the development of grass tetany. Cattle and sheep grazing on grass pastures, on small grains, on highly fertilized fescue in late winter or early spring, or on grass hay diets are more likely to develop tetany. The affected animals become nervous, stagger, and then fall down. Unless they are quickly treated by injection of magnesium, they will often die.

Low levels of available magnesium in grass plants result when the temperature is cool for a period of time and then rises. The new plant growth at this time may be deficient in available magnesium for several days. The specific plant physiological conditions that cause the magnesium to be unavailable are not currently completely understood. Because the magnesium in the plant is unavailable to the animal, the incidence of grass tetany is increased under these conditions. When the potassium content of the grass is more than 5.5 times the combined content of magnesium and calcium, the incidence of grass tetany increases. A properly planned soil-fertilization program with soil and plant analysis will help prevent the problem of grass tetany. Plant magnesium content can be increased by liming fields with dolomitic limestone.

Interrelationships

Adding magnesium to the ration can cause a zinc deficiency, if the ration is not properly supplemented with zinc. And too much magnesium in the ration can interfere with the metabolism of phosphorus and calcium. Higher nitrate levels in the ration require more magnesium for proper utilization of the nitrate.

Sources

If the diet is low in magnesium, the animal will draw upon the magnesium reserve in its bones. While most rations contain enough magnesium, it may be necessary to supplement the ration, especially under conditions described above that may result in grass tetany. A little more than 1 ounce of magnesium per head per day for cattle is enough, with a proportionately smaller amount for sheep. Magnesium sulphate ($MgSO_4$) or magnesium oxide (MgO) may be mixed with salt or supplement and fed free choice in areas where grass tetany is a risk. Magnesium oxide is preferred for feeding because it is not as purgative as magnesium sulphate.

SULPHUR

Functions

Sulphur is an essential part of the amino acids cystine and methionine and is important in the metabolism of lipids (as a part of biotin), carbohydrates (as a part of thiamin), and energy (as a part of coenzyme A).

Ration Considerations

Ruminant rations that are high in nonprotein nitrogen (NPN) may be deficient in sulphur. While a nitrogen:sulphur ratio of 15:1 is considered adequate to supply the sulphur needs, a ratio of 10:1 is more desirable, as this improves the utilization of NPN in the ration. Rations consisting mainly of corn, corn silage, sorghum, sorghum silage, and coastal bermuda grass generally contain nitrogen:sulphur ratios greater than 15:1 and should be supplemented with sulphur to bring the nitrogen:sulphur ratio to 10 or 12 to 1. For ruminants and horses, the sulphur may be supplied in protein, as elemental sulphur, or as sulfate sulphur. For nonruminants, supply the sulphur in protein feeds.

Sources

Forages, especially legumes, which are harvested in the earlier growth stages, should contain enough sulphur for ruminants. If forages are harvested in more mature stages, some sulphur supplementation may be needed to improve the nitrogen utilization.

Water supplies in some areas have a high sulphur content. This should be checked to determine the amount of sulphur being included in the diet through the drinking water.

Deficiency Symptoms

A deficiency of sulphur in the ration will appear as a protein deficiency. Slow growth and a general unthrifty condition are symptoms of a possible sulphur deficiency. Sheep fed NPN as a nitrogen source may have lower wool production unless the ration is supplemented with sulphur. Sulphur toxicity rarely occurs.

IRON

Functions

Iron is necessary for hemoglobin formation and is also involved in oxidation of nutrients in the cells. *Hemoglobin* is that part of the blood that transports oxygen to the cells in the body. Other minerals needed for hemoglobin formation are copper and cobalt. However, copper is needed in only about one-tenth the amount that iron is needed, and only very small amounts of cobalt are required. The amount of iron in the body is small, ranging from 0.01 to 0.03 percent by weight.

Ration Considerations and Deficiency

The amount of iron required in the diet is not known exactly, but it is believed that as little as 80 mg per kg of diet is enough for most animals. Livestock feeds generally contain enough iron to meet the needs of older animals. Milk is low in iron and, therefore, iron deficiency is most likely to occur with young, nursing animals. Iron deficiency occurs more frequently with nursing pigs raised in confinement than with other species of livestock. The young of other species appear to have sufficient reserves of iron in the liver and spleen to carry them through the nursing period.

Adding iron to the ration of sows does not increase the iron content of their milk. Oral doses or injections of iron for young pigs are necessary to prevent anemia, which is the most common symptom of iron deficiency. Symptoms of anemia in young pigs include labored breathing (thumps), listlessness, pale eyelids, ears, and nose, flabby, wrinkled skin, and edema of the head and shoulders.

A concentrated ferrous sulphate or other iron solution may be given to young pigs orally. An iron dextran injection of 150-200 mg at three days of age may be used instead of oral administration.

Sources

Most grains and forages contain enough iron to meet the needs of older animals. It is recommended that trace-mineralized salt containing iron be fed to all livestock to ensure against any possible iron deficiency.

Crops grown on high calcareous or over-limed soils may be deficient in iron. The high pH level in the soil makes the iron less soluble and less available to plants. A lower soil pH increases the availability of iron as do soil deficiencies of zinc, copper, and manganese. Therefore, crops grown on these soils will have a higher iron content.

Interrelationships

Too much iron in the diet will interfere with the assimilation of phosphorus because of the formation of insoluble iron phosphates. Excess iron intake may result in symptoms of phosphorus deficiency in the animal. If this occurs, the iron content of the ration needs to be decreased or the phosphorus content increased (by as much as 50 percent) to overcome the poor assimilation of phosphorus by the animal. Copper is required for the proper metabolism of iron. A deficiency of pyridoxine in the diet reduces the absorption of iron in the body.

MANGANESE

Functions

Manganese is involved in the utilization of phosphorus, in assimilation of iron, in reduction of nitrates, and in enzyme systems that influence estrus, ovulation, fetal development, milk production, and growth. Manganese also functions in amino acid and cholesterol metabolism and the synthesis of fatty acids. Health problems may be associated with too much or too little manganese in the diet.

Deficiency Symptoms

When the level of manganese in the ration drops below 20 ppm (parts per million), it is considered deficient. The symptoms of manganese deficiency are similar to symptoms caused by deficiencies of phosphorus, iron, or vitamin A. Indications of low manganese levels in the diet are: swollen and stiff joints, abnormal bone development, sterility, delayed estrus, reduced ovulation, abortions, deformed young, young born weak or dead, loss of appetite, slow gains, knuckling over in calves, rough hair coats, pinkeye, and greater susceptibility to nitrate poisoning.

A deficiency of manganese may cause reproductive problems such as the degeneration of the testes in males. Other indications of deficiency include poor growth and a shortening of the long bones in the body. In poultry, perosis or slipped tendon may result.

Interrelationships

The normal range of manganese in the ration is 50 to 150 ppm, while a level in excess of 500 ppm will cause problems with phosphorus and iron utilization. Too much phosphorus or iron in the ration will interfere with the utilization of manganese, because of the formation of insoluble forms of manganese. High levels of phosphorus or calcium in the ration require an increase in the manganese content of the diet because of poor absorption of the manganese. Manganese toxicity is rare.

An excessive level of manganese in the diet may be corrected by adding iron and phosphorus or by reducing the manganese content (by blending feeds high in manganese with those lower in manganese).

Sources

While the availability of manganese is not as good from manganese sulphate or manganese oxide, these inorganic sources may be used to add manganese to the ration. Trace-mineralized salt containing 0.25 percent manganese will provide the needed amount.

Ration Considerations

Most livestock rations have enough manganese, with roughages having higher levels than grain and corn being low. Feeds grown on calcareous or high pH soils, on soils high in available iron, zinc, or phosphate, and on soils naturally low in manganese may be deficient in this mineral. Forages grown on acid soils or on soils subject to alternate wetting and drying may be excessively high in manganese.

Beef cattle on all-concentrate diets based on corn and NPN supplements may need manganese supplementation in the diet. Corn and soybean meal rations for swine may be improved with manganese supplementation.

COPPER

Functions

Copper is needed to help form hemoglobin, as an activator in some enzyme systems, for hair development and pigmentation, wool growth, bone development, reproduction, and lactation.

Interrelationships

Only small amounts of copper are needed by the animal; however, the level required is influenced by the levels of iron, manganese, zinc, lead, nitrate, and molybdenum in the ration. Molybdenum has the greatest influence on the amount of copper needed, with an excess of molybdenum in relation to copper causing molybdenosis (interference with copper activiation of enzymes). When forages are fed with an excess molybdenum level, two to three times as much copper as molybdenum is needed in the ration. When the other elements mentioned above are in proper balance, a level of 10 to 20 ppm of copper in the ration is adequate for most livestock.

Sources

Most livestock feeds have three to four times the amount of copper needed by animals. A deficiency of copper may appear in feeds raised in Florida and a few other areas of southeastern United States, as well as in Australia. Copper deficiency may appear in crops raised on soils with over 20 percent organic matter, highly leached sandy soils where rainfall is high, and on calcareous sandy soils under irrigation with high nitrogen fertilization. It is necessary to supplement feeds from these crops with additional copper.

Deficiency Symptoms

Symptoms of copper deficiency include severe diarrhea, slow growth caused by anemia, swelling of joints, bone abnormalities, abortions, weakness at birth, difficulty in breathing, loss of hair color in cattle, abnormal wool growth in sheep, lack of muscle coordination, and possibly sudden death. Thoroughly mixing copper sulphate or copper oxide in the ration will correct ration deficiencies of copper. As insurance, it is common practice to feed trace-mineralized salt containing 0.25 to 0.50 percent copper sulphate in most livestock feeding operations.

Toxicity

Copper levels above 250 ppm are toxic, with death resulting. Levels above 50 ppm are considered potentially dangerous to animal health. Toxic symptoms include anemia and jaundice. Decreasing the level of copper in the ration or increasing the levels of iron, manganese, zinc, and possibly molybdenum will help prevent toxic levels of copper in the diet. Soils seldom have levels of copper high enough to produce toxic levels in the feeds grown on them. Possible sources of excessive copper in the soil include repeated applications of copper as a fungicide and contamination from copper smelter plants in the area.

ZINC

Functions

Zinc is important for the normal development of skin, hair, wool, bones, eyes, preventing parakeratosis, and promoting the healing of wounds. Zinc is necessary in several enzyme systems including peptidases and carbonic anhydrase, for protein synthesis and metabolism, and is a component of insulin. Generally a range of 30 to 50 ppm in the ration is adequate for most livestock.

Interrelationships

There is a relationship between the amount of calcium or phytate in the ration and the amount of zinc required. High levels of calcium or phytate reduces the availability of zinc in the ration. The level of iron, copper, and manganese in the ration also has some influence on the amount of zinc needed.

Animals are generally tolerant of excessive levels of zinc in the ration although high levels will interfere with the utilization of copper and iron and may cause anemia. If the ration is high in zinc, additional copper and iron may be needed. Grains and forages seldom contain excessive levels of zinc, even if grown on soils with high levels of available zinc. A zinc deficiency in feeds is more common than an excessive amount.

Deficiency Symptoms

Symptoms of zinc deficiency in livestock include parakeratosis (rough, thick skin in swine); thickening of skin on the neck, muzzle, and back of ears in cattle; loss of hair; slipping of wool; slow wound healing; poor appetite and slow growth; swelling of hocks and knees; stiff gait; and inflammation of nose and mouth tissues. Zinc deficiency seldom occurs in cattle and sheep on normal rations but sometimes occurs in swine fed in confinement.

Sources

Zinc is often added in trace-mineralized salt for all livestock as insurance against a possible deficiency. Swine rations generally have zinc added at the rate of about 50 mg/kg of air-dry feed in the supplement or in trace-mineralized salt. Zinc sulphate, zinc oxide, zinc carbonate, or zinc chelate are common sources of zinc for the ration.

MOLYBDENUM

Functions

Molybdenum is a component of the enzyme xanthine oxidase, which is found in milk and in body tissues. This enzyme is also important for the formation of uric acid in poultry. Molybdenum is involved in stimulating the action of rumen organisms, and, experimentally, some improvement in the growth rate for lambs has been demonstrated with 2 ppm molybdenum supplementation of the diet.

Ration Considerations

The requirement for molybdenum in the ration is very small, with 1 ppm being considered adequate. Most normal rations contain enough molybdenum, so supplementation in the diet is not recommended.

Toxicity

Molybdenum toxicity is a greater problem than molybdenum deficiency, with toxicity symptoms being observed at levels of 5-10 mg/kg of diet on a dry-weight basis. Increasing the copper level in the ration helps prevent molybdenum toxicity.

Excess molybdenum in the diet interferes with the utilization of copper and the symptoms appear as a copper deficiency. Molybdenum toxicity is referred to as teart disease, molybdenosis, winter dysentery, or peat scours. Symptoms include diarrhea, poor growth, loss of hair coat color, mucous membranes becoming bleached, and, at higher levels, death. Excess molybdenum in the diet over an extended period of time will interfere with phosphorus metabolism, resulting in lameness, abnormalities in the joints, and osteoporosis.

In some areas, forages that are produced on high calcareous soils or high organic soils (such as peat or muck) and on poorly drained soils may contain excess levels of molybdenum. The problem is more severe with cattle than with other classes of livestock. A proper diagnosis of molybdenum toxicity depends on a complete feed analysis to determine the relative levels of molybdenum and copper in the diet.

SELENIUM

Functions

Selenium is needed in small amounts because of its relationship with vitamin E absorption and utilization and it is an essential part of the enzyme glutathione peroxidase. A range of 0.05 to 0.10 ppm appears to be adequate in livestock diets. The current recommendation for pigs under 50 pounds (22 kg) is 0.03 ppm. Toxic symptoms may appear if the dietary level is above 5 mg/kg of dry feed.

Deficiency Symptoms

A deficiency of selenium in the diet causes nutritional muscular dystrophy (white muscle disease) in cattle, sheep, chickens, turkeys, swine, and horses; retained placenta and low fertility in ruminants; loss of condition and diarrhea in cattle and sheep; liver necrosis in swine; exudative diathesis and muscular dystrophy in turkeys; lesions in the gizzards and muscles of ducks; and nutritional pancreatic dystrophy and exudative diathesis in chickens.

Selenium deficient areas in the United States include the southeastern coastal states, states along the Great Lakes, New England states, and the coastal northwest. Feeds grown on acid soils in these areas are often deficient in selenium. It is not considered practical to add selenium to the soil in these areas to raise the selenium content of the forages grown. Young calves and lambs raised in selenium deficient areas are sometimes injected with small amounts of selenium to prevent deficiency symptoms.

FDA Regulation in Diet

The Food and Drug Administration has not authorized a general inclusion of selenium in all livestock diets. In 1974, the FDA authorized inclusion of sodium selenate or sodium selenite at the rate of 0.1 ppm in complete rations for swine, growing chicks up to 16 weeks of age, breeder hens producing only hatching eggs, and other nonfood animals. Complete rations for turkeys may include 0.2 ppm of selenium. The selenium must be carefully mixed in the ration using a premix that contains no more than 90.3 mg of selenium per pound.

Toxicity

Three types of selenium toxicity occur in livestock: (a) acute; (b) chronic blind staggers type; and (c) chronic alkali-disease type. The accumulation of excessive levels of selenium in plants grown in some areas, mainly South Dakota, Montana, and Wyoming, may cause selenium toxicity. The problem is more prevalent on well-drained alkali soils, where selenium oxidizes to selenates that are easily taken up by some plants, called *accumulator plants*. On acid and neutral soils, the selenium tends to form selenites that are relatively insoluble and not available to plants; therefore, plants grown on these soils do not accumulate toxic levels of selenium.

Acute selenium poisoning occurs when animals eat accumulator plants over a short period of time. Symptoms include labored breathing, abnormal movement and posture, prostration and diarrhea, followed in a few hours by death. Acute poisoning does not occur often, because livestock will usually avoid accumulator plants on pasture unless other plants are not available to them.

The intake of a limited amount of accumulator plants over a period of weeks or months will cause chronic poisoning of the blind-staggers type. The symptoms are impaired vision, wandering, stumbling, and finally death from respiratory failure.

Chronic toxicity of the alkali type is caused by the intake of grains or grasses containing 5 to 40 mg selenium/kg over a period of several weeks or months. The symptoms are liver cirrhosis, lameness, hoof malformations, loss of hair, and emaciation.

If symptoms of selenium poisoning occur, animals should be removed from the pasture area or feed suspected of causing the problem. It is sometimes difficult to diagnose selenium poisoning because the symptoms do not always appear until months after the consumption of the feed containing excessive amounts of selenium. Animals will recover if they are removed before the condition has progressed too far. Animals that have been severely affected will die or may have to be destroyed. Experimentally, it has been shown that adding small amounts of arsenic to the ration will counteract the effects of chronic selenium poisoning in cattle. This procedure should be used only with the aid of a veterinarian. Maintaining adequate levels of protein and vitamin A in the diet will help prevent selenium poisoning.

COBALT

Functions

Cobalt, an essential component of the vitamin B_{12} molecule, is used by rumen and cecal bacteria in the synthesis of vitamin B_{12} and in the growth of rumen bacteria. Only traces of cobalt are needed in the ration, with 1 ppm considered adequate for most classes of livestock.

Deficient Areas

Areas of the United States that produce crops deficient in cobalt include Florida, Maine, Massachusetts, Michigan, New Hampshire, New York, Pennsylvania, and Wisconsin. Western Canada and Australia also have cobalt deficient areas. A good indication of the adequacy of cobalt in an area is the amount of cobalt found in the leaves of the catalpa tree.

Legume forages generally contain more cobalt than grass forages; however, legumes may be deficient in cobalt if they are grown on soils deficient in cobalt. Poor quality grass forages and grains are most likely to be deficient in cobalt. A cobalt level below 0.08 ppm in the feed is considered deficient.

Deficiency Symptoms

Cobalt deficiency symptoms are similar to vitamin B_{12} deficiency symptoms. These include poor appetite and general malnutrition, weakness, anemia, slow growth, decreased fertility, and lower milk and wool production. Nonruminants may also develop pernicious anemia.

Ration Considerations

Cobalt deficiency in cattle and sheep can be prevented by feeding 0.05 to 0.10 mg/kg of feed. If there is enough cobalt present, ruminants can synthesize vitamin B_{12} in the rumen. Swine have less ability to synthesize vitamin B_{12} in the intestinal tract, even when sufficient cobalt is present in the ration. If there is enough vitamin B_{12} present in the ration, then additional elemental cobalt is usually not needed.

A cobalt deficiency may be corrected by adding vitamin B_{12} to the ration; supplementing with cobalt sulphate, cobalt chloride, or cobalt carbonate; with a balling gun, orally administering to ruminants pellets made of cobalt oxide and iron; or treating deficient soil with fertilizers containing cobalt. Calcareous or high pH soils depress cobalt uptake by plants and even on other soils the uptake is generally low. Cobalt compounds are often included in protein supplements, vitamin-mineral premixes, trace-mineralized salt, and mineral supplements. Cobalt is usually included in the ration in one of the above-mentioned ways, even in areas that are not cobalt deficient, as a cheap insurance against a possible deficiency.

Toxicity

Cobalt toxicity is rare because animals appear to be tolerant of high levels of dietary cobalt. Levels as high as 10 ppm cobalt are considered nontoxic under normal conditions.

IODINE

Functions

Iodine is necessary for the production of the hormone thyroxine in the thyroid gland. This hormone controls the rate of oxidation of nutrients in the cells and thus controls heat production in the animal's body. A deficiency of iodine reduces the production of thyroxine and leads to the enlargement of the thyroid gland, a condition referred to as *goiter*.

Deficiency Symptoms

Iodine deficiency symptoms include development of goiter, weak or dead young at birth, hairlessness at birth, and infected navels. Older animals rarely show any symptoms of iodine deficiency.

Deficient Areas

Some areas of the United States are iodine deficient, most notably, the Great Lakes region, the Pacific coast, and the far northwest. Iodine deficient areas are found in many other parts of the world.

Ration Considerations

The minimum level of iodine required in the diet is not known, because the presence of plant goitrogens and some other dietary imbalances will increase the requirement for iodine. When pregnant animals are fed diets of corn or corn silage with large amounts of soybean meal, they may give birth to young with enlarged goiters. It is believed that the presence of goitrogens in the soybean meal may be responsible for this condition. Heating soybean meal tends to destroy the goitrogens present, but not all soybean meal is heated to a sufficient level to do this.

An iodine level of 0.2 ppm (0.2 mg/kg) of air-dry feed in the ration is considered adequate for livestock, including pregnant or lactating animals. Feeding iodized salt containing 0.007 percent iodine or 0.01 percent potassium iodide will meet the needs of most livestock. Stable sources of iodine that may be used include calcium and potassium iodates, Ethylenediamine dihydriodide (EDDI), and pentacalcium orthoperiodate.

FLUORINE

Functions

Fluorine helps prevent cavities in teeth and possibly slows down osteoporosis in older animals. An intake below 10 ppm fluorine in ration dry matter or below 1 ppm in drinking water is beneficial, while higher levels may be toxic.

Deficiency

Fluorine deficiency in the diet is rare and supplementation is not recommended. In livestock feeding, toxic levels are more of a concern than is a deficiency. Drinking water and forages in most areas contain enough fluorine to meet the needs of livestock.

Excess and Toxicity

In excessive amounts, fluorine is toxic. The symptoms of too much fluorine in the diet develop slowly over a period of time because fluorine is an accumulative poison. Common symptoms of excess fluorine in the diet include loss of tooth enamel, staining or brown mottling of teeth, uneven or excessive wear of teeth, cavities in the teeth, thickening and softening of the bones, reduced strength in the bones, lameness of animals, poor appetite, rough hair coat, lowered milk production, disturbed energy metabolism, poor gains in young animals, and weight loss in older animals.

Fluorine is normally immobilized in the bones or excreted in the urine. Excessive amounts may become incorporated in some of the enzymes involved in energy metabolism. Thus, citrate cannot be metabolized to isocitrate, which reduces the energy available from the feed. This leads to an increased synthesis of acetoacetic acid, which is converted to acetone for elimination. Chronic fluorosis may be one of the causes of secondary acetonemia (ketosis).

Excess fluorine in the ration may be corrected by removing the source of fluorine from the diet. While excess intake of fluorine is not often a problem, some areas of the United States have too much fluorine in the drinking water or have soils high in fluoride, which results in crops containing excess fluorine. Supplements or mineral mixes made from phosphate high in fluorine may lead to excess amounts in the diet. Industrial operations may contaminate the air with fluorides, which leads to contamination of the crops grown in the vicinity. Fluorine tends to concentrate in the leaves of the plant; therefore, if contamination appears to be a problem, check the forages carefully. There is less danger of fluorine accumulation in the grain produced in these areas.

Mineral supplements made from defluorinated phosphates usually do not contain excessive amounts of fluorine. There is no cure for the effects of high fluorine intake, so check the ration carefully to avoid the problem. The tolerance level for excess fluorine varies with species and use of the animal. Limit fluorine intake for breeding cattle and sheep to no more than 30 mg/kg of air-dry diet and no more than 100 mg/kg for fattening steers and lambs.

CHROMIUM

Chromium has only recently been added to the list of necessary mineral elements. It is believed that some compounds of chromium may activate insulin for sugar metabolism in the body. Little is known about chromium activity in the body. At the present time, chromium supplementation of livestock rations is not recommended, as dietary deficiencies have not been demonstrated. Most forages and grains contain only low levels of chromium. Chromium is not necessary for plant growth and high levels in the soil are toxic to plants.

SILICON, NICKEL, STRONTIUM, TIN, VANADIUM

These minerals have been shown to have beneficial effects in the diets of laboratory animals, but the need for these minerals in livestock feeding is not presently known. Feeds normally contain only small amounts of these minerals, all of which are toxic in excessive concentrations and in some chemical combinations. No recommendation for supplementation of livestock rations with these minerals is made.

ALUMINUM

No beneficial effects from aluminum and boron have been found in livestock feeding. They are normally found in feeds and are no problem at normal concentrations, but can be toxic at excessive levels.

Forages grown on soils with low pH (less than 5.5), under stress conditions such as excess water or when available zinc, copper, manganese, iron, or magnesium are low, may develop high concentrations of aluminum. Excessive amounts of aluminum in livestock diets can cause gastrointestinal irritation and interfere with phosphorus absorption. High concentrations of aluminum in grass forages are suspected as a cause of higher incidences of grass tetany. Problems with excess aluminum in forages may be corrected by adjusting the soil pH and by proper fertilization to ensure correct balance of major and minor elements in the soil.

BORON

Boron is an essential element for plant growth with legumes containing 20-50 ppm, grasses containing 4-12 ppm, and grains containing 1-5 ppm on a dry-matter basis. Occasional boron toxicity may be found in forages in western United States, but it is rare because a high level of soluble boron will interfere with plant growth. Boron toxicity symptoms include frequent urination, dehydration, and digestive and breathing disorders; more acute cases show brain damage, loss of muscular control, and death.

TOXIC MINERALS

Arsenic, cadmium, lead, and mercury are all toxic to livestock. These elements may occur naturally in the soil, be added to the soil through the application of herbicides or insecticides, or come from fumes in air polluted by industrial plant operation.

Arsenic

Arsenic is found in some herbicides, insecticides, and defoliants. Crops do not grow well on soils contaminated with arsenic; however, these crops do not accumulate dangerous levels of arsenic and can be safely fed to livestock. Crops that have been recently sprayed with a material containing arsenic may cause arsenic poisoning if livestock graze on the foliage.

Cadmium

Cadmium is normally not found in soils in excessive amounts; however, contamination from industrial plants or sewage sludge from cities may cause a buildup of cadmium in the soil. It is possible for plants to accumulate levels of cadmium that are toxic to livestock if the zinc level in the soil is low. In this situation, the plant will take up cadmium in place of zinc. Proper fertilization of the soil will help prevent this problem. Symptoms of cadmium poisoning in cattle include poor appetite, slower growth, anemia, retarded testicular development, enlarged joints, scaly skin, liver and kidney damage, and increased mortality. Zinc, cobalt, and selenium may be added to the diet to counteract the toxic effect of cadmium in animal diets.

Lead

The major danger from lead is contamination of the upper parts of plants from lead in the air. Lead uptake by plants tends to remain in the roots and not move to the aerial parts. Under conditions of unusually heavy contamination of the soil, some of the lead will move to the plant foliage. Airborne lead may be inhaled by animals or may get into the feed from contamination from lead pipes; however, the most common source of toxic amounts of lead for livestock is from lead-based paint that the animal licks from discarded paint cans or peeling paint on the walls of buildings. Young calves are especially susceptible to lead poisoning with 200-400 mg lead per kilogram of body weight being toxic. Older cattle can tolerate levels of lead intake about twice as high. Symptoms of lead poisoning in cattle include dullness, lack of appetite, abdominal pain with constipation, sometimes followed by diarrhea, and after two or three days, bellowing, staggering, snapping of eyelids, muscular twitching, frothing at the mouth, and convulsive seizures.

Mercury

Mercury may be discharged into air and water by industrial plants and is used in herbicides and fungicides for seed treatment. Inorganic forms of mercury are not very toxic because they are not readily absorbed by the animal. The major source of mercury poisoning is from ingesting methyl mercury (an organic form) from seed treated with a fungicide. Symptoms of organic mercury poisoning in cattle include muscular incoordination, head-pressing, twitching of eyelids, tetanus-like spasms, excessive salivation, lying down, inability to eat or drink, muscular convulsions, and death.

SUMMARY

The ash left when an organic material is burned is its mineral content. At least 18 mineral elements are needed by animals. Those needed in large amounts are referred to as *major* or *macro*, while those needed in small amounts are called *trace* or *minor minerals*.

Minerals are required for the development of the bones and teeth as well as for many other functions in the body. Mineral deficiencies cause lower production and poor gains, but rarely cause diseases or death.

Commercial feeds and mineral mixes are the most common sources of minerals in livestock rations. Minerals may be mixed in complete feeds or fed free choice. Often a combination of both of these methods is used.

Salt (which contains sodium and chlorine), calcium, and phosphorus are the major minerals most often needed in livestock rations. Trace minerals are often added to the ration by the use of trace-mineralized salt.

The ratio of calcium to phosphorus should be 1:1 or 1:2 for most classes of livestock. The ratio can be as high as 7:1 for ruminants. Lactating animals need a ratio of 1.5:1.

Salt toxicity rarely occurs unless animals are limited in their access to water. Salt starved animals may overeat salt and develop digestive problems as a result.

Excessive amounts of some minerals in the diet may cause toxicity problems or poor utilization of other minerals or nutrients. Care must be taken to properly balance the mineral content of the ration.

Crops grown in some areas of the United States are deficient in some minerals such as cobalt, iodine, and selenium. Some western areas of the United States produce crops that contain excessive amounts of selenium. Selenium toxicity in livestock eating these crops can be a problem.

Arsenic, cadmium, lead, and mercury are all toxic to livestock. Care must be taken to prevent the contamination of feed by these elements.

REVIEW

1. What are minerals?
2. Name the major minerals needed by livestock.
3. Name the trace minerals needed by livestock.
4. Describe the general functions of minerals in nutrition.
5. What are some general symptoms of mineral deficiency in the diet?
6. What are some usual sources of minerals needed to supplement the ration?
7. How are minerals usually included in the livestock ration?
8. Which minerals are most likely to be deficient in livestock feeding?
9. What is a common way to add trace minerals to the livestock ration?
10. What are the functions of calcium?
11. Describe the symptoms of a calcium deficiency.
12. Name some good feed sources of calcium in the ration.
13. What are the common calcium supplement sources?
14. What are the functions of phosphorus?
15. Describe the symptoms of phosphorus deficiency.
16. Name some good feed sources of phosphorus in the ration.
17. What are some common phosphorus supplement sources?
18. What is the recommended calcium/phosphorus ratio?
19. What mineral elements are provided in salt?
20. Why are these elements important in animal nutrition?
21. What are some indications of salt deficiency in the diet?
22. How may salt be added to the ration?
23. Describe the forms of salt available for livestock feeding and list some advantages and disadvantages of each.

24. What are some symptoms of salt toxicity?

25. Describe how to feed salt to animals that have been salt starved.

26. What are the functions of potassium?

27. What are the symptoms of potassium deficiency in the ration?

28. Describe the results of an excessive amount of potassium in the ration.

29. What are the functions of magnesium?

30. What are symptoms of magnesium deficiency in the ration?

31. Describe the cause, symptoms, and treatment of grass tetany.

32. What are the functions of sulphur?

33. What are the symptoms of sulphur deficiency in the ration?

34. What are the functions of iron?

35. Which animals are most likely to have an iron deficiency and how should this be prevented?

36. What are symptoms of iron deficiency?

37. What is the effect of too much iron in the diet?

38. What is the function of manganese?

39. What are the symptoms of manganese deficiency?

40. What is the function of copper?

41. What parts of the United States are considered copper deficient areas?

42. What are the symptoms of copper deficiency?

43. Discuss copper toxicity in livestock nutrition.

44. What are the functions of zinc?

45. What are the symptoms of zinc deficiency?

46. What are the functions of molybdenum?

47. What are the symptoms of molybdenum toxicity?

48. What are the functions of selenium?

49. What are the symptoms of selenium deficiency?

50. What areas of the United States are deficient in selenium?

51. How may selenium be added to the ration?

52. Name the three types of selenium toxicity and describe the symptoms of each.

53. How should selenium toxicity be prevented and/or treated?

54. What are the functions of cobalt?

55. What areas of the United States are deficient in cobalt?

56. What are the symptoms of cobalt deficiency?

57. How may a cobalt deficiency be corrected?

58. What are the functions of iodine?

59. What are the symptoms of iodine deficiency?

60. What areas of the United States are deficient in iodine?

61. What are the functions of fluorine?

62. What are the symptoms of fluorine toxicity?

63. How may excess fluorine in the ration be prevented?

64. Name four toxic elements and discuss each.

6

VITAMINS, FEED ADDITIVES, AND WATER

OBJECTIVES

After completing this chapter you will be able to:

- Describe vitamins and feed additives.
- List sources of vitamins and feed additives.
- Describe the functions of vitamins, feed additives, and water.
- Describe deficiency symptoms caused by lack of vitamins in the ration.
- Discuss the effects of feed additives in the ration.
- Discuss regulations on the use of feed additives in the ration.
- Discuss requirements for vitamins, feed additives, and water in the ration.

VITAMINS

History

Prior to the beginning of the twentieth century, vitamins were unknown, although it was known that certain foods prevented or cured some diseases, such as beriberi and scurvy. Why this happened was not known and it was believed that a satisfactory ration for farm animals consisted of adequate amounts of proteins, carbohydrates, fats, and minerals. However, experimenters (working with laboratory animals fed purified diets) found that these animals did not grow or produce satisfactorily and eventually died if the diet was continued long enough. These investigators realized the existence of substances, unknown to them, essential for nutrition.

Work with experimental animals led to the discovery in 1912 of a water-soluble substance (now known to be the vitamin *thiamine*)

that could prevent or cure beriberi. Because this compound was essential and contained nitrogen it was called *vital amine. Vital* referred to its being essential and *amine* referred to the nitrogen content. Over a period of several years, other substances were discovered that are required in trace amounts and are essential to life. Not all of these substances contained nitrogen and this group of nutrients came to be called *vitamins.*

Defined

Vitamins are organic compounds that are essential for health but are needed in only trace amounts. They are required for normal growth and maintenance of animal life. Only small amounts of vitamins are needed because they function as catalysts (as parts of enzymes or coenzymes) in metabolic processes.

Composition

A total of 16 vitamins have been identified as essential in animal nutrition. These compounds differ from each other in chemical composition as opposed to carbohydrates, fats, and proteins, all of which are chemically related within their group. Chemically, all vitamins contain carbon, hydrogen, and oxygen. All the B complex vitamins, except inositol, also contain nitrogen. Some of the B complex vitamins also have one or more of the mineral elements as a part of their chemical structure. Vitamins differ from each other in their specific functions, but are grouped together because all are organic in nature and each is essential, at least for some species, in trace amounts.

Naming

During the early research on vitamins, they were designated by letters of the alphabet. As time went on, it was discovered that vitamin B was actually a number of different substances; therefore, subscripts were used to differentiate these compounds. Today vitamins are referred to either by letter designations or by their chemical names, with the latter method becoming more common.

Solubility

Some of the vitamins are soluble in water and others in fat. Water-soluble vitamins include the B complex vitamins and vitamin C. Vitamins soluble in fats and fat solvents include vitamins A, D, E, and K. The solubility of a vitamin is related to its function in metabolism, influencing its use in animal nutrition.

Ration Considerations

Some vitamins can be synthesized in the body and, therefore, do not need to be added to the diet. For example, most of the B complex vitamins can be synthesized in the rumen. Vitamins that are not synthesized in the body must be supplied through dietary sources.

Fat-soluble vitamins may be stored in the animal's body, reducing the need for a daily supply in the diet. The water-soluble vitamins are generally not stored in the body and need to be supplied on a regular basis in the diet.

Before the commercial synthesis of vitamins, the animal's needs for vitamins were supplied by the use of feeds known to contain the necessary vitamins. While well-balanced rations may supply the needed vitamins, there are times when the diet may need vitamin supplementation. The vitamin content of feeds varies with the quality of the feed, especially forages. In some cases, a feed may contain an essential vitamin but it may have a low availability in metabolism. Today vitamins are synthesized in the laboratory and are made available in commercial vitamin premixes for incorporation in supplements. It is generally recommended that vitamin premixes be used at the appropriate levels in livestock feeding to assure a ready supply of these essential nutrients.

Specific vitamins have specific functions in the body. A deficiency of a given vitamin may result in specific deficiency symptoms. Vitamins are involved in a number of metabolic processes and the deficiency symptoms are an indication that basic metabolic processes in the body have been disturbed. Deficiency symptoms for specific vitamins are included in the following discussion of each vitamin.

VITAMIN A

Figure 6-1 shows the structural formula for vitamin A. Vitamin A in animals is the product of the conversion in the animal's body of carotene which is found in feeds. True vitamin A is generally not found in feeds. Because vitamin A is converted from carotene, the carotene is regarded as a precursor of vitamin A and is sometimes referred to as *provitamin A*. The amount of carotene in a feed is measured in International Units (IU) of activity. While there are several forms of carotene found in feeds, beta-carotene is the standard used when determining the rate of carotene conversion to vitamin A. When measuring vitamin A activity and beta-carotene, the international standard is 1 IU vitamin A = 1 USP (United States Pharmacopoeia) unit = 0.6 microgram beta-carotene. One IU vitamin A equals the vitamin A activity of 0.300 microgram crystalline vitamin A alcohol or 0.344 microgram vitamin A acetate or 0.550 microgram vitamin A palmitate. One milligram beta-carotene equals 1,667 IU vitamin A. These standards are based on conversion rates for rats. Different animal species convert carotene to vitamin A at different rates, Table 6-1.

When the feed source supplies enough carotene, the animal can usually meet its requirements for vitamin A from the diet.

However, the rate of conversion for an individual animal may vary, depending on the type of carotene in the feed, level of carotene intake, breed of animal, individual differences among animals, viral infections present in an individual, stress conditions, and altered thyroid functions.

Functions

Functions of vitamin A include normal maintenance of the eyes and membrane tissue (epithelium) of the respiratory, digestive, reproductive, nerve, and genitourinary systems. Vitamin A is also needed for normal bone growth. Research has shown that dairy cattle and sows have a requirement of beta-carotene, in addition to their requirement for vitamin A, for normal ovarian function.

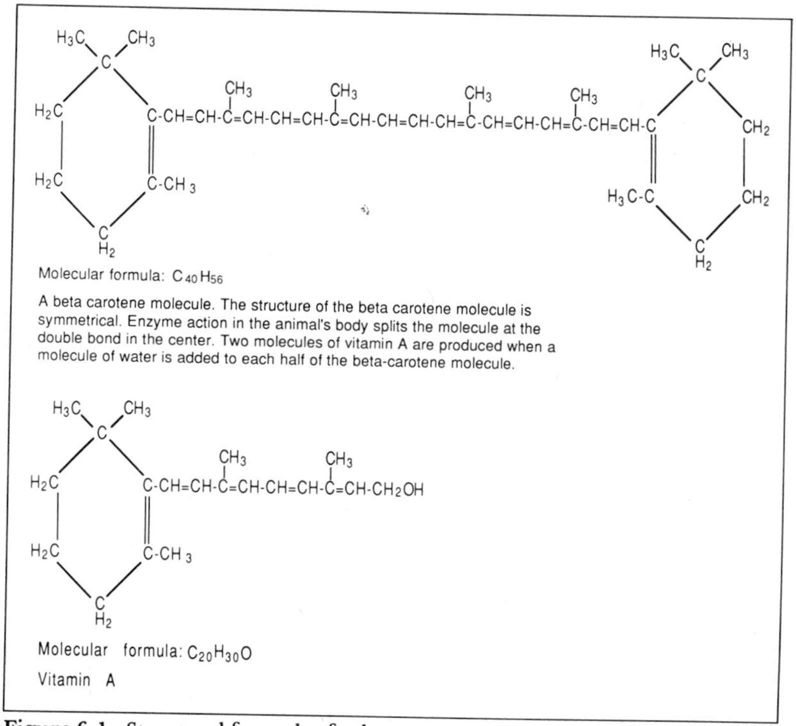

Molecular formula: $C_{40}H_{56}$

A beta carotene molecule. The structure of the beta carotene molecule is symmetrical. Enzyme action in the animal's body splits the molecule at the double bond in the center. Two molecules of vitamin A are produced when a molecule of water is added to each half of the beta-carotene molecule.

Molecular formula: $C_{20}H_{30}O$

Vitamin A

Figure 6-1. Structural formulas for beta-carotene and vitamin A. Vitamin A is fat soluble, not found in plants, and is formed in animals from carotene. Carotene is found in plants and is the precursor of vitamin A.

Table 6-1. Conversion of beta-carotene to vitamin A for different animal species.

Species	Conversion of mg of Beta-Carotene to IU of Vitamin A (mg) (IU)	IU of Vitamin A Activity (Calculated from Carotene) (%)
Standard	1 = 1,667	100
Beef cattle	1 = 400	24
Dairy cattle	1 = 400	24
Sheep	1 = 400-500	24-30
Swine	1 = 500	30
Horses		
Growth	1 = 555	33.3
Pregnancy	1 = 333	20
Poultry	1 = 1,667	100

Source: *United States–Canadian Tables of Feed Composition,* National Research Council, National Academy of Sciences, Washington, D.C., 1982.

Deficiency

Night blindness (the inability to see in dim light) is a symptom of severe vitamin A deficiency. Vitamin A is a part of the visual purple of the eye, which is depleted through the action of light. Visual purple is required for vision in dim light and a deficiency of vitamin A results in a reduction of this substance in the eye. Animals suffering from night blindness will recover when sufficient amounts of vitamin A are added to the diet. Animals may become permanently blind because of a vitamin A deficiency, which can result in eye infections or a constriction in the optic nerves where they pass through the skull. Excessive watering of the eyes and development of cornea ulcerations are indications of possible vitamin A deficiency.

Keratinization of the epithelial tissue, which causes lowered resistance to infections, is common with vitamin A deficiency. Other indications of vitamin A deficiency include diarrhea, reduced appetite, poor growth, and weight loss. The nervous system may be affected by higher cerebrospinal fluid pressure resulting in staggering gait, convulsions, and papilledema. Paralysis of some parts of the body may occur in some species. Reproductive problems such as poor conception rate; reduced fertility in males; shortened ges-

tation period; increased incidence of retained placenta; abortion; or the young being born dead, deformed, blind, or lacking coordination may also result from vitamin A deficiency. Finishing cattle may show lameness in the hock and knee joints and swelling (anasarca) of the brisket. In poultry, discharges from the eyes and nostrils may occur, as well as reduced egg production and hatchability of eggs.

Sources

Carotene is found in good quality, fresh, green forages in amounts generally sufficient to supply the needs of livestock eating the forage. Green, leafy hays that have been in storage less than one year; dehydrated, pelleted legume hays; and good quality grass or legume silages are excellent sources of carotene. Dried, sun-cured forages contain less carotene. Dry summer grass pasture has a lower carotene content than grass pasture during the early stages of rapid growth. Bleached, low quality forages have little carotene content. Carotene content of forages in storage is reduced by exposure to the sun and air, high temperatures, and long storage times. Other feed sources of carotene include yellow corn, fish liver oil, and whole milk. Carotene content of feeds will be reduced by the use of steam or pressure during processing of feeds or by mixing the feeds with oxidizing agents such as some minerals or organic acids.

Storage in the Body

Vitamin A is stored in the liver and fatty tissues of the body. The animal can use this stored vitamin A during periods of feeding when the diet is deficient in carotene. For example, it takes about 200 days for the depletion of the stored vitamin A in sheep and 3-6 months in horses, when the dietary levels of vitamin A are low. High levels of nitrate or nitrite in the ration may increase the need for dietary vitamin A.

Ration Considerations

Conditions of feeding that may indicate a need for supplementation of the diet with vitamin A include (1) poor quality or low levels of forage; (2) limited amounts of colostrum or whole milk in the diet; (3) diets consisting mainly of corn silage and low-carotene concentrates; (4) grazing during drought periods when pastures become dry; and (5) rations made up mainly of cereal grains, except corn. Vitamin A in a stabilized form that is resistant to oxidation may be added to the ration through vitamin premixes. Intramuscular injections of vitamin A may be used to provide the necessary amounts for a period of several months or when a severe deficiency exists, with the diet being adjusted to provide the recommended level of vitamin A on a long-term basis. Tables in the Appendix list the vitamin A requirements for various species of livestock. Stress conditions such as low temperature or exposure to infectious bacteria will increase the vitamin A requirements above those listed in the Appendix tables.

Excess in Diet

Feeding excessive amounts of vitamin A to horses over a long period of time may result in fragile bones, hyperostosis (thickening of bony tissue), and exfoliated epithelium (flaking off of the epithelium). Young poults or chicks may die from extreme overdoses of vitamin A.

Levels above 50,000 IU/kg of diet for the laying hen can result in decreased yolk content of xanthophyll-type pigments. Increased blood spots in the eggs and decreased egg production will result when dietary levels of vitamin A are above 100,000-150,000 IU/kg of diet.

VITAMIN D

Functions

Vitamin D is important for calcium absorption and calcium and phosphorus metabolism in the body. Vitamin D is needed along with calcium and phosphorus to prevent rickets in animals and is sometimes referred to as the *antirachitic* factor.

Solubility

Vitamin D is fat-soluble and stored in the body; therefore, it is less critical in the diets of mature animals as compared to younger animals. It is more important in the diets of pregnant animals where the requirement is higher.

Forms

The plant form of vitamin D is D_2 (ergocalciferol), formed by the irradiation of ergosterol by ultraviolet light. The animal form is D_3 (cholecalciferol) and is formed by the irradiation of 7-dehydrocholesterol by ultraviolet light. Both forms are about equally effective for swine and sheep, but for poultry, vitamin D_2 is only about 1/30 to 1/40 as effective as vitamin D_3. The international standard defines 1 IU of vitamin D as the biological activity of 0.025 micrograms of cholecalciferol. Figure 6-2 shows the structural formulas for vitamin D.

The liver converts cholecalciferol to 25-hydroxy D_3, which is the main form in which the vitamin circulates in the body. The kidney converts 25-hydroxy D_3 (25-OH-D_3) to 1,25-dihydroxy D_3 (1,25(OH)$_2$$D_3$), or 24,25-dihydroxy D_3 (24,25(OH)$_2$ D_3), which is the biologically active form of vitamin D.

Deficiency

A deficiency of vitamin D will result in the development of rickets (because of reduced calcium and phosphorus absorption and metabolism), with insufficient calcification of the bones. Indications are decreased appetite, slower growth, digestive disturbances, stiffness in gait, labored or fast breathing, irritability, weakness, and, sometimes, the development of tetany and convulsions. As the disease progresses, the joints enlarge and become swollen and stiff, a slight arch in the back develops, the legs bow, and the animal has difficulty in walking because of deterioration of the joint surfaces. If the vertebrae fracture, posterior paralysis may occur. Symptoms develop more quickly in younger animals than they do in mature animals.

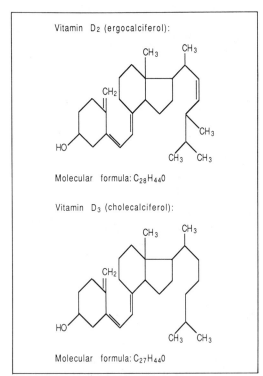

Figure 6-2. Structural formulas for vitamin D_2 (Ergocalciferol) and vitamin D_3 (Cholecalciferol). Vitamin D is fat soluble. It is formed when sterols are irradiated with ultraviolet light.

Pregnant animals may give birth to dead, weak, or deformed young as a result of vitamin D deficiency. Mature swine may show a reduction in bone mineral content with the development of osteomalacia (softening of the bones). If the vitamin D deficiency is severe, pigs may show signs of calcium and phosphorus deficiency with the development of tetany.

Chickens, in addition to rickets, will show poor feathering and growth, while laying hens will produce thin-shelled eggs, show a drop in egg production, and eventually produce shell-less eggs. Chickens may lose the use of the legs entirely. Egg size and hatchability will decrease, with embryos frequently dying at 18 to 19 days.

Sources

Diets that include sun-cured forages generally provide sufficient vitamin D to meet the animal's needs. Animals regularly exposed to sunlight or ultraviolet light also will not develop symptoms of vitamin D deficiency. Green forages, barn-cured hay, and silages have some vitamin D activity. Irradiated yeast, whole milk, and cod-liver and some other fish-liver oils are also sources of vitamin D in the diet. Grains, grain byproducts, and protein feeds have practically no vitamin D activity.

Commercially available forms of vitamin D must be supplied to animals that are not receiving sunlight or are on diets deficient in vitamin D. While vitamin D is more stable than vitamin A, it does oxidize and has poor stability when mixed with minerals, especially calcium carbonate. Dietary requirements of vitamin D for various species of animals are given in the tables in the Appendix. A diet containing the recommended levels of calcium and phosphorus will require less vitamin D than one deficient in these two minerals.

Toxicity

Extremely high levels of vitamin D have been shown to be toxic to poultry, although birds will tolerate levels at least 100 times the requirement level shown in the Appendix tables. Excessive levels of vitamin D can also be toxic to horses, with soft tissue calcification, generalized bone resorption, kidney damage, and death occurring in young ponies fed 3,300 IU/kg body weight daily for a period of four months. While the toxic level for horses is not definitely established, levels above 50 times the recommendations may be harmful. A high level of calcium in the diet increases the toxicity of vitamin D. Young pigs fed excessive levels (250,000 IU daily for 4 weeks) of vitamin D had poor appetites and slower gains.

While feeding massive doses of vitamin D during the last few days of pregnancy in dairy cattle has reduced the incidence of milk fever, continued feeding of massive doses after calving has produced toxic effects. Dairy cows with a previous history of milk fever have shown a reduction in this disease when fed vitamin D at levels of 70,000 IU/kg of concentrate on a year-round basis. Further study is needed before recommendations can be made on the use of vitamin D to reduce the incidence of milk fever.

VITAMIN E

Functions

Vitamin E functions as an antioxidant, which helps in the absorption and storage of vitamin A. As an antioxidant, it also acts in other metabolic functions in the cell. There is some storage of vitamin E in the liver and other fatty tissues. It is a fairly stable compound but is rapidly destroyed in the presence of rancid fat.

Forms

Vitamin E occurs in several forms of complex organic compounds called *tocopherols*, the alpha, beta, and gamma forms being most common. Figure 6-3 shows the structural formula for alpha-tocopherol. There is a wide variation in the biological activity of the tocopherols with *d*-alpha-tocopherol being the most active. One IU of vitamin E activity is defined as the biological activity of 1 mg *dl*-alpha-tocopherol acetate.

Deficiency Symptoms

Vitamin E deficiency results in symptoms similar to selenium deficiency, that is, white muscle disease or nutritional muscular dystrophy. Hatchability of eggs is reduced, although production does not appear to be affected. Extended vitamin E deficiency in poultry will cause permanent sterility in the male and reproductive failure in the female. Vitamin E deficiency will cause eye problems in turkey embryos.

The addition of selenium to the diet can prevent some of the symptoms of vitamin E deficiency. Alpha-tocopherol added to the diet will increase the tocopherol level of the blood and body tissues and reduce the susceptibility of body fats to oxidative rancidity. The addition of cystine to poultry diet will prevent nutritional muscular dystrophy caused by a vitamin E deficiency.

Sources

Good dietary sources of vitamin E include whole cereal grains, the germ or germ oils of cereal grains, green forages, and good quality hay. The vitamin E level in feeds will decline after a long storage period. Vitamin E is seldom deficient in the diet unless feeds being used were produced on selenium deficient soils.

Vitamin E is produced commercially and may be added to the diet when needed, by using a vitamin premix, or it may be injected intramuscularly. Excessive levels of nitrite in the diet may increase the need for vitamin E. Dietary recommendations for vitamin E are found in the tables in the Appendix. No toxic effects from excessive levels of vitamin E in the diet have been reported.

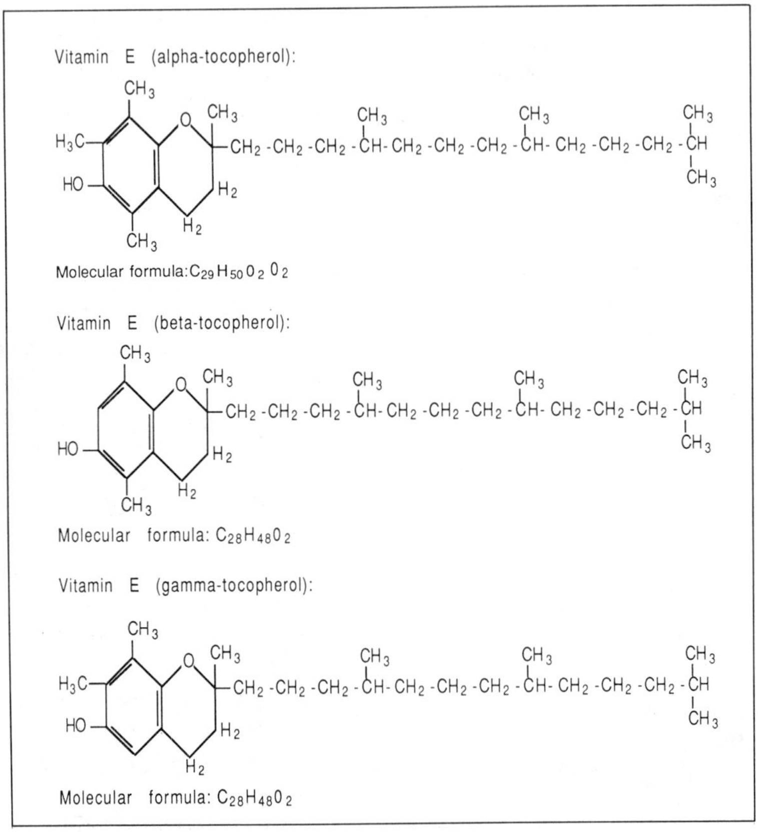

Vitamin E (alpha-tocopherol):

Molecular formula: $C_{29}H_{50}O_2O_2$

Vitamin E (beta-tocopherol):

Molecular formula: $C_{28}H_{48}O_2$

Vitamin E (gamma-tocopherol):

Molecular formula: $C_{28}H_{48}O_2$

Figure 6-3. Structural formula of vitamin E. Vitamin E is fat soluble. Three isomeric tocopherols are shown: alpha-tocopherol, beta-tocopheral, and gamma-tocopherol. The alpha-tocopherol form is the most active.

VITAMIN K

Functions

Vitamin K is necessary for the formation of prothrombin in the blood. Prothrombin is the material from which the enzyme that causes blood to clot is formed. A deficiency of vitamin K prevents blood clotting and can lead to internal hemorrhaging and ultimately death.

Deficiency

Deficiencies of vitamin K rarely occur because it is synthesized in the rumen and in the intestinal tract of monogastric animals. The feeding of moldy feeds, such as moldy sweetclover, with a high dicoumarin content may cause a vitamin K deficiency leading to a bleeding syndrome called *sweetclover poisoning* or *bleeding disease*. Moldy feeds with a high dicoumarin content fed to swine or poultry will also cause internal bleeding and death. Feeding vitamin K or water-soluble synthetic forms of the vitamin will counteract the effects of the deficiency.

Sources

Dietary sources of vitamin K include green, leafy feeds (either fresh or dry), fish meal, liver, and soybeans. Because vitamin K is fat-soluble, solvent process soybean meal contains only a small amount of the vitamin.

Generally, vitamin K is very stable in feeds. Figure 6-4 shows the structural formula for vitamin K_1 (phylloquinone), which is found in green plants and menadione.

Common synthetic forms of material with vitamin K activity are menadione sodium bisulfite, menadione sodium bisulfite complex, and menadione dimethyl pyrimidinol bisulfite. These are water-soluble forms that depend on their menadione content for their biological activity. Menadione has a higher antihemorrhagic effect than vitamin K. It is common to add vitamin K to swine diets and chick starter rations; however, supplementation of ruminant or horse diets is generally not considered to be necessary.

B COMPLEX VITAMINS

There are eleven identified B complex vitamins with ten of them known to be necessary in animal nutrition. There is some evidence that there may be more in this group that have not yet been identified. All of the B complex vitamins are soluble in water.

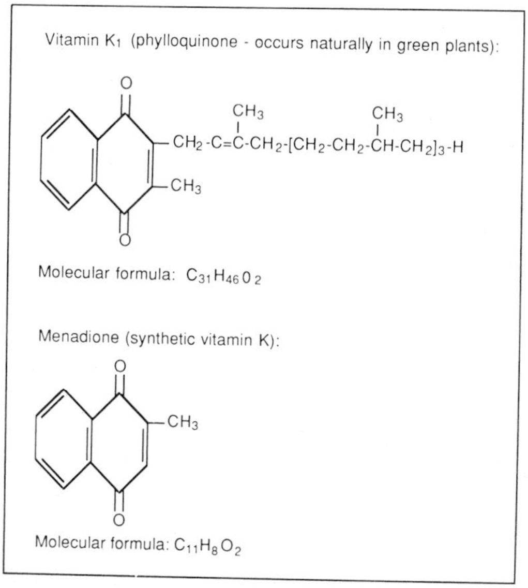

Ruminants (cattle, sheep, goats) generally do not need vitamin B complex supplementation of the diet. Microorganisms in the functioning rumen synthesize many of these vitamins, and feeds used in ruminant nutrition also supply many of the B complex vitamins. However, the rumen does not function in young animals; therefore, supplementation is recommended if the young animal is not nursing. The mother's milk contains many of the B complex vitamins that will supply the dietary needs of the nursing young. A deficiency of cobalt may cause a vitamin B_{12} deficiency in mature ruminants. Animals that are sick or have disorders upsetting the normal function of the rumen may also need additional B complex vitamins.

Swine fed on unsupplemented grain-soybean meal diets may be deficient in niacin, pantothenic acid, riboflavin, choline, and vitamin B_{12}. The other B complex vitamins are usually present in feeds in sufficient amounts to meet the needs of swine.

Horses generally do not need additional vitamin B complex supplementation in their diet. Adequate amounts are secured from the feed and from synthesis in the intestine. When horses are under severe stress, they may need more vitamins than those required for maintenance and growth. Additional research is needed to determine more accurately the vitamin needs of horses.

Poultry diets need to be supplemented with B complex vitamins. The B complex vitamins often included in poultry diets include thiamine, riboflavin, pantothenic acid, niacin, pyridoxine, biotin, choline, folacin, and B_{12}.

Thiamine (B_1)

Thiamine functions as a coenzyme in energy metabolism. It is unstable when heated; therefore, grain drying or cooking soybeans reduces the amount of available thiamine in these feeds. Figure 6-5 shows the structural formula of thiamine.

Vitamin K_1 (phylloquinone - occurs naturally in green plants):

Molecular formula: $C_{31}H_{46}O_2$

Menadione (synthetic vitamin K):

Molecular formula: $C_{11}H_8O_2$

Figure 6-4. Structural formulas for vitamin K (natural and synthetic). Natural vitamin K is fat soluble. Solubility of menadione depends on its form.

Indications of thiamine deficiency are anorexia, slow growth, weakness, increased irritability, lowered body temperatures, slower heartbeat, and enlargement of the heart. Calves develop polyneuritis, which is characterized by incoordination of the legs (especially the front legs), inability to rise and stand, and retraction of the head when lying down. These symptoms are often accompanied by anorexia, diarrhea, dehydration, and death. Swine develop anorexia, have poor growth, and are prone to sudden death. Horses show anorexia, weight loss, and lack of coordination (especially in the hind legs). Poultry develop polyneuritis characterized by lethargy, head tremors, and retraction of the head as the deficiency continues. Other indications in poultry include anorexia, loss of weight, weakness, and convulsions.

Dietary sources of thiamine include cereal grains; grain byproducts; brans; germ meals; green, leafy hay; green pastures; brewer's yeast; and milk. Infection of grains with certain kinds of mold may destroy the thiamine content. When cattle and lambs are on a high-concentrate feeding system, a thiaminase may become active in the rumen and cause a possible deficiency of thiamine. Normally, farm diets for animals provide enough thiamine. Thiamine is commercially available in vitamin premixes.

Riboflavin (B₂)

Riboflavin is a part of two coenzymes (flavin mononucleotide and flavin-adenine dinucleotide), which are involved in energy and protein metabolism. Figure 6-6 shows the structural formula of riboflavin.

Indications of riboflavin deficiency include slow growth in most species; poor reproduction and lower milk production in sows; anemia, diarrhea, vomiting, eye cataracts, stiffness of gait, seborrhea (fatty discharge on the skin), and alopecia (loss of hair) in young pigs; lesions around the mouth and loss of hair in calves; and diarrhea and curled-toe paralysis in young chicks.

Dietary sources of riboflavin include milk; skimmed milk; meat scraps; fish meal; green, leafy hay (especially alfalfa); green pastures; grass silage; and brewer's yeast. Riboflavin is commercially available in vitamin premixes. Ruminant rations may need riboflavin supplementation and swine and poultry rations generally have riboflavin added.

Pantothenic Acid

Pantothenic acid is a component of coenzyme A, which is important in carbohydrate and fatty acid metabolism. Figure 6-7 shows the structural formula of pantothenic acid.

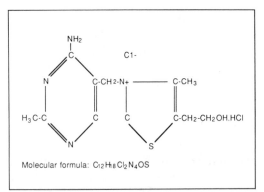

Molecular formula: $C_{12}H_8Cl_2N_4OS$

Figure 6-5. Structural formula for thiamine. Thiamine is a water soluble vitamin and is seldom deficient in livestock rations.

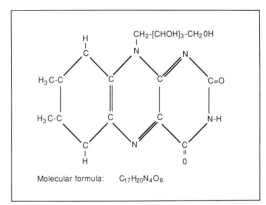

Molecular formula: $C_{17}H_{20}N_4O_6$

Figure 6-6. Structural formula for riboflavin. Riboflavin is a water soluble vitamin. Swine rations which do not contain milk and/or alfalfa meal will generally be deficient in riboflavin.

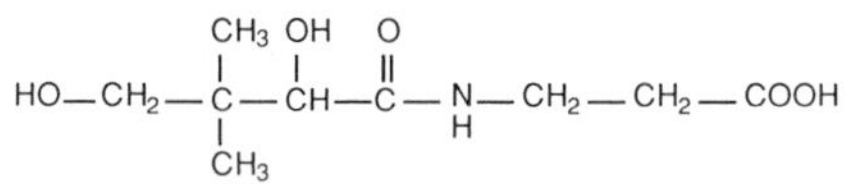

CH₃ OH O

HO—CH₂—C—CH—C—N—CH₂—CH₂—COOH
             | | || |
            CH₃   H

Molecular formula: $C_9H_{17}O_5N$

Figure 6-7. Structural formula for pantothenic acid. Pantothenic acid is a water soluble vitamin. It may be deficient in swine rations.

Deficiency symptoms common to all species include slow growth, loss of hair, and enteritis (inflammation of the intestine). In young pigs, a deficiency of pantothenic acid results in stiff legs and lack of coordination when walking (called *goose-stepping*) and poor condition of the hair and skin. Pantothenic acid deficiency in calves results in dermatitis around the eyes and muzzle, loss of appetite, diarrhea, weakness, increased susceptibility to respiratory infection, difficulty in standing, and, eventually, convulsions. Young chicks will have extremely ragged feather development and lesions on the mouth, eyelids, and feet. When breeder hen diets are deficient in pantothenic acid, newly hatched chicks die shortly after hatching and, eventually, hatchability of eggs declines.

Good feed sources of pantothenic acid include brewer's yeast, cane molasses, dried milk and whey, and fish solubles. Grains are deficient in this vitamin and deficiencies are most likely to appear when animals are fed high grain diets in confinement. The supplemental form commercially available is calcium pantothenate (*d* and *l* forms), which is a calcium salt. Only the *d* form is biologically active and the guaranteed amount should be stated in terms of this form.

Pyridoxine (B₆)

Pyridoxine is important as a component of coenzymes in amino acid and essential fatty acid metabolism, in the production of red blood cells, and in the endocrine system.

Figure 6-8 shows the structural formula for pyridoxine.

Indications of pyridoxine deficiency are anorexia, slow growth, and convulsions in all species; abnormal feathering, and development of nervous symptoms in chicks; and, in hens, lower egg production, poor hatchability, rapid weight loss, and eventually death.

Dietary sources of pyridoxine include cereal grains and grain byproducts; rice bran and polished rice; green pastures; green, leafy hay (especially alfalfa); yeast; and meat and liver meals. Pyridoxine is commercially available in vitamin premixes. Livestock rations generally do not need pyridoxine supplementation.

Niacin (Nicotinic Acid)

Niacin is an essential part of enzyme systems involved in lipid, carbohydrate, and protein metabolism and is needed by all living cells. Figure 6-9 shows the structural formula of niacin.

General indications of deficiency include poor appetite (anorexia), slow growth, and unthriftiness. In addition, swine show signs of diarrhea, vomiting, dermatitis, and loss of hair. Chicks develop an inflammation of the tongue, mouth cavity, and upper esophagus; show reduced feed consumption, slow growth, poor feather development; and develop scaly

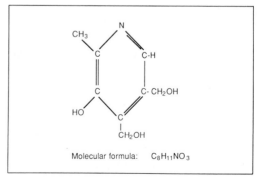

Molecular formula: $C_8H_{11}NO_3$

Figure 6-8. Structural formula for pyridoxine. Pyridoxine is a water soluble vitamin. It is generally found in adequate quantities in livestock rations.

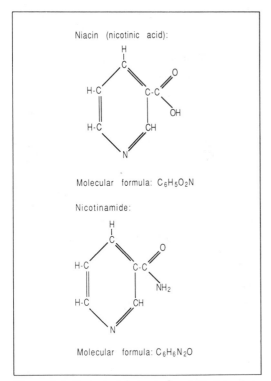

Niacin (nicotinic acid):

Molecular formula: $C_6H_5O_2N$

Nicotinamide:

Molecular formula: $C_6H_6N_2O$

Figure 6-9. Structural formula for niacin (nicotinic acid) and its amide (nicotinamide). Niacin is a water soluble vitamin. It can be synthesized in the body from surplus tryptophan. Swine rations may be deficient in available niacin.

dermatitis of the feet and skin. Turkey poults develop a hock disorder similar to perosis.

Some niacin is available in feeds, with meat and bone meal being a good source and green alfalfa being a fair source. However, the niacin in cereal grains is largely unavailable, therefore, swine and poultry diets need to be supplemented with this vitamin. Surplus tryptophan in the diet can be converted to niacin, so the level of this amino acid in the diet is important when determining the amount of niacin needed. In swine, each 50 mg of surplus tryptophan can be converted to 1 mg of niacin. Niacin supplementation is normally not needed for horses and ruminants, however, niacin supplementation has been beneficial to high-producing dairy cows.

Choline

Choline is a structural component of fat and nerve tissue and is required in the diet at levels higher than the other vitamins. Figure 6-10 shows the structural formula for choline.

It functions as a part of the cell structure, lipid transport, nerve impulse transmission, fat metabolism in the liver, and as a donor of readily available methyl groups involved in several one-carbon transfer reactions called *transmethylation*. Most of the dietary requirement for choline is used for transmethylation. In swine, methionine can replace approximately one-half the choline needed for transmethylation.

Choline deficiency symptoms include slow growth rate, unthriftiness, fatty livers, poor coordination, reproductive problems, lower milk production, higher death rate in the young, and perosis (slipped tendons) in poultry. Most farm rations contain enough choline, with deficiencies more likely to occur on low protein diets. Choline is synthesized in the body when there is a sufficient supply of other vitamins, especially vitamin B_{12}.

Good dietary sources of choline include meat scraps, oil meals, brewer's dried yeast, fish meal, and distiller's solubles. Some grains, forages, and dairy byproducts contain lesser amounts of choline. Choline is available in synthetic form, although general supplementation of the diet is usually not needed.

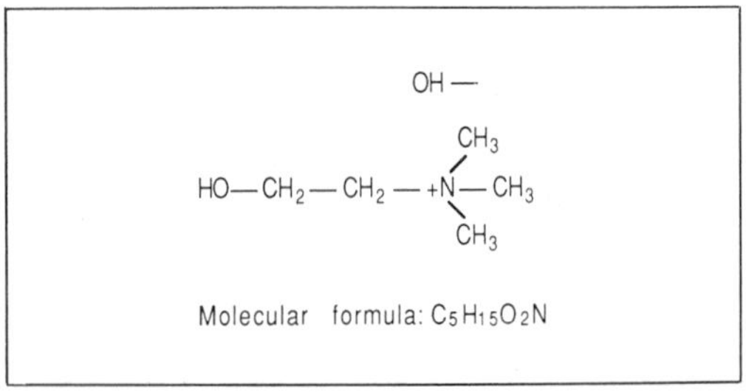

OH —

$$HO-CH_2-CH_2- \overset{\overset{\displaystyle CH_3}{|}}{\underset{\underset{\displaystyle CH_3}{|}}{N}} -CH_3$$

Molecular formula: $C_5H_{15}O_2N$

Figure 6-10. Structural formula for choline. Choline is a water soluble vitamin. It is generally not deficient in livestock rations. Surplus choline may be used as a methyl donor to spare methionine.

Biotin

Biotin is a part of the enzymes involved in carboxylation reactions such as the synthesis of fatty acids. Figure 6-11 shows the structural formula of biotin.

Deficiency symptoms include dermatitis, loss of hair, cracks in the feet, slow growth, spasticity of the hind legs in swine, and reduced hatchability of eggs. Raw egg white contains the protein avidin, which makes biotin unavailable to the animal. Egg whites that have been heat treated may be fed safely because the heat treatment inactivates the avidin.

Feed sources of biotin include most grains (wheat and barley are low in available biotin), soybean meal, green forages, yeast, and cane molasses. Animals can readily synthesize biotin and it is generally not deficient in normal farm rations. Biotin is quite stable as a compound and is available in commercial vitamin premixes. It is sometimes added to the ration of brood sows as insurance against deficiency.

Folic Acid (Folacin)

Folic acid is a part of the folate coenzymes and is essential to the normal metabolic function of body cells. It is involved in the combining of single carbon units into larger molecules. Folic acid is closely related to vitamin B_{12} metabolism and deficiencies of either one affect the function of the other. Figure 6-12 shows the structural formula of folic acid.

Deficiencies of folic acid result in weakness, slow growth, and anemia. Additional signs in young chicks include poor feathering, and loss of feather pigment; and in breeding hens, lower egg production and hatchability. Turkey poults have shown signs of nervousness, droopy wings, and cervical paralysis (stiff and extended neck). Cervical paralysis usually leads to death within two days after symptoms appear. Turkey breeder hens show reduced hatchability.

Green pasture; green, leafy alfalfa hay; and some animal proteins (particularly those made from body organs) are good dietary sources of folic acid. Synthetic forms are readily available when dietary supplementation is indicated, although for most animals there is generally enough available through the diet and by synthesis in the body. Supplementation is common for turkey breeder and pre-starter diets and chicken starter and breeding diets, especially for broiler chicks.

B_{12}

Vitamin B_{12} functions as a coenzyme in a variety of metabolic reactions and is necessary for the maturation of red blood cells. Cobalt is found in vitamin B_{12} and must be present for synthesis of the vitamin to occur. This is the only known function of cobalt in the animal's body. A cobalt deficiency can lead to a vitamin B_{12} deficiency in all species. Figure 6-13 shows the structural formula of vitamin B_{12}.

A deficiency of vitamin B_{12} is indicated by slow growth in all species; lack of coordination in the hind legs in young pigs; reduced litter size and higher pig death rate in breeding swine; and lower egg hatchability in breeding chickens. Dietary deficiencies may occur in swine and breeding hen rations and

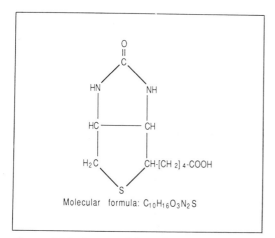

Molecular formula: $C_{10}H_{16}O_3N_2S$

Figure 6-11. Structural formula for biotin. Biotin is a water soluble vitamin. It is generally not deficient. Biotin is synthesized in the intestine in fairly large quantities.

Molecular formula: $C_{19}H_{19}N_7O_2$

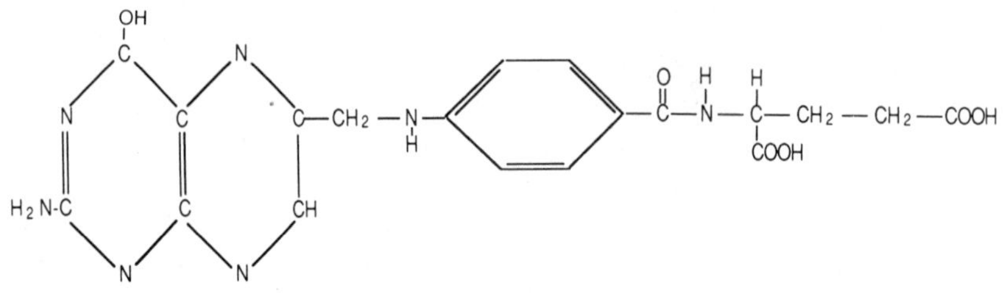

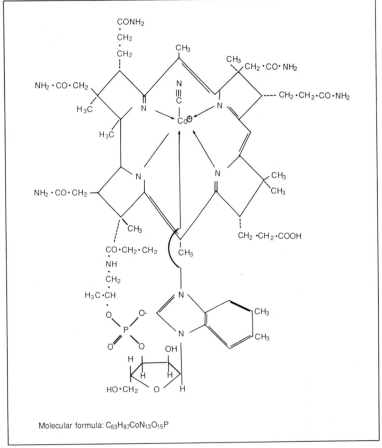

Figure 6-12. Structural formula for folic acid. Folic acid is a water soluble vitamin. It is usually not deficient for livestock. Folic acid is synthesized to a limited degree in the intestine.

Molecular formula: $C_{63}H_{87}CoN_{13}O_{15}P$

Figure 6-13. Structural formula for vitamin B_{12}. Vitamin B_{12} is a water soluble vitamin. It may be deficient for swine and early weaned calves.

for other species if a cobalt deficiency is present.

Good dietary sources of vitamin B_{12} include animal proteins and fermentation products. Vitamin B_{12} is available in commercial vitamin premixes.

Inositol

The function of inositol is not known and deficiency symptoms have not been described. It is available in synthetic form and is found widely in commonly used farm feeds. Inositol is synthesized in the intestinal tract and is generally not deficient in livestock feeding. Figure 6-14 shows the structural formula of inositol.

Para-aminobenzoic Acid

Para-aminobenzoic acid is essential for the growth of some microorganisms but deficiency symptoms have not been observed in animals. Its distribution in farm feeds is not clearly established but it is synthesized readily in the intestine and is usually not deficient in livestock rations. It is available in synthetic form. Figure 6-15 shows the structural formula of para-aminobenzoic acid.

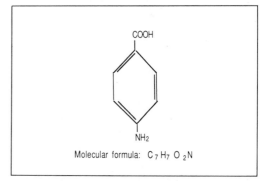

Molecular formula: $C_7H_7O_2N$

Figure 6-15. Structural formula for para-aminobenzoic acid (PABA). Para-aminobenzoic acid is a water soluble vitamin. It is usually not deficient for livestock.

VITAMIN C (ASCORBIC ACID)

Vitamin C is necessary for the formation of collagen, a gelatinlike protein that is the main constituent of the fibrils of connective tissue and bones. Vitamin C deficiency symptoms have not been observed in farm animals. Normally they synthesize sufficient amounts in body tissues to meet their needs; therefore, it is not required as a supplement to their diet. There is some evidence that under stress conditions, poultry, and possibly swine, may need some supplementation of the diet with vitamin C. Figure 6-16 shows the structural formula of vitamin C.

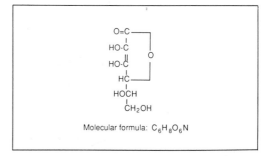

Molecular formula: $C_6H_8O_6N$

Figure 6-16. Structural formula for vitamin C (ascorbic acid). Vitamin C is a water soluble vitamin. It is synthesized in body tissues and is never deficient for livestock under normal conditions.

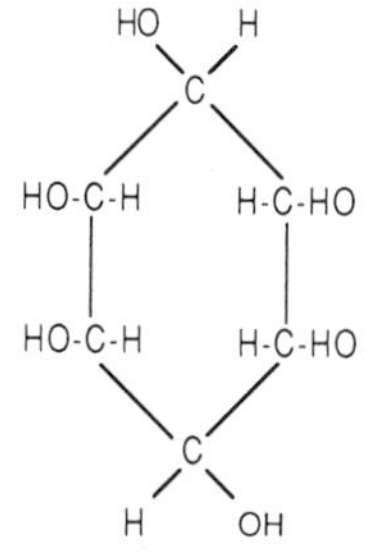

Molecular formula: $C_6H_{12}O_6$

Figure 6-14. Structural formula for inositol. Inositol is a water soluble vitamin. It is usually not deficient for livestock.

FEED ADDITIVES

Feed additives are products used in animal nutrition that are not nutrients in the usual sense of the word. Additives are used because they promote greater feed efficiency, produce more rapid gains or higher production. It is not known for sure why additives have the effect they do, but some appear to control disease organisms and parasites, others affect metabolism, some improve digestion, and others increase appetite. Types of feed additives include antibiotics and antibacterials, hormones, anthelmintics, and other miscellaneous compounds.

Some additives are incorporated in the feed and others in the form of a drench or bolus. In some cases, more than one form of an additive is available, giving the feeder a choice of methods for administering the additive.

While additives have been developed over the years for most species of livestock, the major uses today are with growing-finishing beef cattle, swine, and poultry. Some additives are used with dairy calves and lambs. Additives do not replace good management practices, including proper sanitation and balanced rations.

Health Concerns

There has been a growing concern in recent years that the use of some types of additives for livestock feeding may have an impact on human health as well as on animal health. Much of this concern centers around the use of antibiotics at subtherapeutic levels in animal feeding which, it is feared, might lead to the rise of resistant strains of bacteria with a negative impact on animal and human health. Bacterial resistance to drugs has been observed almost from the time antibiotics were first used in animal feeding. The impact of a ban on the use of antibiotics in animal feeding would vary from species to species, but the overall effect would be to raise the cost of animal products to the consumer. To date, there is no persuasive evidence of animal or human health problems arising after nearly 30 years of the use of antibiotics in animal feeding. There is a need for critical experimental studies on the effect of low-level antibiotic feeding on animal and human health.

The Food and Drug Administration (FDA) has also expressed concern about the possible carcinogenic (cancer causing) effects of some feed additives. The 1958 Food Additives Amendments to the Food, Drug and Cosmetic Act of 1938 include prohibitions against adding to food any substance that is a known carcinogenic. The hormone diethylstilbestrol, which was introduced as a beef cattle additive in 1954, has demonstrated carcinogenic effects in mice, when fed at levels several thousand times as high as any residues found in beef. Approval for the use of diethylstilbestrol was withdrawn. Other additives are under investigation for possible carcinogenic effects and will be withdrawn if these effects are found.

Regulation of Use

There are strict regulations governing the use of feed additives. These regulations change from time to time as new information on effects and dangers becomes available. Many additives may be used only within certain specified levels and for specific species or types of animals. Many additives must also be withdrawn from use in the feed within specified times of marketing the animal. Current detailed information on feed additive use and regulations may be found in the *Feed Additive Compendium*, published annually by the Miller Publishing Company, 2501 Wayzata Boulevard, Minneapolis, Minnesota 55440. The Compendium is updated during the year on a regular basis. In Canada, information on use and regulations is available in the compendium of *Medicating Ingredient Brochure*, available from the Plant Products Division, Canada Department of Agriculture, Ottawa, Canada. The *Code of Federal Regulations (CFR)* Title 21 gives official information from the Food and Drug Administration concerning approval of antibiotics and other animal drugs. Revisions of Title 21 are made annually as of April 1 and the *CFR* is updated in individual issues of the *Federal Register*. The *CFR* and the *Federal Register* must be used together to determine the current regulations concerning the use of animal drugs. Title 21 (part 500-599 covering animal drugs, feeds, and related products) is available from the Superintendent of Documents, U.S. Government Printing Office, Washington, D.C. 20402 (inquire for current price). The *Federal Register* is available from the Superintendent of Documents on an annual subscription basis (inquire for current price) and includes monthly issues of the *List of CFR Sections Affected* and *The Federal Register Index*.

Antibiotics and Antibacterials

Antibiotics are organic in nature (being produced by organisms) and slow or stop the growth of other organisms. Antibacterials are chemicals that also slow or stop the growth of organisms. Sometimes an antibiotic and an antibacterial are combined into one compound, called *chemobiotic*, to combat a problem that is not susceptible to either one individually. In the discussion which follows, the term *antibiotic* is used generically to refer to all these compounds.

During the past 30 years, there has been a remarkable increase in confinement operations in the swine, poultry, and beef cattle industries. Concurrent with this change, and probably facilitating it, has been the increased use of antibiotics in feeding. The regular use of antibiotics at subtherapeutic levels is more common in swine feeding than with other classes of livestock, accounting for about 40 percent of the feed-additive antibiotics produced. The FDA has defined subtherapeutic level as that which is lower than the therapeutic level needed to cure disease.

The subtherapeutic use of antibiotics in cattle feeding is regional, with little use in the arid regions of the Southwest, and routine use in other cattle feeding regions. There is no response in terms of growth stimulation to the use of antibiotics in the Southwest, an effect believed to be related to the climate of the region. In other areas, 50-60 percent of all feedlot cattle are fed antibiotics at subtherapeutic levels, with an estimated 40 percent of the total beef supply in the U.S. being so fed.

With the increased use of confinement production, the most common disease problems reported with swine are enteric (intestinal) and respiratory. Antibiotics and antibacterials commonly used at subtherapeutic levels in swine feeding include bacitracin, bambermycins, carbadox, chlortetracycline (Aureomycin), lincomycin, oleandomycin, oxytetracycline (Terramycin), penicillin, streptomycin, tylosin (Tylan), virginiamycin, arsenicals, nitrofurans, and sulfa drugs. **CAUTION:** Current regulations should be consulted, as noted above, to determine current approvals for the use of these materials.

Poultry producers make only limited use of penicillin and the tetracyclines for subtherapeutic feeding. These antibiotics are not approved by the FDA for use in combination with monensin (an anticoccidial drug). About 85 percent of the broiler feed manufactured in the U.S. contains monensin. Only the antibiotics lincomycin, bacitracin, and the bambermycins can be combined with monensin for subtherapeutic feeding. Penicillin is incorporated in turkey feed for the control of erysipelas, but is rarely used in feeds for chickens. Tetracycline or a combination of oxytetracycline and neomycin is used in starter feeds for turkeys and broilers. In the case of broilers, the anticoccidial drug is removed because of the prohibition on the combination. Tetracyclines are also used periodically in the feeds for laying hens to improve shell quality and production, especially during the latter part of the laying cycle. **CAUTION:** Current regulations should be consulted, as noted above, to determine current approvals for the use of these materials.

The tetracyclines are commonly used in beef cattle feeding to accomplish the following four purposes: (1) controlling liver abscesses; (2) controlling respiratory diseases associated with shipping fever; (3) controlling foot rot; and (4) improving gains and feed efficiency. High-concentrate finishing diets for feeder cattle increase the incidence of liver abscesses, and the use of the tetracyclines at subtherapeutic levels has shown as much as a 50 percent decline in liver abscesses, with a concomitant increase in daily gains of as much as 5-10 percent. Cattle coming into the feedlot are fed chlorotetracycline in combination with sulfamethazine for a period of 7-28 days for control of respiratory infections related to shipping fever. After this initial feeding period, the level is reduced for the remainder of the feeding period, for the purpose of promoting growth and improving feed efficiency. Monensin (an antibiotic additive) in combination with tylosin has also been approved for use with feedlot cattle. Lasolacid (Bovatec) plus monensin (Rumensin) increases the proportion of propionic acid and decreases the proportions of acetic and butyric acids in the rumen. This action increases energy utilization by the animal, thus improving feed efficiency. Tylosin helps reduce liver abscesses. The only other antibiotics approved at the present time for feedlot cattle are bacitracin and erythromycin, although neither have been field-tested sufficiently to produce substantive information on their effectiveness. The use of antibiotics in feeding cattle does not always result in these benefits, and it is believed that the variation in response may be related to the disease level present in the feedlot. CAUTION: Current regulations should be consulted, as noted above, to determine current approvals for the use of these materials.

The tetracyclines (especially chlorotetracycline and oxytetracycline) have been shown to improve the growth rate and reduce enterotoxemia (intestinal disorders), in lamb creep and finishing diets, when used at subtherapeutic levels of feeding. The degree of response appears to be related to the level of management and the amount of stress to which the animals are subjected. The response has also been more variable with western feeder lambs. CAUTION: Current regulations should be consulted, as noted above, to determine current approvals for the use of these materials.

The use of tetracyclines (especially chlorotetracycline and oxytetracycline) in milk replacers and starter feeds for dairy calves has been shown to be effective in promoting growth and controlling bacterial scours. There has been little evidence of continued benefit from feeding antibiotics after four months of age. The inclusion of antibiotics in the rations of lactating dairy cows is not recommended, mainly because little benefit in increased milk production has been shown and because of the danger of residues in the milk. CAUTION: Current regulations should be consulted, as noted above, to determine current approvals for the use of these materials.

Dairy calf starters and milk replacers containing antibiotics may be used when raising kids (young goats). There are no recommendations for the inclusion of antibiotics at subtherapeutic levels in the rations of horses.

Hormones

Hormones are substances secreted by the endocrine glands into the body fluids and transported to other locations, where they produce specific effects on cell activity. Hormones and hormonelike substances are produced commercially to be used as feed additives, primarily in beef nutrition.

The first synthetic hormone feed additive was stilbestrol (diethylstilbestrol <DES>), which was approved by the FDA in 1954 for use in beef finishing rations. The FDA approved stilbestrol implants for steers in 1956. As noted earlier, however, DES was found to be carcinogenic when fed in massive doses to mice and has been withdrawn from use as a feed additive.

A number of other hormones and hormonelike products are available for use, including melengestrol acetate (MGA), Ralgro (zeranol, a derivative of corn mold), Synovex S, and Synovex H. Melengestrol acetate is a progestin that suppresses estrus, increasing gain and improving feed efficiency. It is used with unspayed or nonpregnant feedlot heifers but not with steer, or bulls. Ralgro is an alcohol, rather than a hormone, but it mimics the action of a hormone. MGA has been approved for use with Rumensin and can be included in a complete mixed feed.

Ralgro is used as an implant in male and female nursing calves, grazing cattle, and finishing cattle. Synovex S implants are used for steers weighing 400 (181 kg) to 1,000 pounds (454 kg) and contain progesterone and estradiol benzoate. Synovex H implants are used for heifers weighing 400 (181 kg) to 800 pounds (363 kg) and contain testosterone and estradiol benzoate. **CAUTION:** Current regulations should be consulted, as noted above, to determine current approvals for the use of these materials.

The use of hormone additives is not recommended for swine, dairy cattle, poultry, or horses. Some benefits from hormone use have been observed in lambs.

Anthelmintics

These compounds are used to control stomach and intestinal worms and may be administered in feed or water. Because the presence of worms in the system reduces feed efficiency and gain, the use of anthelmintics can improve performance if there is a problem with worms. If worms are not present, then these materials should not be used. Some anthelmintics available for use include hygromycin, loxon, phenothiazine, piperzine, thiabendazole, and tramisol.

Miscellaneous Additives

Coccidiostats are added to some poultry rations to prevent the disease *coccidiosis*. Sodium bicarbonate and ground limestone affect the pH level in the digestive tract and sometimes are used to improve performance. The thyroid

gland functions to control the rate of metabolism and an additive such as iodinated casein (thyroprotein) may be used to increase the amount of the hormone thyroxin in an animal's system. If too much thyroxin is being produced in the body, there are several inhibitors such as thiourea, thiouracil, and methimazole (tapazole), which may be used to regulate the amount of thyroxin present. Poloxaline is a bloat preventative, which may be added to ruminant feeds when bloat is a problem. Animals under stress in the feedlot may be calmed by the use of tranquilizers such as hydroxyzine, reserpine, or trifluomeprazine. Enzyme feed additives are available but their cost on a continuous use basis may outweigh the benefits realized from their use. Antioxidants may be added to feeds to prevent rancidity, and sodium bentonite is sometimes used as a pellet binder when feeds are pelleted.

WATER

Functions

Water is a vital factor in animal nutrition, being one of the largest single constituents of the body. It performs many vital physiological functions including:

1. Involved in many of the biochemical reactions of digestion and metabolism in the body.
2. Transports nutrients and wastes in the body.
3. Helps regulate body temperature.
4. Helps give the body form by filling the cells, giving them shape (turgor pressure).
5. A constituent of many important fluids in the body and helps provide lubrication.
6. Milk production.

Animals can survive longer without feed than they can without water. The normal intake of water ranges from three to eight times the intake of dry matter in the diet. All feeds contain water with a range of approximately 10 percent for air-dry feeds to more than 80 percent in fresh green forages.

As Percent of Body Weight

Younger animal bodies contain a higher percent of water in relation to their weight than do older animal bodies. As animals fatten, the fat replaces some of the water in their bodies; therefore the bodies of fat animals contain a lower percent of water in relation to their weight, as compared to thinner animals. Some comparative percentages of water in the body compared to weight are: newborn calf, 70 percent; 1,000 lb (454 kg) steer, 50 percent; fat dairy cow, 50 percent; thin dairy cow, 70 percent; newborn pig, 80 percent; finished market hog, 50 percent; newborn lamb, 80 percent; fat lamb, 50 percent; young horse, 70-80 percent; goat, 60-76 percent.

Requirements

Clean water should be readily available to livestock at all times, preferably near feeding areas. In addition to water that they drink, animals get water from the metabolic breakdown of carbohydrate, fat, and protein, and from the feed they consume. Limited information is available on water requirements for livestock, with the assumption being that the amount consumed when freely available is roughly equivalent to the requirement. The amount of water an animal consumes will vary with the species, breed, body size, rate and composition of gain, age, pregnancy, lactation, activity, type of diet, feed intake, frequency of watering, humidity, ambient temperature, and mineral content of the water.

Animals that are gaining weight require more water than those that are losing weight. Animals with diarrhea or fever will have a higher water intake as will those with a high salt or protein intake. Pregnant and lactating animals require more water than those not pregnant or milking. Increased activity increases the water intake of animals. Animals consuming high levels of dry matter will drink more water than those consuming feeds high in water content.

Temperature and Water Intake

The temperature of the water also affects intake. When ambient temperatures are below freezing, animals will drink more water if it is warmed slightly above freezing. The major benefit, however, of warming water under these conditions is to keep it from freezing. High humidity decreases water intake, while high ambient temperatures significantly increase water consumption.

Minerals in Water

Water supplies sometimes contain salt, sulphates, nitrates, or other minerals. The effect of these substances varies with species.

Salt concentrations below 1,000 ppm are generally safe for all livestock. Animals unaccustomed to the range of 1,000 to 5,000 ppm may develop diarrhea, with possible refusal of the water at the upper level of concentration. When the salt concentration is from 5,000 to 7,000 ppm, the water is considered safe for dairy, beef, and swine but should not be used for pregnant or lactating animals. At a concentration of 7,000 to 10,000 ppm, the water is considered unsafe for swine, pregnant or lactating cows, young calves, or animals under heat stress. Water containing more than 10,000 ppm salt is not considered safe for any livestock.

Salt concentrations up to 1.2 percent cause an increase in water consumption, with higher concentrations causing water consumption to decrease. Excessive levels of nitrates or high alkalinity may make water unpalatable.

Hard (33 ppm calcium and magnesium) or soft (1 ppm) water has no effect on lactation in dairy cattle. The presence of elements such as iron, aluminum, zinc, manganese, and strontium can also increase the hardness of water. Some of these elements can be toxic if present in high enough concentrations, Table 6-2.

Table 6-2. Recommended limits of concentration of some potentially toxic substances in drinking water for livestock.

Element or Compound	Safe upper limit of Concentration (ppm)
Arsenic	0.2
Barium	not established
Cadmium	0.05
Chromium	1.0
Cobalt	1.0
Copper	0.5
Cyanide	not established
Fluoride	2.0
Lead	0.1
Mercury	0.01
Nickel	1.0
Nitrate–N	100.0
Nitrite–N	10.0
Vanadium	0.1
Zinc	25.0

Source: Ankerman, D. and Petersen, L., eds. 1980. *Feed analysis and livestock feeding guide.* Fort Wayne, Indiana: A & L Agricultural Laboratories.

pH Level

The pH of most water ranges from 6 to 9, which appears to be satisfactory for livestock. Water that has pH values above or below this level may not be suitable for use with livestock.

Contamination

Lake and other surface waters may become contaminated with blue-green algae, which has resulted in poisoning of cattle drinking this water. Water sources containing heavy growths of algae should not be used for livestock.

Administering Medications in Water

Medications are sometimes administered through water, especially for poultry and swine. Turkeys can detect minor differences in salt and pH levels of water and may not consume medicated water because of this difference.

Water Consumption

Water consumption varies with species, Table 6-3. Swine will consume from 2.0-5.0 kg (0.5-1.3 gals.) water per kilogram (2.2 lbs.) of dry feed and 7-20 kg (1.8-5.3 gals.) water per 100 kg (220 lbs.) body weight daily.

Beef cattle will consume from 3-8 kg (0.8-2.1 gals.) of water per kilogram (2.2 lbs.) of dry matter with the higher level of consumption occurring at higher temperatures.

Nonlactating dairy cattle will consume from 3.1-15.6 kg (0.8-4.1 gals.) of water per kilogram (2.2 lbs.) of dry matter with increasing ambient temperatures causing the increase in consumption. Lactating dairy cattle will consume from 2.08-3.83 kg (0.55-1.0 gals.) of water for each kilogram (2.2 lbs.) of milk produced in addition to the amount consumed for maintenance purposes. Again, the higher amounts consumed were at higher temperature levels.

The water consumption of poultry varies with the age, type of poultry, and kind of production.

It is recommended that goats receive 145.6 grams (0.04 gals.) water per $w_{kg}^{0.75}$ (body weight in kilograms to the power of 0.75) for maintenance and 1.43 kg (0.4 gal) water per kilogram (2.2 lbs.) of milk produced. In tropical areas meat goats have been reported to have an average daily water intake of 680 grams (0.2 gal).

Horses need from 2-4 litres (0.5-1.0 gal) of water per kilogram (2.2 lbs.) of dry matter intake with the requirement increasing by 15-20 percent as the ambient temperature rises.

Water Losses from the Body

Water losses from the body are through urine, feces, sweat, lungs, and production of milk and eggs. Animals have a limited capacity to reduce the amount of water excreted through the urine when suffering from water deprivation. Loss of water through the feces varies with species, with cattle feces having a higher moisture content than sheep feces.

Water Deprivation

Water deprivation will result in reduced feed intake, digestive upset, lower production, and eventually starvation. Dairy cattle suffer more quickly from a lack of water than from a deficiency of any other nutrient. Because of their milk production, dairy cows require more water for their size than do any other farm animals.

Toxic Elements in Water

Toxic elements such as lead, cadmium, and mercury are sometimes found in sufficient concentration in water to pose a problem for livestock health (see Table 6-2). Even if the health of animals is not threatened by these substances, they may accumulate in the meat, milk, or eggs produced by the livestock. Products with high levels of these elements are not fit for human consumption, thus presenting a human health hazard and an economic loss for the livestock producer if the products are condemned.

Occasionally, pesticides get into the water supply for livestock, either through runoff from the soil, drift, direct application, accidental spill, or careless handling in waste disposal. The major danger is from the organophosphorus insecticides, which are lethal to animals at very low dosages. Insecticides such as DDT, Methoxychlor, Dilan, Perthane, and DDD present some danger as their lethal dosages are also low, though not as low as the organophosphorus insecticides. The pyrethrins and rotenones, being of vegetable origin, are practically nontoxic to livestock and so present little danger if found in the water supply.

Even if the pesticide is not lethal to livestock, it may present a hazard to human health if dangerous levels accumulate in meat, milk, or eggs. There is also danger of possible interaction between the pesticide and drugs used in feed, which could be harmful to animals.

SUMMARY

Vitamins are organic compounds essential for life, but needed by the animal in only trace amounts. Sixteen vitamins have been identified as essential in animal nutrition. The B complex vitamins and vitamin C are water soluble. Vitamins A, D, E, and K are soluble in fat or fat solvents.

Most of the B complex vitamins are synthesized in the rumen. Fat soluble vitamins may be stored in the body, reducing the need for dietary sources. The water-soluble vitamins are generally not stored in the body and need to be supplied in the diet of monogastric animals. Vitamins necessary in the diet are available through vitamin premixes.

Vitamin A is associated with vision and other functions and requirements can usually be met through the diet of the animal, with fresh green forages and yellow corn being good sources.

Vitamin D is associated mainly with bone development and the prevention of rickets. Sun-cured forages and irradiated yeast are good sources of this vitamin.

Vitamin E is associated with the prevention of the oxidation of lipids within the body and the absorption and storage of vitamin A and other metabolic functions in the cell. A deficiency results in symptoms similar to selenium deficiency. Whole grains, green forages, and good quality hay are good sources of this vitamin.

Vitamin K is associated with the ability of the blood to clot. Green leafy feeds and fish meal are good sources of vitamin K.

There are eleven identified B complex vitamins, with ten of them known to be essential in livestock nutrition. In general, ruminants (except the very young) do not need dietary sources of B complex vitamins because of their ability to synthesize these vitamins in the rumen. Swine and poultry diets generally need supplementation with the B complex vitamins.

Feed additives are substances that are not nutrients but promote feed efficiency, more rapid gains, or higher production. Some feed additives control disease organisms and parasites, others affect metabolism, some improve digestion, and others increase appetite. The most commonly used additives are antibiotics and hormones.

Feed additives may be incorporated in the feed or used as implants, drenches, or boluses. The major use of feed additives is with growing-finishing cattle, swine, and poultry. Additives are not a substitute for good management practices.

There are strict regulations governing the use of feed additives. These regulations are updated periodically and need to be consulted for current approved practices.

Water is a vital factor in animal nutrition, being one of the largest single constituents of the body. A clean, readily available supply of water is vital to good animal nutrition. Animals can survive longer without feed than they can without water. Water is involved in metabolic processes, transport of nutrients and waste in the body, temperature regulation, giving form to the body, as a part of many fluids in the body, and in milk production.

Care must be taken to ensure that the water supply is safe for use. Toxic elements such as lead, cadmium, and mercury sometimes contaminate water supplies leading to human and animal health problems. Surface water contaminated with heavy algae growth can be dangerous to livestock. Pesticide contamination, especially the organophosphorus insecticides, may be toxic to animals or contaminate the meat, milk, and eggs produced.

REVIEW

1. What are vitamins?
2. Discuss the historical development of vitamins.
3. How many vitamins are essential in animal nutrition?
4. What chemical elements are found in vitamins?
5. Which vitamins are soluble in water and which are soluble in fat or fat solvents?
6. Which vitamins are commonly synthesized in the rumen?
7. How does the solubility of vitamins affect the need for supplying them in the diet?
8. How may vitamins be supplied other than through natural feed sources?
9. What is the source of vitamin A in feeds?
10. What is the international standard for measuring vitamin A activity?
11. Discuss the functions of vitamin A.
12. What are the symptoms of a deficiency of vitamin A?
13. What are some good feed sources of vitamin A?

14. List the conditions that might indicate a need for supplementing the ration with vitamin A.

15. What are effects of feeding excessive amounts of vitamin A?

16. What is the function of vitamin D?

17. Compare the effectiveness of the animal and plant forms of vitamin D for different species of livestock.

18. Describe the symptoms of a deficiency of vitamin D.

19. List some feed sources of vitamin D.

20. What is the function of vitamin E?

21. What are the symptoms of vitamin E deficiency?

22. What is the relationship between vitamin E and selenium?

23. List some dietary sources of vitamin E.

24. What is the function of vitamin K?

25. What are the symptoms of vitamin K deficiency?

26. List some dietary sources of vitamin K.

27. Name some synthetic materials with vitamin K activity.

28. How many of the B complex vitamins are necessary in animal nutrition?

29. Discuss the need for supplementing ruminant rations with the B complex vitamins.

30. Which of the B complex vitamins are usually needed in swine rations?

31. Which of the B complex vitamins are usually needed in poultry rations?

32. List each of the B complex vitamins including functions, deficiency symptoms, dietary sources, and need for each in livestock rations.

33. What is the function of vitamin C?

34. Discuss the need for vitamin C supplementation in the ration.

35. Define and list the types of feed additives used in animal nutrition.

36. With which classes of livestock are feed additives mainly used?

37. Discuss the concern about the use of feed additives as it relates to human and to animal health.

38. Where can current information on regulations for feed additive use be secured?

39. What are antibiotics and antibacterials?

40. Discuss the increased growth in use of antibiotics during the past 30 years.

41. List antibiotics currently in common use for the different classes of livestock and the purpose of each.

42. Define hormone and hormonelike additives and discuss their functions.

43. Describe the hormone additives commonly used for beef cattle.

44. Define anthelmintics and discuss their function.
45. List and describe the functions of other feed additives sometimes used in livestock nutrition.
46. What are the functions of water in an animal's body?
47. Discuss the relationship between age and fat content of the body and the percent of water it contains.
48. In addition to drinking water, what are some other sources of water for the animal?
49. List and discuss factors affecting the amount of water an animal will consume.
50. Discuss the effects of salt concentrations in water as they relate to animal nutrition.
51. Discuss other factors affecting the suitability of water for livestock.
52. What are typical water intakes for various classes of livestock?
53. What are ways by which animals lose water from the body?
54. What are symptoms of water deprivation in livestock?
55. Discuss the dangers of toxic elements and pesticides in the water supply.

7

CLASSIFICATION AND USE OF FEEDS

OBJECTIVES

After completing this chapter you will be able to:

- Describe the classification of feeds.
- Describe the uses of feed in the animal's body.
- Describe underutilized and byproduct sources of feeds.

CLASSIFICATION OF FEEDS

Materials fed to animals for the purpose of meeting their nutritional needs are called *feeds (or feedstuffs)*. The two general classes of feeds are roughages and concentrates. In general, air-dry roughages are feeds that contain more than 18 percent crude fiber and less than 60 percent total digestible nutrients (TDN). Air-dry concentrates generally contain less than 18 percent crude fiber and more than 60 percent TDN.

Both roughages and concentrates are further divided by moisture content. Feeds that have more than 80 percent dry matter (DM) are considered air-dry, while feeds with less than 80 percent dry matter are called *high-moisture feeds*.

INTERNATIONAL FEED NOMENCLATURE

The feed composition tables in the Appendix are adapted from materials published by the National Research Council. Feedstuffs are identified by a nomenclature system that follows the International Feed Vocabulary of Harris et al. (1980, 1981). The Association of American Feed Control Officials (AAFCO) have designated official names and definitions for many of the feeds used in the United States. In many cases, these names are common or trade names and the origin of the feed name does not follow a standardized naming system.

Feed names are determined by using descriptors based on one or more of the following:

1. Origin of the feed (includes scientific and common name).
2. Part eaten by the animal.
3. Process(es) and treatment(s) applied to the feed before it is fed to the animal.
4. Stage of maturity (forages).
5. Cutting or crop (forages).
6. Grade or quality designations.

Table 7-1 shows examples of complete feed names made up of all the applicable descriptors.

Table 7-1. Feed names using applicable descriptors.

International Feed Number	1-00-022	1-03-619	5-20-637	3-03-296	5-02-009	4-02-935
Origin						
Genus	Medicago	Arachis	Glycine	Avena	Brevoortia	Zea
species	sativa	hypogaea	max	sativa	tyrannus	mays
variety						indentata
common name	alfalfa	peanut	soybean	oats	fish	corn, dent yellow
kind					menhaden	
Part fed		hay	seeds			seeds
Process	meal dehydrated	sun-cured	meal solvent extracted	silage	meal mechanical extracted	
Maturity				dough stage		
Cut						
Grade	15% protein		44% protein			

The nutritive value of some feeds (especially forages and silages) is affected by the stage of maturity at which the feed is harvested. A comparison of various stages of maturity for a forage listed in the feed composition table illustrates this effect. Definitions of terms for stages of maturity are given in Table 7-2.

INTERNATIONAL FEED CLASSES

There are eight classes of feeds, based on composition and use. Table 7-3 (page 104) lists these eight classes with their definitions. In some cases, it might be expected from the definitions given that a feed would fall into a different class from that listed. In borderline cases, the feed is placed in a class according to its most common use in feeding practices.

INTERNATIONAL FEED NUMBER

Feeds are identified in the feed composition table by a six-digit number. The first number is the feed class number and the remaining five digits are the International Feed Number (IFN). The IFN links the International Feed Name with chemical and biological data in the USA and World databanks. The IFN is useful for computer formulation of diets and is used in examples of diet formulation in this text.

FORMS OF FEED

The three general forms of feeds are dry, green, and high-moisture. Typical examples of dry feeds include hays, grains, oilseed meals, straw, stover, corn cobs, corn husks, soybean hulls, soybean mill feeds, cottonseed hulls, peanut hulls, oat hulls, and rice hulls.

Green feeds include pasture and green chop. Green chop feeds are roughages chopped daily in the field and brought to livestock for feeding.

Some typical examples of high-moisture feeds include high-moisture grain, haylage, wet byproduct feeds, roots and tubers, fresh milk, and silages.

Table 7-2. Stage of maturity terms for plants.

Preferred term	Definition	Comparable Terms
For Plants that Bloom		
Germinated	Stage in which the embryo in a seed resumes growth after a dormant period.	Sprouted
Early vegetative	Stage at which the plant is vegetative and before the stems elongate.	Fresh new growth, before heading out, before inflorescence emergence, immature prebud stage, very immature, young.
Late vegatative	Stage at which stems are beginning to elongate to just before blooming; first bud to first flowers.	Before bloom, bud stage, budding plants, heading to in-bloom, heads just showing, jointing and boot (grasses), prebloom, preflowering, stems elongated.
Early bloom	Stage between initiation of bloom and stage in which 1/10 of the plants are in bloom; some grass heads are in anthesis.	Early anthesis, first flower, headed out in head, up to 1/10 bloom.
Midbloom	Stage in which 1/10 to 2/3 of the plants are in bloom; most grass heads are in midanthesis.	Bloom, flowering, flowering plants, half bloom, in bloom, midanthesis.
Full bloom	Stage in which 2/3 or more of the plants are in bloom.	3/4 to full bloom, late antheses.
Late bloom	Stage in which blossoms begin to dry and fall and seeds begin to form.	15 days after silking, before milk, in bloom to early pod, late- to post-anthesis.
Milk stage	Stage in which seeds are well formed but soft and immature.	After anthesis, early seed, fruiting, in tassel, late bloom to early seed, past bloom, pod stage, postanthesis, postbloom, seed developing, seed forming, soft, soft immature.
Dough stage	Stage in which the seeds are of doughlike consistency.	Dough stage, nearly mature, seeds dough, seeds well developed, soft dent.
Mature	Stage in which plants are normally harvested for seed.	Dent, dough to glazing, fruiting, fruiting plants, in seed, kernels ripe, ripe seed.
Postripe	Stage that follows maturity; some seeds cast and plants have begun to weather (applies mostly to range plants).	Late seed, overripe, very mature.
Stem cured	Stage in which plants are cured on the stem; seeds have been cast and weathering has taken place (applies mostly to range plants).	Dormant, mature and weathered, seeds cast.

continued

Table 7-2. (continued).

Preferred term	Definition	Comparable Terms
For Plants that Bloom		
Regrowth early vegetative	Stage in which regrowth occurs without flowering activity; vegetative crop aftermath; regrowth in stubble (applies primarily to fall regrowth in temperate climates); early dry season regrowth.	Vegetative recovery growth.
Regrowth late vegetative	Stage in which stems begin to elongate to just before blooming; first bud to first flowers; regrowth in stubble with stem elongation (applies primarily to fall regrowth in temperate climates).	Recovery growth, stems elongating, jointing and boot (grasses).

Source: *United States-Canadian Tables of Feed Composition*, third revision, National Research Council, National Academy of Sciences, Washington, D.C., 1982

ROUGHAGES

The two general kinds of roughages are legume and grass. Legumes have, on their roots, nodules containing bacteria that fix nitrogen from the air. Some typical legume roughages used for animal feeds include alfalfa, the clovers, birdsfoot trefoil, lespedeza, peanut hay, cowpea hay, and soybean hay, Figure 7-1. Grass roughages include timothy, brome grass, orchard grass, coastal Bermuda grass, common Bermuda grass, bluegrass, redtop, reed canary grass, Bahia grass, fescue, Dallis grass, Johnson grass, Sudan grass, millet hay, oat hay, prairie hay, and meadow hay.

CONCENTRATES

The three kinds of concentrates include grains, supplements, and byproduct feeds. Common grains used for livestock feeds include corn, oats, barley, grain sorghum (milo), and wheat, Figure 7-2 and Figure 7-3.

Supplements include protein feeds, minerals, and vitamins. Protein supplements are of either animal or vegetable origin. Animal protein feeds include tankage, meat scraps, meat and bonemeal, fish meal, dried skim milk, dried whole milk, blood meal, feather meal, and poultry byproduct meal.

Vegetable proteins include soybean meal, soybeans, cottonseed meal, peanut oil meal, corn gluten meal, safflower meal, sesame oil meal, sunflower meal, copra meal, and linseed meal.

Urea and other nonprotein products are also protein supplements, neither animal nor vegetable, but manufactured from other sources. They are used primarily in feeding ruminants.

Mineral supplements are used to provide the mineral needs of the animal. They are generally some combination of calcium and phosphorus with trace minerals added. Salt (which provides sodium and chlorine) may also have trace minerals added. Trace mineral elements are available as premixes, which may be added to complete rations.

Vitamin supplements are available to provide necessary vitamins in the ration.

Byproduct feeds from the milling and brewing industries, as well as other sources, are used in feeding. Feeds included in this group are wheat bran, rice bran, wheat mid-

Table 7-3. Feed classes.

Class Number	Class Denominations and Explanations
1	DRY FORAGES AND ROUGHAGES All forages and roughages cut and cured and other products with more than 18 percent crude fiber or containing more than 35 percent cell wall (dry basis). Forages and roughages are low in net energy per unit weight, usually because of the high cell wall content. *Example forages:* hay; straw; stover (aerial part without ears and without husks (for corn) or aerial part without heads (for sorghum)). *Example roughages:* hulls, pods.
2	PASTURE, RANGE PLANTS, AND FORAGES FED FRESH This group comprises all forage feeds either not cut (including feeds cured on the stem) or cut and fed fresh.
3	SILAGES This class comprises ensiled forages (corn, alfalfa, grass, etc.), but not ensiled fish, grain, roots, and tubers.
4	ENERGY FEEDS Products with less than 20 percent protein and less than 18 percent crude fiber or less than 35 percent cell wall (dry basis), for example, grain, mill byproducts, fruit, nuts, roots, and tubers. When these feeds are ensiled they are classified as energy feeds.
5	PROTEIN SUPPLEMENTS Products that contain 20 percent or more protein (dry basis) from animal origin (including ensiled products) as well as oil meals, gluten, etc.
6	MINERAL SUPPLEMENTS
7	VITAMIN SUPPLEMENTS Including ensiled yeast.
8	ADDITIVES Feed supplements such as antibiotics, coloring material, flavors, hormones, and medicants.

Source: *United States-Canadian Tables of Feed Composition*, third revision, National Research Council, National Academy of Sciences, Washington, D.C., 1982

dlings, rye middlings, rice polish, molasses, brewer's grain, beet pulp, screenings, malt sprouts, and whey.

USES OF FEED IN AN ANIMAL'S BODY

An animal uses feed for several purposes, some of which are basic to all animals and some of which are specific to the purpose of the animal. Basic uses include reproduction, growth, and maintenance. Uses reflecting the purpose of the animal include finishing, fitting for show, production, and work.

Reproduction

If the livestock farmer is to make a profit, animals must reproduce. Poor nutrition is a major contributing factor to reproductive failure in livestock. The production of sperm and the quality of semen is influenced by the quality of the ration. Males that are too fat may become temporarily or permanently sterile. Rations balanced with the proper nutrients improve fertility. Nutrient requirements for pregnant females are most critical during the last third of the gestation period, when the developing fetus grows most. Especially critical are nutrient requirements for young females

Figure 7-1. Alfalfa-bromegrass hay raked into windrows ready to be baled.

Figure 7-2. Corn is the most common grain used in livestock feeding.

during their first pregnancy. Poor nutrition at this stage will result in a poorly developed fetus and poor growth of the mother.

Growth

During the growth period, there is a substantial increase in the size of muscles, bones, internal organs, and other body parts. Animals need nutrients to grow properly so that they will be efficient producers when mature. The full genetic potential of an animal cannot be reached if it is not fed a well-balanced ration during the growing period. The production of meat, milk, eggs, and wood is severely restricted by poor growth, as is reproductive ability. Horses will not perform to their capacity if they are stunted or suffer from skeletonal injuries during growth.

Figure 7-3. Grain sorghum headed out and ready to be combined.

Young, growing animals have more rigorous nutritional requirements than do mature animals. Nutritional deficiencies are reflected more quickly and more seriously in young animals than in mature animals. Nutrients for growing animals are needed in larger quantities than for mature animals. Protein quality must be higher and calcium and phosphorus needs are greater. The younger the animal, the greater is its need for good quality protein because of the rapid growth of muscle at this time. A shortage of total digestible nutrients or net energy during growth will result in slower growth or smaller size when animals mature. Mineral requirements are higher during growth because of the rapidly developing skeleton and the formation of muscle tissue. Mineral or vitamin deficiencies in the ration may result in rickets, which can leave the animal permanently deformed or crippled. A shortage of calcium and phosphorus can result in weakened bones, which will more easily break under stress. A lack of iron in the ration can cause anemia in young animals.

Maintenance

A maintenance ration is one that maintains basic life processes without any work or production being done. A maintenance ration must supply (1) heat to maintain body temperature; (2) energy for vital functions and a mininum amount of movement; and (3) small amounts of protein, minerals, and vitamins. Maintenance requirements for various species of livestock are listed in the tables in the Appendix.

Oxidation of food in the digestive tract and of nutrients in the muscles and other tissues produces heat. The rate of oxidation is relatively constant and occurs at a low temperature. Increasing the intake of oxygen does not cause an increase in the rate of oxidation. In easily digested feeds such as corn, about one-third of the total energy available is converted into heat by the digestive process. A much higher proportion of the total energy found in roughages is converted into heat by digestion. The body temperature of animals varies by species, Table 7-4.

An animal at rest must still maintain activity in such vital organs as the heart and lungs, as well as other internal organs. Animals that are standing require a higher rate of oxidation of nutrients than those lying down, because of the muscle tension needed to remain standing. Even more energy is needed when the animal moves around. An animal lying down at rest will require less energy than one standing or moving. Maintenance rations for mature animals, except swine and poultry, can be mostly roughages, which yield enough heat but are relatively low in energy value. A

Table 7-4. Normal temperatures of farm animals.

Species	Average		Range	
	(Fahrenheit)	(Celsius)	(Fahrenheit)	(Celsius)
Beef Cattle	101.5	38.6	100.0-102.4	37.8-39.1
Dairy Cattle	101.5	38.6	100.4-102.8	38.0-39.3
Goats	103.8	38.9	101.7-105.3	38.7-40.7
Horse	100.5	38.0	99.0-100.8	37.2-38.2
Poultry	107.1	41.7	105.0-109.4	40.6-43.0
Sheep	102.3	39.1	100.9-103.8	38.3-39.9
Swine	102.0	38.9	101.6-103.6	38.7-39.8

roughage containing little net energy is not satisfactory for a maintenance ration. Generally, swine and poultry, being simple-stomached animals, cannot use enough roughage to meet their needs for maintenance. However, adult swine during gestation can be maintained on high quality pasture.

An animal's maintenance requirement is relative to its amount of body surface, as well as its weight. An animal weighing 1,200 pounds (544 kg) does not require twice as much total digestible nutrients for maintenance as two animals, each weighing 600 pounds (272 kg). This is because the major heat loss is from radiation and conduction from the body surface and the 1,200 pound (544 kg) animal does not have twice as much body surface as the two 600 pound (272 kg) animals. Also, the weight of the internal organs, glands, etc. tends to be in proportion to body surface rather than to weight. However, the heavier animal will require more energy for standing and moving than the smaller animals, thus weight is a factor in determining the maintenance requirements for activity.

Some protein is required in the maintenance ration because there is some breakdown of protein in the body tissues each day. Also, hair, wool, skin, and hoofs all contain protein and continue to grow. The amount of protein needed in the maintenance ration is small (see the tables in the Appendix).

There is some daily loss of calcium and phosphorus from the body of an animal. For this reason, a small amount is required in the maintenance ration to replace the loss. Salt and a small amount of vitamins A and D are also needed in maintenance rations for most species of livestock. Maintenance requirements of both minerals and vitamins are listed in the tables in the Appendix.

Finishing for Market or Fitting for Show

Animals such as beef, swine, lambs, and broilers that are fed for meat need nutrients above the maintenance requirements for fattening. The juiciness, flavor, digestibility, and nutritive value of meat is improved by *marbling*, the stored fat in the lean meat tissues of the animal.

Nutrients for fattening animals come mainly from carbohydrates and lipids in the ration. Protein that is not needed for repair of body tissues and other needs of the animal may also be converted into body fat. However, because protein is higher in price than other nutrients, it is not fed primarily for fattening purposes. Most animals are finished for market while they are still young and growing. Rapid gains in fattening livestock depend on a ready supply of TDN or energy, protein, minerals, and vitamins in the ration. Feed efficiency is higher when the ration is properly balanced for fattening livestock.

Animals being fitted for show purposes are fed a liberal supply of carbohydrates and fats, usually through the addition of more grain to the ration. Mature animals being fitted for show need little additional protein, minerals, and vitamins above the maintenance requirements. However, young, growing animals must be fed more liberal amounts to reach the desired level of finish for show purposes.

Production

The production of milk requires a liberal supply of TDN or energy, protein, minerals, and vitamins in the ration. The quantity of milk and its fat content influence the amount of nutrients needed above the maintenance level for milk production. Heavy milkers with a high fat content require proportionately more nutrients than do light milkers with low fat content. When balancing rations for dairy cattle, the requirements for milk production at a given fat content are added to the maintenance and growth requirements to arrive at the total nutrient needs of the animal. Tables in the Appendix give the requirements for milk production.

Wool and mohair have a high protein content; therefore, rations for sheep and goats must contain an adequate protein level to meet this production need. TDN or energy requirements are also higher because of the wool and mohair production. When the ration is deficient in protein, weak spots appear in the wool fiber, and total wool production is reduced.

Work

The production of work is limited mainly to horses. While horses are seldom used for draft animal purposes, they are used for riding and racing, which require additional nutrients for efficient production. As the amount of work being done increases, so does the TDN or energy requirements of the animal. The protein, mineral, and vitamin requirements of mature working horses are only a little greater than the requirements needed for maintenance. Young, growing horses used for work have higher nutrient needs than do mature horses. Tables in the Appendix give recommendations for meeting the nutrient needs of horses of various ages and at different levels of work.

Animals require additional nutrients (especially energy) in relation to the kind of terrain they are on and whether they are in confinement feeding or not. Also, when animals are on pastures, the distance to feed and water influences the amount of nutrients they need. Rough terrain and pasture feeding require more nutrients, as does greater distances to feed and water. Animals on level terrain or in confinement feeding do not expend as much energy in their daily movements.

UNDERUTILIZED SOURCES OF FEED

Most of the feedstuffs used in animal nutrition are the conventional grains and forages grown on the farm. However, currently there are a number of potential sources of feedstuffs not being utilized extensively in animal feeding. These include industrial food-processing wastes, nonfood industrial wastes, forest residues, animal wastes, crop residues, and aquatic plants. Some materials included in these groups are readily available, relatively inexpensive, and have substantial nutritive value. Others are available in limited quantity, are expensive to process, and have limited nutritive value. Some of these materials have alternate uses such as soil conditioners, fuel, or fertilizer. The use of underutilized resources for animal feeding depends primarily on the cost of the nutrients they contain, relative to more conventional feeds, and their safety insofar as human and animal health is concerned.

Industrial Food-Processing Wastes

The three general classes of food-processing wastes are fruit and nut, vegetable, and animal. Fruit and nut processing wastes include apple, citrus, peach, pear, fruit-cannery activated sludge, winery wastes, cacao processing wastes, fruit pits, fruit pit kernels, nut hulls, and nut shells. Vegetable-processing wastes include potato, sweet potato, tomato, and other vegetables. Animal byproducts include dairy whey, seafood processing wastes, poultry processing wastes, and red meat processing wastes.

As a class, food-processing wastes have a high water content, are perishable, and require rapid processing. From a nutritive viewpoint, the dry matter of animal processing wastes is generally high in good quality protein and low in carbohydrates. The dry matter of fruit and vegetable processing wastes, on the other hand, tends to be low in protein and high in carbohydrates. Animal processing wastes are generally available all year, while fruit and vegetable processing wastes are seasonally available.

As a class, fruit and vegetable processing wastes are more suitable for use in feeding ruminants rather than nonruminants. Animal processing wastes may be used for both ruminants and nonruminants. The use of food-processing wastes in the animal diet does not lower rate of gain or production if they are fed at appropriate levels. The use of some food-processing wastes at too high a level in the diet may adversely affect digestibility and/or palatability of the feed.

The most economically feasible method of utilizing high-moisture food-processing wastes as animal feed appears to be by ensiling with drier material, such as poor quality hay or crop residues. Dehydration for inclusion in the diet is often too expensive to be a viable processing method.

Pesticide levels in fruit and vegetable processing wastes may pose a threat to human and animal health unless they are carefully monitored. Processing fruit and vegetables often concentrates any pesticides present in the wastes. Tomato wastes may have pesticide levels that are higher than permitted for feeds.

Apple pomace should not be fed with nonprotein nitrogen, such as urea or biuret. Problems such as dead, weak, or deformed calves have resulted from feeding this combination. Also, feed consumption has been lower with a resultant increase in body weight loss as compared to feeding corn silage with nonprotein nitrogen. Apple pomace may be safely fed with other protein supplements.

Cacao processing wastes are limited for use as a feed because of their theobromine and caffeine content. Too high a theobromine intake has adverse effects on the animal. Low levels (0.027 g/kg body weight) of theobromine intake has caused death in horses. Feeding theobromine to racehorses may cause reactions similar to doping.

Feeding cacao shells to pigs, poultry, or calves has an accumulative effect that is detrimental to their health. Limit the use of cacao shells to mature cattle with a maximum rate of 2.5 percent of the diet or a maximum of 0.907 kg/day (2 lbs./day).

Nonfood Industrial Wastes

Three sources of nonfood industrial wastes are the organic chemical industry, the fermentation industry, and municipal solid waste. Most nonfood industrial wastes must be processed before being used for animal feeding.

Wastes from the organic chemical industry may be processed into single-cell protein feed for livestock use. Another possible way for organic chemical wastes to be used as calorie sources is by processing it into carbohydrate or fat material. Processing costs are relatively high, with most work currently being done on an experimental basis. There are only a few commercial processing plants in the world producing single-cell protein from organic chemical wastes. Because the product is used as a protein supplement, it must be priced competitively with other protein supplements.

The fermentation process can result in pathogenic microorganisms being in the final product. Guidelines for the allowable limits of certain microorganisms in single cell protein feeds have been recommended by the International Union of Pure and Applied Chemistry.

Toxic organic chemicals and heavy metals

may be found in organic chemical wastes. These materials are concentrated by the microorganisms used to convert the waste to animal feed. The removal of toxic organic chemicals and heavy metals from the waste increases the cost of processing.

The primary waste from the fermentation industry producing antibiotics and organic acids is the spent fungal mycelia. This contains from 20-50 percent protein, about 10-30 percent minerals, and typically 15 percent solids. There is some experimental evidence to indicate that mycelia waste is useful in supplementing protein feeds that are low in the amino acids lysine and threonine. A major problem in the use of mycelia waste is its low digestibility. Processing with acid or enzymatic hydrolysis helps to make the mycelia more soluble and thus increases digestibility. Experimentally, mycelia waste has been blended with other feed components to be fed directly to livestock. There are problems with palatability of feed mixed by this method. There is widespread use of fermentation wastes from the brewing and distilling industries in animal feeding. Experience with these wastes will help in the evaluation and utilization of other fermentation waste materials if they become available for use in livestock diets.

The presence of residual antibiotics in the mycelia waste presents a possible health hazard in their use as feed. Processing to remove the residual antibiotics from the mycelia waste is necessary before it can be utilized in animal diets. Securing approval from the appropriate regulatory agencies for the use of feeds containing mycelia waste may be difficult because of the residual antibiotic problem. Establishing standards is hard to do because of the diversity of the raw material used. To date, there is little economic incentive for the industry to develop processes resulting in high-quality mycelia waste for inclusion in animal feeds.

Municipal solid waste contains large quantities of foreign material, which reduces its potential value as an animal feed source. Processing out undesirable elements, collecting, and storing this waste material are major stumbling blocks to its being utilized extensively for animal feeding. Municipal solid wastes could provide minerals and fiber for livestock diets. However, there are some potentially harmful elements found in municipal solid wastes that may be present at levels unacceptable for livestock feeding. The removal of these elements is expensive and may make the use of this waste material in animal diets economically unfeasible in the foreseeable future.

Forest Residues

Residues from the lumbering industry have some potential as an animal feed source, primarily for their carbohydrate value. As an energy source, they are comparable to an average to low quality hay. Their major use would be with ruminants such as overwintering beef cows, ewes, dry cows, and larger size replacement dairy and beef heifers. Typically, forest residues are low in protein and other essential nutrients and, therefore, require heavy supplementation if used as an animal feed.

Low digestibility is a major problem with using the forest residues in livestock diets. Coniferous species are practically indigestible by rumen microorganisms. Deciduous species generally have low digestibility. Aspen is the most digestible of the forest residues, but it must be treated to improve digestibility. Chemical and physical treatments may be used to improve digestibility, with the hardwoods responding better to treatment than the softwoods. The cost of treatment to improve digestibility is the major barrier to the use of forest residues in livestock diets. Because conventional feedstuffs are generally priced low enough to make treated forest residues uneconomical as a feed, there has been little interest in commercial development of this source of carbohydrates for animal feeding.

Forest residues can be used as a roughage in the diet of ruminant animals. Five to 15 percent sawdust in the ration for beef cattle has proved to be practical. Lactating dairy cattle can be fed aspen sawdust as a roughage substitute in high-grain diets; however, some long hay is necessary to stabilize feed intake.

Forest residues are not generally harmful to the health of animals. The major nutritional problem is proper balancing of the diet including them. Care must be taken with pine needles, as they can be toxic and have caused abortion in cattle. Residues from the pulp and papermaking industries have potential as an animal feed, but may contain toxic chemicals.

Animal Wastes

Animal wastes are best used in the diets of ruminant animals, although some utilization by swine and poultry has been shown to be possible. Generally, animal wastes are high in nonprotein nitrogen and fiber, with a low to fair energy value, and high in mineral content.

Layer waste and broiler litter have the highest nutritive value. Successful processing by dehydration or ensiling has been demonstrated with these animal wastes. They contain as much true protein as common feed grains and are also good sources of calcium and phosphorus. Poultry waste has a metabolizable energy value about one-half to one-third that of common feed grains.

Animal wastes are used most successfully by growing and finishing beef animals, beef breeding stock, growing dairy heifers, dry dairy cows, and sheep. Palatability of feed containing animal wastes has not proven to be a problem. Care must be taken to balance the ration, as too high a level of animal waste can result in reduced animal performance. High energy diets can utilize up to 10-20 percent waste, while beef cow diets can contain as much as 80 percent animal waste.

Processing by dehydration or ensiling appears to minimize the danger of disease being transmitted through the animal waste. The major risk is spore-forming bacteria, which are not destroyed by dehydration or ensiling. There is some problem with excess copper in animal wastes used in the diet of sheep. This requires careful control on the part of the feeder. There do not appear to be any problems with drug or chemical residues in animal wastes used for feeding. It is recommended that a 15-day withdrawal period be observed when animals are to be used for producing milk, meat, or eggs for human consumption, although the feeding of animal waste does not affect the quality of the food produced.

Crop Residues

The amount of vegetative material produced by feed grain plants is at least equal to, and often greater than, the amount of grain produced. This residue from the production of grain is a valuable potential source of feed for livestock. It is estimated that there are at least 300 million tons of straw, stalks, and stubble available in the United States each year with an additional 40 million tons produced in Canada. The world supply of crop residues that might be used for feeding livestock is estimated at more than 1.5 billion tons annually. Because of its low digestible energy content, this residue does require additional treatment or processing to make it practical for livestock feeding.

More than one-half of the total residue supply comes from the production of corn. Soybean and wheat residues each account for about 15 percent of the total supply. Grain sorghum and barley each supply about 5-6 percent of the total crop residues available. The remainder come from barley, oats, cotton, rice, peanuts, flax, rye, sugar beets, and other miscellaneous feed crops.

Because crop residues are bulky, they are difficult to transport economically. Some, such as small grain straw, can be collected relatively easily after the grain is combined and stored dry. Corn and grain sorghum residues tend to have higher moisture content at harvest than small grain straw and will keep better when stored as silage. Care must taken when harvesting crop residues to avoid contamination with soil. Crop residues are usually harvested in the late summer and fall and used for livestock feed during the winter. Cotton and rice milling byproducts are collected at central processing points and are thus readily available for treatment or processing into livestock feed.

The use of crop residues for livestock feed may increase problems from soil erosion and loss of organic matter from the soil. To reduce these losses, it is recommended that no more than one-half of the crop residues be removed. When manure from livestock is returned to the soil, loss of organic matter is reduced.

The nutritional value of crop residues is low. The vegetative portion of the plant at the normal stage of maturity for grain harvest is high in cell-wall and lignin content and low in protein and digestible dry matter. Lignin reduces the digestibility of cellulose and hemicellulose. The digestibility of crop residues is highly variable. A major problem in the practical use of crop residues for livestock feed is controlling the digestibility of the material, which is done by treating crop residues in various ways. Chemical treatments have been used experimentally to break down the ether linkages between lignin and cellulose or hemicellulose. The use of sodium hydroxide has shown promise for increasing the digestibility; cornstalks treated with sodium hydroxide have been pelleted and fed to beef calves, with promising results. Putting the treated material in the silo has also shown promise as a method of utilizing crop residue for livestock feeding. The increase in the sodium content of the diet may create a mineral imbalance, necessitating the feeding of additional potassium and chloride.

Experimental work has also shown that treating crop residues with ammonia can improve their feeding value. The material must be ensiled in an airtight silo to prevent the loss of ammonia. This method provides nonprotein nitrogen in the diet from the ammonia and does not cause a mineral imbalance. Ammonia-treated residues must be aerated or mixed with a fermented feed to make them palatable to livestock.

Calcium hydroxide may be used to improve digestibility of crop residues, but a greater amount is required as compared to sodium hydroxide. However, the use of calcium hydroxide is less likely to cause a mineral imbalance in the diet.

Treatment of crop residues with high pressure steam, with or without chemicals, will also increase the digestibility of the material. A continuous-flow digester is commercially available for this type of treatment.

Residues from soybeans do not respond well to chemical treatment to improve digestibility. More research is needed with soybean residues to determine the feasibility of their use as livestock feed.

The use of low-quality crop residues is most feasible with ruminants on maintenance rations. Grazing cornstalks and milo stubble is the most economical method of utilization of these crop residues. However, conditions such as muddy fields or snow cover make it difficult for livestock to graze crop residues. Harvesting and storing some of the crop residues can provide feed during times when grazing is not practical.

To avoid problems with insecticide and herbicide residues on crop residues being fed, it is necessary to strictly observe label restrictions when applying these chemicals to the growing crop. As long as the label restrictions are observed, there do not appear to be any health hazards to livestock or humans from using crop residues for animal feeding.

The economical use of crop residues for livestock feeding depends upon the development of effective methods of collecting, storing, treating, and feeding the material. More research is needed in all of these areas.

Aquatic Plants

Aquatic plants have potential for use as animal feed. Because these plants grow in water, they have a high moisture content, so harvesting and processing present major economic obstacles to their use as feed. Because potential yield is quite high, it is hoped that more research will provide effective methods of harvesting and processing these plants.

Algae have high protein and mineral content. They have potential as swine and poultry feeds on a limited basis if economical methods of harvesting and drying can be developed.

Other aquatic plants such as kelp, water hyacinth, and duckweed are low in protein and energy value. Their major use appears to be in feeding ruminant animals. They do contain minerals and vitamins that are of value to feeding ruminants. The best method of utilizing these plants appears to be by ensiling rather than drying.

There do not seem to be any major health problems in the use of aquatic plants in livestock diets. These plants do not appear to have so much heavy metal content that they adversely affect animal health.

SUMMARY

Feeds or feedstuffs are materials fed to animals for the purpose of meeting their nutritional needs. Generally, feeds are either roughages (high-fiber content) or concentrates (low-fiber content). Roughages are usually legumes or grasses and are often fed in the form of hay. Concentrates are usually grains, supplements, or byproducts and usually have more total digestible nutrients than do roughages.

Feeds are identified and classified by names and numbers based on descriptors of origin, part used, treatment, maturity, cutting, and grade.

Animals use feed for reproduction, growth, maintenance, finishing, fitting for show, production, and work. Nutrient requirements vary according to the use made in animal bodies.

There are a number of underutilized sources of feedstuffs available for animal nutrition. Sources include industrial food processing wastes, forest residues, animal wastes, crop residues, and aquatic plants. Availability, nutrient value, and economic feasibility of use vary among these underutilized sources of feedstuffs. The use of these feedstuffs for animal feeding depends primarily on the cost of utilizing their nutrients as compared to the cost of nutrients from more conventional feedstuff sources. In some instances, their safety for animal and human health is also a factor in their use in animal nutrition.

REVIEW

1. Name and briefly tell the difference between the two general classes of feeds used for animal nutrition.
2. List the eight descriptors used in determining International Feed Names.
3. Name and define five different stages of maturity used when naming feeds.
4. Name and briefly describe each of the eight feed classes.
5. Name and give examples of the three general forms of feed.
6. What is the difference between legume and grass roughages?
7. Give three examples of a legume and three examples of a grass roughage.
8. Name three common grains used in animal nutrition.
9. Name the two kinds of protein supplements and give two examples of each.
10. Which class of livestock are fed urea and other nonprotein nitrogen sources?

11. Describe how nutrition affects reproduction in livestock.

12. Describe nutrient needs of young, growing animals as compared to more mature animals.

13. A maintenance ration requires approximately how much of the total feed consumed by an animal?

14. Describe the life processes that are supported by a maintenance ration.

15. Why is the amount of an animal's body surface more closely related to its maintenance needs than is its weight?

16. Which part of the ration supplies most of the nutrients needed for finishing or fattening an animal?

17. Why is a balanced ration important for finishing an animal?

18. How does milk production affect the nutrient requirements of an animal?

19. How does wool and mohair production affect the nutrient requirements of sheep and goats?

20. Describe the effect of work on nutrient requirements of horses.

21. List six underutilized sources of feedstuffs.

22. Name three factors that affect the use of underutilized sources of feedstuffs.

23. Select one of the underutilized sources of feedstuffs and describe its potential for animal nutrition.

8

FEED QUALITY AND FEED ANALYSIS

OBJECTIVES

After completing this chapter you will be able to:

- Describe quality characteristics of grains and other feeds.
- Describe methods of harvesting quality hay and silage.
- Describe the analysis of feedstuffs.
- Describe methods of measuring the value of feeds.
- Describe regulations relating to the manufacture and sale of feedstuffs.

FEED QUALITY

The quality of feed affects its value for animal nutrition. Factors of feed quality include palatability and nutrient content. Improper harvesting or handling methods will reduce feed quality.

Palatability refers to how well the animal will accept the feed. If the animal does not like the feed, it will not eat enough to grow properly, make good gains, or produce meat, milk, eggs, and wool in a profitable manner. Factors that reduce the palatability of feeds include mold, dust, loss of leaves, mustiness, or other foreign matter. Anything that adversely affects the smell, taste or mouth-feel of the feed reduces its palatability.

Feeds (especially roughages) grown on poorly fertilized soils are of lower quality than feeds produced on well-fertilized soils. The mineral content of feeds is lower when they are produced on poor soils.

Feed quality is also affected by its vitamin and amino acid content. Feeds low in necessary vitamins and amino acids are of poorer quality than those high in these nutrients.

HARVESTING HAY TO MAINTAIN QUALITY

The quality of hay is affected by the harvesting and handling methods used. Hay that is stemmy, with few leaves, is of lower quality than good leafy hay because the leaves of hay contain most of the nutrients and are the more easily digested part of the plant, Figure 8-1. *Shattering* refers to the loss of leaves during harvesting and handling. Legume hays tend to shatter more readily than do grass hays.

When you are harvesting hay, the moisture content must be reduced enough so it can be safely stored without spoiling or serious loss of nutrients. Field-dried forages must not contain more than 18-22 percent moisture for proper baling. Factors affecting the allowable moisture content for baling include the humidity and movement of the air, the fineness of

115

Figure 8-1. High quality hay is leafy. Most of the nutrients in hay are found in the leaves which are the more easily digested part of the plant.

the forage, and the tightness of the bale. Hay baled when it is too wet may ignite spontaneously or mold in storage.

Three causes of loss of quality during harvesting include shattering, leaching, and bleaching. When the forage gets too dry, the leaves tend to shatter when it is handled. Forage that is rained on while it is curing in the field will suffer loss of quality by *leaching* (the loss of water-soluble nutrients). *Bleaching* refers to the loss of nutrients from exposure to sunlight. Forages must have some exposure to sunlight to cure properly for baling; however, excessive exposure will cause a loss of nutrients. Carotene loss is especially high from excessive exposure to sunlight.

To secure the highest quality of hay, the forage should be cut as soon as possible after reaching the early bloom stage of maturity. Listening to weather forecasts and selecting harvesting periods when rain is not expected for several days will improve the quality of the hay. Unless the hay is conditioned, it requires about two days of drying in the field before it is ready to bale. *Conditioning* is a process of passing the forage through smooth or fluted rollers to break the stems. This operation reduces the drying time before baling by as much as 50 percent. Equipment is available to cut, condition, and windrow the hay in one operation, Figure 8-2. This practice reduces the number of times the hay is handled before baling, thus reducing losses from shattering and improving the quality of the hay.

To reduce losses from shattering and bleaching, the hay is raked into windrows before it is completely dry. If the crop is rained on, it may be necessary to turn the windrow with a rake, so it can dry for baling. Turning the windrow should be done while the hay is slightly damp from dew, in order to reduce shattering losses.

Figure 8-2. Hay which is cut, conditioned, and windrowed in one operation does not have to be handled as much before baling. This reduces losses from shattering and improves the quality of the hay.

Baling Hay

Hay should be baled as soon as possible after it reaches a safe moisture level for storage. Delays in baling at this stage will increase the likelihood of losses from bleaching or rain. Square bales should be stored inside or in a covered stack outside to protect them from the weather. Rain will easily soak a square bale, causing loss of quality from molding and mustiness.

Hay may be stored in large round bales or mechanically formed stacks, but special equipment is necessary for these operations. This equipment is usually more expensive than that needed for conventional square bales; however, the labor required for harvesting is generally lower.

Large round bales may weigh from 500-3,000 pounds (237-1361 kg) and are handled with special equipment. They may be left in the field for feeding because they tend to shed water and do not lose quality as readily as square bales. Sometimes round bales are moved to an outside storage area and later moved to the feeding area, Figure 8-3. There are mechanical unrollers available to unroll the round bale for feeding. Losses from livestock trampling on the hay may be high if the animals are fed on the ground. Feed losses are

Figure 8-4. Placing a large round bale in a rack for feeding reduces waste.

reduced if the bales are placed in racks for feeding, Figure 8-4.

Mechanically formed stacks may contain as much as 3 tons or more of hay. These stacks are stored outside and may be covered to help maintain the quality of the hay. Restricting livestock access to the stack, using some type of restraining fence or rack, will reduce losses from trampling.

PRODUCTION OF QUALITY SILAGE

When high-moisture crops are stored under anaerobic conditions, a fermentation process

Figure 8-3. Large round bales stored outside. These bales will later be moved to the feeding area.

Figure 8-5. A cement stave upright silo. Vertical silos require less land for construction, have less spoilage, and require less labor for unloading than do horizontal silos.

occurs that produces silage. Crops may be stored in several types of silos that may be vertical or horizontal. Horizontal silos are usually cheaper to construct than vertical silos. However, vertical silos require less land for construction, have less spoilage, and require less labor for unloading, Figure 8-5.

Vertical Silos

Vertical silos are either conventional or gas-tight. Conventional silos are constructed of concrete staves, metal, or cast-in-place concrete. To achieve good preservation, the crop should contain 25-35 percent dry matter at the time of ensiling. If the crop is too wet or too dry when put in the silo, spoilage will be high. Conventional silos may be unloaded by

hand or with the use of mechanical unloaders. Mechanically unloaded silos require less labor for unloading than do those unloaded by hand or than horizontal silos do, Figure 8-6.

Gas-tight silos are made of glass-lined steel or cast-in-place concrete. The construction of this type of silo eliminates air, ensuring the preservation and storage of forages with a range of approximately 25-75 percent dry matter. Some brands of airtight silos unload mechanically from the bottom and others unload from the top. The initial cost of an airtight silo is higher than that of the conventional silo. Gas-tight silos cannot be unloaded by hand, which makes the farmer dependent upon the proper functioning of the mechanical unloading system. Some brands use a breather bag in the top to permit interior gas expansion and contraction. Others use a two-way valve from the inside to the outside of the silo.

Horizontal Silos

The three types of horizontal silos are trench, bunker, and stack. The *trench silo* is usually dug into the side of a hill and may have dirt sides and bottom or may be constructed of concrete. It is recommended that at least the bottom be constructed of concrete to avoid problems with mud in wet weather. The floor of the silo is sloped to the front for good drainage and the sides tilt out slightly so the forage may be more effectively packed during filling. Filling is accomplished with dump trucks or wagons and a tractor is often used to pack the material during this operation. After filling, the silage is usually covered with a polyethylene sheet and weighted with dirt, boards, or some other material to hold it in place. The purpose of covering is to reduce spoilage losses from exposure of the top material to the air. Spoilage in properly covered horizontal silos will not be much more than in vertical silos. Cane molasses may also be used as a seal to help prevent spoilage. The silage may be removed for feeding by using a tractor and front-end loader or a movable barrier may be used to permit self-feeding. Spoilage is reduced during feeding if at least 3 inches of silage is removed daily.

Bunker silos are used where the terrain or soil conditions are not suitable for trench silos, Figure 8-7. Construction is of posts and boards with a side lining of building paper or plastic. The bottom is usually of concrete. The major difference between a trench silo and a bunker silo is that the former is below ground, while the latter is built on the surface.

Stack silos require only a small investment and may be used for emergency storage of surplus silage. While losses from spoilage are generally higher than in trench, bunker, or vertical silos, they can be reduced through good management practices. The stack should be placed in a well-drained area, protected from animals. Silage is spread in shallow layers using unloading wagons, dump trucks, or a silage blower. The sides should slope about 60 degrees. During filling, the silage is

Figure 8-6. Mechanical equipment may be used to unload upright silos. A circular feeder for the silage is located around the base of this silo.

Figure 8-7. A horizontal bunker silo. After filling, this silo will be covered with plastic to help reduce spoilage.

packed continuously with a tractor. When filling is completed, the top of the stack is rounded off. Black plastic (4 to 6 mil thick) is used to cover the stack. The plastic should be sealed well around the edges. This can be done by burying the plastic in a furrow about 8-10 inches (20-25 cm) deep around the edges. The film is held tightly to the silage by hanging rope or cord weighted with old automobile tires or poles over the stack. Spoilage is reduced by (when unloading) keeping weighted rope on the top edge to prevent air from getting into the stack. Spoilage is also reduced by quickly filling and sealing the stack by the end of the first day. No more than two days, maximum, should be used to fill and seal the stack.

Physiological Changes in the Silo

The most commonly used crops for silage are corn and sorghum. Other crops, including small grains, grass-legume forages, grasses, and legumes, may be used for silage. To produce good quality silage, it is usually necessary to allow the crop to field dry to approximately 62-68 percent moisture before ensiling. If the crop is too wet when ensiled, a lack of air smothers the plant cells before the temperature reaches the proper level for lactic acid formation. Butyric acid forms, causing a sour feed with a high level of nutrient loss. When the silage is too dry, it does not pack well and forms mold more readily. Material that is cut 1/4 to 3/8 inch (0.6-1 cm) in length will pack better than more coarsely cut material.

It normally takes about 21 days for the fermentation process that produces silage to complete its action. After the chopped material is put in the silo, the living cells continue to burn plant sugars, use oxygen, and give off carbon dioxide, water, and heat. During this time, the temperature of the silage increases and bacteria present on the material continue to grow and multiply. Acetic acid is produced early in the process and the plant pH changes from approximately 6.0 to about 4.2. After approximately 3 days, the formation of acetic acid slows down and lactic acid is formed.

Formation of lactic acid continues for about another 2 weeks and the temperature of the silage decreases. Bacterial action stops when the pH level reaches 4.0. If conditions have been right, the silage is now fairly stable and will remain so for a long period of time.

A temperature in the range of 80-100°F (27-38°C) gives the best results during fermentation. When the temperature is lower, butyric acid is formed, giving the silage a foul odor, breaking down the protein, and increasing spoilage. At higher temperatures, caramelized silage forms, which is sweet, tobacco-smelling, and dark brown in color. This results in a loss of dry matter and crude protein. Caramelized silage is highly palatable but has a lower nutrient value than normal silage.

Harvesting Crops for Silage

Corn for silage should be harvested at the dough to hard-dent stage (32-38 percent dry matter) for optimum feed value and total dry-matter intake. A black layer forming at the tip of the kernel will indicate that the kernels are physiologically mature and at the right stage for ensiling.

Grasses and legumes should be wilted to below 65 percent moisture (35 percent dry matter) to make good silage. Grasses should be ensiled in the early heading stage and legumes at about one-tenth bloom. When they are ensiled at more mature stages of growth, the palatability and digestibility of the silage is reduced.

Sudan grass and sorghum-sudan hybrids are harvested for silage before the heads emerge from the boot. Delaying harvest for silage will reduce digestibility as much as 50 percent, if the crop is fully headed.

Small grains are harvested for silage in the boot to early milk stage for best results. Small grains pass quickly through this stage of growth, making a short time frame for harvesting quality silage. Soybeans are harvested for silage when the beans are forming in the pod.

Grasses and legumes need to be wilted in the field to reach the desired moisture level before being put in the silo. The use of a conditioner will speed up the drying process. Grasses and legumes should be chopped at 1/4 inch (0.6 cm) for silage. Dry-matter losses are kept to a minimum when grasses and legumes are stored in the silo at 50-65 percent moisture. Losses may be as high as 10 percent if the moisture content is below 50 percent. Direct-cut forage, which has a higher moisture content, will have harvest losses of 1 to 3 percent and seepage losses after ensiling as high as 10 percent of the dry matter.

Adding Dry Matter and Preservatives

Instead of wilting the forage before chopping, dry matter may be added as it is ensiled. Ground dry grain may be added as the forage is put into the silo. It will require 430 pounds (195 kg) of dry grain added to 1,570 pounds (712 kg) of direct cut forage to reduce the moisture level from 80 percent to 65 percent. If this method is used, the cost of the grain as well as the extra time and labor must be considered. A poor mix as the material is put into the silo may result in more molding of the silage. Adding 100 to 200 pounds (45-90 kg) of dry feed per ton of direct cut forage will increase the feeding value of the silage because of the higher dry-matter content. Using this amount of dry feed reduces the cost, time, and labor involved (as compared to adding enough to reduce the moisture content to 65 percent) while still gaining some of the benefits of higher dry matter content.

Chemical or biological preservatives may be added to wilted or direct cut forage silage. These have no feeding value of their own and may be difficult to justify on a cost basis. Some of these materials slightly increase the palatability of the silage, but do not reduce seepage losses of direct-cut forages.

Other Practices

To save the maximum amount of feed, the following practices are important: rapid filling, even distribution of material, packing, and covering the top of the silage. Grasses and legumes are higher in protein than corn or sorghum and will, therefore, spoil quicker. Putting a plastic film covering on the silage after filling and weighting it down with a few loads of direct cut forage will add weight that helps packing and reduces spoilage.

Grass silage does not keep well in trench or bunker silos. Too much surface area is exposed and storage in warm weather results in high spoilage losses.

Corn silage should be chopped 1/4 to 3/8 inch (0.6-1 cm) in length. When corn is harvested at 35 percent dry matter, field losses will range from 1 to 7 percent of the total dry matter harvested. Harvesting corn for silage when it is wetter than optimum may increase losses to 18 percent of dry matter. Harvesting when it is too dry can increase losses of dry matter up to 30 percent.

To make good quality corn silage, filling should be done as rapidly as possible. Proper distribution of the material in the silo will result in better packing, with less loss from spoiling. The use of some type of mechanical distributor during filling will give a more even distribution of particles of varying sizes. Sealing the top of the silo with plastic will also help reduce spoilage losses.

COLLECTING FEED SAMPLES FOR ANALYSIS

A feed analysis can be no better than the sample that is submitted for analysis. Care must be taken in the collection, preparation, and sending of the sample to the laboratory, in order to obtain the most accurate results. The sample must be carefully secured so as to accurately represent a much larger quantity of feed. Less than 1 quart of feed sample is usually submitted for analysis. The container must be accurately labeled with the name and address of the person submitting the sample. The sample number, feed name, type of feed, variety name, stage of maturity, and harvest date must also be placed on the sample label. Any unusual circumstances relating to the feedstuff being sampled may also be included. This information may prove to be useful in developing rations to utilize this feed.

Moisture Content

At the time of taking samples of silage, high-moisture grain, haylage, and fresh forage, the moisture content must be low enough so that seepage is not occurring. If each crop is harvested at the recommended moisture content, seepage should not be a problem at harvest time and samples can be taken then. Usually, there will be no seepage if corn is ensiled at the dent stage. If the material is wet enough to cause seepage, samples for analysis should not be taken until the seepage stops. Samples can also be taken at the time the crop is being fed.

Sampling Procedures

At harvest time, samples are taken from several representative loads of the crop by taking handsful at random and placing the material in a large plastic bucket or other container. This large sample is then mixed thoroughly and a smaller sample is taken from it and placed in a plastic bag. The plastic bag is immediately sealed to prevent loss of moisture from the sample. The sample may be sent immediately to the laboratory for analysis or may be preserved by freezing and submitted at a later time. Freezing prevents decomposition of the material if it is to be held for a period of time before being sent to the laboratory.

If the material is to be sampled after it is ensiled, random samples are again taken and mixed together. A smaller composite sample for analysis is then taken from this larger amount. The random samples should be taken from at least ten different locations around the top area of the silo. Material should be secured from a depth of about 1 foot (30.5 cm) from the top. If samples are taken as the material is being unloaded, allow the unloader to make one or two revolutions before secur-ing the samples. Do not take the samples directly from the top of the ensiled material and do not include spoiled silage in the sample, unless it is mixed in the fed mixture sufficiently so that the animals do not separate it during feeding.

Core Sampling

Hay samples may be taken at the time of harvest if the moisture content is low enough for safe storage. Sample different cuttings separately if they are to be fed separately. If they are being mixed at feeding, then secure samples in the same proportion as they appear in the mix. A core sampler is used to secure representative samples of hay from bales, chopped, or loose hay. The core sampler must be inserted far enough into the hay to secure representative samples. The sampler should be inserted at least 12-15 inches into the end of square bales. Stacked hay should be sampled to full depth and round bales should be sampled at least to the center of the bale. Take only one sample from each bale when sampling baled hay. Take at least 12 samples from random locations in chopped or loose hay or from random bales. Mix the cores in a clean pail and take a smaller composite sample to be submitted to the laboratory. Be careful not to lose leaves or stems when getting the composite sample.

Hand Sampling

If a core sampler is not available, hand samples may be taken by breaking open 10-12 bales and getting a handful of hay from the center of each. Take care not to lose leaves or stems from the area being sampled. The samples are then mixed and a composite sample is secured in the same manner as if a core sampler had been used.

Sampling Grain or Bulk Feed

Grain or bulk feed samples are taken at random from several locations in the bin. Secure at least 12-15 different samples. Sacked feed samples are taken at random from several bags. A sack probe may be used to secure these samples if one is available. If a sack probe is not available, then sacks must be opened to secure the samples. Grind ear corn before taking random samples. Mix the random samples in a clean pail to make a smaller composite sample to send to the laboratory. Check the composite to be sure that the feed has not become segregated by particle size during the mixing process.

PROXIMATE ANALYSIS OF FEEDSTUFFS

The approximate nutritional value of a feed may be determined by chemical analysis. More than 100 years ago, the Weende Experiment Station in Germany developed a method for chemically analyzing feedstuffs to give an approximate indication of their value in livestock feeding. The feedstuff is separated into the following six components: (a) water; (b) crude protein; (c) crude fat; (d) crude fiber; (e) nitrogen-free extract; and, (f) ash.

Five of the components are separated from the feed sample by chemical or physical methods. The nitrogen-free extract is then determined by subtracting the amount of the other five from the weight of the total sample. With the exception of water, each of these components is a combination of similar substances. Not all of these substances have nutritional value for an animal.

Determining the Water Content

The water content of the feed sample is determined by physically drying a sample in an oven. A small amount is weighed and placed in a container which is then placed in the drying oven. It is heated to a temperature just above the boiling point of water (212°F or 100°C) and held at that temperature until a constant weight is obtained. This will take about 8-24 hours. The dried sample is then weighed and the weight loss is noted. The per-

cent of water content is found by dividing the amount of weight lost by the weight of the original sample and multiplying by 100. The dry matter content may be calculated directly by dividing the weight of dried sample by the weight of the original sample and multiplying by 100. The water content is then calculated by subtracting the percent of dry matter content from 100.

The formulas are:

$$\frac{\text{Weight loss}}{\text{Original weight}} \times 100 = \text{percent water}$$

$$\frac{\text{Weight of dried sample}}{\text{Original weight}} \times 100 = \frac{\text{percent dry}}{\text{matter}}$$

$$100 - \text{percent dry matter} = \text{percent water}$$

Feed analysis reports usually are made on the basis of both the moisture-free weight and the moisture content when received at the laboratory. Knowing the water content of a feed makes it easier to calculate the weight of feed necessary to balance a ration properly. Proper storage of feed also depends on a knowledge of the moisture content. Feeds stored at too a high a moisture content may result in molding or heating. Moisture content also affects fermentation of silages. All of these factors affect palatability and feeding value.

Measuring Crude Protein

The crude protein content of a feedstuff is determined by a chemical analysis that determines the nitrogen content of the sample. Proteins will average 16 percent nitrogen; therefore, the crude protein content is found by multiplying the nitrogen content by 6.25. In some cases, the nonprotein nitrogen content (such as nitrates, nitrite, ammonia, or urea) is subtracted from the total nitrogen content before multiplying by 6.25.

Kjeldahl Process

Chemical analysis of the feed sample for nitrogen content is carried out by the Kjeldahl process. The sample is digested in concentrated sulfuric acid, which destroys all the organic matter and converts the nitrogen to ammonium sulfate. The material that remains is neutralized with sodium hydroxide and then distilled to drive the ammonia over into a standard acid solution. Titration is then used to determine the amount of nitrogen in the sample. Because protein contains an average of 16 percent nitrogen, the nitrogen content is multiplied by 6.25 to determine the crude protein content. The percentage of crude protein in the feedstuff is then found by dividing the crude protein content of the sample by the weight of the sample and multiplying by 100.

The formula for this computation is:

$$\frac{\text{Crude protein in sample}}{\text{Weight of sample}} \times 100 = \text{percent crude protein}$$

Amino acids, enzymes, certain vitamins, urea, and other nonprotein compounds are found in crude protein but not in true proteins. The determination of the true protein content requires a more complex analysis. The Kjeldahl process of analysis does not distinguish between true protein and nonprotein nitrogen in the feed. Nonprotein materials are normally present in only small quantities in grains with a higher level in forages. Ruminants can make use of these materials to meet their protein needs.

Determining the Crude Fat Content

The portion of a feed sample that is soluble in ether is called *crude fat*. Crude fat is often referred to as *ether extract*. Ether extract contains many substances in addition to true fat, including the fat soluble vitamins A, D, E, and K, free fatty acids, sterols, xanthophyll, phospholipids, chlorophyll, lecithin, resins, volatile oils, and waxes. Chlorophylls, resins, and volatile oils are not classified as nutrients, although they are found in the ether extract. Most of the ether extract from grains is fat; however, as much as one-half of the ether extract of forages may be these other (nonnutrient) substances.

To determine the crude fat content of the feed sample, the sample is dried in a drying oven. Ether is used to extract the sample. The ether is then evaporated from the extract and the remaining material is weighed. The resulting value is the weight of the crude fat in the sample. The percentage of crude fat in the feed is found by dividing the weight of the crude fat by the weight of the sample and multiplying by 100.

The formula for this is:

$$\frac{\text{Weight of crude fat}}{\text{Weight of sample}} \times 100 = \text{percent crude fat}$$

Determining Crude Fiber Content

Crude fiber consists of cellulose, more insoluble hemicellulose, and lignin, which is not soluble in weak acid or alkali solutions. Some of the more soluble hemicellulose and lignin in a feed may be found in the nitrogen-free extract. These materials are relatively low in digestibility, with hemicellulose being more digestible than cellulose. Lignin is almost completely indigestible. A high level of crude fiber in a feed indicates a relatively low level of digestible energy.

The first steps in determining crude fiber content of the sample are the same as the determination of crude fat. The same sample can be used after the crude fat has been determined. The protein, sugars, starches, and soluble hemicelluloses, minerals, and lignins are removed from the sample by boiling in dilute (1.25 percent) sulfuric acid for 30 minutes, followed by boiling in dilute (1.25 percent) sodium hydroxide for 30 minutes and then passing the solution through a filter. The material (which contains the crude fiber and insoluble mineral matter) is then dried and weighed. The material is then heated in a furnace to oxidize the crude fiber. The material that is left is the ash or mineral matter. This material is weighed and the amount of crude fiber is calculated by subtracting the weight of the ash from the weight of the material left after boiling and drying. The percentage of crude fiber is calculated by dividing the weight of the crude fiber by the weight of the original sample and multiplying by 100.

The formula for this is:

$$\frac{\text{Weight of crude fiber}}{\text{Weight of original sample}} \times 100 = \begin{array}{l}\text{percent} \\ \text{crude fiber}\end{array}$$

Determining the Mineral Content

The inorganic, or mineral, content of the feed sample is found in the ash. Grains and concentrates usually contain from 1-4 percent ash on a moisture-free basis. Forages generally contain 3-12 percent ash (also on a moisture-free basis). The percentage of ash in the sample is found by weighing a small sample of the feedstuff and placing it in a crucible. The sample is then heated in a furnace at 600°C (1,112°F) until the sample is reduced to ash. This usually takes several hours. The ash is then weighed and the percentage of ash or mineral matter is calculated by dividing the weight of the ash by the weight of the original sample and multiplying by 100.

The formula for this is:

$$\frac{\text{Weight of ash}}{\text{Weight of original sample}} \times 100 = \begin{array}{l}\text{percent of} \\ \text{mineral matter}\end{array}$$

Determining the Amount of Nitrogen-free Extract

Nitrogen-free extract is often referred to as NFE and contains the easily digested carbohydrates such as sugars, starches, organic acids, and the more soluble cellulose, hemicelluloses, and lignin. The presence of lignin in the feedstuff reduces the digestibility of the NFE because lignin is indigestible. In approximately 30 percent of the analyses, the NFE is determined to have a lower digestibility than the crude fiber because of the presence of lignin. For this reason, the feed composition tables do not show the NFE, and it is recommended that NFE not be used as a determinant of nutritive value.

The NFE content of the feedstuff is found by difference rather than by direct analysis. The total percentages of the other five components are added together and subtracted from 100. The remaining amount is the NFE content of the feedstuff.

The method for doing this is:

100 − (percentage of water + percentage of crude protein + percentage of crude fat + percentage of crude fiber + percentage of ash) = percentage of nitrogen-free extract

LIMITATIONS OF PROXIMATE ANALYSIS

The chemical analysis of a feedstuff is a good starting point in determining its value. However, this procedure does not give any indication of its palatability, digestibility, toxicity, or nutritional value. Other methods must be utilized to make these determinations. Digestion (metabolism) and feeding trials with animals are necessary to determine the actual worth of a feed for practical use.

Feed composition tables contain information about average amounts of the various components found in feedstuffs. Individual feeds found on farms may vary widely from the table values. Thus, the feed composition tables should be used only as general guidelines. The organic components of the feed such as crude protein, amino acids, ether extract, and cell wall constituents may vary in individual feeds as much as 15 percent more or less than the table values. Inorganic components such as mineral matter may vary as much as 30 percent more or less than the table values. Energy values of individual feeds may vary as much as 10 percent more or less than the table values.

The crop variety, climate, and fertility of the soil upon which the feed was grown all have a major effect on its composition. The time and conditions of harvest and storage also affect the composition of the feed. For more accurate ration balancing, samples should be taken of the actual feeds to be used and a laboratory analysis made. Only in this way can the livestock feeder know for sure what the feed contains and how much is needed to get the most economical gains or production.

When analyzing feeds with a high-fiber content, crude fiber as identified by proximate analysis is not chemically uniform, containing cellulose, hemicellulose, and lignin. The feeding value of cellulose and hemicellulose is greater for ruminants than for nonruminants. Lignin is mostly indigestible for all species. Proximate analysis does not reflect this difference in feeding value by species.

Some of the hemicellulose and lignin are reported in the NFE fraction when proximate analysis is done. This fraction is designed to show the more digestible nutrients such as sugars and starches. When significant amounts of hemicellulose and lignin appear in the NFE, it has a lower average digestibility than if it contained only the sugars and starches. Concurrently, the crude fiber fraction does not contain as much of the more indigestible nutrients as it should.

Proximate analysis does not provide any information on specific amino-acid content or the mineral composition of the feed. These must be determined by more complicated analysis procedures.

Proximate analysis does not give the total vitamin content or the amount of individual vitamins in a feedstuff. There are methods of doing this when it appears to be necessary. In proximate analysis, vitamins appear as a part of the crude protein, crude fat, and NFE components. They are not a significant part of any of these components since they are usually less than one-tenth of 1 percent. The vitamin choline may be a more significant part of the crude protein component but it does have a protein value for most animals.

FEED COMPOSITION BASIS

All feeds contain some moisture. The amount varies with the form of the feed, the stage of growth at which it was harvested, the length of time in storage, and the conditions of storage.

Feed composition tables often list feeds on both an as-fed and 100 percent dry matter basis. The term *as-fed* means the data is calculated on the basis of the average amount of moisture found in the feed as it is used on the farm. Sometimes the term *air-dry* is used. For most feeds, air-dry means the same thing as as-fed or 90 percent dry-matter basis. The term *100 percent dry-matter basis* means that the data presented is calculated on the basis of all the moisture removed from the feed.

One of the major reasons for the variation in the composition of feedstuffs is the dry-matter content. For this reason, it is recommended that rations be formulated on a 100 percent dry-matter basis and then the figures converted to an as-fed basis to determine the quantities to be fed.

Converting from One Basis to Another

The method of converting from one basis to the other is as follows:

Let a = pounds (kilograms) of feed on 100 percent dry-matter basis
 b = pounds (kilograms) of feed on an as-fed basis
 c = the percent of dry matter in the feed

To convert from as-fed basis to 100 percent dry-matter basis:

$$a = b \times c$$

The pounds (kilograms) of feed on a 100 percent dry-matter basis equals the pounds (kilograms) of feed on an as-fed basis multiplied by the percent of dry matter in the feed.

Example 1: A ration calls for 8.9 pounds (4.04 kilograms) of #2 dent corn on an as-fed basis. The feed composition table shows that #2 dent corn has 89 percent dry matter. Therefore: $8.9 \times .89 = 7.9$ pounds (or $4.04 \times .89 = 3.6$ kilograms) on a 100 percent dry-matter basis. (Amounts have been rounded to the nearest tenth of a pound or kilogram.)

Example 2: A ration calls for 15 pounds (6.8 kilograms) of alfalfa hay on an as-fed basis. The alfalfa has been cut in the midbloom stage. The feed composition table shows that midbloom alfalfa hay has 89.2 percent dry matter. Therefore: $15 \times .892 = 13.4$ pounds (or $6.8 \times .892 = 6.1$ kilograms) on a 100 percent dry-matter basis.

Example 3: A ration calls for 23.6 pounds (10.7 kilograms) of corn silage on an as-fed basis. The feed composition table shows that mature corn silage has a dry-matter content of 55 percent. Therefore: $23.6 \times .55 = 13$ pounds (or $10.7 \times .55 = 5.9$ kilograms) on a 100 percent dry-matter basis.

To convert from 100 percent dry-matter basis to as-fed basis:

$$b = \frac{a}{c}$$

That is, the pounds (kilograms) of feed on an as-fed basis equals the pounds (kilograms) of feed on a 100 percent dry-matter basis divided by the percent of dry matter in the feed.

Example 4: A ration calls for 4.7 pounds (2.1 kilograms) of #2 dent corn on a 100 percent dry-matter basis. The feed composition table shows that #2 dent corn has 89 percent dry matter. Therefore:

$$\frac{4.7}{.89} = 5.3 \text{ pounds (or } \frac{2.1}{.89} = 2.4 \text{ kilograms)}$$

on an as-fed basis.

Example 5: A ration calls for 18 pounds (8.2 kilograms) of alfalfa hay on a 100 percent dry-matter basis. The alfalfa has been cut in the midbloom stage. The feed composition table shows that midbloom alfalfa hay has 89.2 percent dry matter. Therefore:

$$\frac{18}{.892} = 20.2 \text{ pounds (or } \frac{8.2}{.892} = 9.2 \text{ kilograms)}$$

on an as-fed basis.

Example 6: A ration calls for 14 pounds (6.4 kilograms) of corn silage on a 100 percent dry-matter basis. The feed composition table shows that mature corn silage has a dry-matter content of 55 percent. Therefore:

$$\frac{14}{.55} = 25.5 \text{ pounds (or } \frac{6.4}{.55} = 11.6 \text{ kilograms)}$$

on an as-fed basis.

When using nutrient requirement and feed composition tables given on a 100 percent dry-matter basis, it is easier to work out the ration on the dry-matter basis and then convert the final figures to an as-fed basis.

FORAGE EVALUATION BY VAN SOEST METHOD

As mentioned earlier, the proximate analysis of feedstuffs does not give an accurate analysis of crude fiber and NFE content in roughages. An alternate method of determining crude fiber content has been developed by Van Soest and associates, working at the USDA's Agricultural Research Service laboratory in Beltsville, Maryland.

The feed sample is boiled in a neutral detergent solution for 1 hour and then filtered. This separates the sample into a high digestible fraction and a low digestible fraction. The highly digestible fraction consists of the lipids, sugars, starches, and proteins. This fraction is called the *neutral detergent solubles* (NDS).

The low digestible fraction is called the *neutral detergent fiber* (NDF) and consists of the more insoluble material found in the cell wall. This fraction is made up mainly of cellulose, lignin, silica, hemicellulose, and some protein. In the Van Soest method, almost all of the lignin and hemicellulose are contained in the NDF fraction. In the Weende method, a considerable amount of these two materials are present in the NFE fraction. The Van Soest method of analysis results in a higher NDF amount being reported for roughages than does the Weende method.

The amount of lignin present in the NDF influences its digestibility. The acid detergent lignin procedure is used to determine the amount of lignin present in the fraction. The first step is to determine the amount of *acid detergent fiber* (ADF) present in the sample. This fraction consists mainly of cellulose, lignin, and some silica. It does not contain hemicellulose and protein. The difference between the amount of NDF and the amount of ADF gives an estimate of the hemicellulose content of the sample. The ADF is then digested in sulphuric acid and filtered. The residue is then washed, dried, weighed, and burned to ash. The ash that is left is composed of silica and the weight loss from burning is the amount of lignin present in the sample. The lignin is called the *acid detergent lignin* or the *acid insoluble lignin*.

The amount of lignin found to be present is higher if the roughage has been processed at temperatures above 122°F (50°C). The amount of nitrogen found in the ADF is a good indication of the amount of damage that has been done to the feed by excessive heat during processing.

An estimate of the true digestibility of the roughage is made by using the following formula:

$$TD = 0.98 \text{ NDS} + (1.473 - 0.789 \log_{10} \text{lignin}) \text{ NDF}.$$

The NDF and NDS are used as percentages of the dry matter in the roughage. The lignin is the percentage of acid insoluble lignin in the ADF. Approximately 12.9 percent of the dry-matter intake will be passed through the animal and is found in the feces. An apparent digestibility estimate for the feed may be made by subtracting this amount of dry matter from the true digestibility figure. The apparent digestibility of feeds that are heat damaged or that have an ash content above 2 percent of the feed dry matter will have to be adjusted to allow for the presence of additional silica or lignin.

DIGESTION TRIALS

The value of a feed depends upon its digestibility. Nutrients in the ration that are not digested are of no value to the animal, though not all of the nutrients in a feed can be digested and used.

An analysis of the feed is made to determine the percentage of each nutrient present. The ration to be tested is fed for a period of time to allow all residues of feeds that have previously been fed to pass out of the animal's digestive tract. This usually requires about 7-10 days for ruminants and 3-5 days for nonruminants. Weighed amounts of the ration are then fed to the test animal for a period of time. The feces passed during the test period are collected, weighed, and analyzed for nutrient content. The difference between the amount of the nutrient fed and the amount collected in the feces is determined, and the percentage of the nutrient digested is calculated. The digestion coefficient for the nutrient is the calculated percentage that is digested.

The digestion coefficient determined for a nutrient will vary from trial to trial, depending upon: the kind of digestive tract (ruminant/nonruminant); the manner in which the feed was processed; its chemical makeup; and the individual animal being tested. For these reasons, a number of digestion trials are necessary to determine the average digestibility of a feed for a given species of animal.

DETERMINING NET ENERGY VALUES OF FEED

Net energy is a valuable indicator of the true value of a feed. Two methods may be used to find the net energy value of a feed: measuring the amount of energy lost by the animal or measuring the amount of energy retained by the animal.

Measuring Energy Losses

The amount of energy lost by an animal may be determined by either direct calorimetry or indirect calorimetry. These two methods give comparable results, however, direct measurement is more expensive. Therefore, indirect methods of measuring energy loss are more commonly used when measuring the energy value of feeds.

Direct calorimetry requires the use of an insulated chamber in which the animal is confined. All heat losses by radiation, conduction, and convection from the body surface, by water evaporation from the lungs and skin, and by excretion of urine and feces are then measured. Measurements are taken by determining the temperature rise in a known volume of water or by measuring the amount of electricity generated when the heat is sensed by thermocouples.

Indirect calorimetry is done by measuring the exchange of O_2 and CO_2 as the animal breathes. The amount of O_2 used and the amount of CO_2 produced are proportional to the amount of heat produced. The *respiratory quotient* (RQ) is the ratio of CO_2 produced to the O_2 used. There is a specific RQ for each nutrient present in the feed. Thus, the RQ gives an indication of which nutrient is being metabolized. By measuring the RQ and oxygen, it is possible to determine the total heat production by using tables or by utilizing a formula that relates heat production to the respiratory quotient.

Measuring Energy Retained

Measurements of the amount of energy retained by the animal are made by the comparative slaughter method. At the beginning of an experiment, a control group of animals is slaughtered and the body composition is determined. At one time this determination was made by chemical analysis. However, a modified method is currently a more common practice. The carcass density is determined by weighing in water and making use of known relationships between carcass density and composition. The energy content of the carcass can be estimated quite accurately from its composition.

When the experiment is done, the remaining animals are slaughtered and their body composition is determined. The difference in calorie content is the amount of energy stored in the gain on the animals. This gives a more accurate measure of the energy value of the feed than does the amount of liveweight gain.

This method of determining energy value is most effective when used with growing and fattening animals. It is not usable for dairy animals. Based on this type of experiment, feeds are given NE_m (net energy for maintenance) and NE_g (net energy for gain) values.

OTHER MEASURES OF RATION VALUE

Palatability

Palatability refers to how well the animal likes the feed and how readily it eats the feed. Factors affecting palatability include taste, odor, appearance, texture, and temperature of the feed. A feed that has low palatability is of little value for feeding livestock no matter how nutritious it may be—if the animals will not eat it, it does them no good. Care must be taken when making changes in the ration. Livestock sometimes initially hesitate to eat a ration when its composition has been changed. Usually it is wise to make gradual adjustments to the ration so the animals become used to the change.

Palatability of feed is especially important when feeding animals for production, be it meat, milk, wool, or eggs. Feeds of low palatability will not be consumed in sufficient quantities to allow economical production. Feeds with lower palatability may be used when feeding livestock maintenance rations. Mixing limited amounts of low-palatability feeds with feeds of higher palatability may induce the livestock to eat sufficient quantities of the ration to make efficient gains or production.

Digestive Disturbances

Some kinds of feeds tend to induce digestive disturbances in livestock. For example, lush legume pastures may cause bloat in cattle. The regrowth of sorghum that has been frosted may contain enough prussic acid to be poisonous to livestock. Generally, livestock will not eat poisonous plants unless forced to by hunger. Frosted sorghum, however, is palatable and livestock will eat it. Some kinds of feed may cause constipation, and others will cause scours. The livestock feeder must watch for these problems and take corrective action (by adjusting the ration).

Other Considerations

In order to improve the palatability and nutritive value of the ration, it is usually a good idea to include a variety of feeds. This is particularly true when feeding nonruminants in order to assure a proper balance of the essential amino acids in the ration.

The amount of bulk desirable in the ration depends on the class and age of livestock being fed. Young animals and nonruminants should be fed less bulk than mature animals and ruminants. Animals being fattened for market should have less bulk in their rations than breeding animals.

FEEDING TRIALS

Feeding trials are conducted by many of the Agricultural Experiment Stations in the United States. Results of these experiments relate to the value of feeds and rations under controlled feeding conditions. The information obtained is valuable in determining practical rations for farm use.

FEED REGULATIONS

The regulation of the manufacture and sale of feeds is generally left to the states. The major exception to this is regulations relating to feed additives, which are made by the federal government. The Association of the American Feed Control Officials has published the "Uniform State Feed Bill," followed by many states in the preparation of their regulations of feed manufacture and sale. This has resulted in a fairly high degree of uniformity across state lines in matters relating to the manufacture and sale of feed. A copy of the "Uniform State Feed Bill" is found in the Appendix of this text.

Major provisions of most state feed laws, rules, and regulations include (1) registration of feed manufacturers; (2) labeling requirements; (3) prohibited acts; (4) definitions of misbranding and adulterations of feed; (5) schedule of inspection fees and reports; (6) inspection, sampling, and analysis procedures; and, (7) penalties for violations.

Feed Labeling

The feed tag or label found on a bag of commercial feed is of importance to the livestock feeder because it contains information about the content of the feed and its proper use. The format and content of the feed tag or label is regulated by state laws.

The tag or label generally contains the following information:

1. net weight
2. product name and brand name
3. guaranteed analysis of the feed
   a. minimum percentage of crude protein
   b. maximum or minimum percentage of equivalent protein from nonprotein nitrogen
   c. minimum percentage of crude fat
   d. maximum percentage of crude fiber
   e. minimum and maximum percentages of calcium and salt
   f. minimum percentage of phosphorus
   g. other minerals
   h. vitamin content
4. when drugs are used as an additive
   a. the word "medicated" must be on the label
   b. purpose of the medication must be stated
   c. directions for use and precautionary statements must be included
   d. list of active drug ingredients

Certain exemptions on labeling are common:

1. No mineral guarantee is needed if no label claims concerning minerals are made and the total mineral content is less than 6.5 percent of the total contents.
2. No vitamin information is required when the feed contains no claims concerning vitamins or is not being sold as a vitamin supplement.
3. Crude protein, crude fat, and crude fiber guarantees are not needed if the feed is not intended to furnish these substances or if they are a minor part of the total ingredients, i.e, in drug premixes, mineral or vitamin supplements, and molasses.

This general description of feed tag labeling is not intended to be a specific guide for a given state. The feed laws, rules, and regulations of each state should be consulted to determine the specific requirements for that state.

SUMMARY

Palatability and nutrient content are major factors affecting the quality of feeds. Poor quality feed may reduce the profitability of the livestock enterprise.

Harvesting methods affect the quality of hay. Improper methods of harvesting will reduce the palatability and nutrient content of the hay crop. The leaves of hay contain most of the nutrients and are the more easily digested part of the plant. Most of the losses in hay quality result from shattering, leaching, and bleaching.

To produce good quality silage, the crop should contain 25-35 percent dry matter at the time of ensiling. Silage may be stored in upright or horizontal silos. Corn and sorghum are the most commonly used crops for silage. Forages and small grains may also be used for silage.

To secure a more accurate feed analysis, representative samples of the feedstuff must be carefully secured. Poor sampling techniques will result in inaccurate analysis.

Chemical analysis may be used to secure a proximate analysis of the feedstuff. A proximate analysis will separate the feedstuff into these six components: (a) water, (b) crude protein, (c) crude fat, (d) crude fiber, (e) nitrogen-free extract, and (f) ash. Proximate analysis does not give an indication of the palatability, digestibility, toxicity, or nutritional value of the feed to the animal. Digestion and feeding trials are used to determine the actual worth of a feed for practical use. Forage evaluation by the Van Soest method will give a better indication of the crude fiber and NFE content of roughages.

Feed regulation is generally left to the states. The use of feed additives is regulated by the Federal government. Many states follow the "Uniform State Feed Bill" published by the Association of the American Feed Control Officials when preparing regulations governing the manufacture and sale of feedstuffs.

REVIEW

1. Name and describe the factors that affect feed quality.
2. Describe a good quality hay.
3. How can shattering be reduced when harvesting hay?
4. Describe the proper methods of harvesting good quality hay.
5. Describe and compare the forms in which hay may be harvested.
6. Describe the types of structures which may be used for the storage of silage.
7. Describe methods of harvesting good quality silage.

8. Which crops are most commonly used for silage?

9. Describe the changes that occur in the ensiled material after it is placed in the silo.

10. Describe the stage of maturity at which corn, grasses, legumes, sorghum, and small grains should be harvested for good quality silage.

11. Describe methods of collecting grain, sacked feed, silage, and forage samples for analysis.

12. List the six components into which a feedstuff is separated by proximate analysis.

13. Briefly describe the method of proximate analysis for each of these six components.

14. What are the limitations of using proximate analysis to determine feed value?

15. Describe and give examples of how feeds may be converted from one composition basis to another.

16. Why is the Van Soest method of forage analysis sometimes used?

17. Briefly describe the Van Soest method of forage analysis.

18. Why are digestion trials of importance when determining the value of a feedstuff?

19. Briefly describe how net energy values of feed may be determined.

20. List and briefly describe some other measures of feed value.

21. Why are feeding trials of value in developing rations?

22. Briefly describe the major provisions found in most state feed laws.

9

RATIONS: SELECTING, BALANCING, MIXING

OBJECTIVES

After completing this chapter you will be able to:

- Describe general principles for formulating rations.
- Discuss the economics of ration selection.
- Describe general principles for balancing rations.
- Describe the steps in balancing a ration.
- Use feeding standards and feed composition tables to help balance rations.
- Use the Pearson Square or algebraic equations to balance a ration.
- Discuss the use of computers to balance rations.
- Describe methods of feed preparation and storage of feed.

FORMULATING RATIONS

An animal must receive the proper amounts of nutrients in the right proportion in order to efficiently produce meat, milk, eggs, wool, work, etc. A ration is said to be balanced when it provides the nutrient needs of the animal in the proper proportions. Strictly speaking, a *ration* is the amount of feed given to an animal to meet its needs during a 24-hour period; however, in common practice, the term may refer to feed provided without reference to a time period. The term *diet* refers to the ration without reference to a specific time period.

Feed accounts for approximately 50-80 percent of the total cost of raising livestock. In order to feed livestock efficiently, it is necessary to develop diets that are as economical as possible, yet still meet palatability and nutritional requirements of the animals. Home-grown feeds are used as much as possible because they are generally less expensive than purchased feeds. Commercial feeds are used when homegrown feeds are not available. Every effort is made to formulate diets balanced to provide the best nutrients at the lowest possible cost, assuring an efficient and profitable ration.

Feeds used in rations must not be harmful to the animal's health or lower the quality of the product. Poisonous plants should not be included in diets for livestock. Poisonous plants sometimes grow in hay and pasture fields. Eradicate these plants before harvesting the hay or allowing animals to graze the pasture. Usually, animals will not eat poisonous plants, but if they are in the hay the animal may not sort them out. If the pasture is sparse, animals may eat poisonous plants that are growing there. Animals get sick every year from eating poisonous plants because farmers do not take care to keep the poisonous plants out of the animals' diet.

It is necessary to balance the intake of roughage and concentrates for the particular species and age of livestock being fed. Ruminants can use more roughage in their diets than nonruminants. Also, younger animals cannot use as much roughage in their diets as can more mature animals. The purpose for which the animal is being fed must also be considered when including roughage in the diet. For example, fattening animals generally should be fed less roughage than breeding animals.

Micronutrients and feed additives are used in small quantities in the diet. Care must be taken to thoroughly mix these materials to assure uniform distribution in the feed. Failure to do so may result in one animal getting too much of the micronutrient or additive while another animal may get too little. Excessive amounts of some additives may be harmful to an animal. Feed only the recommended amounts of these materials and make sure they are well mixed with the rest of the feed ingredients.

The functions of a ration, as discussed earlier, must be considered when determining the nutrient requirements of livestock. These functions include maintenance, growth, fattening, production, reproduction, and work.

ECONOMICS OF RATION SELECTION

Because the cost of feed ranges from 50 to 80 percent of the cost of raising livestock or producing livestock products, it is necessary to give careful consideration to economic factors when formulating rations. When possible, use homegrown feeds such as grains and roughages in the ration—this is usually lower in cost than buying these feeds. Supplements to provide additional protein, vitamins, minerals, and other additives are generally pur-

chased. Because of processing and mixing costs, it is usually not economical to produce these on the farm.

The price of a purchased feed does not necessarily reflect its true nutritive value. Comparisons among possible feed sources must be made on the basis of their nutritive content. Price per bushel or per hundredweight is not an adequate indicator of the relative worth of a feed.

The most commonly purchased feeds are the high protein supplements. These should be compared on the basis of the cost per pound or kilogram of crude or digestible protein content. For example, several dairy feeds of various protein content might be compared as follows:

Percent Protein	Price per Ton	Pounds Protein per Ton	Price per Pound Protein
14%	$186	280	$0.66
16%	200	320	0.625
18%	213	360	0.59
20%	219	400	0.5475

In this case the 20 percent protein dairy feed has the lowest cost per pound of protein content.

In a similar manner, high-energy feeds may be compared on the basis of energy content or TDN. For example corn, oats, and grain sorghum might be compared as follows:

Feed Name	Price	Price/ CWT	TDN/ CWT	Cost/ lb TDN
Shelled Corn	$3.20/bu.	$5.71	77	$0.074
Oats	2.01/bu.	6.28	68	0.092
Grain Sorghum	5.21/Cwt	5.21	78	0.067

In this example, grain sorghum is slightly lower in cost per pound of TDN than is corn and both are significantly lower than oats.

When comparing prices, transportation costs, if any, must also be taken into consideration. The suitability of the particular feed for the class of animal being fed must also be considered.

Another economic factor to consider when selecting feedstuffs is the changing prices of feeds over a period of time. It may be profitable to vary the feedstuffs used in a diet as prices change. However, the availability of the feedstuff and the effect of a diet change on the performance of animals being fed must be considered before making major changes. Some species of animals do not respond favorably when major changes are made in their diet. Relative palatability of the feedstuff may make it necessary to introduce a change in the diet gradually in order to avoid a reduction in feed intake and a subsequent reduction in rate of gain or production.

When the livestock feeder knows that a particular feed will be needed in large quantity during the year, it may be profitable to purchase a supply when the price is lower. Some feeds are lower in price at certain times of the year. If the feeder has the capital to invest and can arrange suitable storage, it may pay to purchase needed feeds in larger quantities. However, alternate uses of the capital so invested must be considered when making this decision. Using an electronic spreadsheet on a microcomputer to make various kinds of projections concerning alternate uses of capital may prove helpful in this decision-making process.

GENERAL PRINCIPLES FOR BALANCING RATIONS

The diet must meet the nutritional needs of the animal. Nutrient requirements are listed in tables in the Appendix. Nutrient allowances figured in balanced rations should not be more than 3 percent below the animal's requirement. For best results, the nutrient requirements must be met as closely as possible.

Diets must include a minimum level of dry matter or the digestive tract will not func-

tion properly, but there is an upper limit on the amount of dry matter that can be included in an animal's diet. This varies with the kind, age, and size of animal being fed. The amount of dry matter in the diet should not be more than 3 percent above the recommended level for an animal on full feed. The permitted variation in levels of dry matter can be wider for animals not on full feed.

Diets are commonly balanced to meet the protein, energy, calcium, phosphorus, and vitamin A requirements of the animal. Methods of measuring protein and energy content of feedstuffs are discussed in earlier chapters in this text.

Protein

The amount of protein in the diet may be measured by *crude protein* (CP) or *digestible protein* (DP) content. When diets are balanced for nonruminants, essential amino acids must be considered. Often the first limiting amino acid is used to determine if the amino acid requirements of the nonruminant are being met by the diet. Feed composition tables list average nutrient content for feedstuffs. Actual nutrient content of the feed being used may vary somewhat from table values. Feed analysis is a more accurate way to determine feedstuff nutrient content but may not always be practical for the livestock feeder. It is acceptable to have the protein content of the formulated diet as much as 5 to 10 percent above the nutrient requirements listed in the tables. However, it must be remembered that protein is an expensive part of the diet and using excessive amounts will substantially raise the cost of the ration.

Energy

Four measures of energy commonly used when formulating diets are *digestible energy* (DE), *total digestible nutrients* (TDN), *metabolizable energy* (ME), and *net energy* (NE). The energy provided in the diet should not be more than about 5 percent above requirements, because animals are limited in the total amount of energy they can use.

Minerals

Calcium and phosphorus are the two minerals generally needed in larger amounts. The ratio of calcium to phosphorus is just as important as the total amount being fed. The ratio of calcium to phosphorus should be between 1:1 and 2:1. When other nutrient needs are met by the diet, the total amount of calcium and phosphorus provided may be more than required by the animal.

The other mineral needs of the animal are generally not considered when balancing rations. There are usually enough minerals provided in the ingredients used or by the addition of trace-mineralized salt.

Vitamins

A vitamin supplement is usually added to the ration to meet the vitamin needs of the animal. However, vitamin A requirements may be considered in balancing rations. The amount of vitamin A in a ration may be in excess of the animal's needs but this will not harm the animal. When low-quality legume hay is included in the diets of pregnant cattle or sheep, a vitamin deficiency may occur. Always add a vitamin supplement to a gestation ration.

Cost of Nutrients

The cost of the nutrients included in the diet is another factor to consider when balancing rations. The cost per pound (or kilogram) of each nutrient must be considered when developing least-cost rations for maximum efficiency. Energy and protein nutrients are the major ones to consider when making nutrient cost comparisons among feeds.

Moisture Basis

Nutrient requirements and feed composition tables often list nutrient content of feeds on a 100 percent dry-matter basis. It is easier to compare feeds having different moisture contents on a 100 percent dry-matter basis as compared to an as-fed basis. When using tables that give the nutrient values on a 100 percent dry-matter basis, it is easier to work out the ration on this basis and then convert the final figures to an as-fed basis to determine the quantities of feed to use. The method for making these conversions is given in Chapter 8.

Feeding Guidelines

General guidelines for feeding specific species of animals are found in the chapters for those species later in this text. These guidelines should be considered when formulating rations.

STEPS IN BALANCING A RATION

Step 1. Identify the kind, age, weight, and function of the animal(s) for which the ration is being formulated. In this text, suggested rations and feeding programs are found in the chapters referring to specific species of animals. These may be used as general guides in formulating rations.

Step 2. Consult a table of nutrient requirements to determine the nutrient needs of the animal(s). These requirements are called *feeding standards.* Feeding standards are based on average requirements and may not meet the needs under specific feeding conditions. If unusual conditions such as weather stress are present, adjustments in the diet may be needed.

Step 3. Choose the feeds to be used in the ration and consult a feed composition table to determine the nutrient content of the selected feeds. Note that the nutrient content of a feed may be different for different species. Values given in a feed composition table are average values and may not represent the actual composition values of the feeds being used. An analysis of feeds being used is a more accurate method of determining feed composition.

Step 4. Calculate the amounts of each feed to use in the ration. Several methods may be used to do this. The Pearson Square or algebraic equations may be used to balance a ration using two or more feeds. Computer programs may also be used to balance rations.

Step 5. Check the ration formulated against the needs of the animal(s). Be sure it meets the requirements for minerals and vitamins. If there is an excessive amount of a nutrient present, it may be necessary to recalculate the ration to bring it more closely in line with the requirements. Check the cost of the nutrients in the ration to determine if this is the most economical ration that is practical to feed. Calculate the cost of the ration per pound (kilogram) or ton (tonne). The daily cost of feeding the animal may also be calculated if a daily consumption rate is known or assumed. In some cases it may be necessary to feed certain nutrients, such as salt or other minerals, on a free-choice basis in addition to the amounts provided in the formulated ration.

USE OF THE PEARSON SQUARE TO BALANCE A RATION

Balancing a ration by trial and error methods is difficult and time consuming. The Pearson Square is a simple method of balancing a ration consisting of two feeds or feed mixtures. The Pearson Square shows the proportions or percentages of two feeds that need to be combined to give the desired percent or amount of a given nutrient. While most of the examples given in this chapter use protein as the nutrient to be balanced, any desired nutrient may be balanced with this method.

The Pearson Square can be used to balance rations using more than two feeds and this method is explained later in this chapter.

One Nutrient and Two Feeds

The first example demonstrates the use of the Pearson Square to balance a ration for one nutrient using two feeds. The ration will consist of corn and soybean oil meal and will be balanced to a desired level of protein content. Assume that one ton (2,000 lbs) of mix containing 15 percent total protein is needed. The feed composition table shows that corn grain (IFN 4-02-935) contains 9.6 percent crude protein and that soybean meal, solvent process, contains 44.6 percent crude protein (as-fed basis).

Draw a square and place the desired level of protein for the final mix in the center of the square. The number in the center of the square must lie between the values for the feed ingredients, which are placed in the upper and lower left hand corners of the square. If it is not in this range, the Pearson Square will give a solution but the solution will not be correct. Draw diagonal lines to each corner of the square. Next, place the protein content of the corn next to the upper left corner of the square and the protein content of the soybean oil meal next to the lower left corner of the square, Figure 9-1.

Subtract the smaller number from the larger number along the diagonal lines in the square. Write the difference at the opposite end of the diagonals. The difference between the percent protein in the soybean meal (44.6) and the needed percent protein in the ration (15) is the parts of corn needed (29.6). The difference between the percent protein in the corn (9.6) and the percent protein needed in the ration (15) is the parts of soybean oil meal needed (5.4). The sum of the numbers on the right equals the difference of the numbers on the left. This fact is used as a check to see if the

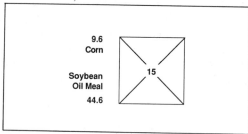

Figure 9-1. Setting up the Pearson Square.

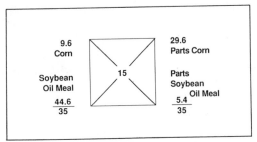

Figure 9-2. Completing the Pearson Square.

square is set up correctly, Figure 9-2.

Divide the parts of each feed by the total parts to find the percent of each feed in the ration.

Corn (29.6/35) × 100 = 84.6%
Soybean oil meal (5.4/35) × 100 = 15.4%

One ton (2,000 lbs or 907 kg) of mix is needed. The amount of corn needed is 1,692 pounds (767 kilograms). This is found by multiplying the percent of corn in the mix by the total pounds (kilograms) of the mix.

2,000 × 0.846 = 1,692 lbs.

or

907 × 0.846 = 767 kilograms

The amount of soybean meal needed is 308 pounds (140 kilograms). This is found by multiplying the percent of soybean meal in the mix by the total pounds (kilograms) of the mix.

2,000 × 0.154 = 308 lbs.

or

907 × 0.154 = 140 kilograms

The calculated results need to be checked to be sure the protein need is met by the ration. Multiply the pounds (kilograms) of corn by the percent of protein in the corn (1,692 x 0.096 = 162; or 767 × 0.096 = 74). Multiply the pounds (kilograms) of soybean meal by the percent of protein in the soybean meal (308 × 0.446 = 137; or 140 × 0.446 = 62). Add the pounds (kilograms) of protein together and divide by the total weight of the mix. The result should equal the required

percent of total protein in the ration. If it does the ration is balanced for this nutrient.

162 + 137 = 299
(299/2,000) × 100 = 15%

74 + 62 = 136
(136/907) × 100 = 15%

(Numbers in this example are rounded off to full pounds or kilograms.)

Mix Two Grains with a Supplement

The Pearson Square can be used to mix more than two feeds in the desired proportions. In this example, two grains will be mixed with a supplement to produce 1 ton (2,000 pounds) of feed. The mix will be balanced to contain 16 percent crude protein. As-fed values will be used in balancing this ration. Corn (IFN 4-02-935) oats (IFN 4-03-309), and soybean meal (IFN 5-20-637) are the feeds to be used in this example.

When using the Pearson Square to mix more than two feeds, the proportion of each grain or each supplement to be used must be known or decided upon in advance. In this example the proportion of corn to oats will be 3:1. How many pounds of corn, oats, and soybean meal will be needed?

The first step in solving this problem is to find the weighted average percent of crude protein in the mix of corn and oats to be used in the ration. This is found by multiplying the proportion of corn (3) by the percent crude protein in corn (9.6). Do the same for oats. Add the two answers together and divide by the total parts (4). The answer is the weighted average percent of crude protein in the corn-oats mix.

3 × 9.6 = 28.8
1 × 11.8 = 11.8
            40.6

40.6/4 = 10.2% crude protein in the corn-oats mix.

The Pearson Square is then used to find the pounds of corn-oats mix and soybean

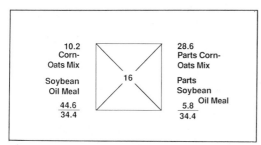

Figure 9-3. Mixing two grains and one supplement using the Pearson Square.

meal needed, Figure 9-3.

    28.6/34.4 × 100 = 83.1% corn-oats mix
    0.831 × 2,000 = 1,662 pounds corn-oats
                    mix.
    1,662 × 3/4 = 1,246 pounds corn needed
    1,662 × 1/4 = 416 pounds oats needed
    100 − 83.1 = 16.9% soybean meal
    0.169 × 2,000 = 338 pounds of soybean
                    meal needed

This method may also be used to mix two protein supplements with one grain or two supplements and two grains. In either case, the feedstuffs are grouped in like groups (such as two grains) and the proportions of each in each group must be decided before calculating the ration. The weighted average percent of the nutrient being used to balance the ration is then found. Finally, the Pearson Square is used to balance the mix. Any of the measures of nutrients in feeds may be used. To balance on energy needs, use TDN, NE, ME, or DE. To balance on protein needs, use total (crude) or digestible protein. When balancing swine rations, the first limiting essential amino acid (lysine) is sometimes used to balance for protein requirements.

USE OF ALGEBRAIC EQUATIONS TO BALANCE RATIONS

Algebraic equations may be used in place of the Pearson Square to balance rations. This may be illustrated by using the same problem as the first Pearson Square example shown earlier in this chapter. The mix of 2,000 pounds

is to be balanced for protein using two feeds. The basic equations are:

    X = pounds (or kilograms) of grain needed
    Y = pounds (or kilograms) of supplement
        needed

Equation 1:

    X + Y = total pounds (or kilograms) of mix
            needed

Equation 2:

    (percent nutrient in grain) × (X) + (percent
    nutrient in supplement) × (Y) = pounds (or
    kilograms) of nutrient desired in mix

Place the desired values in equation 2:

    0.096X + 0.446Y = 300

    (The quantity 300 is found by multiplying the
    quantity of feed by the percent [or the amount/
    kg or lb] of the nutrient desired.)

Either X or Y must be canceled by multiplication of equation 1 by the percent of nutrient for either X or Y, and the resulting equation 3 is subtracted from equation 2:

Equation 3:

    0.096X + 0.096Y =   192

    0.096X + 0.096Y =   300
    −0.096X − 0.096Y = −192
    ─────────────────────────
          0.35 Y =   108
               Y =   308

The value of X may be found by substituting the value of Y in equation 1 and solving for X:

    X + 308 = 2,000
    X = 2,000 − 308
    X = 1,692

Answers in the above example have been rounded to whole numbers. Note that this method gives the same results as the use of the Pearson Square demonstrated earlier in this unit. The same restriction as mentioned with the Pearson Square regarding the value of the desired nutrient lying between the values of the two feeds being used also applies when using algebraic equations.

Algebraic equations may also be used to balance rations using three or more feeds. The same initial step must be taken as when using the Pearson Square, i.e., group like feeds into two groups and determine the proportions of each to be used in each group. After this is done, the same procedure as outlined above is followed to balance the ration.

Balancing Rations with Simultaneous Algebraic Equations

Simultaneous algebraic equations may be used to balance a ration using two feeds or groups of feeds and balancing for two desired nutrients. Assume that one tonne, or metric ton, (1000 kg) of feed mix is desired for growing-finishing pigs weighing 20-35 kg (44-77 lbs). The mix is to be balanced for lysine and metabolizable energy (ME) requirements using corn and soybean meal. The nutrient and energy composition of the feeds to be used are found in the Appendix tables of feed composition. When using the tables, note that energy values are different for different classes of livestock. Be sure to secure the value for the type of livestock for which the diet is being formulated. In this example corn (IFN 4-02-935) and soybean meal (IFN 5-04-600) are to be used. Lysine is often the first limiting amino acid in swine diets; therefore it is used to balance this example diet. If the lysine requirement is not met, it makes no difference how much of the other essential amino acids are present. The pigs will not grow any faster than the amount of lysine in the diet will permit.

Step 1. Set up the requirements and composition of the feeds:

	Lysine	ME
Req/kg of diet	0.70%	3,175 kcal
corn	0.25%	3,300 kcal/kg
SBM	2.79%	2,972 kcal/kg

Step 2. Set up the algebraic equations and solve:

X = amount of grain needed per kilogram of diet

Y = amount of supplement needed per kilogram of diet

Equation 1:

$0.25X + 2.79Y = 0.70$ (lysine equation)

Equation 2:

$3,300X + 2,972Y = 3,175$ (energy equation)

Divide 3,300 by 0.25 to get a factor which is multiplied times equation 1. The resulting equation is then subtracted from equation 2 to eliminate the X unknown and solve for Y. Alternatively, the Y unknown could be eliminated by dividing 2,972 by 2.79 and then solving for X:

$$3,300/0.25 = 13,200$$

$$\begin{array}{r} 3,300X + \quad 2,972Y = \quad 3,175 \\ -3,300X - \quad 36,828Y = -9,240 \\ \hline -33,856Y = -6,065 \\ Y = 0.179 \end{array}$$

Substitute the value of Y in equation 1 and solve for X:

$$\begin{array}{r} 0.25X + (2.79 \times 0.179) = 0.70 \\ 0.25X + 0.5 \quad = 0.70 \\ 0.25X = 0.70 - 0.5 \\ 0.25X = 0.2 \\ X = 0.8 \end{array}$$

Step 3. The accuracy of the solution may be checked by comparing the computed amounts of lysine and ME provided by this diet with the original requirements:

Lysine requirement
Corn 0.25 × 0.8   = 0.2
SBM 2.79 × 0.179 = 0.5
Total lysine = 0.7 per kilogram of diet

ME requirement
Corn 3,300 × 0.8   = 2,640
SBM 2,972 × 0.179 = 532
Total ME = 3,172 per kilogram of diet

Step 4. As one tonne (1,000 kg) of the mix is needed, the computed amounts of corn and SBM for one kilogram of diet are multiplied by 1,000 to find the amounts of each to put in the mix:

Corn 0.8 × 1,000 = 800 kilograms
SBM 0.2 × 1,000 = 200 kilograms
Total mix = 1,000 kg (one ton)

(Decimals are rounded to one place in the final calculation to make the mix total 1,000 kg.)

The feed mix as formulated probably will not meet the needs of the animal for minerals and vitamins. These ingredients will have to be added to have a properly balanced diet. The addition of fixed ingredients to meet these needs is discussed in the next section of this chapter.

USING FIXED INGREDIENTS WHEN FORMULATING DIETS

Feed mixes formulated to provide a complete diet for the animal normally have small amounts of minerals, vitamins, and/or antibiotics added. Generally these total less than 10 percent of the total mix and provide little of the protein or energy needed in the diet. However, these fixed ingredients must be taken into account when formulating diets if the final computed protein and energy needs of the animals are to be met.

The first step in formulating a diet using fixed ingredients is to determine what these ingredients are and how much of each is to be in the final mix. Next, determine if any of these fixed ingredients provide any of the nutrients for which the ration is being balanced. If they do, then these amounts must be calculated and subtracted from the amount to be provided by the major ingredients in the diet. After this is done, then the procedures outlined above may be followed to balance the major ingredients for the mix.

This procedure is demonstrated using algebraic equations. Assume a one ton (2,000 lb) mix is needed to feed finishing hogs weighing 125 pounds (57 kg). The major ingredients selected are corn (IFN 4-02-935) and soybean meal (IFN 5-04-600). The fixed ingredients that provide additional minerals and vitamins

do not add either energy or protein to the ration and total 55 pounds. The ration is to be balanced for daily requirements of lysine and ME.

Step 1. Set up the requirements and composition of the feeds:

	Lysine	ME
Daily req. (kg)	0.0122	6,320 kcal
corn (4-02-935)	0.0025	3,300 kcal/kg
SBM (5-04-600)	0.0279	2,972 kcal/kg

Step 2. Set up the algebraic equations and solve:

X = amount of grain needed per day
Y = amount of supplement needed per day

Equation 1:

$$0.0025X + 0.0279Y = 0.0122 \text{ (lysine equation)}$$

Equation 2:

$$3,300X + 2,972Y = 6,320 \text{ (energy equation)}$$

Divide 3,300 by 0.0025 to get a factor which is multiplied times equation 1. The resulting equation is then subtracted from equation 2 to eliminate the X unknown and solve for Y. Alternatively, the Y unknown could be eliminated by dividing 2,972 by 0.0279 and then solving for X:

$$3,300/0.0025 = 1,320,000$$

$$
\begin{aligned}
3,300X + 2,972Y &= 6,320 \\
-3,300X - 36,828Y &= -16,104 \\
\hline
-33,856Y &= -9,784 \\
Y &= 0.289
\end{aligned}
$$

Substitute the value of Y in equation 1 and solve for X:

$$
\begin{aligned}
0.0025X + (0.0279 \times 0.289) &= 0.0122 \\
0.0025X + 0.008 &= 0.0122 \\
0.0025X &= 0.0122 - 0.008 \\
0.0025X &= 0.0042 \\
X &= 1.68
\end{aligned}
$$

Step 3: The accuracy of the solution may be checked by comparing the computed amounts of lysine and ME provided by this diet with the original requirements:

Lysine requirement (0.0122 kg)
Corn 0.0025 × 1.68 = 0.0042
SBM 0.0279 × 0.289 = 0.008
                    Total lysine = 0.0122 kg per day

ME requirement (6320 kcal)
Corn 3,300 × 1.68 = 5,544
SBM 2,972 × 0.289 = 859
Total ME   = 6,403 kcal per day

Step 4: Determine the amount of corn and soybean meal to mix together to make 1,945 pounds of mix. The amounts of corn and soybean meal needed daily are added together and each amount is divided by the total to determine the percent of each ingredient in the ration. This percent is then multiplied times 1,945 pounds to determine how many pounds of corn and soybean meal are necessary in the total mix. The balance of the 2,000 pounds is composed of the fixed ingredients previously determined to provide the added minerals and vitamins needed in the ration.

	kg/day	% diet	lb/ton
Corn	1.68	85.3	1,654
SBM	0.289	14.7	286
Total	1.969	100	1,945

USING COMPUTERS TO FORMULATE RATIONS

A computer may be used to formulate and balance rations more quickly than can be done using other methods. Large, mainframe computers utilizing complex programs may be used to develop least-cost rations for livestock feeding. Many universities and commercial feed companies offer computer services for ration formulation.

Using Mainframe Computers to Balance Rations

Complex programs on mainframe computers usually use a data base, which incorporates information relating to predetermined maximum and minimum amounts of the major nutrients needed in the diet. The data base also includes information on available feedstuffs, feed composition, restrictions relating to the kind of livestock being fed, and feed prices. When correctly programmed, the computer can then make many calculations at high speed, resulting in the formulation of a least-cost ration. The major advantage in the use of such a program is its ability to make many trial runs with the information given in a much shorter time than could be done by hand.

While such a ration will in fact be the lowest in cost possible with the data given and will be balanced for the nutrients involved, it may not be the best ration for the livestock feeder to use. It is difficult to assign numerical values to all of the variables, such as palatability and acceptability, involved in ration formulation, and the computer can only deal with numerical values. Feeding values of some feeds vary with the amount being fed or when they are used in conjunction with other feeds. While it is not impossible to program these factors into the computer, given a sufficiently large data base, it does increase the complexity of the program.

Another factor to be considered when formulating least-cost rations is the effect of major ration changes on the livestock being fed. In many cases, changes in diet must be made over a period of time to avoid digestive upset and a resultant reduction in rate of gain or production. This must be considered when a computer program designed to develop a least-cost ration indicates a major change in the diet based solely on the cost of the ingredients. It may be necessary to formulate several rations to make a gradual change in order to avoid the problems mentioned above.

Using Microcomputers to Balance Rations

A microcomputer may be used to formulate and balance a ration using the same basic techniques as outlined in this unit. A program may be written that simply does the mathematical calculations after the operator enters the information found in nutrient requirement and feed composition tables. Such a program is relatively simply and does not utilize a data base to formulate the diet.

Because the amount of available memory is limited in most microcomputers, the development and use of a large data base that would not require the operator to enter information is more difficult but not impossible. Such a program would require the storage of the data base on a disk, which would then be accessed by the computer as the program is running. Commercial programs available for use on microcomputers will formulate and balance rations. Generally, such programs are fairly expensive and the livestock feeder must determine if their use is cost effective for a particular feeding operation.

Using Electronic Spreadsheets to Balance Rations

Another possible method of utilizing a microcomputer to balance rations is by using an electronic spreadsheet program. There are many electronic spreadsheet programs on the market available for microcomputer use and some are powerful enough to permit the use of a limited data base to provide information for the balancing of the ration.

To use an electronic spreadsheet, one must develop a template, which is a series of formulas entered on the spreadsheet in the appropriate places to perform the mathematical calculations required. The simplest template would require the operator to enter basic information from nutrient requirement and feed composition tables as well as the prices of the feedstuffs being considered. With a little practice, the electronic spreadsheet permits the operator to try many different possible formulations fairly quickly, thus saving much time as compared to making the necessary calculations by hand.

PREPARATION OF FEEDS

The value of various methods of preparing feeds for livestock depends on the particular kind of feed and the class of livestock to which it is being fed. Processed feeds are generally easier to handle with mechanized equipment. Processing often increases the palatability of the feed, especially of lower quality roughages and some grains such as rye. The digestibility of some feeds is increased when the method of processing breaks it up into smaller size particles, which increases the surface area of the feed. The increased surface area improves bacterial and/or enzyme activity, which leads to greater efficiency of feed use. Feed efficiency may be increased by 5 to 15 percent by some methods of processing. However, rarely is the rate of gain improved by some methods of feed processing. When considering various methods of feed processing, the cost/benefit ratio of the particular method must be considered. This will vary depending upon the price of the feed and the cost of processing.

MECHANICAL PROCESSING

Feeds are mechanically processed by grinding, rolling, crimping, cracking, pelleting, or cubing. Most grains are processed by one of the above methods before being fed to livestock. Roughages are sometimes processed by mechanical means before feeding.

Grinding

Grinding is a mechanical process that reduces the size of the feed particle, usually done by using either a hammer mill or a burr mill.

Hammer Mill

A hammer mill is constructed with a series of swinging or fixed metal hammers attached to a rotor that turns in a grinding chamber with a perforated screen at its bottom. The screen is removable and can be replaced with screens of different sizes. The size of the holes in the screen determines the size of the resulting product. The feed material fed into the chamber is hit by the hammers as the rotor turns until its size is reduced enough to pass through the holes in the screen. A hammer mill can grind feed into very fine particles. Coarse grinding usually results in nonuniform particle size in the final product. The hammer mill can grind grain, ear corn, hay, and mixtures of feedstuffs. Maintenance costs are generally lower with a hammer mill than with a burr mill.

Burr Mill

A burr mill is constructed with two plates that have burrs on their surfaces. Generally one of the plates is stationary and the other rotates as feed is fed into the grinding chamber. The distance between the two plates can be adjusted to determine the fineness of the finished product. The feed is ground into relatively uniform size particles at all settings. A burr mill is not designed to be used for grinding roughages, but works well for ear corn, small grains, and grain mixtures.

Fineness of Grind

Feeds should be ground to a medium fineness for most classes of livestock. Grinding feeds too finely (flour fine) results in the following problems: (1) excessive dustiness with resulting loss in the wind; (2) reduced palatability; (3) reduced digestibility; (4) digestive problems such as stomach ulcers in swine or ruminal parakeratosis in feedlot cattle; (5) lowered milk fat production with dairy cows; and (6) bridging (failure to feed down) in a self-feeder.

Grinding Grains

Grains should be ground for young animals because their teeth are not developed enough to properly chew the feed. Older animals with worn teeth also need to have grain ground for them.

Sorghum grains (milo) have a hard seed coat and should be ground for all classes of livestock. When corn is included in beef cattle rations that contain more than 15-20 percent roughage, it should be coarsely ground. When the roughage level is below 15-20 percent, there is no increase in rate of gain as a result of grinding the grain. When silage is the principal roughage in beef cattle rations, the corn should be ground to a medium fineness. Grains in dairy cattle rations should be coarsely ground.

Grinding Hay

There is little or no advantage in grinding good quality hay for livestock feeding. The intake of poor quality hay may be improved by grinding but the digestibility is not affected. If the hay is to be handled by mechanized feeding equipment or incorporated into a complete ration mixture, it must be ground.

Coarse grinding of hay is recommended. A fine grind increases the speed of passage through the digestive tract and thus reduces digestibility. Ground hay fed to dairy cattle reduces the butterfat content of the milk.

Grinding hay increases the cost of the ration; therefore, the benefits of grinding must be carefully considered to determine if they outweigh this additional expense. Grinding hay also increases the dustiness of the ration, which may decrease feed intake.

Rolling and Crimping

Rolling is done by passing the grain between rollers with smooth surfaces set close together. This action compresses the grain, leaving it in the shape of a flake. This process is sometimes called *cracking* or *flaking*. *Crimping* is the same process except that the rollers have corrugated surfaces; however, the grain is left in about the same form as is accomplished by rolling. All of these processing methods are similar to coarse grinding and leave the grain with about the same feeding value. Cattle appear to prefer grain prepared by these methods and seem to gain a little better as compared to being fed ground grain.

Pelleting

Pelleting is done by grinding the concentrate or roughage and then forcing it through die openings, usually using some combination of heat, pressure, and moisture. Pellets can be made in different lengths, diameters, and degrees of hardness depending upon the equipment and process used.

Pelleting reduces the amount of storage space needed for the feed (as much as 75 percent for roughages), its dustiness, and the loss of fine particles during transport. The palatability of the feed is increased, which leads to increased consumption. Digestibility of the feed is improved because of the partial gelatinization of the starch in the grain, which increases susceptibility to the action of enzymes or bacteria in the digestive tract. Feeds with a high fiber content are better utilized by some classes of livestock after pelleting.

There is less waste with pelleted feeds because livestock cannot be selective about which portion of the ration they eat. Pelleted feeds are in a free-flowing state, thus lending themselves more easily to mechanized feeding systems and to use in self-feeders.

Pelleting feeds increases the cost of the diet and this must be taken into consideration when weighing the benefits of this practice. Pelleting roughages is about twice as expensive as pelleting concentrates. Pelleted roughages have a higher density and thus lose much of their value as roughages, especially for ruminants.

If care is not taken during the pelleting process, the pelleted feed may be more susceptible to spoiling while in storage. Improper pelleting procedures may reduce the quality of the pelleted feed. It is also difficult to pellet feeds which have a high fat content.

Complete Pelleted Rations

Sometimes complete rations are prepared by pelleting for livestock. Some cautions need to be observed about complete pelleted rations for dairy or beef cattle. Pelleting roughages for dairy cattle reduces rumen acetate production and lowers the milk fat test. High roughage or all roughage rations for finishing beef cattle may be pelleted, which will result in increased feed consumption and up to 25 percent faster gains. For finishing beef cattle, high concentrate rations that are pelleted reduce feed intake and gains, but increase feed efficiency by 5 to 10 percent. Pelleting complete high concentrate rations for finishing beef cattle probably will not pay.

Probably sheep also need some roughage in the diet; therefore, a completed pelleted ration is of questionable value for them.

Swine and poultry diets may be made up of complete pelleted rations. This usually shows improved feed efficiency (5-10 percent) and rate of gain (0-5 percent), likely because of reduced feed waste and increased feed intake.

Complete pelleted diets are most useful for young horses because they are more likely to sort out the various parts of the feed available to them. The complete pelleted diet should include at least 60 percent coarsely ground hay. Feeding some grass hay along with the pelleted ration may reduce wood chewing, or mane and tail chewing when several horses are penned together.

Cubing

Cubing is a variation of the pelleting process in which dry hay is forced through large (usually 1-1¼ inch) square dies. The resulting cubes are approximately 1-1¼ inches square and 2 inches long. They have a bulk density of 30-32 pounds per cubic foot. The best cubes are normally produced by using alfalfa hay. The hay is normally not ground prior to cubing, but water may be sprayed on it during the process. Cubing hay makes it easier to handle with mechanized feeding equipment and cubes may be fed on the ground with little waste. There is less nutrient loss as compared to baled hay and hay cubes require less storage space. Although cubes are more expensive than baled hay, the advantages may more than offset the additional cost.

Granules and Crumbles

Granules or *crumbles* are produced by breaking up pelleted feed and removing the fine

Figure 9-4. Control center and scales for a commercial feed mixing center. Note the diagram showing a schematic of the bins and flow of materials.

particles, which are then repelleted. Poultry feeds are often produced by this method.

HEAT TREATMENT PROCESSING OF GRAINS

Heat treatment processing of grains includes cooking, steam rolling, steam flaking, pressure flaking, roasting, extruding, popping, exploding, and micronizing. The various methods of heat treatment may improve the palatability and nutritive qualities of some feeds; however, the high cost of the processes generally makes heat processing an unprofitable practice.

Most states have laws which require that garbage fed to swine be cooked. This requirement is the result of an outbreak of *Vesicular Exanthema* (VE), a swine disease, which occurred in garbage-fed hogs a number of years ago. The feeding of garbage to swine is sometimes practiced near large cities where a ready supply of garbage is available.

FEED MIXING

The two general types of commercially available feed mixers are batch and continuous flow (percentage volumetric). Either system will mix feed ingredients satisfactorily. In the batch mixing system, all the ingredients are measured into a common container (either vertical or horizontal) and then mixed together in batches. In the continuous flow system, all the ingredients are metered into a blender on a percentage basis and mixed in a continuous process.

Scales and metering equipment used to measure ingredients must be accurate. Scales should be able to weigh ingredients within a 2 percent accuracy range. If small amounts of ingredients (such as a premix) are to be measured, the scales should be capable of giving readings in tenths of a pound, Figure 9-4.

Batch Mixing

The batch mixing system requires the ingredients to be processed separately before they are mixed. For the most efficient operation, this system requires a high capacity grinder and conveying system. The operator must be present while the system is in use.

Vertical batch mixers are generally cylindrical in shape, with an inverted cone-shaped bottom. The material is moved from the bottom to the top of the mixer by high-speed vertical augers, which spread it by centrifugal force over the top of the mix. The mixing time is relatively long (15-20 minutes); however, the initial investment and power cost to operate are generally low. The vertical mixer cannot handle molasses, silage, or ground hay.

Horizontal batch mixers are U-shaped with paddle agitators on a horizontal shaft in the bottom of the mixer, which mixes the batch as it turns. This type of mixer can handle molasses, silage, and ground hay. The initial investment and power requirements are generally higher than for a vertical mixer; however, the horizontal batch mixer can do a thorough job of mixing the batch in a shorter period of time (7-10 minutes).

Auger wagons may also be used for batch mixing. The vertical auger can be positioned to return the material to the top of the wagon, allowing the mixing process to continue until a thorough mix is obtained. This usually requires more time than with either the vertical or horizontal mixers. An auger wagon can handle molasses, silage, and ground hay in the mix. It is possible to get some mixing of the batch by layering the ingredients in the wagon, causing them to be mixed as the wagon unloads into the feeder. However, with this method, small quantities of ingredients cannot be thoroughly mixed in the batch.

The stationary batch mixer is usually a separate unit from the grinder. The portable batch mixer is generally a combination of grinder and mixer. The stationary batch mixer does a good job of mixing the ingredients; however, it does require electrical service and transporting of the feed to storage or the point of use. The portable batch mixer provides more flexibility in grinding, mixing, and delivering the feed anywhere on the farm. Tractor power is generally used so a separate electrical service is not required; however, this does tie up a tractor during feed processing and mixing. Portable batch mixers are well-adapted to use with fenceline bunk feeders.

Continuous Flow Mixing

The continuous flow system is better suited to automation, utilizing timers and switches to control the processing and mixing of the feed, Figure 9-5 and Figure 9-6. It can be low capacity because it can operate automatically over long periods of time without an operator being present. Automatic electric mills may operate as much as 30 hours or more per week. Grinding, mixing, and conveying operations are done simultaneously and continuously from start to finish. Operating costs are generally low for this type of system; however, a package feed center with automatic mill will have a high initial cost because of the need for overhead bins. Continuous flow systems will not handle roughages and the finished product must be either stored or conveyed to the point of use. Electrical service must be provided for the equipment and routine calibration of the metering system is necessary.

Housekeeping in Feed Mixing Areas

Feed mixing areas must be kept clean, well-lighted, and secure from weather and rodents. Good ventilation is important to prevent dangerous dust accumulation. Dust from feed grinding and mixing operations is highly explosive and serious damage to equipment and buildings or injury to the operator may result if a spark ignites a dust buildup.

Feed additives and premixes must be kept in a separate room or area in tightly closed, labeled containers. Care must be taken when handling premixes containing drugs to prevent contamination of rations in which the drug is not to be included. The operator should wash after using premixes containing feed additive drugs. Good inventory records

Figure 9-5. A continuous flow feed processing center on a swine farm in the Midwest. Feed storage bins and a distribution system are part of the system.

Figure 9-6. A continuous flow feed mixing system as part of a feed distribution system on a Midwest swine farm. Note the controls for the mix mill.

are essential to the efficient mixing and use of feed additives. All bags and bins of mixed feeds must be labeled to prevent accidental misuse.

The floor of the feed mixing area should be kept swept clean. Smooth, solid floors make this job easier. Spilled feed materials that are swept up off the floor should be immediately added to the batch to which they belong or be thrown away. Adding floor sweepings to subsequent feed batches may contaminate the new feed mix with undesirable feed additive drugs or other unwanted material.

FEED STORAGE

The proper storage of feed after harvest is necessary to preserve its palatability and nutritive value, and prevent possible harm to

animals. Three sources of feed storage problems are moisture, insects, and rodents.

Moisture and Feed Storage

When the moisture content of stored feed is too high, bacteria and molds may grow, causing excessive heat to develop. Ambient temperature, humidity, and air circulation in the stored feed all may affect the amount of moisture that can be safely tolerated. Additionally, safe moisture levels in stored forage are affected by the coarseness of the material and the tightness of the stack or bale. Heating of hay in storage may result in browning or heat damage to the protein content. Heat buildup may become high enough to cause spontaneous combustion in the stored feed. Excessive moisture can cause spoilage and nutrient loss in both forages and grains.

Palatability is reduced when fungi develop in the presence of excess moisture. In some cases, toxic substances are produced that can be harmful to livestock. Feeds that are too wet in storage will also tend to cake together, reducing their value.

Safe moisture levels for storage of baled or stacked air dry forages is around 18-22 percent. Whole grains can be safely stored at or below 13 percent, with the exception of whole shelled corn, which can be safely stored at moisture levels as high as 15.5 percent. Ground or rolled feeds should be at or below 11 percent moisture for safe storage.

Artificial drying may be used to bring feeds down to a safe storage level. With the increased cost of fossil fuels, there has developed an interest in the use of solar energy for drying of feeds. Experimental work has shown that this can be a cost-effective method of drying feeds for safe storage.

The use of propionic, acetic, formic, and isobutyric acids for treatment of grain to prevent the development of mold has proven to be effective. Grains treated with acid preservatives have a feeding value equal to or slightly better than those dried naturally before being placed in storage.

High-moisture grains can be successfully stored in airtight silos. For successful storage the moisture content should be around 22-30 percent at the time of ensiling. Feeding value is retained or slightly improved with this method of storage.

Forages containing 60-75 percent moisture can be stored as silage. At 40-60 percent moisture levels, forages can be stored as haylage in conventional upright silos, if the material is well-packed. Haylage can also be stored in airtight silos. The storing of haylage in trench or bunker silos is not recommended because of the difficulty of packing the material sufficiently to prevent the formation of mold.

Insect Problems and Feed Storage

Insect infestation is generally not a problem with forages but can be a serious storage problem in grains. Damage may be caused directly to the grain by the insect, or contamination may result from insect eggs or excrement. Moths, beetles, and weevils in both the adult and larval stages can damage stored grain. The longer grain is kept in storage, the more likely insect damage is to occur, especially at higher ambient temperatures.

Insect damage can be reduced by treating grains with a protectant at the time of harvest and taking care to minimize the amount of infestation in the initial stages of storage. Thoroughly cleaning grain storage bins prior to use will help prevent insect infestation. Regular inspection for signs of infestation should be made when grain is held in storage for several months, especially during warm weather. At the first sign of infestation, fumigation procedures should be initiated to prevent further infestation and damage. **CAUTION:** Grain fumigation can be a dangerous procedure and should be conducted only by someone with proper training and equipment, using appropriate safety procedures.

Rodent Problems and Feed Storage

The most common rodent damage to stored grain is from mice and rats. The greatest economic loss from rodent damage is from the contamination of the feed rather than from the amount actually eaten. It is estimated that contamination damage from rodent feces and urine is approximately ten times the loss which comes from their eating the grain.

Contamination of feed makes it less palatable to livestock and therefore reduces feed intake. Rodents also carry many livestock diseases that may be spread through grain contamination.

Rodent control is an ongoing process, which is necessary to reduce problems caused by infestation. Keeping feed spills cleaned up, cleaning bins prior to their use, and rodent-proofing bins are all recommended procedures to follow for effective rodent control. The use of rodenticides may also be necessary to keep mice and rats out of stored grain.

CAUTION: Rodenticides must be used with care. Always follow label directions when using these materials.

SUMMARY

It is important that animals get the right nutrients in the proper proportions so they may efficiently produce meat, milk, eggs, wool, work, etc. Balanced rations or diets are those that meet the nutrient needs of animals in the correct proportions.

Feed costs range from 50-80 percent of the total cost of raising livestock or producing livestock products. The use of homegrown feeds to the maximum extent possible helps to hold down the cost of feeding livestock. The major purchased feeds used for livestock are protein supplements, minerals, and vitamins. Energy needs are generally met by the use of grains and roughages grown on the farm. The cost per pound or kilogram of nutrient content is one method of comparing feeds to secure the lowest cost rations.

Diets are generally balanced for energy, protein, calcium, phosphorus, and vitamin A requirements. Tables of nutrient needs and feed composition are used to provide the data necessary to balance diets. After the nutrient requirements are determined for the class, age, weight, and function of animal to be fed, the composition of the feeds available for the diet is determined. The Pearson Square or algebraic equations may then be used to balance the diet to meet the needs of the animals.

Both mainframe and microcomputers may be used to help balance rations. The major advantage of using computers is the speed with which they perform mathematical operations, permitting trial runs on many different formulations to secure the best least-cost ration.

Many feeds are processed before use in animal rations. Grinding is a common processing method, although feeds may also be rolled, crimped, cracked, pelleted, or cubed. Processing may improve the palatability of the feed and make it easier to handle with mechanized equipment. Rates of gain and feed efficiency are sometimes improved by processing of feed.

Rations may be mixed in either batch mixers or continuous flow mixers. A wide range of equipment at various prices is available to grind and mix feeds. Initial investments and operating costs vary with the complexity of the system. Continuous-flow systems generally have lower operating costs than batch-mixing systems.

Feeds must often be stored for long periods of time after harvest before they are used. Moisture, insects, and rodents can present problems with feed storage. Moisture levels must be low enough to prevent feed spoilage or spontaneous combustion in storage. Good housekeeping practices and attention to control measures will help reduce insect and rodent problems in feed storage areas.

REVIEW

1. Define the terms *ration, balanced ration,* and *diet*.
2. Why is a balanced ration important in livestock feeding?
3. Why should homegrown feeds be used as much as possible in livestock diets?
4. Why is it necessary to thoroughly mix micronutrients and feed additives in the ration?

5. Why should feeds to be purchased be compared on the basis of cost per pound or kilogram of nutrient?

6. Compare several locally available protein feeds and energy feeds on the basis of cost of nutrient content.

7. Why should changes in diet be made gradually?

8. Describe the permitted variations in levels of dry matter in the ration for different classes of animals.

9. What are the five nutrients usually considered when rations are balanced for livestock?

10. List the steps followed when rations are balanced.

11. Describe how computers may be used to help balance rations for livestock.

12. Why are feeds often processed before being fed to livestock?

13. Name and describe several methods of mechanically processing feeds.

14. Discuss the use of complete pelleted rations for livestock.

15. Name the two general systems of feed mixing.

16. Briefly describe each system.

17. Describe good housekeeping methods that should be used in feed-mixing areas.

18. Discuss moisture problems that may be encountered when storing feeds and describe how these problems can be reduced.

19. How can insect infestation problems be reduced in feed storage areas?

20. How can rodent problems be reduced in feed storage areas?

RATION BALANCING PROBLEMS

1. Balance a ration for a growing-finishing hog weighing 50 kg (110 lbs). Balance the ration for lysine content using the Pearson Square method. Use feeds commonly available in your area.

2. Balance a ration for a 400 kg (882 lbs) growing-finishing steer with an expected daily gain of 1.2 kg (2.6 lbs) per day. Balance the ration for crude protein content using algebraic equations. Use feeds commonly available in your area.

3. Balance a ration for a 25 kg (55 lbs) growing-finishing hog. Balance the ration on lysine and ME requirements using simultaneous algebraic equations. Use feeds commonly available in your area.

4. In each problem above, calculate the amount of each feed needed to make 2,000 lbs of mix.

5. In each problem above, use two grains as energy sources in a ratio of 3:1.

6. In problem 3 above, assume that mineral, vitamin, and additive requirements are to be met by 3 percent of the ton mix. Recalculate the ration with two energy sources (ratio 4:1) and two protein sources (ratio 1:1). Determine the total amount of each feed needed to make 2,000 lbs of mix. Use any of the ration balancing methods described in this chapter.

10

ENVIRONMENT AND NUTRITION

OBJECTIVES

After completing this chapter you will be able to:

- Describe effects of temperature on nutritional requirements.
- Describe the effects of environment on feed intake.
- Discuss forage/temperature interaction on feed intake.
- Describe the effect of temperature on water intake.
- Describe the effect of the environment on feed efficiency.

The livestock nutrient requirements included in the tables in the Appendix have, for the most part, been established without considering the effects of environmental stress on the animal. When animals are subjected to extremes of temperature or other stress, their feed and nutritional requirements may need to be altered to achieve desired levels of production.

EFFECTIVE AMBIENT TEMPERATURE

The efficiency of energy use by farm animals is primarily affected by air temperature, with humidity, precipitation, wind, and heat radiation being secondary influences. The term *effective ambient temperature (EAT)* describes the combined effects of these factors. As the EAT varies, animals attempt to compensate by altering feed intake, metabolism, and heat dissipation. While there are limits to the ability of animals to make such compensation, their attempts do change the efficiency with which they use energy, which may make

changes in diet necessary as the EAT varies.

Temperature Changes

Farm animals maintain a fairly constant internal temperature by balancing heat generated from metabolism with heat gained from or lost to the environment. When body heat is lost too rapidly, the animal suffers from *hypo*thermia (temperature too low) and when heat is lost too slowly, *hyper*thermia (temperature too high) results. Either condition affects production, and if continued for too long a period of time, can lead to death.

Normally, an animal has a continuous loss of heat from the surface of the body by conduction, convection, and radiation and through evaporation from the respiratory tract and skin surface. The rate of this heat loss is influenced by the environment and by species variation in the resistance to heat flow of the skin, body tissues, and body covering (hair, wool, etc.). Under extremely hot conditions, animals may actually gain heat from the environment; however, they must then expend energy to lose heat by evaporation.

153

Farm animals receive heat radiation from two sources: (a) the sun, and (b) their surroundings. The net amount of heat radiation an animal receives is the difference between the amount received from these two sources and the amount lost by radiation from its own body. Available shade, ground cover, clouds, buildings in the area, other animals in the area, and the body characteristics of the animal all affect the net amount of heat gained or lost by radiation. When animals are in direct sunlight, there is usually a net gain in heat, which has the effect of raising the EAT by 5.4 to 9°F (3 to 5°C). This is an advantage in cold weather, but may present problems in keeping animals cool during hot weather.

During periods of high humidity, it is more difficult for animals to lose heat by evaporation. Animals such as cattle (which depend more upon sweating to lose excess heat) are more affected by high humidity than are those such as swine (which do not sweat but lose heat by panting).

Wind Chill Index

Heat loss from the body by convection and evaporation is affected by the movement of air surrounding the animal. The rate of change in heat transfer is greatest at lower air velocities. When the air movement is above 3.7 mph (6 km/hour) the rate of change of heat transfer is relatively small. *Wind chill index* is a measure of the combined effect of air temperature and speed of air movement. Cold air in motion has a greater adverse impact on animals than cold air that is still. The wind chill index is often given in weather reports.

Other Factors Affecting Heat Gain or Loss

Animals can lose or gain heat by conduction from surfaces they come into contact with. Normally, there is little heat exchange by this method. However, this can be a significant factor in the case of young pigs on concrete floors, which have a high thermal conductivity.

Precipitation (in the form of rain or wet snow) combined with low temperature and wind can cause animals to lose heat at a rapid rate. The insulation value of an animal's hair or wool coat is reduced when it is wet or becomes matted by rain or snow, and the animal loses heat more rapidly by conduction. When the hair or wool coat dries, the animal loses heat by evaporation.

Thermoneutral Zone

The term *thermoneutral zone* refers to that range of effective ambient temperatures in which an animal does not have to increase normal metabolic heat production to offset heat loss to the environment. It could be viewed as the range of temperatures in which an animal is most comfortable, uses feed with maximum efficiency, and does not feel any temperature stress. The thermoneutral zone varies with livestock species and is not necessarily the temperatures at which a human is most comfortable. It is possible for the thermoneutral zone to shift up or down as an animal becomes accustomed to warmer or colder temperatures. For example, as cattle become accustomed to the winter season, their thermoneutral zone may shift downward as much as 27°F (15°C).

Critical Temperatures

Below the thermoneutral zone, animals will show symptoms of cold stress, and above the thermoneutral zone, they will show symptoms of heat stress. The points at which these events begin to occur are called the *lower critical temperature* and the *upper critical temperature*.

As lower temperatures produce cold stress, animals will increase metabolic heat production. Over a period of continued cold stress, animals make gradual physiological adjustments such as adding insulative layers of fat or thickening the hair coat.

A number of factors affect the point at which animals will become temperature stressed. These include housing conditions, age, breed, stage of lactation, level of nutrition, time after feeding, length of time exposed to the temperature change, number of animals in the group, and amount of hair or thickness of wool coat.

Large ruminants at high feeding and production levels produce a lot of metabolic heat, have small surface areas relative to total body mass, and have a large amount of insulative tissue. For these reasons, they have significantly lower critical temperatures than do smaller animals such as swine, poultry, or young animals. Predicted measures of lower critical temperature are more useful in determining nutritive requirements and housing needs of swine, poultry, and young animals than they are for large ruminants, because the lower critical temperature for the latter are lower than normal winter temperatures where they are usually raised. Large ruminants generally adjust to lower temperatures by gradually becoming accustomed to them and through metabolic and digestive adjustments.

Adjustments animals can make as they pass the upper critical temperature are limited. The first reaction is generally a lowering of feed intake as the animal attempts to reduce the rate of metabolic heat production. High producing animals such as large ruminants have greater difficulty in adjusting to high temperatures than they do to low temperatures because of their higher metabolic rate. They become heat stressed more easily than they become cold stressed. The primary method of heat loss in hot environments is through evaporation from the surface of the skin or from the respiratory tract. In extreme heat, animals may actually gain heat from the environment.

FEED INTAKE

The amount of feed an animal will eat is influenced by its environment. Generally, feed intake is increased as temperatures go lower and decreased when temperatures rise. Summer temperatures also affect the quality of the forage available and this, in turn, affects the intake of these forages.

EFFECT OF TEMPERATURE ON FORAGE INTAKE

Forages mature rapidly when the temperature is high and have a higher cell-wall content. This causes a decrease in the digestibility of the forage, which results in decreased intake by ruminants grazing on pastures.

The cell-wall content and, thus, the digestibility of a forage are affected by both the temperature and the amount of light received. In the spring, when temperatures are lower and the light intensity is high, the cell-wall content is lower and the digestibility is higher. Forages harvested at second and third cuttings have a higher cell-wall content and also a higher stem-to-leaf ratio, resulting in lower digestibility. The highest quality forage comes from the first cutting, with second and third cuttings being of lower quality. This is true for both legumes and grasses.

Lower quality forages result in lower feed intake, digestibility, and efficiency of metabolizable energy utilization. This makes it more difficult to estimate the direct effect of temperature on feed intake of grazing ruminants.

WATER AND THE ENVIRONMENT

Animals get water by (a) drinking it; (b) from the feed; and (c) metabolic processes in the body. The amount of water in the feed varies from 5 percent in dry grains to approximately 90 percent in young, fast-growing grasses. Swine and poultry are normally fed diets consisting mainly of dry feeds, and therefore, get a smaller part of their water from the feed.

Animals lose water through (a) urine; (b) feces; (c) evaporation from the body surface and respiratory tract; and (d) milk. Ruminants lose about as much water through the feces as they do through the urine. In all animals, the amount of water lost through the respiratory tract is dependent on relative humidity and rate of respiration. When the relative humidity is high, the rate of water loss is lower than when relative humidity is low. Animals breathe faster during heat stress, which causes a faster rate of loss through the respiratory system.

Water losses from sweating vary greatly among animal species. The relative importance of sweating as a source of water loss, in descending order, is horses, cattle, goats, sheep, and swine. Both swine and poultry lose more water from the respiratory tract than they do from sweating. As the temperature rises the importance of respiratory tract losses of water in poultry become increasingly larger.

Sheep and swine normally consume about 40 percent less water per unit of dry matter consumed than do cattle. Young calves need more water than older cattle. Lactation and pregnancy increase water requirements over the requirements for nonlactating and nonpregnant animals. Heavier swine use less water than young pigs.

The availability of water also influences the amount animals will consume, Figure 10-1.

When the distance to water increases, the amount consumed by animals decreases.

The physical form of the feed in the diet has an effect on the intake of water. For example, dairy heifers fed diets containing higher levels of silage have a higher water intake than those on a diet of only hay.

When air temperature is cold, cattle drink more water if it is heated. Heating water for sheep when the air temperature is cold does not appear to increase water intake. The rate of digestion in ruminants is not affected by the temperature of the water.

When the air temperature rises above 80°F (27°C), lactating cows significantly increase the amount of water they consume. In all cases, rising air temperatures cause cattle to increase their water consumption. As the relative humidity increases, the water intake of cattle decreases.

Sheep follow the same general pattern as cattle in the intake of water under various temperature and humidity conditions. Sheep consume about 50 percent less water when the air temperature is 10°F (−12°C) compared

Figure 10-1. Animals will consume more water when it is readily available.

to an air temperature of 59°F (15°C). Experimental evidence indicates that Merino sheep are better able to withstand a lack of water in the diet than are European breeds. The type of body covering may also influence the need for water in the diet of sheep. Breeds with hair covering appear to have the least need for water, followed by coarse-wool sheep and then fine-wool breeds.

Swine reduce water intake as the air temperature rises to about 77°F (25°C), with water consumption increasing when air temperatures go above this level, Figure 10-2. The amount of water that swine will consume below 50°F (10°C) is not currently known.

As air temperatures rise, poultry require increasing amounts of water. Laying hens consume more water on days when an egg is laid. A major use of water by laying hens is the formation of the egg, which is about 66 percent water. Poultry cannot survive for any significant length of time without water. It is important, during times of heat stress, to provide plenty of free water in containers that allow the dunking of the head during drinking.

This seems to be a method by which poultry cool themselves.

EFFECT OF ENVIRONMENT ON NUTRITIONAL EFFICIENCY

Temperature extremes above or below optimum levels reduce feed efficiency relative to growth and fattening. Rates of feed intake and maintenance energy requirements are affected by the environment. The reduction in feed efficiency results in an economic loss to the livestock producer. Hence, the benefit of controlling environmental conditions must be weighed against this economic loss, when making management decisions concerning the expenditure of capital assets.

The efficiency of producing milk and eggs increases during periods of higher air temperatures. Animals utilize body tissue to produce these products during periods of reduced feed intake caused by the higher air temperatures. The calculation of this apparent improved efficiency does not take into

Figure 10-2. During hot weather swine will consume more water when it is available.

account the cost of the original deposition of body tissue prior to the production of milk or eggs. Another reason for the higher efficiency of egg production during times of heat stress is the increased use of body fat as an energy source as compared to utilization of energy from the diet.

ADJUSTING NUTRITION IN RELATION TO ENVIRONMENT

Dairy Calves

Dairy calves have a wide tolerance to variations in temperature. High humidity generally causes more problems with calves than does temperature. Feeding dairy heifers to maintain good health is more important than feeding for maximum growth rate. When feeding dairy bulls for maximum growth rate, suitable guidelines for adjusting diet to environment are the same as those that apply to beef cattle.

Dairy Heifers

Some experimental studies have indicated that the temperature does have an influence on the feed intake and growth rate of dairy heifers. However, they do make compensatory growth after temperatures return to a more optimum range. Temperature effects on dairy heifers are of less economic importance than they are for lactating cows or feedlot cattle.

Dairy heifers from six months of age to first calving should be fed sufficiently to maintain a normal heat balance regardless of the season. Dairy heifers exposed to heat stress have a lower feed intake, which results in smaller fetuses and a slower rate of growth. If the last three months of pregnancy come during midwinter, the energy level of the diet needs to be increased by approximately 30 percent to provide for normal fetal development and the development of a reserve of body fat.

Dairy Cows

Using a baseline of 50-68°F (10-20°C), lactating dairy cows being fed a free choice diet of 60-65 percent high quality roughage and 35-40 percent concentrates will increase feed intake by approximately 35 percent when the temperature drops to −4°F (−20°C). During continuous hot weather, lactating dairy cows will decrease feed intake at 77-81°F (25-27°C) with a greater decrease as the temperature goes above 86°F (30°C). The amount of milk being produced and the breed affect the rate of decrease or increase in feed intake at extreme temperatures.

When summer temperatures are above 77°F (25°C), lactating cows on pasture will decrease dry-matter intake at a fairly rapid rate, Figure 10-3. This is attributable in part to reduced activity because of heat stress and, in part, to the lower quality of the forage during hot weather. The lower forage quality may have greater effect than heat stress on the amount of dry matter taken in by the cow. Feeding more concentrate to grazing cattle during heat stress will help maintain the energy intake.

The optimum temperature range for production of milk by dairy cattle is 55-64°F (13-18°C). During the first two months of lactation, the influence of temperature on production is greater than during the latter stages of lactation.

Lactating dairy cows show an increase in dry-matter intake, a reduction in water intake, and a decrease in milk yield when the temperature drops below a range of 64-68°F (18-20°C). When temperatures rise above this level, dry-matter intake decreases, water intake increases, and milk yield decreases. The environmental effect on ME (metabolizable energy) intake may be reduced by providing shelter for lactating cows during cold weather (below 32°F (0°C)) or by increasing the proportion of concentrate in the diet (20 percent roughage/80 percent concentrate). A combination of these two methods may also be

Figure 10-3. During hot weather lactating dairy cows reduce their dry matter intake at a fairly rapid rate.

used. Higher concentrate levels in the diet during cold weather help to maintain milk yield without excessive loss of body weight.

Heat stress reduces appetite in lactating dairy cows. At 95°F (35°C), milk yield may be reduced by as much as 33 percent, with more than 50 percent reduction in yield when temperatures rise above 104°F (40°C). Raising the level of concentrate in the ration helps to some degree to alleviate this problem; however, feed intake will still drop at the higher temperatures. Feed efficiency also drops when the temperature is high, especially over an extended period of time (40-87 days).

Utilizing confined housing for lactating dairy cows will help reduce the impact of environmental extremes on production. Care must be taken that other problems, such as high humidity or foot problems from wet floors, do not develop, as these can offset gains made by protecting the cows from temperature extremes.

Feedlot Cattle

Feedlot cattle react much as lactating dairy cattle do to temperature extremes. Crossbred cattle appear to be less affected by hot weather than are purebreds. When the temperature is below 14°F (−10°C) or above 77°F (25°C), the type of ration fed and the temperature do have a significant effect on feed intake. At temperatures between 32-77°F (0-25°C), the digestibility of the ration has a more important effect than the temperature, on feed intake. Other important factors affecting feed

intake of feedlot cattle are the type of feedlot surface and the amount of space allowed per animal.

Beef Cattle

Recommendations for feeding beef cattle are generally based on an assumed environment with little temperature stress. Adjustments are based on a temperature of 68°F (20°C), the midpoint of an optimum temperature range for beef cattle.

When temperatures are above 95°F (35°C), feed intake for beef cattle on full feed may be reduced as much as 10-35 percent. High humidity, little shade, and little night cooling tend to make the reduction in feed intake closer to the high end of the range. If shade or other methods of cooling are provided, if night temperatures are more moderate, and if low fiber diets are fed, the reduction in feed intake is closer to the low end of the range.

In a temperature range of 77-95°F (25-35°C), feed intake is depressed 3-10 percent. The temperature range of 59-77°F (15-25°C) is considered to be optimum for feeding beef cattle. When temperatures are in a range of 41-59°F (5-15°C), feed intake is increased by 2-5 percent. A temperature range of 23-41°F (−5-5°C) results in feed intake being increased by 3-8 percent. Feed intake is increased by 5-10 percent in a temperature range of 5-25°F (−15-−5°C). Below 5°F (−15°C), feed intake is increased by 8-25 percent. When temperatures go below −13°F

($-25°$C) or during blizzards, there may be a temporary reduction in feed intake, especially of roughage feeds.

Other kinds of stress, such as rain, mud, or illness, will also reduce feed intake. Rain reduces feed intake by 10-30 percent and mud by 5-30 percent, depending upon its depth and the amount of bedded area available. Deeper mud and less bedded area will result in reductions nearer the upper end of the range. Illness normally results in severe reduction in feed intake.

Cattle become acclimated to variations in temperature. Adjustments to energy requirements listed in the Appendix tables for beef cattle can be made for extended exposure to temperatures above or below $68°$F ($20°$C). For each $1.8°$F ($1°$C) that cattle have been exposed to for an extended period of time, the NE_m or maintenance energy requirements for ME or TDN may be adjusted by 0.91 percent. The adjustment is a decrease for temperatures above $68°$F ($20°$C) and an increase for temperatures below this level.

For short periods of heat stress, the maintenance energy requirement may increase from 7 to 25 percent. However, appetite is usually reduced during heat stress, which results in lower productivity. During hot weather, feeding less roughage to beef cattle is advantageous. Providing shade or methods of cooling for cattle during heat stress may be a more practical approach than attempting to adjust the diet.

Beef cattle, especially when they have had the opportunity to become acclimatized to the environment, are not greatly affected by colder temperatures, Figure 10-4. However, survival of beef cattle during cold temperatures may depend upon the degree of acclimatization they have undergone. Those with little opportunity to become acclimatized may not survive conditions that acclimatized cattle will survive. Moisture and wind combined with low temperatures will have a greater detrimental effect on young stock than on older cattle. Providing protection from the wind

and cold, along with dry bedding, may be more practical in preventing losses during cold weather than making adjustments in the diet.

During cold weather, it is advantageous to increase the amount of roughage in diets of cattle on restricted feed intake. For cattle on full feed, increasing the amount of roughage in the diet during cold weather may decrease the amount of energy available and thus reduce productivity.

Sheep

There is little experimental data available that describes the interaction of temperature and feed intake for sheep. It is known that the length of the fleece and the level of feeding will affect feed intake as temperatures change. Unshorn sheep generally respond to high temperatures in the same manner as do lactating cows. Shorn sheep tend to respond in a manner similar to dry cows. It is known that heat-stressed sheep require more metabolizable energy per unit of gain than do sheep under more optimum temperature conditions.

Sheep can tolerate colder climatic extremes than other farm animals. The age of the animal, length of fleece, and whether it is wet or dry influence its ability to survive in extremes of cold. Wind combined with wet fleece raises the lower critical temperature of sheep.

High temperatures combined with high humidity lowers feed intake of sheep, especially when the diet is high in roughage. Feed intake increases as the temperature drops to $23°$F ($-5°$C), but does not increase below that temperature. Shorn lambs do not increase feed intake over a temperature range of 32-$86°$F ($0-30°$C), compared to feed intake at $59°$F ($15°$C).

Sheep need a higher energy intake during cold stress. This higher energy requirement can be met most economically by increasing the amount of roughage in the ration.

During hot weather, sheep diets should contain less roughage, which lowers the

Figure 10-4. When beef cattle have had an opportunity to become acclimated to the climate they are not greatly affected by cold weather.

amount of heat produced by the feed. Adding dried beet pulp and increasing the amount of fat in the diet maintains net energy at a constant level when the proportion of roughage is decreased. During warm weather, lambs gain faster when the diet contains 60 percent concentrate as compared to the rate of gain when the diet contains 40 percent concentrate.

Swine

Swine also increase their feed intake as the temperature declines and decrease it as the temperature rises. Heavier hogs (above 154 lb (70 kg)) are affected by heat stress quicker than are lighter hogs. There is some evidence to indicate that pigs under eight weeks of age may actually have a higher feed intake at 77°F (25°C) than they do at 68°F (20°C). Young pigs grow slower under extremes of either cold or hot weather.

A temperature range of 64-70°F (18-21°C) is considered to be optimum for growing-finishing swine. Growing-finishing swine need additional feed when the temperature drops below 61°F (16.5°C), to maintain approximately the same rate of gain. Leaner hogs are more sensitive to reductions in temperature than are fat hogs. For each 1.8°F (1°C)

of temperature drop, the feed requirement increases 1-1.4 oz (30-40 grams) per day for swine in the weight range of 44-220 pounds (20-100 kg). The crude protein requirement does not appear to vary with environmental temperature. Because feed intake increases, the percent of crude protein in the diet may be reduced when pigs are cold stressed, as long as the actual amount of crude protein intake remains the same.

Heat stress reduces feed intake and the rate of gain in swine. However, humidity has a greater effect on feed intake during hot weather as compared to cold weather. The weight of the pig influences the effect temperature has on feed intake and rate of gain, with heavier hogs being more sensitive to hot weather than hogs of lighter weights.

Fat has a lower heat increment than carbohydrate or protein and, for this reason, has an advantage as a part of the diet of hogs during heat stress. Fat added to the diet also helps to maintain caloric intake during hot weather because of its higher caloric density, thus offsetting some of the effects of lowered feed intake.

The percent of protein in hog rations may need to be increased during heat stress because of the lowered feed intake. The amount of protein intake needs to be kept at a constant level, in order to reduce the decrease in the rate of gain. It is important to check the calorie:lysine ratio of the diet during hot weather to ensure an adequate amino-acid intake during these times of reduced feed intake.

Poultry

Laying hens have the ability to become acclimated to a fairly wide range of temperatures. When temperature changes occur, feed intake will drop temporarily and then return to approximately the level observed prior to the temperature change. When temperatures drop below 41°F (5°C), feed intake must be increased to meet a higher maintenance requirement. During times of heat stress, the level of protein in the diet affects feed intake. When temperatures are high, the crude protein level of the ration should be approximately 25 percent or higher. When temperatures drop below 77°F (25°C), the crude protein level of the ration may be less than 25 percent.

Poultry become acclimated to changes in the environment, which results in less impact on feed intake and production. Confinement systems of poultry production generally result in less environmental impact, because the environment is more closely controlled. A confinement system is generally designed to maximize production with the greatest efficiency that is economically feasible.

Temperature appears to have little effect on the efficiency of feed use by poultry. Feed intake increases when the temperature is lower than 68-70°F (20-21°C) and decreases with higher temperatures. The rate of change is about 1.5 percent for each 1.8°F (1°C) change in temperature.

During periods of heat stress, it may be beneficial to feed diets with less heat increment. This is accomplished by adding fat to the diet. Experimental evidence is not consistent regarding the success of this type of dietary adjustment with poultry.

Increases in nutrients in the diet of laying hens has helped to reduce the decrease in egg production; however, it has not reduced the loss of shell quality. When calcium intake levels are kept constant, higher temperatures still result in a loss of shell quality. After a period of acclimatization to the higher temperatures, laying hens have returned to near normal levels of egg production and shell quality, without an increase in nutrients in the diet.

The availability of adequate drinking water is critical to maintaining growth or production in poultry. Compared to water intake at 70°F (21°C), water intake is doubled at 90°F (32°C) and is 2.5 times greater at 98°F (37°C).

SUMMARY

Nutrient requirement tables for livestock are generally based on the assumption that no environmental stress is present. Feed and nutritional requirements of farm animals may need to be adjusted to reduce the impact of environmental stress on production.

Efficiency of energy use by animals is affected by air temperature, air movement, humidity, precipitation, and heat radiation. The term *effective ambient temperature* (EAT) is used to describe the combined effect of these factors.

The thermoneutral zone is that range of effective ambient temperatures in which animals are most comfortable, use feed with maximum efficiency, and do not feel any temperature stress. Some animals, such as cattle and poultry, become acclimated to changes in temperature, which results in a change in their thermoneutral zone. Temperatures above or below the thermoneutral zone result in changes in feed intake and efficiency of feed use.

Most animals increase feed intake when temperatures go below their thermoneutral zone and decrease feed intake when temperatures are above their thermoneutral zone. During cold weather animals generally decrease their intake of water, and they increase their intake during hot weather. Heating water during cold weather may increase water intake for some species.

Feed efficiency generally decreases when temperatures are below or above the thermoneutral zone for animals. Adjustments in energy levels in the diet may help overcome some of the impact on production caused by changes in feed intake and feed efficiency.

REVIEW

1. Define the term *effective ambient temperature*.
2. Define the terms *hypothermia* and *hyperthermia* and describe how they affect livestock.
3. Describe how animals maintain body heat balance.
4. Define the term *wind chill index* and describe its significance for livestock producers.
5. Define the term *thermoneutral zone*.
6. Define the terms *upper critical temperature* and *lower critical temperature* and discuss their significance for livestock producers.
7. Why do large ruminants have lower critical temperatures than other farm animals?
8. How do animals generally react when they pass the upper critical temperature?
9. Discuss the effects of temperature on forage quality and intake.
10. What are the three major sources of water for livestock?
11. What are the three major ways livestock lose water?
12. What effect does humidity have on water losses by livestock?
13. Rank the common farm animals according to their relative utilization of sweating as a method of losing water and cooling the body.
14. Discuss the factors that influence the intake of water by farm animals.
15. What effect does temperature have on feed efficiency?
16. Why does the efficiency of egg production increase during periods of high temperature?
17. What adjustments in diet may be beneficial for lactating dairy cows when temperatures are above or below the thermoneutral zone?
18. What adjustments in diet may be beneficial for beef cattle when temperatures are above or below the thermoneutral zone?
19. What adjustments in diet may be beneficial for sheep when temperatures are above or below the thermoneutral zone?

20. What adjustments in diet may be beneficial for swine when temperatures are above or below the thermoneutral zone?

21. What adjustments in diet may be beneficial for poultry when temperatures are above or below the thermoneutral zone?

11

FEEDING BEEF CATTLE

OBJECTIVES

After completing this chapter you will be able to:

- Describe nutrient requirements for beef cattle.
- Creep-feed calves.
- Feed replacement heifers.
- Feed bulls and weaned calves.
- Select appropriate feeds for feeder cattle.
- Start cattle on feed.
- Develop balanced rations for beef cattle.
- Calculate the total feed needed for feeder cattle.

GENERAL GUIDELINES FOR BALANCING BEEF RATIONS

Dry Pregnant Beef Cows— Maintenance

The major part of a maintenance ration for dry beef cows is roughage. In general, feed an amount of air-dry roughage equal to 1¾ percent of the animal's body weight. This may be increased to 2½ percent for thin cows and decreased to 1½ percent for cows in good flesh. When poor quality roughage is used, some additional supplement may be required to meet the nutritive needs of the cows. An allowance of 1½ to 2 pounds (0.68-0.90 kg) of a 35-40 percent supplement per day is adequate to meet the needs of the dry cow, even if the roughage is of poor quality. Sample rations for wintering dry, pregnant beef cows are given in Table 11-1.

When haylage or silage is used as the roughage in the ration, an adjustment in the amount fed must be made to account for the higher moisture content. Three pounds of silage at 70 percent moisture content or 1.5 pounds of haylage at 40 percent moisture content are roughly equal to 1 pound of air-dry roughage.

Additional vitamin A should be included in the ration if poor quality roughage is fed. This may be provided at the rate of 20-30 thousand IU of vitamin A per head each day. Provide a mineral supplement on a free-choice basis. The use of a mix of 1 part trace-mineralized salt to 1 part dicalcium phosphate is recommended. Defluorinated phosphate or steamed bone meal may be substituted for the dicalcium phosphate in the mineral mix.

Overfeeding beef cows results in cows becoming too fat, increased maintenance

Table 11-1. Rations for wintering dry, pregnant, beef cows (wt. 1,000 to 1,100 lb; 453 to 499 kg).

Ration	Amount per day	
	(lb)	**(kg)**
Legume hay	16-25	7.26-11.3
Mixed legume-grass hay (1/3 legume)	18-22	8.2-9.98
Legume hay	5-10	2.3-4.5
Straw or low quality grass hay	10-15	4.5-6.8
Legume-grass haylage	30	13.6
Corn or grain sorghum silage	35-50	15.9-22.7
Protein supplement (40% total protein)	0.5-1	0.23-0.45
Legume-grass hay	10	4.5
Straw or cobs	10	4.5
Corn or sorghum silage	30	13.6
Legume hay	5	2.3
Straw, low-quality grass hay, cottonseed hulls, ground corncobs, or other low-quality roughage	Unlimited	Unlimited
Prairie or grass hay	Unlimited	Unlimited
Protein supplement	0.5-1	0.23-0.45
Grass silage	30-40	13.6-18
Straw or low quality grass hay	Unlimited	Unlimited
Grazing crop residue	Unlimited	Unlimited
Cornstalk silage	40	18
Legume hay or hay silage	4-5	1.8-2.3
Corn silage	30	13.6
Mixed hay	4	1.8
Grass silage	25-35	11.3-15.9
Grass or mixed hay	10	4.5
Mixed or grass hay	10	4.5
Pea-vine silage	25-35	11.3-15.9
Husklage	11	4.99
Alfalfa-brome hay	7	3.2
Corn stover	12	5.4
Alfalfa-brome hay	7	3.2
Corn stover	15	6.8
Shelled corn	4	1.8
Protein supplement (35% total protein)	1.1	0.5
Husklage	15	6.8
Shelled corn	3	1.4
Protein supplement (35% total protein)	1.1	0.5
Corn silage	43	19.5
Soybean stover	20	9
Alfalfa-brome hay	7	3.2

costs, increased difficulty calving, higher calf losses, and decreased milk production. Underfeeding results in cows becoming too thin and also lowers the percentage of surviving calves. Cows of normal weight in the fall should not lose more than 10 percent of their body weight (including weight lost at calving) from fall through spring calving.

Young cows and heifers are still growing and need more feed than mature cows. The size and condition of the heifers or cows affects the amount of feed they need. Feeding about 20 pounds (9 kg) of legume or mixed hay per head per day will meet the needs of the average mature beef cow. Young cows and heifers may need a little more. Sample wintering rations for bred heifers are given in Table 11-2. In all cases, rations are formulated to meet the nutrient requirements for the particular class of animal being fed.

Beef Cows Nursing Calves

Protein requirements for lactating beef cows are 160-268 percent greater than for dry cows. Energy needs are 36-68 percent higher. Calcium and phosphorus needs are 100-250 percent higher, while vitamin A needs are 18-88 percent greater.

High quality pastures can usually meet the needs of the lactating beef cow. Provide salt and minerals on a free-choice basis. Grain needs to be added to the ration only if the roughage is of poor quality. Protein supplement is needed if there is a limited amount of legume roughage in the ration.

Cows nursing calves in drylot require about 50 percent more roughage than dry cows. The protein requirement is roughly twice that for dry cows, while the energy requirement is approximately 50 percent more. Several sample rations for lactating

Table 11-2. Wintering rations for bred heifers (wt. 800-900 lb; 363-408 kg).

Ration	Amount per day	
	(lb)	(kg)
Legume-grass hay	20	9
Corn silage	50	22.7
Soybean meal	1	0.45
Corn silage	25	11
Legume-grass silage	10	4.5
Corn silage	45	20.4
Protein supplement (48% total protein)	1.5	0.68
Legume-grass haylage	35	15.9
Legume-grass hay	20	9
Corn (shelled)	5.6	2.5
Alfalfa hay	22.8	10.3
Corn (shelled)	3.4	1.5
(Feed vitamin-mineral mix free choice)		

cows in drylot are given in Table 11-3. Table 11-4 gives some sample lactation rations for first-calf heifers being fed in drylot.

Cows that calve in the fall and nurse during the winter need additional amounts of concentrate in the ration unless the quality of the roughage is unusually high. The addition of 3-5 pounds (1.4-2.3 kg) of a 16 percent protein supplement to a roughage ration will generally supply the necessary nutrients.

The nutritive requirements of cows nursing calves may be met by feeding the same amount of roughage as is fed to dry cows and increasing the concentrate in the ration to 4-6 pounds (1.8-2.7 kg).

Mineral requirements are higher for cows nursing calves. The suggested mineral mix offered on a free-choice basis may be adjusted to meet these needs by doubling the dicalcium phosphate (or steamed bone meal or defluorinated phosphate) in the mineral mix. The vitamin A requirements of nursing cows may be met by providing 30-40 thousand IU of vitamin A per head daily.

Feeding Protein Supplements

When high quality roughages are fed, additional protein supplement is usually not needed in beef cow rations. However, poor quality roughages (such as dry, mature range grasses) require the feeding of additional amounts of protein supplement. Cows on western range may need as much as one-third of the protein in the ration supplied by additional supplement, Figure 11-1.

A convenient way to feed protein supplement to beef cows is by using protein blocks. The amount the cow eats is controlled by the

Table 11-3. Lactating rations for cows in drylot (wt. 1,000 lb; 499 kg).

Ration	Amount per day	
	(lb)	(kg)
Legume-grass hay	10	4.5
Corn silage	40	18
Vitamin A	40,000 IU	
Legume-grass hay	30	13.6
Legume-grass hay	20	9
Corn	4	1.8
Legume-grass silage	74	33.6
Legume-grass haylage	50	22.7
Corn or grain sorghum silage	60	27.2
Protein supplement (48% total protein)	1.5	0.68
Mixed hay	20	9
Shelled corn	5	2.3
Corn stover	20	9
Shelled corn	5	2.3
Protein supplement (35-40% total protein)	2.8	1.3
Alfalfa haylage (55% dry matter)	29.8	13.5
Shelled corn	8	3.6
Corn silage	50-55	22.7—24.9
Legume hay	4-5	1.8-2.3

Figure 11-1. Protein supplement may be needed in the diet, especially when poor quality roughages are fed.

salt content and the hardness of the block. Liquid protein supplements may be fed in lick tanks. Cubed or pelleted protein supplements are convenient for hand feeding. Provide enough bunk space so "boss" cows do not get more than their share.

Beef cows that are hungry or thirsty will eat too much protein. Feeding plenty of roughage in the diet helps to control the amount of protein intake. It is important to have plenty of water available to flush out excess protein if the cow overeats, Figure 11-2.

Limit the intake of urea in beef cow rations. Urea can be toxic; therefore, self-feeding is not recommended. When corn silage is fed, urea can supply all the protein supplement needed in the ration. If poor quality roughage that is low in protein content is used, then some grain or molasses needs to be fed, along with the urea, to provide a source of easily digested carbohydrate, which will help in the utilization of the urea by the rumen microbes. Urea needs to be limited to not more than 50 percent of the protein value in the supplement in a ration containing low quality roughage.

There is no experimental evidence to indicate that the use of urea in the rations of breeding animals has any effect on reproductive efficiency, as long as the protein needs of the animals are being met. If urea is used with a low energy ration and the protein needs of the animal are not being met, then some negative impact on reproductive efficiency would be expected. In experiments at Oklahoma State University, massive amounts of urea fed to cows that were 3-4½ months pregnant did not produce abortion. After recovery, cows had normal pregnancies and rebred as well as control animals.

Table 11-4. First-calf heifers—lactation rations, drylot.

Ration	Amount per day	
	(lb)	(kg)
Legume-grass hay	25	11.3
Ground shelled corn	3	1.4
Corn silage	60	27.2
Soybean meal	1.5	0.68
Corn silage	30	13.6
Legume-grass hay	13	5.9
Legume-grass silage	65	29.5
Ground shelled corn	3	1.4

Courtesy of the University of Illinois

Figure 11-2. Plenty of fresh, clean water is essential when feeding cattle.

PASTURE FOR THE BEEF BREEDING HERD

Most of the nutrients needed by the beef breeding herd may be supplied by good quality pasture, Figure 11-3. Legume crops contain higher levels of protein and minerals than do grass roughages. Provide additional mineral supplement on a free-choice basis. The same mineral mix as suggested for dry cows in drylot may be used. Additional vitamin A should be supplied in the ration if grass pastures are mature. Beef cows will bloat on some fresh legume forages; however, feeding hay to cattle on fresh legume pasture will help prevent bloat, Figure 11-4.

During periods of drought or when pastures are overgrazed, it may be necessary to feed additional amounts of roughage or other feed to maintain the cows in good condition. Hay or silage may be fed alone or along with grain and protein supplement. Fifteen to 30 pounds (6.8-13.6 kg) of corn silage will provide enough nutrients to substitute for one- to two-thirds of the normal intake of nutrients from pasture.

Temporary pasture crops, such as oats, rye, ryegrass, or mixtures of these with crimson clover, will supply nutrients needed by the beef cow herd during the winter months in southern areas of the United States. Sudan grass is a popular temporary pasture in the western states. This kind of temporary pasture is more expensive than regular pastures used for beef cattle. Beef cows tend to get too fat when allowed unlimited access to these kinds of temporary pastures. Limiting the grazing time to 2-4 hours per day and feeding supple-

Figure 11-3. Good quality pasture can supply most of the nutrients needed by the beef breeding herd.

Figure 11-4. Feeding hay to cattle on legume pasture will help prevent bloat.

mental roughage for the rest of the ration will help to prevent too much weight gain.

CREEP-FEEDING CALVES

Creep-feeding is a method of providing extra feed for calves in a feeder in an area that cannot be reached by the cows. Grain, commercial feeds, and/or roughage may be fed in this manner to calves.

Advantages of creep-feeding

- produces heavier calves at weaning.
- produces higher grade and more finish at weaning.
- calves go on feedlot rations better at weaning time.
- reduces feedlot stress.
- cows and calves can stay on poorer quality pasture for a longer time.
- cows lose less weight.

Creep-feeding is often used if

- calves are to be sold at weaning.
- calves are to be fed out on high-energy rations.
- cows are milking poorly.
- calves are from first-calf heifers.
- calves were born late in the season.
- calves have above-average inherited growth potential.
- calves were born in the fall.
- calves are kept in drylot.
- calves are to be weaned early (45-90 days).

Disadvantages of creep-feeding

- requires more feed.
- results in higher feed cost.
- weight advantage of creep-feeding is lost if calves are well fed after weaning.
- when production testing, it is harder to detect differences in inherited gaining ability.
- replacement heifers may become too fat.

Creep-feeding is generally not used if

- calves are to be fed through the winter on roughages.
- cows are above-average milk producers.
- the calf-feed price ratio is poor.
- calves are on good pasture.

Creep-feeding Guidelines

When calves are about four weeks of age, they will start to eat grain. The amount they eat is relatively small until they reach six to eight weeks of age.

Each pound (0.45 kg) of additional gain requires approximately 6-9 pounds (2.7-4.0 kg) of feed. A total of 280-480 pounds (127-218 kg) of feed are required for 40-60 pounds (18-27 kg) of additional gain.

The energy needs of calves can often be met with only grains in the diet. Protein, mineral, and vitamin needs are met by the milk and pasture in the diet. A good simple creep feed consists of whole oats and cracked corn mixed 50-50. The palatability of the ration is improved and the calf will eat more, if molasses is added to the mix. Calves prefer rolled shelled corn and linseed meal pellets over whole oats or whole shelled corn. Table 11-5 gives some suggested creep-feed rations for self-feeding calves.

Pelleted commercial creep feeds are available for creep-feeding calves. These have a high palatability, are easy to handle, and do not blow away in the feeder. Calves eat more and gain more rapidly on these feeds. Commercial mixes generally contain minerals, vitamins, and often antibiotics. If antibiotics are included in the creep feed, they should provide 40-60 mg per head per day.

Locate the creep-feeder near the area where the cows loaf, in the shade, if possible. Locate waterers, salt, and mineral feeders for the cows to attract them to the area near the creep-feeder. This will bring the calves into the area, so they are more likely to use the creep-feeder.

Provide 3-4 inches (7.6-10.0 cm) of feeder space for each calf. Use self-feeders or bunks in the creep and include a hay self-feeder if energy intake is to be limited. The feeders or bunks are covered to protect the feed from the weather.

The opening into the creep-feeding area should be about 16 inches (40.6 cm) wide and 36 inches (91.4 cm) high. This will allow the calves access, but keep the cows out. An adjustable opening is best.

GROWTH IMPLANTS

Implants are growth stimulants, which are implanted in the calf at birth to 60-90 days of age. The implant contains a hormone that is released slowly in the calf's body over a period of time. Growth implants will increase the rate of gain approximately 8-12 percent. Both heifers and steers will grow faster if implants are used. It is recommended that heifers

Table 11-5. Mixtures for creep feeding calves—self-fed (100 lb. (45 kg) mix).

Ration	Amount per day	
	(lb)	(kg)
Shelled corn	100	45
Shelled corn	50	22.7
Oats	50	22.7
Ground ear corn	90	40.8
Soybean meal	10	4.5
Corn or barley (rolled, cracked, or coarsely ground)	50	22.7
Whole oats	30	13.6
Protein supplement (6.7 to 26% total protein)	10	4.5
Molasses (dried or liquid)	10	4.5
Shelled corn	90	40.8
Soybean meal	10	4.5

being kept for breeding purposes not be implanted with growth hormones. However, research has shown that if the implant is used at the correct time (3 months of age) and at the recommended level, there are no harmful effects on breeding performance.

CAUTION: Regulations on the use of implants change from time to time. Current regulations must be followed concerning the kind and dosage level of implants approved for use with any given class of livestock. Directions for the amount to use and age at which to implant must be carefully followed.

BACKGROUNDING CALVES

Backgrounding is the growing and feeding of calves from weaning until they are ready to enter the feedlot. The feeding period is generally from 120-150 days with an expected daily gain of 1.5 to 2.0 pounds (0.68-0.90 kg). Heifer calves generally gain about 10 percent slower than steer calves fed the same ration. Calves must not be allowed to become too fat because fat calves bring lower prices when going into the feedlot for finishing.

The basic ration for backgrounding is made up of forages, including corn silage. A ration of 23 pounds (10.4 kg) of corn silage (35 percent dry matter) and 1.5 pounds (0.68 kg) of a 40 percent supplement might be used with calves weighing 400 pounds (181 kg). Increase the amount of corn silage by approximately 5 pounds (2.26 kg) for each 100 pounds (45 kg) of gain in weight, but keep the amount of supplement constant. Mixed hay, corn, and supplement could be used as the ration for backgrounding. Feed 400 pound (181 kg) calves 8 pounds (3.6 kg) of mixed hay, 2 pounds (0.90 kg) of corn, and 0.5 pound (0.22 kg) of 40 percent supplement. Increase the amount of hay by 2 pounds (0.90 kg) for each 100 pounds (45 kg) of gain in weight. Increase the corn to 2.5 pounds (1.13 kg) when the calves weigh 500 pounds (227 kg) and keep the supplement level at 0.5 pound (0.22 kg). Thereafter, increase the amount of corn by 0.2 pound (0.09 kg) for each 100 pounds (45 kg) of gain in weight, but eliminate the supplement from the ration at 600 pounds (272 kg). In all cases it is expected that calves will be moved into the feedlot for finishing at approximately 700 pounds (318 kg).

A wintering ration made up of 5-6 pounds (2.26-2.72 kg) of corn, 9-10 pounds (4.1-4.5 kg) of hay, and 0.75 pound (0.34 kg) of 44 percent supplement will produce 1.5-2.0 pounds (0.68-0.90 kg) of gain daily. If calves are to be put on pasture for grazing after winter feeding, use a ration of 14-17 pounds (6.35-7.7 kg) of a grass-legume hay. Expected daily gain on such a ration is 1.0-1.25 pounds (0.45-0.56 kg). Feed a mineral mix free-choice with any of the above rations.

Daily water intake on the above rations will be 5-7 gallons (18.9-26.5 litres) per head daily. Silage rations require lower amounts of water and hay rations require more.

During the first two or three weeks of feeding calves, use a ration that is highly digestible and contains high-energy feed. Commercial companies have stress rations that may be used when starting calves on feed. Keep calves confined and provide a good quality hay, along with 0.5-1.0 percent of their body weight in concentrate feeds. Corn silage and concentrates may be used as the starting feed. Keep fresh feed in front of the calves and feed limited amounts several times daily. Remove feed that is not eaten during a 24-hour period and start with fresh feed the next day. Do not feed moldy silage to calves.

Urea or other nonprotein nitrogen sources may be used when feeding calves on backgrounding rations, but the rate of gain will be slower than when soybean meal supplements are used. When using urea or other NPN sources, combine them with the grain or silage part of the ration.

Growth stimulants that are implanted are recommended for backgrounding calves. Several are available but current regulations regarding dosage and withdrawal times must be followed.

Calves may be fed on pasture but gains will decline in late summer when pastures begin to dry up, Figure 11-5. A limited intake of grain may be used to supplement the pasture. Additional protein supplementation is generally not profitable for backgrounding calves on well-fertilized, grass-legume pastures when grain is limit-fed. If pasture is of poor quality because of drought or maturity, then a mixture of 14 parts corn or milo and 1 part of 44-percent protein supplement may

Figure 11-5. Calves may be fed on pasture but gains are slower in late summer as pastures begin to dry up.

be used to provide the protein requirement. A crude protein level of 10-12 percent in the concentrate is enough under most pasture feeding conditions.

FEEDING REPLACEMENT HEIFERS

To maintain the size of an established beef breeding herd, it is necessary to keep 30-40 percent of the heifers calved to provide replacements. Heifers are selected as herd replacements on the basis of soundness, body conformation, high weaning weight, and rate of gain. Smaller breed heifers should gain 1.0-1.25 pounds (0.45-0.56 kg) per day from weaning to breeding and weigh 575-700 pounds (261-318 kg) at the time of breeding.

Heifers from the larger breeds should gain 1.25-1.75 pounds (0.56-0.79 kg) per day from weaning to breeding and weigh 650-750 pounds (295-340 kg) at breeding. Heifers are usually bred at 14-16 months of age to calve at two years of age, when they weigh approximately 900-1,000 pounds (405-450 kg). Breed heifers according to weight, rather than age.

The cheapest feed for growing breeding heifers after weaning is pasture. However, most heifers are weaned in the fall and, in many parts of the United States, must be fed through the winter in drylot. Rations that may be used for feeding heifers in drylot are given in Table 11-6. Feed a vitamin-mineral mix free choice with these rations. In some areas, permanent pastures are available for wintering

Table 11-6. Rations for growing replacement heifers (450-500 lb; 204-227 kg, expected gain 1 to 1.25 lb (0.45-0.56 kg) daily).

Ration	Amount per day	
	(lb)	(kg)
Legume-grass hay	10	4.5
Oats	3	1.36
Legume-grass haylage	25	11.3
Legume-grass hay	10	4.5
Ground ear corn	4	1.8
Corn silage	30	13.6
Soybean meal	1.5	0.68
Corn silage	20	9.1
Legume-grass hay	6	2.7
Alfalfa hay	12.5	5.7
Shelled corn or ground ear corn	2.2	1
Sorghum silage	43.4	19.6
Protein supplement (35% total protein)	1.7	0.77
Alfalfa hay	4.6	2.1
Corn silage	25.7	11.6
Grass hay (brome, orchardgrass, canarygrass)	11.2	5.1
Shelled corn or ground ear corn	3.4	1.5
Alfalfa hay	12.5	5.7
Oats	2.6	1.2

heifers. If green forage is not available, add protein supplement to the ration at the rate of 1-2 pounds (0.45-0.90 kg) per head per day. Use an all-natural protein supplement rather than urea, for young heifer calves. Young ruminants lack the ability to efficiently utilize a nonprotein nitrogen source such as urea. They may have a protein shortage in their ration if a nonprotein nitrogen source is used.

Provide high quality, palatable roughages when possible. Coarse, poor quality roughages will not produce the desired growth without additional concentrates added to the ration. During cold weather, the energy requirements are higher in order to maintain body heat. A rule of thumb is to add 1 percent more energy to the ration for each degree of temperature below freezing. Energy needs also increase when there is no protection from the weather during periods of low temperatures and rapid air movement. The wind chill index should be followed as a guide for additional energy needs under these conditions.

An allowance for feed waste must be included when determining the amount of feed to provide. As much as 15 percent of the feed may be wasted, with even greater amounts being wasted during wet, muddy weather. As heifers grow larger, energy needs increase and adjustments must be made in the amount of feed allowed in the ration. Do not feed bred heifers the same rations as steers.

FEEDING BULLS

Feeding Replacement Bulls

Bulls are weaned at six to eight months of age. A high-energy ration is fed for about six months after weaning to determine which bulls gain best. Liberal feeding during this period results in rapid development and permits earlier use of the bull for breeding purposes. Keep the best gaining bulls for use in the herd or for sale as herd sires.

Keep bulls on full feed during the winter months. In the spring, bulls may be put on pasture to complete their growth. Young bulls

Table 11-7. Feed mixtures for weaned bull calves (weaning to about 700 lb (318 kg)).

Feed	Amount in mix	
	(lb)	(kg)
Corn	1,200	544
Alfalfa-grass hay	600	272
Soybean meal	200	91
Corn	1,200	544
Oats	600	272
Soybean meal	200	91
Ground ear corn	1,700	771
Soybean meal	300	136
Corn silage	1,860	844
Protein supplement (32-35% total protein) (includes vitamin mix and trace minerals)	156	71
Shelled corn	460	209
Corn silage	1,431	649
Protein supplement (32-35% total protein) (includes vitamin mix and trace minerals)	156	71

gain more rapidly than heifers and, consequently, need more feed. Do not allow bulls to become too fat as this wastes feed and reduces fertility. However, thin bulls also have breeding difficulties.

Young bulls on a corn silage ration are fed grain at the rate of 1 percent of body weight. When hay or haylage is used, feed grain at the rate of 1.5 percent of body weight. Supplement poor quality roughage with additional protein. Feed minerals and salt free-choice and provide vitamin A at the rate of 30,000-50,000 IU per day if the ration is mostly corn silage or limited hay. If the bull is self-fed, include plenty of roughage in the ration so the bull will not become too fat.

Rations for bulls from weaning to about 700 pounds (318 kg) are given in Table 11-7. Rations for bulls over 700 pounds (318 kg) are given in Table 11-8. Trace-mineralized salt is fed free choice with these rations.

Feeding Yearling Bulls

Yearling bulls may need grain added to the ration at the rate of 0.5-1.0 percent of body weight. If corn silage is included in the ration, no additional grain is needed. The expected rate of gain for yearling bulls is 1.5-2 pounds (0.68-0.90 kg) per day.

During the winter two- to four-year-old bulls need more energy and protein in the ration than does the cow herd. A satisfactory ration is a full feed of silage plus 2 pounds (0.90 kg) of 40-percent protein supplement per day. Feeding a good quality legume hay at the rate of 16 to 20 pounds (7.25-9.0 kg) plus 10 pounds (4.5 kg) of grain per day will also provide a satisfactory ration for bulls of this age. A mineral supplement is fed free choice. The expected daily rate of gain is 1.0-1.5 pounds (0.45-0.68 kg).

Feeding Mature Bulls

Mature bulls that are in good condition may be fed the same ration as the cow herd. Thin bulls require 5-6 pounds (2.3-2.7 kg) of grain per day above the amount fed to the cow herd. Vitamin A requirements for mature bulls are met when at least one-half of the ration is legume hay, corn silage, sorghum silage, or grass hay. If the ration does not include legume hay, add 1-2 pounds (0.45-0.90 kg) of protein supplement to the ration. Provide a mineral supplement free choice.

Bulls will lose 50-100 pounds (22.5-45 kg) during the breeding season. They must be fed enough during the rest of the year to gain this weight back. It may be necessary to provide

Table 11-8. Feed mixtures for bulls over 700 lb (318 kg).

Feed	Amount in mix	
	(lb)	(kg)
Shelled corn	1,200	544
Alfalfa-brome hay	756	343
Corn silage	1,940	880
Urea	50	23
Dicalcium phosphate	10	4.5
Shelled corn	440	200
Corn silage	1,500	680
Urea	50	23
Dicalcium phosphate	10	4.5
(Feed trace—mineralized salt free choice)		

additional feed for 6-8 weeks prior to the start of the breeding season, if the bull is too thin. Providing good pasture during the summer and high quality roughage during the winter will usually keep breeding bulls in good condition. Provide plenty of exercise and keep the bull in medium flesh (not too fat and not too thin). Bulls that are too fat or too thin have low fertility.

It may be necessary to feed 1 pound (0.45 kg) of protein supplement and 5 pounds (2.3 kg) of grain per day 30-60 days prior to the breeding season and during the breeding season. Base the amount of additional feed to provide on the condition of the bull.

FINISHING CATTLE

Starting Cattle On Feed

Cattle brought into the feedlot may or may not have had grain in the growing ration. It is desirable to bring cattle up to a full feed of grain in the finishing ration as quickly as possible without causing them to go off feed. The microorganisms in the rumen cannot adjust to the ration if the cattle are put on a full feed of grain too quickly so it is necessary to acclimate cattle to grain to prevent digestive problems.

A good quality grass hay or first-cutting alfalfa-brome hay mix are good roughages to use when starting cattle on feed. Second- or third-cutting alfalfa hay may cause scours (diarrhea). Oat hay, Sudan hay, or green chop may be used as roughages when starting cattle on feed. Corn silage with a protein supplement and hay fed separately may be used to start cattle on feed.

A starting mixture of 80 percent concentrate and 20 percent roughage may be used when cattle have been receiving some grain prior to being put on a finishing ration. The amount of grain may be increased at the rate of 0.5 pound (0.22 kg) per day until they are on full feed. Take care that cattle do not overeat as they start full feed. Provide no more feed than they will clean up each day. By the end of the first week the ration should consist of 90 percent concentrate and 10 percent roughage. This ration may be continued throughout the feeding period, Figure 11-6.

When cattle are not used to grain in the ration, it may be advisable to start with an all-roughage ration to avoid digestive disorders. For example, start with no concentrates and 100 percent roughages and increase the percent of concentrate gradually over a four to six week period until they are on full feed. This type of starting ration allows the rumen

Courtesy of the University of Illinois

Figure 11-6. The use of a self-unloading wagon to feed a mixture of roughages and concentrates can reduce labor requirements for feeding cattle.

microorganisms to gradually adjust to the higher energy level of the concentrate in the finishing ration.

If cattle are started on 90 percent concentrate and 10 percent roughage in the ration, limit the daily feed intake to 1 percent of body weight. Gradually increase the amount fed until they are receiving all they will eat by the end of two weeks.

Another method is to feed 1-3 pounds (0.45-1.36 kg) of grain the first day, along with 5 pounds (2.28 kg) of hay or its equivalent in silage. The grain and hay is mixed together before feeding. After the first day, gradually increase the amount of grain in the ration by 1 pound (0.45 kg) per day per head until they are receiving 1 percent of their body weight. Thereafter, increase the amount of grain by 0.5 pound (0.22 kg) per day for heavy feeders and every other day for light feeders until they are on full feed. Make sure cattle are cleaning up the amount of feed being given them. The amount of roughage is gradually reduced as the grain allowance is increased. Protein supplement is added to the ration as required to meet nutritional needs. Cattle are on a full feed of grain when they are consuming 1.5 percent of their body weight in grain and supplement, in addition to the roughage in the ration. If there is limited roughage in the ration, cattle will consume 2.0-2.5 percent of their body weight in grain and supplement, Figure 11-7.

During the first three weeks of starting cattle on a finishing ration, add 50,000 IU of vitamin A per day to the ration. Adding a broad-spectrum antibiotic at the rate of 350 milligrams per head per day during the first three weeks will reduce sickness and improve the rate of gain. Feed a mineral mix of 60 percent dicalcium phosphate and 40 percent trace-mineralized salt on a free-choice basis.

Rations for Finishing Cattle

Nutrient requirements for finishing cattle are given in Appendix tables. An animal on a finishing ration will consume from 2-2.5 percent of its body weight in air-dry grain and protein supplement. When calculating the ration on a 100 percent dry-matter basis, use 1.8-2.25 percent of its body weight. About 0.5-1.0 percent of body weight should be fed as air-dry roughage or 0.45-0.9 percent of body weight when calculating the ration on a 100 percent dry matter basis.

The ratio of protein supplement to grain should be 1:8 to 1:12. The total ration should have about 10-15 percent air-dry roughage or

Courtesy of the University of Illinois

Figure 11-7. Cattle in the feedlot on a full feed of grain.

Table 11-9. Beef cattle rations for cattle on full feed[1]. Steer calves—initial weight 400-450 pounds (181-204 kg).

Dry Rations	lb/day	kg/day	Silage in Ration	lb/day	kg/day
Grain	11-14	5-6	Grain	10-15	4.5-6.4
High quality legume hay	5-6	2.3-2.7	Protein supplement	0.5-1.5	0.23-0.68
			High quality legume hay	2-4	0.9-1.8
			Corn or sorghum silage	8-12	3.6-5.4
Grain	11-14	5-6	Grain	10-14	4.5-6.4
Protein supplement	1.25-1.75	0.57-0.79	Protein supplement	1-1.5	0.45-0.68
Mixed hay or low quality			Mixed hay or low quality		
legume hay	4-6	1.8-2.7	legume hay	2-4	0.9-1.8
			Corn or sorghum silage	8-12	3.6-5.4

Steers-Initial Weight 600-750 pounds (272-340 kg)

Dry Rations	lb/day	kg/day	Silage in Ration	lb/day	kg/day
Grain	13-16	5.9-7.3	Grain	12-15	5.4-6.8
High quality legume hay	6-8	2.7-3.6	Protein supplement	0.75-1.25	0.34-0.57
			High quality legume hay	3-5	1.4-2.3
			Corn or sorghum silage	14-18	6.4-8.2
Grain	13-16	5.9-7.3	Grain	12-15	5.4-6.8
Protein supplement	1.25-1.75	0.57-0.79	Protein supplement	1-1.5	0.45-0.68
Mixed or low quality			Mixed hay	2-4	0.9-1.8
legume hay	5-7	2.3-3.2	Corn or sorghum silage	14-18	6.4-8.2
Corn and cob meal	14-17	6.4-7.7	Ground snapped corn	13-17	5.9-7.7
Protein supplement	1-2	0.45-0.9	Protein supplement	2-2.25	0.9-1.1
Mixed hay	4-6	1.8-2.7	Corn or sorghum silage	12-16	5.4-7.3
Ground snapped corn	18-20	8.2-9.1			
Protein supplement	2-2.5	0.9-1.1			
Peanut hay	3-5	1.4-2.3			

Steers-Initial Weight 800-900 pounds (363-408 kg)

Dry Rations	lb/day	kg/day	Silage in Ration	lb/day	kg/day
Grain	16-18	7.3-8.2	Grain	12-16	5.4-7.3
Legume hay	6-8	2.7-3.6	Protein supplement	1.25-1.75	0.57-0.7
			Legume hay	1-3	0.45-1.4
			Corn or sorghum silage	16-20	7.3-9.1
Grain	13-17	5.9-7.7	Grain	12-16	5.4-7.3
Protein supplement	1.5-2	0.68-0.9	Protein supplement	2.5-3.0	1.1-1.4
Mixed hay	7-9	3.2-4.1	Grass hay	3-5	1.4-2.3
			Corn or sorghum silage	16-20	7.3-9.1
Ground snapped corn	14-18	6.4-6.8	Ground snapped corn	15-17	6.8-7.7
Protein supplement	3.0-3.5	1.4-1.6	Protein supplement	2.75-3.25	1.2-1.5
Cottonseed hulls	6-8	2.7-3.6	Corn or sorghum silage	16-18	7.3-8.2

[1]Quantities of feed are averages for entire feeding period. Less feed is required in the early part of the feeding period and more in the later part of the feeding period. More roughage is used early in the feeding period and less in the later part of the feeding period.

Source: USDA *Finishing Beef Cattle, Farmers Bulletin No. 2196,* May 1973.

about 9-13.5 percent on a 100 percent dry-matter basis. The fastest and most efficient weight gains are ensured by using rations with a high grain content.

A mineral supplement may be fed free choice to fattening cattle. When feeding a high grain diet, use a mix of 2 parts dicalcium phosphate, 2 parts limestone, and 6 parts trace-mineralized salt. When feeding a high roughage ration, use 2 parts trace-mineralized salt and 1 part dicalcium phosphate.

Examples of rations that may be fed to cattle on full feed are given in Table 11-9. While these tables serve as general guidelines for rations for finishing cattle, the feeder must consider the quality of available feeds and relative feed prices. These considerations may make it necessary to vary the ration from these examples in order to meet the nutrient needs of cattle.

Rations that are high in crude fiber may be used to finish cattle; however, gains are a little slower as compared to high-concentrate rations. High crude fiber rations may contain as much as 70 percent pelleted or cubed alfalfa hay. In feeding experiments, this type of ration has reduced feed requirements by as much as 21 percent; however, fewer cattle on high-fiber rations will be graded USDA Choice as compared to cattle on high-concentrate rations.

Calculating Total Feed Needed

The total amount of feed needed to finish cattle varies with age and condition of the cattle, kind of feed used, weather conditions, and management practices of the feeder. Some general guidelines may be used to estimate feed needs during the finishing period.

Fattening cattle will eat about 2.5-3 percent (dry-matter basis) of their body weight in feed each day. Multiplying the average body weight of the animal by 2.5-3 percent and then multiplying by the number of days in the feeding period gives an indication of the amount of feed needed.

Example:

Starting weight of steer	600	lb	272	kg
Length of feeding period	210	days		
Average daily gain		2.4 lb	1.1	kg
Ending weight	1,104	lb	500	kg

Daily ration:

Shelled corn	14.8 lb	6.7 kg
Protein supplement	1.5 lb	0.68 kg
Mixed hay	5 lb	2.27 kg
Feed consumption	2.5% of average body weight	

((starting weight + ending weight) / 2)
x Feed consumption as percent of body
weight x days on feed = total pounds (kg)
of feed per head

English:

((600 + 1,104) / 2) x 0.025 x 210 =
4,473 pounds of feed per head

Metric:

((272 + 500) / 2) x 0.025 x 210 = 2026.5 kg
of feed per head

Consumption of each feed in ration:

	lb		kg	
Corn	14.8 × 210 = 3,108	6.7 × 210 = 1407		
Protein supplement	1.5 × 210 = 315	0.68 × 210 = 142.8		
Mixed hay	5 × 210 = 1,050	2.27 × 210 = 476.7		
Total feed consumption	4,473	2026.5		

UREA IN FINISHING RATIONS

Urea can be used by ruminants such as cattle to meet part of the protein requirement of the diet. Urea is broken down in the rumen to carbon dioxide and ammonia. The ammonia is used by microorganisms in the rumen to make microbial protein. These microorganisms move through the digestive tract and are, in turn, digested, which releases the protein in their cells for use by the animal.

Urea decomposes rapidly in the rumen and the ammonia that is released may be used by the microorganisms or may be absorbed into the bloodstream, where it is carried to the liver. If the microorganisms in the rumen cannot utilize the ammonia rapidly enough, the excess goes into the bloodstream. Too much ammonia in the bloodstream can be toxic to the animal. If there is enough ammonia in the rumen from all protein sources to meet the metabolic needs of the microorganisms, additional urea in the ration is not needed to meet the protein needs of the animal.

Microorganisms in the rumen require energy, minerals, and vitamins for rapid growth. If these nutrients are not present in sufficient quantity, then less urea can be utilized in the diet. High-energy rations are better adapted to the use of urea as compared to low-energy rations.

Urea Fermentation Potential

A method of estimating the value of urea in cattle finishing diets by determining the urea fermentation potential (UFP) of feeds has been developed by researchers at Iowa State University (Iowa State University A.S. Leaflet R190, 1974). The UFP value of a feed is an estimate of the amount of fermentable energy it contains, as determined by its TDN content and the amount of ammonia released when the protein is broken down in the rumen by fermentation. The UFP value is expressed as grams of urea per kilogram of feed dry matter available for use by the microorganisms in the rumen.

A feed with a positive UFP value contains more fermentable energy than is needed by rumen microorganisms to break down the ammonia produced from the protein in the feed (when it undergoes rumen fermentation). Urea in amounts up to the UFP value of the feed, when added to a ration containing feeds with positive UFP values, can be utilized by the rumen microorganisms to supply part of the metabolizable protein requirement of the animal.

On the other hand, a feed with a negative UFP value does not contain enough fermentable energy to enable rumen microorganisms to utilize ammonia released during rumen fermentation of that feed. Therefore, additional urea added to the ration cannot be utilized to supply part of the metabolizable protein requirements of the animal. When feeds with both positive and negative UFP values are used in the ration, the UFP value of the ration is the total of the positive and negative values of the individual feeds.

Corn, corncobs, corn silage, and milo are examples of feeds with positive UFP values. Soybean meal, linseed meal, alfalfa, meat scraps, and milk are examples of feeds with negative UFP values. A high-protein value in relation to the amount of energy in the feed will tend to produce negative UFP values, while lower-protein values with higher-energy values tend to produce positive UFP values.

Amounts of Urea to Feed

No more than one-third of the total protein requirement of the ration should be supplied from urea. When cattle are fed high-energy rations, urea use should not exceed 0.23 pounds per day per head. With lower-energy rations, the amount of urea fed per head per day should not exceed 0.15 pounds. A minimum of 3 pounds of grain should be fed per head per day when sorghum silage or other similar energy roughages are used in rations containing urea. Feeding less grain than this with this type of ration results in unsatisfactory urea utilization.

When the crude protein requirement is higher than 12-13 percent in a high-energy ration for young cattle, plant protein sources are more satisfactory than urea. The rumen bacteria in young cattle cannot effectively convert more ammonia to metabolizable protein than that supplied in the ration. Urea will not give an increase in rate of gain for young cattle.

Other Considerations in Utilizing Urea

Grains provide a better source of readily available energy than do roughages. Therefore, high-grain rations can better utilize urea than can high-roughage rations.

Do not feed excessive amounts of urea at one time. This will cause more ammonia to be released in the rumen at one time than can be utilized by the rumen microorganisms. It is better to feed rations containing urea twice a day to provide a more uniform entry of urea into the rumen. It is also important to thoroughly mix urea in the ration to prevent overconsumption by some animals in the feedlot. Mixing the urea into the ration with a scoop shovel does not produce the required uniform mix. For the same reason, urea should not be top-dressed over feed in the feed bunk.

Cattle should not be started in the feedlot on rations containing urea. Use plant protein sources for the first 20 to 30 days of feeding, and then gradually change to a urea supplement. Cattle started on urea supplements will not gain as rapidly during the early part of the feeding period. This reduced rate of gain early in the feeding period lasts longer with lightweight cattle than with heavier cattle.

The calcium, phosphorus, potassium, cobalt, zinc, and sulphur requirements of the cattle must be met in order for rumen microorganisms to do the best job of utilizing urea. A nitrogen to sulphur ratio for the total ration of 15:1 or 10:1 is recommended for rations using urea.

SUMMARY

Roughages may be used as the major part of a maintenance ration for dry, pregnant beef cows. If the roughage is of poor quality, some supplementation with protein may be necessary. Young cows and heifers are still growing and need more feed than mature cows.

When beef cows are nursing calves, their protein, energy, mineral, and vitamin needs are greater. High-quality pastures may meet the nutrient needs of lactating beef cows. Grain and supplement may need to be added to the ration if the roughage is of poor quality.

Beef calves will start eating grain in small amounts when they are about four weeks old. They may be creep-fed to get heavier calves at weaning. Energy needs of calves are met with grain, while protein, mineral, and vitamin needs are met by the milk and pasture in the diet.

Backgrounding is the growing and feeding of calves from weaning until they are ready to enter the feedlot. The basic ration is forages, including corn silage. Do not allow calves to become too fat during this feeding period.

Pasture is the cheapest feed for raising replacement heifers for the beef herd. If the calves are born in the fall, then they must be fed in drylot during the winter. High-quality, palatable forages make a good basic ration for raising replacement heifers.

A variety of methods are used to start cattle on feed in the finishing lot. If they have not had grain, they must be acclimated to it to prevent digestive problems. Use high-quality roughages when starting cattle on feed.

Cattle on finishing rations will consume from 2.5-3 percent (dry-matter basis) of their body weight in feed. Corn, silage, hay, grain, and protein supplement with minerals and vitamins added as needed make up the basic finishing ration widely used for feeding cattle.

Urea may be used in beef cattle rations but care must be taken to get the animals used to it and not to use too much urea in the diet. It must be mixed thoroughly in the ration. No more than one-third of the total protein requirements in finishing rations should come from urea. Less urea should be used on low-energy rations.

REVIEW

1. Describe the nutrient requirements for maintenance of dry, pregnant beef cows.
2. What adjustments must be made in maintenance rations for dry, pregnant beef cows if poor quality roughage is a part of the ration?
3. Why is it important to properly balance the ration for dry, pregnant beef cows?
4. How do the nutrient needs of young cows and heifers differ from those of mature beef cows?
5. Describe the energy, protein, mineral, and vitamin requirements for lactating beef cows.
6. How can these requirements for lactating beef cows be met?
7. Discuss the use of protein supplements in feeding beef cows.
8. To what extent should urea be used in the rations of beef cows?
9. Discuss the use of pasture for the beef breeding herd.
10. Describe the advantages and disadvantages of creep feeding beef calves.
11. What kind of rations may be used for creep feeding beef calves?
12. Discuss the use of growth implants with beef calves.
13. What is backgrounding?
14. Describe rations that might be used for background beef calves.
15. Discuss the use of urea when backgrounding beef calves.
16. What rations may be used to meet the nutrient needs of replacement heifers in the beef herd?
17. Discuss the feeding of replacement bulls for the beef herd.
18. Describe a method of starting finishing cattle on feed.
19. Describe the nutrient needs of finishing cattle.
20. Describe a ration that might be used to finish cattle.
21. Describe and give an example of how to calculate the total amount of feed needed to finish cattle.
22. Discuss the use of urea in finishing rations.
23. Describe the use of the urea fermentation potential method of evaluating feeds for beef cattle.

12

FEEDING DAIRY CATTLE

OBJECTIVES

After completing this chapter you will be able to:

- Describe nutrient requirements for dairy cattle.
- Describe methods of feeding dairy cattle.
- Select the right feeds for dairy cattle.
- Balance rations for lactating cows, dry cows, and herd replacements.

NUTRIENT REQUIREMENTS

The importance of balancing rations for dairy cattle to maintain a high level of production is indicated by the fact that feed costs represent approximately 50-60 percent of the total cost of producing milk. The ability of a dairy cow to produce milk is limited by heredity. Differences in the level of milk production among cows are determined by approximately 25 percent heredity and 75 percent environment. A large part of the environmental factor that influences milk production is the feed available to the cow.

Nutrients are used by dairy animals to meet the following needs:

- Growth of the immature animal.
- Pregnancy. Needs are small during the first six months and large during the last two or three months of the gestation period.
- Fattening or regaining of normal body weight lost during lactation.
- Maintenance of the mature animal. Needs vary according to the size of the animal.

- Milk production. Needs vary with the pounds (kilograms) of milk produced and its composition.

Maintenance and milk production needs are the two most important to consider when balancing rations for dairy cows. With high-producing cows, nutrient requirements for milk production may be several times that needed for maintenance. Rations for lactating dairy cattle are generally balanced for energy (either $NE_{Lactation}$ or TDN) and crude protein.

Dairy cattle are ruminants and can use fairly large amounts of roughages (pasture, hay, silage, haylage) in the ration. Use high quality roughages and supplement the ration with concentrates (grain and protein feeds). Mineral needs of dairy cattle are met by adding some mineral supplement to the ration and also providing a mineral supplement free choice.

The nutrient requirements for dairy cattle are found in the Appendix. These requirements may be used as guides when balancing rations for dairy animals.

185

METHODS OF FEEDING DAIRY COWS

Traditional

Roughages are fed free choice in bunks to the entire herd or in mangers in the stanchion barn. Concentrates are fed to cows individually, according to milk production. The concentrate mixture is fed in mangers in the stanchion barn or in the milking parlor during milking, Figure 12-1.

Advantages of traditional feeding:

- Less specialized equipment is needed.
- Theoretically feeds each cow according to individual needs based on production.
- Permits adjusting concentrate feeding to the stage of lactation.
- Permits challenge feeding of each cow.

Disadvantages of traditional feeding:

- It is hard to measure the amount of forage (roughage) each cow eats. Therefore, it is hard to balance the ration with the right amount of concentrate for each cow.
- Low producers are often overfed concentrates.
- High producers are often underfed concentrates.
- Grain feeding facilities are needed in the milking parlor.
- The level of dust in the milking parlor increases.
- Cleanup of uneaten grain in the milking parlor is required.
- Milking in the parlor may be delayed while waiting for the cow to finish eating the grain mix.
- Feeding in the parlor slows down the milking.
- Cows do not stand as quietly and they defecate more during milking.
- More labor is necessary.
- There is less control over the total feeding program.
- Cost for equipment is higher.

Figure 12-1. The concentrate mix may be fed individually to cows in a stanchion barn.

- Careful records of individual production and continual adjustment of concentrate feeding to match production are required.

A feeding practice once common was feeding concentrate on a grain-to-milk ratio. That is, a pound of concentrate was fed for so many pounds of milk produced. This is not currently considered a good practice because it tends to overfeed concentrate to the low-producing cows and underfeed the high-producing cows.

Challenge or Lead Feeding

Challenge, or *lead feeding*, is the practice of feeding higher levels of concentrate to challenge the cow to reach her maximum potential milk production. Challenge feeding gives more concentrates to the cow early in the lactation period and less during the later part of the lactation period. Feed dry cows in good condition about 1/2 pound (0.22 kg) of concentrate per 100 pounds (45 kg) of body weight. Two or three weeks before calving, increase the amount of concentrate by about 1 pound (0.45 kg) per day until the cow is eating from 1-1½ pounds (0.45-0.68 kg) per 100 pounds (45 kg) of body weight at calving. About three days after calving, increase the concentrate by 1-2 pounds (0.45-0.90 kg) per day for two to three weeks after calving. At this time, the cow should be eating about 2 pounds (0.90 kg) of concentrate per 100 pounds (45 kg) of body weight daily. Continue to increase the level of concentrate feeding until milk production levels off. At maximum milk production, hold the level of concentrate feeding constant.

Keep accurate production records and when the level of production drops, decrease the level of concentrate feeding by about 1 pound (0.45 kg) of concentrate for each 3 pounds (1.36 kg) of milk-production drop.

Increase the level of concentrate feeding during the early part of the lactation period if the cow is losing weight. Cows that do not respond to challenge feeding should be culled from the milking herd.

Feeding Complete Rations

The most common method of feeding dairy cows today is by using a complete ration and feeding by groups of cows. The complete ration has all or almost all of the ingredients blended together and is fed free choice to all the cows of a given group.

Methods of grouping cows in large herds for the purpose of feeding complete rations is discussed later in this chapter.

The roughages and concentrates required to meet the energy, protein, mineral, vitamin, and crude fiber needs of the cows are blended together in the complete ration. When feeding a complete ration free choice, no concentrates are fed in the milking parlor.

Advantages of feeding a complete ration:

1. Each cow receives a balanced ration.
2. Each cow is challenged to produce to her maximum genetic potential.
3. Feeds are used more efficiently.
4. NPN (nonprotein nitrogen) is utilized more effectively.
5. Fewer cows have digestive problems or go off feed.
6. It is not necessary to feed minerals free choice (separately) from the rest of the ration.
7. Rations can be changed easier without affecting consumption.
8. Less labor is required for feeding.
9. Problems with low milkfat test are reduced.
10. Cost of cow housing and feeding facilities are lowered.
11. Cows that gain weight early in the lactation period are quickly identified and may be culled.
12. Weight gains usually come later in the lactation period.
13. It is possible to substitute cheaper grains and urea in the ration, because silage tends to mask the taste and dustiness of feed ingredients.
14. There are several advantages of not feeding concentrates in the milking parlor:

a. Cost of parlor construction and maintenance of feeding equipment is less.
b. Cows stand more quietly in the parlor during milking.
c. There is less dust and manure in the parlor.
d. No concentrate cleanup and no clogging of the drain system by concentrates in the parlor.
e. It takes less time to milk the cows.
f. There is less wasted concentrate.

Disadvantages of feeding complete rations:

1. Special equipment for weighing and mixing the ration is necessary.
2. It may be more difficult at first to get cows to enter the milking parlor when they are not fed there.
3. It is hard to include hay in the complete ration.
4. If hay is fed separately, the ration may not be balanced for some cows.
5. Low-producing cows tend to get too fat.
6. Cows need to be divided into groups for most efficient feeding.

Special Equipment for Feeding Complete Rations

The successful use of complete rations requires two pieces of special equipment. These are a mixer-blender unit and a weighing device. The mixer-blender unit may be mobile or stationary and is used to get the ration uniformly and completely mixed.

The feed ingredients used in a complete ration must be accurately weighed to ensure a balanced ration. Weighing devices may be stationary scales or electronic scales on the mixer-blender unit. A complete discussion of this type of equipment and its use is found in Chapter 9.

Feed Analysis

The analysis of feeds used in complete dairy rations is especially important for success. Proper analysis will help in balancing the ration and reducing feed costs.

The most important feeds to be analyzed are the roughages used in the ration. Roughages tend to have a greater variation in nutrient content than do concentrates.

Analysis of feeds is discussed more completely in Chapter 8.

Grouping Cows for Feeding Complete Rations

The purpose of grouping cows for feeding complete rations is to match more nearly the ration to the nutrient needs of the cows in the group. Herd size, facilities, and the time the cow must wait in the holding area for the milking parlor are factors to consider when making up the groups.

Large herds are divided into several groups while it may be more practical to have only one group if the herd is small. The kind of facilities available may be a limiting factor on the number of groups that is practical. Groups should be small enough so that an individual cow will not be in the holding area waiting for entry into the milking parlor for more than two hours. Cows should not be kept from rest for more than two hours.

The most common method of grouping cows is by production level. Keep all dry cows in a separate group. Some dairy farmers prefer to keep all first calf heifers in a separate group. Cows with mastitis or other health problems may also be kept in a separate group.

In large dairy herds, a minimum of three groups are used. These are high-, medium-, and low-producing cows. Higher overall milk production and increased profits will usually result from this type of grouping.

High-producing cows need more concentrates in their ration than do medium- or low-producing cows. High producers are more efficient in converting feed into milk. Low-producing cows tend to get too fat when fed at the same level as high-producing cows.

If the herd is too small to make it practical to divide the cows into groups, magnetic feeders may be used to permit selected cows

Figure 12-2. Feed hay in separate mangers when it is not included in the complete ration.

to secure more feed. This creates another group without physically dividing the herd.

Fresh cows are moved into the high-producing group about three days after calving. Leave them in this group for at least two months to challenge them to reach their maximum potential of production. At the end of two months, move any cow whose production is not high enough to a lower-producing group.

Move cows from one group to another at no more than monthly intervals. This shift is based on Dairy Herd Improvement Association (DHIA) production records for individual cows. Cows are shifted in small groups rather than one cow at a time. Also consider such factors as physical condition of the cow, age, stage of lactation, stage of gestation, and individual cow temperament, when making decisions about moving cows from one group to another.

The ration for each group is determined by the average production and milkfat test of the group and by the size of cows. Adjust the ration when the quality of the forage changes.

Forages should be analyzed at least once a month.

Make sure plenty of clean, fresh water is always available for the group. Feed hay in a separate manger if it is not included in the complete ration, Figure 12-2.

The group may be fed once a day or several times a day. No significant difference in production has been found for frequency of feeding. Spoilage of feed is reduced when feeding occurs several times a day during warm weather.

Make sure there is enough bunk space so cows may eat free choice whenever they want to. Keep feed in the bunk at all times, especially for the high-producing group.

Problems Associated with Grouping Cows

It requires more labor to change groups at milking time and to sort cows when moving them from one group to another. Proper facility design reduces this problem. Some drop in milk production may occur when cows are moved from one group to another, because of the change in ration and the fighting that may occur when new cows come into the group. Usually this is not a major problem. Groups may become mixed together if gates are left open, but care in handling the groups and proper attention to detail will prevent mixing.

Automatic Concentrate Feeders

There are three types of automatic concentrate feeders currently in use on dairy farms. These are magnetic, electronic, and transponder, all of which automatically control access to concentrate feed by individual cows. A magnetic or electronic device is attached to each cow, which allows access to feed. The transponder controls the amount of feed each cow receives. The purpose is to allow high-producing cows to have more feed, while limiting the amount received by low-producing cows. Good management is needed to successfully use these systems, Figure 12-3.

GUIDELINES FOR FEEDING LACTATING DAIRY COWS

A shortage of energy is usually the most limiting factor in milk production. The total ration for lactating dairy cows should contain 60-70 percent total digestible nutrients (TDN). This is equivalent to 0.60-0.80 megacalories of net energy per pound of feed (a megacalorie is 1,000,000 calories).

Forage is the foundation for a dairy feeding program. Feed 1.5 to 2.8 pounds (0.68-1.27 kg) of forage dry matter per 100 pounds (45 kg) of body weight. A minimum level of fiber in the ration is necessary to maintain milkfat percentage in the milk. The ration should contain a minimum of 15 percent crude fiber. When roughage is fed free choice, daily intake (dry-matter basis) as a percent of body weight will vary from 1 percent for low quality to 3 percent for high quality forage.

The total ration should contain from 12-22 percent crude protein on an as-fed basis. Approximately 75-80 percent of the crude

Figure 12-3. Access to extra feed may be controlled by electronic devices which permit only selected cows to get at the feed.

Figure 12-4. Oats may be used as a part of the dairy cows ration.

protein in the ration is digestible and available for use by the dairy cow.

Use 0.5 to 1 percent mineralized salt in the concentrate mix. One to two percent of the concentrate mix should be a calcium-phosphorus supplement. These levels will provide the major and trace minerals needed by the dairy cow.

Grains and forages are often processed before feeding; use a coarse to medium grind for dairy cows.

The most expensive part of the ration is the grain and protein supplement. The use of home grown grains whenever possible will generally help lower the cost of the ration, Figure 12-4. Buy feeds on the basis of the least cost per unit of nutrient provided.

Feed requirements vary with the stage of lactation and gestation. The most critical feeding period is the first ten weeks of lactation, when milk production is increasing rapidly.

During the first ten weeks of lactation, increase the amount of grain in the ration by 1-2 pounds (0.45-0.90 kg) per day. Maintain the fiber level in the ration above 15 percent to keep the rumen working properly. Limit the grain to no more than 65 percent of the total dry matter in the ration. Do not increase the rate of grain feeding too fast or feed more than the cow will eat. Increasing the rate of grain feeding too fast may result in digestive problems such as acidosis, displaced abomasum, or going off feed. On the other hand, if the nutrient needs of the cow are not met, problems with low-peak production and ketosis may develop.

Younger cows (2- and 3-year olds) need extra amounts of feed nutrients for continued growth during lactation. If their nutrient needs are not met, they often will not reach their full milk producing potential.

Milk production peaks approximately six to eight weeks after calving. During the second ten weeks of lactation, feed a ration to maintain peak milk production. Grain intake will equal about 2½ percent of body weight and roughage intake will be about 1 percent of body weight. It is important to maintain an adequate level of roughage in the ration to maintain rumen function and milkfat test.

During the latter part of lactation (140-305 days after calving) milk production is usually dropping. Generally, the grain intake should be matched to the level of milk production. However, thin cows will need extra grain in the ration to restore the body condition they had prior to the dry period. It is easier to improve condition during lactation than during the dry period. Young cows are fed more nutrients to meet their growth needs. Two-year olds should get 20 percent more and three-year olds 10 percent more nutrients.

GUIDELINES FOR FEEDING DRY COWS

Dry cows normally need fewer nutrients than lactating cows. Nutrients are needed for the developing calf and to replace body weight lost during lactation. Care must be taken not to overfeed dry cows or they will become too fat. Limit the intake of corn silage and grain during the dry period, because of their high energy content. This will reduce problems with excess fat deposit in the liver area and with displaced abomasum.

Total dry-matter intake during the dry period should be limited to 2 percent of body weight. Roughage intake must be at least 1 percent of body weight. The amount of grain needed depends upon the quality and type of forage in the ration. Grain intake should be limited to no more than .50 percent of body weight daily.

If no grain is fed during the dry period, begin some grain feeding during the last two weeks before calving. This will prepare the rumen for digesting grain during the lactation period. Begin grain feeding with about 4 pounds (1.8 kg) daily and slowly increase the daily amount fed to a maximum of 12-16 pounds (5.44-7.25 kg) per head at calving time.

Limit calcium intake to 3.5 ounces (100 grams) daily. A minimum of 1.4 ounces (40 grams) of phosphorus should be fed daily. The calcium-phosphorus ratio should be about 2:1. Reducing the level of calcium in the diet to less than 0.2 percent about two weeks before calving will reduce the chances of milk fever developing at calving time.

When poor quality feeds are included in the ration, add vitamins A and D. The addition of these vitamins increases the calf survival rate and reduces problems with retained placenta and milk fever at calving time.

Trace minerals, especially iodine and cobalt, are included in the ration. If urea is to be fed during lactation, begin feeding it about two weeks before calving, so the rumen has time to adjust to the urea in the ration.

FEEDING HERD REPLACEMENTS

Feeding Calves from Birth to Weaning

Colostrum milk is the first milk secreted by the cow after calving. Feed the colostrum milk within 15 minutes after birth if possible, and certainly no later than 4 hours after birth. Colostrum milk, as compared to regular milk, is high in fat, solids not fat, total protein, and antibodies, which protect against disease.

The newborn calf does not have antibodies to protect it from disease until it receives the colostrum milk. The ability of the calf to absorb the disease-protection component (gammaglobulin) of colostrum milk is sharply reduced after 24-36 hours. The danger of infection from disease-causing bacteria is high during the first hours after the calf is born. Early feeding of colostrum milk will reduce death losses of newborn calves.

Wash the cow's udder and teats before allowing the calf to nurse. This will reduce the chances of the calf picking up germs and dirt while nursing.

Colostrum milk may be fed from a bucket or nipple pail. While this is more work and requires attention to cleanliness to prevent scours, it does make it easier to keep track of the amount of milk the calf is drinking.

Extra colostrum milk is a nutritious feed and may be stored for use at a later time. It may be frozen for storage and then thawed for use or it may be stored as sour (fermented) colostrum. During the summer months fermented colostrum may turn putrid and be unfit for feeding. This may be prevented by using organic acids to reduce the pH to 4.6 for storage. Formic, acetic, and propionic acids may be used for this purpose. Formic acid at a concentration of 0.3 percent by weight may be added to colostrum at the rate of ¼ cup per 5 gallons of colostrum milk. Acetic acid at a concentration of 0.7 percent by weight may be added at the rate of ½ cup per 4¼ gallons of colostrum. Propionic acid at a concentration of 1.0 percent by weight may be added at the rate of 1 cup per 6 gallons of colostrum.

The calf should be fed colostrum milk for the first three days after birth. During this time, the colostrum milk is higher in nutrients than regular whole milk or milk replacers. After three days, the calf may be fed whole milk, excess colostrum or milk replacer. Small breed calves are fed about 5 pounds (2.3 kg) and large breed calves are fed about 7 pounds (3.2 kg) of whole milk, excess colostrum, or milk replacer. These amounts are fed for three weeks on an early weaning program and for four or five weeks on a liberal milk feeding program.

Feed only what the calf will drink in three to five minutes. Overfeeding a liquid diet may cause scours (diarrhea). If scours should develop, reduce the amount of milk or milk replacer being fed. When the calf stops scouring, gradually increase feeding to the recommended level.

Use milk replacers of high quality, made from dairy products rather than from plant products. Milk replacers that have the following protein sources are preferred:

- skim milk powder
- buttermilk powder
- dried whole whey
- delactosed whey
- casein
- milk albumin

Plant protein sources have lower digestibility than milk protein sources and are not

as desirable in milk replacers. The following protein sources in milk replacers are considered to be of poorer quality:

- meat solubles
- fish protein concentrate
- soy flour
- distiller's dried solubles
- brewer's dried yeast
- oat flour
- wheat flour

The following protein sources are acceptable in a milk replacer but are not considered to be of as good a quality as those from dairy product sources:

- chemically modified soy protein
- soy concentrate
- soy isolates

When all protein sources are from dairy products, the milk replacer should contain 20 percent protein. When plant protein sources are used in the milk replacer, the protein content should be 22-24 percent.

A minimum fat level of 10 percent is recommended in a milk replacer and the level may be as high as 30 percent. Fat reduces scours and provides needed energy in the diet. Animal-fat sources of good quality are better than plant-fat sources. Homogenized soy lecithin is a good fat source for milk replacers, and it improves the mixing properties of the replacer.

Lactose (milk sugar) and dextrose are good carbohydrate sources in milk replacers. Starch and sucrose (table sugar) are not good carbohydrate sources for use in milk replacers.

It is important that calves get enough dry matter diet for proper growth. An 80-100 pound (36-45 kg) Holstein calf will need about 1 pound (0.45 kg) of dry matter during the first four weeks. Smaller breed (Jersey or Guernsey) calves require about 0.75-0.80 pounds (0.34-0.36 kg) of dry matter in the diet during this period. Determine the percent of dry matter in the liquid diet and dilute with water to get the appropriate dry-matter level. When using a milk replacer, follow the directions on the feed tag for mixing with water. Typical dry-matter levels are 28 percent for first milk colostrum, 16 percent for pooled excess colostrum, and 12 percent for whole milk.

A large breed calf is fed about seven pounds (3.2 kg) of milk or milk replacer per day. This amount needs to contain about 1 pound (0.45 kg) of dry matter. To calculate the dilution rate with water use these formulas:

a. (% dry matter) $\times$ (lb/feeding) = 1 pound
b. (lb/feeding from formula A) + (water) = 7 pounds

Using first milk colostrum, the calculations are as follows:

a. (0.28) $\times$ (lb/feeding) = 1 pound
   lb/feeding = 1/.28
   lb/feeding = 3.57
b. (3.57) + (water) = 7 pounds
   water = 7 − 3.57
   water = 3.43 pounds

Milk or milk replacer may be fed with a bucket or nipple pail. Take care to keep the bucket or pail clean for each feeding. A dirty bucket or pail will increase the chances of the calf getting scours.

Calves may be fed either twice a day or once a day. Once a day feeding is successful unless the weather is very cold or there are other stress conditions present in the environment. If calves are fed once a day, it is necessary to watch carefully for sickness and provide enough nutrients without overfeeding.

The cost of raising herd replacements can be reduced by weaning calves at three or four weeks of age. Calves must be healthy and not subjected to other stress conditions such as cold weather if they are to be successfully weaned before four weeks of age.

Calves may be weaned at four to five weeks of age if they are eating well and are healthy. During the last week, reduce the amount of milk fed to 3 pounds (1.4 kg) for small breeds and 4 pounds (1.8 kg) for large breeds.

Milk or milk replacer may be fed up to six or eight weeks of age, but this will increase the cost of raising the calves. Reduce the amount being fed during the last two or three weeks before weaning.

Calf starter should be fed beginning when the calf is about four days of age. Teach the calf to eat the starter by rubbing a little on its nose after each milk feeding. The calf will soon start to eat the starter.

After the calf begins to eat the starter, it may be fed free choice. The calf should be eating 1-2 pounds (0.45-0.9 kg) by the time it is four weeks old. Feed up to 3 pounds (1.4 kg) to small breeds and 4 pounds (1.8 kg) to large breeds per day. Feed calf starter for three or four months.

Recommended commercial calf starters are palatable and contain about 16 percent crude protein. If the calf is to be weaned early (before four weeks), the calf starter should contain 18 percent crude protein. Home-mixed calf starters may be used instead of commercial starters. Good calf starters contain whole, coarsely ground, cracked, or rolled grains. Palatability may be improved and dust reduced by adding up to 5 percent molasses in the mixture. Whole grains such as oats may be fed with starter rations until calves are 12 weeks of age.

High quality hay may be fed beginning at about four weeks of age; however, forages are not necessary in the ration until 8 to 10 weeks of age. If forage is not fed early, it is recommended that additional fiber be included in the starter ration. Silage and pasture may be used as roughage if enough starter is fed to meet the nutrient needs of the calf. Limit the amount of corn silage fed when the calf is under 12 weeks of age because of its high-moisture content, which may limit intake and growth. Hay is a better roughage for young calves. Do not feed moldy or damaged hay to calves. Bright, leafy, early cut legume-grass hay makes a good roughage for young calves. Feed the hay free choice.

Preventing Calf Scours

Death losses are reduced when care is taken to prevent scours in young calves. Scours may be caused by overcrowding, poor ventilation, calves getting wet, overfeeding liquid diet, not getting colostrum milk, and dirty feeding pails. Preventing these problems will reduce the incidence of calf scours and lower the cost of raising herd replacements.

Baby calves that get scours dehydrate rapidly, which often causes death. The calf becomes weak and refuses to drink. The calf may be force-fed by putting a tube down the esophagus. Equipment for force-feeding calves may be purchased at feed and livestock supply stores.

A commercial solution may be used for force-feeding calves with scours or a home-mixed solution may be used. A home-mixed solution may be made by mixing the following:

½ cup (118 cm³) of light corn syrup
3 teaspoons (15 cm³) of baking soda
4 teaspoons (20 cm³) of salt
1 gallon (3.8 litres) of warm water

Do not force the tube when putting it down the esophagus. Be sure it is in the esophagus and not the trachea. The tube can be felt going down the esophagus by placing a hand on the left side of the calf's neck. After the tube is in place, hold the bag containing the solution above the calf's head to allow the liquid to drain into the stomach.

The calf may be fed 2 pounds (0.9 kg) of solution four times per day for two days. On the third day, replace one of the feedings with milk or milk replacer. Replace two of the feedings with milk or milk replacer on the fourth day. Continue to replace with milk or milk replacer till the calf is entirely off the special solution.

Feeding the Calf from Weaning to One Year

Proper diet is necessary if heifers are to be ready for breeding at the right time. Improper diet during this period of growth will result in heifers not producing milk to their potential when they are in the milking herd.

The ration can be forage fed free choice with 4-5 pounds (1.8-2.3 kg) of grain per day. The amount of protein supplement needed depends on the protein content and amount of forage fed. Pasture may supply some of the roughage needed but will not supply all of the nutrients needed for growing heifers. It is recommended that heifers on pasture be fed some grain and stored forage to supplement the nutrients provided by the pasture. Trace-mineralized salt and a calcium-phosphate mineral mix may be fed free choice if the grain mix does not meet these needs. A supply of clean, fresh water is also essential.

Care must be taken not to allow heifers to become too fat. It may be necessary to limit the amount of grain in the diet to prevent this. Heifers which become too fat during this period will not produce as well during lactation.

A 12-16 percent protein level in the forage will eliminate the need for a protein supplement fed with the grain mix. The same grain mix that is fed to the milking herd may be used if it contains enough minerals and vitamins.

Feeding Heifers One to Two Years of Age

A good quality forage may supply all the nutrients needed during this period of growth. Feed trace-mineralized salt and a calcium-phosphorus mineral mix on a free choice basis. A gain of 1.5-1.8 pounds (0.7-0.8 kg) per day is desirable. It may be necessary to feed some additional grain to maintain this rate of gain.

If pastures are mature or heavily grazed it will be necessary to provide additional feed. A diet deficient in energy, phosphorus, or vitamin A will prevent the onset of estrus.

BALANCING RATIONS FOR DAIRY CATTLE

The steps necessary to balance rations for dairy cattle are the same as for any other kind of farm animal. A general discussion of balancing rations is found in Chapter 9. The Appendix contains nutrient requirement and feed composition tables for dairy cattle.

Selecting Feeds for Balancing Rations—Roughages

Alfalfa is one of the most commonly used roughages for dairy cattle. Other roughages used include red clover hay, lespedeza hay, timothy, millet, oat hay, prairie grass, Johnson grass, and coastal Bermuda grass. Mixtures of legume-grass hay are also popular roughages for dairy cattle and include alfalfa-brome grass, alfalfa-timothy, and alfalfa-orchard grass, Figure 12-5. Proper fertilization and harvest methods increase the feeding value of forage crops. Any of the forage crops may be used as dry hay, green chop, haylage, or silage.

There is a trend toward feeding a combination of alfalfa haylage and corn silage to dairy cattle to provide the roughage needed. Advantages include easy mechanization of feeding, reduction in handling bulk, greater uniformity in feedstuff, higher yield per acre as compared to pasture, and lower field losses when compared to hay.

Pastures are an excellent source of roughage for dry cows and growing heifers; however, it is difficult for lactating cows to get enough dry matter in the diet from pasture alone. Because of the high moisture content of pasture, it usually cannot meet the energy needs of high-producing cows.

Grains

The most commonly used grain in dairy cattle rations is corn. Other grains used include oats, barley, wheat, and rye. Corn is high in energy value and is generally a cheaper source of energy than other grains. The high-fiber content of corn and cob meal helps keep the percent of milkfat higher. High moisture corn may be used for dairy cattle; however, more must be fed (as compared to dry corn) to get the same nutrient value.

Oats are lower in energy but higher in protein value than corn. Oats should replace no more than one-half of the corn in the ration.

Figure 12-5. Alfalfa-brome hay is a popular roughage for dairy cattle.

Limit barley, wheat, or rye to no more than 25 percent of the grain mix. Barley should be rolled when fed to dairy cattle. Wheat is usually priced too high to be included in the ration. Rye is not very palatable and is seldom used in dairy rations.

Protein Supplements

Some protein supplements commonly used in dairy rations include soybean meal; ground, unprocessed soybeans; linseed meal; and cottonseed meal. Urea and other nonprotein nitrogen sources may also be used to provide some of the protein needed in dairy rations.

Soybean meal is found in many commercial dairy protein supplements because it is palatable and highly digestible. Ground, unprocessed soybeans should not make up more than 20 percent of the concentrate mix and should not be used in conjunction with urea in the ration. Usually soybeans are too high priced to be used in dairy rations as a protein source. Linseed meal is palatable and slightly laxative and is usually higher priced than other protein supplements. It is sometimes used for fitting show cattle or sale cattle, because it adds a shine to the hair coat. Cottonseed meal is palatable and slightly constipative and is often used in dairy cattle rations.

Urea will lower the cost of the ration. It has a bitter taste and must be mixed completely in the feed for cattle to eat it. Too much urea in the diet can be toxic to cattle, so limit the amount of urea in the ration to no more than 0.4-0.5 pounds (0.18-0.23 kg) per head per day. The concentrate mix should be no more than 1 percent urea. Limit the use of urea during the early stages of lactation and gradually increase the amount in the diet so the rumen bacteria can adapt to its use.

Monoammonium phosphate (11 percent nitrogen, 68.75 percent crude protein equivalent) and diammonium phosphate (18 percent nitrogen, 112.5 percent crude protein equivalent) are two examples of other nonprotein nitrogen sources for dairy cattle.

Liquid protein supplements are made by putting a nonprotein nitrogen source (usually urea) in a liquid carrier such as molasses. Other carriers, such as fermentation liquors or distiller's solubles, may be used. Most liquid protein supplements contain 32 to 33 percent crude protein equivalent and may be fed free choice, top dressed, or blended into a complete ration. There is usually no economic advantage to using a liquid protein supplement. They may be most useful in rations for low-producing cows, dry cows, yearlings, and older heifers when corn silage is the major forage in the ration.

Byproducts

Byproduct and other processed feeds may be used in dairy cattle rations. These include alfalfa meal, alfalfa leaf meal, beet pulp, brewer's grain, citrus pulp, hominy feed, molasses, potatoes, wheat bran, and whey. Typically, byproduct feed sources make up only a small part of the ration for dairy cattle. Each must be evaluated in terms of the cost of the nutrients provided compared to more conventional nutrient sources, as well as the convenience of use and palatability for dairy cattle.

Minerals

Calcium and phosphorus are the two major minerals needed by dairy cattle. Steamed bone meal, dicalcium phosphate, and limestone may be used to supply the calcium. Dicalcium phosphate, monocalcium phosphate, and monosodium phosphate may be used to supply phosphorus. Salt supplies sodium and chloride, and the use of trace-mineralized salt will supply most of the minor minerals needed. Commercial mineral mixes fed at the recommended levels will generally supply the mineral needs of dairy cattle.

Vitamins

Some supplementation of vitamins A, D, and E may be needed in dairy cattle rations. Poor quality forages, high levels of grain feeding, or lack of sunshine may cause a need for additional vitamins above those supplied by the forages and grains being fed. A commercial vitamin premix may be added to the concentrate mix to supply the needed vitamins.

Water

Milk is 85 to 87 percent water, therefore lactating dairy cows require more water in relation to their size than any other farm animal. Always keep plenty of fresh water available for dairy cattle. Make sure the water supply is free of high levels of nitrates and sulphates. A high level of blue-green algae in the water may be toxic to cattle, therefore, do not allow cattle to drink from ponds or lakes which contain a heavy growth of algae.

Example Ration Balancing Problem

Balance a ration for a group of cows with an average body weight of 1,400 pounds and an average milk production of 50 pounds per day testing 3.5 percent fat.

Step 1: Find and list the daily requirements from the dairy cattle nutrient requirement table, Table 39, in the Appendix:

Daily requirements	Crude Protein	NE_l	Ca	P
	(lb)	(Mcal)	(lb)	(lb)
Maintenance	1.12	10.12	.048	.039
Milk production	4.1	15.5	.13	.09
Total daily requirements	5.22	25.62	.178	.129

The requirements for milk production are listed in the Appendix table per one pound of milk produced; therefore, this number must be multiplied by the pounds of milk being produced:

Crude protein	$50 \times 0.082 = 4.1$
NE_l	$50 \times 0.31 = 15.5$
Calcium	$50 \times 0.0026 = 0.13$
Phosphorus	$50 \times 0.0018 = 0.09$

Step 2: Calculate the nutrients provided by the forage fed. Daily forage intake on a dry-matter basis is 1.5-2.0 percent of body weight. A 1,400 pound cow will eat 21-28 pounds (dry-matter basis) of forage per day.

Forage intake may need to be limited for high-producing cows so they will eat enough concentrate to meet their needs. If forage intake is limited to less than 1 percent of body weight, milkfat test may be lowered.

This example will use 2 percent of body weight for forage intake. The total amount of forage is, therefore, 28 pounds. Assume one-half the forage is alfalfa hay (reference #1-00-063) and one-half is corn silage (reference #3-02-912). All calculations in this example are done on a dry-matter basis, with the final figures converted to an as-fed basis. The ration will contain 14 pounds of alfalfa hay and 14 pounds of corn silage (dry-matter basis). The amounts of nutrients provided by the forage are calculated from the feed composition table in the Appendix. Multiply the percentage or amount of the nutrient found in the feed by the pounds of feed to be fed to find the amount of the nutrient provided in the ration. Note that NE_l is given in Mcal/kg and must be converted to Mcal/lb by multiplying by 0.453592.

Calculation of nutrients provided by the forage:

Alfalfa

Crude protein	$14 \times 0.17 =$	4.1
NE_l	$14 \times 1.30 \times 0.453592 =$	8.2554
Calcium	$14 \times 0.0141 =$	0.1974
Phosphorus	$14 \times 0.0024 =$	0.0336

Silage

Crude protein	$14 \times 0.083 =$	1.162
NE_l	$14 \times 1.57 \times 0.453592 =$	9.97
Calcium	$14 \times 0.0029 =$	0.0406
Phosphorus	$14 \times 0.0026 =$	0.0364

Nutrients from the forage:

Feed	lb fed	Crude Protein (lb)	NE$_l$ (Mcal)	Ca (lb)	P (lb)
Hay					
Alfalfa (INF 1-00-063)	14	2.38	8.2554	0.1974	0.0336
Corn silage (INF 3-02-912)	14	1.162	9.97	0.0406	0.0364
Total nutrients from forage	28	3.542	18.2254	0.238	0.07

Step 3: Determine the nutrients needed in the concentrate mix by subtracting the amounts provided by the forage from the requirements:

	Crude Protein (lb)	NE$_l$ (Mcal)	Ca (lb)	P (lb)
Total nutrients needed	5.22	25.62	0.178	0.129
From forage	3.542	18.2254	0.238	0.07
To be supplied by concentrate mix	1.678	7.3946	−0.06	0.059

Step 4: Determine the amount of concentrate mix required based on the need for NE$_l$. This is done by dividing the Mcal of NE$_l$ needed by the Mcal supplied by the concentrate mix. The average value of Mcal/lb of NE$_l$ provided by a mix of ground ear corn (reference #4-28-238) and soybean meal (reference #5-20-637) is approximately 0.87 on a dry-matter basis. This is found by the following calculations:

Ground ear corn NE$_l$ = 1.91 Mcal/kg
Soybean meal NE$_l$ = 1.94 Mcal/kg

$(1.91 \times 0.453592) + (1.94 \times 0.453592) / 2$
$= 0.87$

The pounds of concentrate (ground ear corn plus soybean meal) mix needed: (Mcal of NE$_l$ needed from concentrate mix divided by average Mcal/lb of NE$_l$ provided by feeds in concentrate mix)

7.3946 / 0.87 = 8.5

The total pounds of concentrate mix needed (ground ear corn + soybean meal + dicalcium phosphate + salt) is calculated (based on the assumption that the minerals make up 2% of the total):

8.5 / 98% = 8.67

Step 5: Determine the percent of protein needed in the concentrate mix by dividing the amount of protein needed by the pounds of concentrate mix and multiplying by 100. Therefore, 1.678 divided by 8.5 times 100 equals 19.7 percent protein.

The concentrate mix in this example will use ground ear corn and soybean meal plus 1 percent dicalcium phosphate (calcium phosphate) (reference #6-01-080) and 1 percent salt (sodium chloride) (reference #6-04-152). The protein in the concentrate mix will come from the ground ear corn and soybean meal, which together make up 98 percent of the total concentrate mix. Therefore, 19.7 divided by 98 times 100 equals 20.1 percent protein needed in the ground ear corn and soybean meal mix.

Step 6: Use the Pearson Square method to mix the grain and supplement in the proper proportions. Set up the Pearson Square as shown in Figure 12-6.

The percent of ground ear corn is 29.8 divided by 40.9 times 100 equals 72.86. The percent of soybean meal is 11.1 divided by 40.9 times 100 equals 27.14.

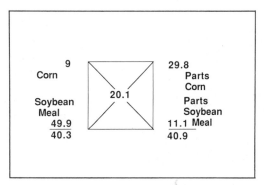

Figure 12-6. Using the Pearson Square to determine the proper mix for the corn and soybean meal.

The percent corn and soybean meal determined by the Pearson Square represents only 98 percent of the final concentrate. Therefore, the actual percentages in the final concentrate are 71.4 percent ($0.7286 \times 0.98 \times 100$) corn and 26.6 percent ($0.2714 \times 0.98 \times 100$) soybean meal.

The nutrients provided in the final concentrate mix are calculated:

NE$_l$:

	lb each feed	NE$_l$ Mcal/ lb	NE$_l$ Mcal (total)
Ground ear corn	6.19	$\times$ 0.87	= 5.39
Soybean meal	2.3	$\times$ 0.88	= 2.02
Dical Phosphate	0.0867	$\times$ 0	= 0
Salt	0.0867	$\times$ 0	= 0

Pounds each feed in concentrate mix:

	lb conc. mix	% conc. mix	lb each feed
Ground ear corn	8.67	$\times$ 71.4%	= 6.19
Soybean meal	8.67	$\times$ 26.6%	= 2.3
Dical Phosphate	8.67	$\times$ 1.0%	= 0.0867
Salt	8.67	$\times$ 1.0%	= 0.0867

Calcium:

	lb each feed	% calc- ium	lb calc- ium
Ground ear corn	6.19	$\times$ 0.07%	= 0.0043
Soybean meal	2.3	$\times$ 0.34%	= 0.0078
Dical Phosphate	0.0867	$\times$ 22%	= 0.019
Salt	0.0867	$\times$ 0	= 0

Protein:

	lb. each feed	% crude protein	lb crude protein
Ground ear corn	= 6.19	$\times$ 9%	= 0.5571
Soybean meal	= 2.3	$\times$ 49.9%	= 1.1477
Dical Phosphate	= 0.0867	$\times$ 0	= 0
Salt	= 0.0867	$\times$ 0	= 0

Phosphorus:

	lb conc. mix	% phos- phorus	lb phos- phorus
Ground ear corn	6.19	$\times$ 0.27%	= 0.0167
Soybean meal	2.3	$\times$ 0.70%	= 0.0161
Dical Phosphate	0.0867	$\times$ 19.3%	= 0.0167
Salt	0.0867	$\times$ 0	= 0

Step 7: Compare the nutrients provided to the requirements:

Feed	lb fed	crude protein (lb)	NE$_l$ (Mcal)	Ca (lb)	P (lb)
Forage	28	3.542	18.2254	0.238	0.07
Ground ear corn	6.19	0.5571	5.39	0.0043	0.0167
Soybean meal	2.3	1.1477	2.02	0.0078	0.0161
Dicalcium phosphate	0.0867	0	0	0.019	0.0167
Salt	0.0867	0	0	0	0
Nutrients from ration	36.7	5.25	25.63	0.27	0.1195
Requirements	36.4	5.22	25.62	0.178	0.129
Surplus	0.3	0.03	0.01	0.092	
Deficit					0.0095

The total dry-matter intake of cows weighing 1,400 pounds and producing 50 pounds of 3.5 percent milk per day is approximately 2.6 percent of body weight. Therefore, 1,400 times 0.026 equals 36.4 pounds of dry-matter intake per day.

The ration as calculated slightly exceeds the requirements for crude protein, Mcal of NE_l, and calcium. It is slightly deficient in phosphorus, which can be corrected by adding a little more phosphorus to the ration. It is recommended that 2,000 IU of vitamin A per pound of concentrate mix be added to the ration. This will ensure that the need for vitamin A will be met by the ration.

Step 8: Determine the amount of each feed needed on an as-fed basis by dividing the amount of feed on a dry-matter basis by the percent of dry matter in each feed:

Converting dry-matter basis to as-fed basis:

	lb dry basis	% dry matter	lb as-fed basis
Alfalfa hay	14	/ 90	= 15.6
Silage	14	/ 30	= 46.7
Ground ear corn	6.19	/ 87	= 7.1
Soybean meal	2.3	/ 89	= 2.6
Dical Phosphate	0.0867	/ 97	= 0.089
Salt	0.0867	/ 100	= 0.0867
Total			72.1757

Percent of each feed in ration:

	Amount fed	Percent total ration	Percent concentrate mix
Alfalfa hay	15.6	21.61	
Corn silage	46.7	64.7	
Ground ear corn	7.1	9.84	71.89
Soybean meal	2.6	3.6	26.33
Dicalcium phosphate	0.089	0.13	0.9
Salt	0.0867	0.12	0.88
Total	72.1757	100	100

FEEDING AND REPRODUCTION

There is a relationship between the amount of energy in the ration of developing heifers and reproduction. A shortage of energy in the ration delays the time the heifer reaches first heat, which causes a delay in breeding and shortens the productive life of the cow.

High-energy intake during growth will cause the heifer to reach first heat earlier, but may result in breeding problems when the cow is mature. Cows overfed energy have a higher sterility rate than cows fed the right amounts of energy feeds. Overfeeding energy also shortens the productive life of the cow, and bulls that become too fat have problems with sperm production.

A protein shortage in the ration may cause silent heats or discontinued heats. Protein shortage is more common when corn silage is the main forage fed in the ration. Greater care must be taken to properly balance the ration for protein when corn silage is the main forage fed.

Urea feeding has been blamed for reproductive problems. Studies done at several universities have shown that feeding urea at the recommended levels has no apparent effect on calving interval or sterility.

A shortage of phosphorus in the ration appears to cause irregular heat cycles; therefore, it is important to meet the cow's needs for phosphorus.

Breeding problems can also be caused by a shortage of vitamin A in the diet. The addition of vitamin A to the ration will help prevent these problems.

SUMMARY

The diet of dairy cows has a major influence on their ability to produce milk according to genetic potential. Maintenance and milk production are the two most important nutritional needs to consider when formulating rations for dairy cows. Because they are ruminants, dairy cows can utilize large amounts of roughage in their diet. Use high-quality roughages and supplement as needed with grain and protein feeds.

The most common method of feeding dairy cows today is by using a complete ration and feeding by groups of cows. Feeding complete rations has a number of advantages, but does require the use of a mixer-blender unit and a weighing device.

Dividing large herds into groups for feeding will more nearly match the ration to the nutrient needs of the cows in the group. The herd is grouped by levels of production, with dry cows and those with mastitis or other health problems being kept in separate groups. Herd size and facilities are factors to consider when choosing this method of feeding.

Feed requirements of lactating dairy cows vary with the stage of production. There are four distinct stages of production to consider. During the first ten weeks of lactation, challenge the cows to produce to maximum levels. During the second ten weeks of lactation, feed to maintain that high level of production. During the rest of the lactation period, milk production is dropping and feeding levels are adjusted accordingly. Feed dry cows in a separate group because they need fewer nutrients than lactating cows. Take care not to overfeed dry cows or they will get too fat.

Calves need to get colostrum milk during the first few hours after birth. This will help prevent disease problems. Milk replacers may be used when feeding calves. Begin feeding calf-starter rations when the calf is a few days old. Roughage feeding may begin when the calf is about four weeks old.

From weaning to one year of age, feed forage free choice with 4-5 pounds (1.8-2.3 kg) of grain per day. Some protein supplementation may be needed depending on the quality of the forage. High-quality forage with mineral supplement may provide all the nutrients needed.

Feeding balanced rations with proper energy levels to replacement heifers will reduce breeding problems and increase the probability that they will produce milk to their genetic potential when they come into production.

REVIEW

1. List the needs that must be met by rations for dairy cattle.
2. What are the two most important needs that must be met by the ration?
3. Describe the traditional method of feeding dairy cows and list its advantages and disadvantages.
4. Describe challenge, or lead, feeding.
5. What are the advantages and disadvantages of feeding complete rations to dairy cows?
6. What special equipment is needed to feed complete rations?
7. Why is feed analysis important when feeding complete rations?
8. Describe how cows are grouped for feeding complete rations.
9. Discuss the management of feeding in groups.
10. What are some of the problems associated with feeding complete rations in groups?

11. Describe the use of automatic concentrate feeders with dairy cows in group feeding.

12. What is the most critical nutrient for lactating dairy cows?

13. How much forage should be fed to lactating dairy cows?

14. What is the minimum level of fiber needed in the ration of lactating dairy cows and why is it needed?

15. What are the four feeding periods for dairy cows?

16. Describe feeding practices during each of these feeding periods.

17. Why is colostrum milk important for newborn calves?

18. Describe a good quality milk replacer.

19. How should calves be fed after weaning?

20. Describe a good calf-starter feed.

21. Discuss the prevention and treatment of calf scours.

22. How should the calf be fed from weaning to one year of age?

23. Discuss the selection of feeds for balancing rations for dairy cattle.

24. Discuss the use of urea in dairy cattle rations.

25. Discuss the relationship between feeding and reproduction.

13

FEEDING SWINE

OBJECTIVES

After completing this chapter you will be able to:

- Describe nutrient requirements for swine.
- Select proper feeds for swine.
- Balance rations for the breeding herd, baby pigs, and growing-finishing pigs.

NUTRIENT REQUIREMENTS

Energy

Energy in swine rations is provided by carbohydrates, lipids (fats and oils), and, to a lesser extent, by protein. Carbohydrates and lipids are the primary sources of energy. Only the excess protein that is not used for other purposes is available as an energy source.

Energy intake above that needed for maintenance, growth, or reproduction is deposited as body fat. The excess amount of energy intake, therefore, affects the lean-to-fat ratio of the pig. High amounts of energy intake result in more fat in relation to lean meat.

The energy intake of breeding sows must be limited or they will become too fat. Lactating sows require enough energy for maintenance and milk production. The energy level in the diet of lactating sows should be adjusted to the size of the litter. Energy is used more efficiently when sows are allowed moderate weight gains during gestation and moderate weight losses during lactation. There should be a small amount of overall weight gain in each reproductive cycle.

As the size and capacity of the digestive tract increase with age, the digestibility of fiber also increases. Thus, mature hogs can use more fiber in the diet than can young, growing pigs.

During the first two weeks of life, pigs cannot effectively digest starch. Carbohydrate in the diet during this time should come from glucose and lactose during the first week and from fructose and sucrose during the second week. After pigs reach two weeks of age, they have the necessary enzymes in the digestive tract to digest starch. Pig starters fed during the first two weeks after the birth of the pig should provide carbohydrate in a form that the pig can utilize.

The digestibility of dry matter, protein, and ether extract is reduced when large amounts of fiber are included in the diet. High intake of crude fiber slows the growth rate of pigs. Growing-finishing pigs from 40 pounds (4.5 kg) to market weight can use up to 5 percent of a high-fiber feed, such as alfalfa, in the ration without a noticeable effect upon rate of gain. Levels of 15-25 percent of high-fiber feed, such as ground alfalfa hay or meal, are sometimes included in rations for brood sows, to reduce energy intake and so prevent them from becoming too fat. Even higher levels may be included if the brood sows are self-fed.

When fats or oils are used to replace corn or other grain as an energy source in the diet, there is usually a reduction in total feed intake. Therefore, the amount of protein, minerals, and vitamins in the diet must be increased to compensate for the reduced feed intake. The use of fats and oils in the ration will increase the energy level in the diet and result in faster gains and more fat in the carcass. The effect is greater when low-protein diets are fed as compared to high-protein diets. Feed conversion is more efficient on high-fat diets, but the additional cost of the feed may offset any benefits gained from this increased efficiency.

Protein

Pigs are simple-stomached animals and cannot synthesize the essential amino acids in the digestive tract. The ten essential amino acids, which must be included in swine diets, are:

- Arginine
- Histidine
- Isoleucine
- Leucine
- Lysine
- Methionine
- Phenylalanine
- Threonine
- Tryptophan
- Valine

At least one-half of the methionine requirement can be met by cystine, and one-half of the phenylalanine requirement can be met by tyrosine. Lysine is the first limiting amino acid in swine growth and is often considered when balancing rations. As long as the amino acid requirements are met, some variation in the total protein level of the diet will not adversely affect rate of gain.

Minerals

Major minerals needed in swine diets include calcium, phosphorus, potassium, sodium, and chlorine. Trace minerals that need to be supplied in the diet include copper, iodine, iron, magnesium, manganese, selenium, sulfur, and zinc. When swine are raised in confinement, it is especially important to include these essential minerals in the ration.

The ratio of calcium to phosphorus needs to be 1-1.5:1 in a grain-soybean meal diet. This can be provided by using a mineral supplement. A 0.20-0.25 percent level of salt in the diet will provide the needed sodium chloride. Trace-mineralized salt or a trace-mineral supplement may be used to provide the necessary trace minerals in the diet.

Vitamins

Vitamins A, D, E, K, niacin, pantothenic acid, riboflavin, choline, and vitamin B_{12} need to be added to the diet of swine. This is done by adding a vitamin supplement to the ration. Other vitamins needed are usually found in the normal feeds used in swine rations.

Water

A supply of fresh, clean water located near feeders is needed for optimum growth and weight gains, Figure 13-1. Lactating sows will experience a drop in milk production if water intake is restricted. Young pigs have higher water needs than more mature pigs. Market hogs require 1-2.5 gallons (3.78-9.45 litres) of water per head per day. Sows, plus the litter, require 4.5-6 gallons (17-22.7 litres) of water per head per day.

Water intake is increased by high air temperature, lactation, fever, diarrhea, and high salt or protein intake. The amount of water intake is affected by water temperature. More energy is required to warm water, which is consumed at temperatures below body temperature. In the summer when environmental temperature is high, water intake increases when the temperature of the water is around 48-52°F.

Figure 13-1. A supply for fresh water is always available to growing-finishing hogs in this confinement feeding system.

SELECTING FEEDS FOR SWINE

Energy Feeds

Corn is the basic energy feed used in swine rations because it is high in digestible carbohydrates, is low in fiber, and is palatable. Other feeds used as energy sources in swine rations are compared to corn when determining their feeding value.

Number two (#2) dent corn contains 8.9 percent protein but does lack some of the essential amino acids needed in swine diets. Lysine and tryptophan are the two amino acids that are not found in corn in large enough amounts. Corn must be supplemented with protein, minerals, and vitamins when fed to swine.

Yellow dent corn contains 0.24 percent lysine. High-lysine corn (HLC) containing as much as 0.55 percent lysine and higher levels of tryptophan has been developed, which can reduce the amount of protein that must be supplemented in swine rations. Total protein and other amino acid content is about the same as normal corn. High-lysine corn needs to be analyzed when used in swine rations. The ration may then be balanced on the lysine content rather than total protein content.

The dry matter in high-moisture corn has the same feeding value as the dry matter in corn at normal moisture levels; however, a greater amount must be fed because of the higher moisture level. A ration of high-moisture corn and protein supplement fed free choice may result in a lower rate of gain and higher feed requirement per pound of gain as compared to complete rations of dry corn or high-moisture corn. Because of its higher palatability, pigs tend to eat more of the high-moisture corn in relation to the supplement and, thus, do not get a balanced ration. If a complete ration utilizing high-moisture corn is used, equipment must be available for daily mixing because it cannot be stored for later use. Processing (grinding, rolling, cracking, or crimping) high-moisture corn does not result in better performance with growing-finishing pigs.

Whole, shelled corn and a protein supplement can be fed free choice. However, this system is not as efficient as using complete rations that are self-fed. Gains are generally slower when grain and supplement are fed free choice.

Complete rations should be ground and the protein supplement mixed in the ration. Self-fed complete rations are more commonly used with growing-finishing pigs because of the better control of nutrient intake, resulting in more efficient gains. This system also lends itself to automation, which reduces the labor required for feeding, Figures 13-2 and 13-3.

Figure 13-2. An open front confinement hog feeding system which utilizes self-feeders protected from the weather.

Figure 13-3. A confinement feeding system which is completely enclosed and uses automated equipment to distribute feed to self-feeders in the pens.

Ground ear corn should not be used in diets for growing-finishing pigs, because of its high-fiber content. It may be used in rations for sows during the gestation period.

Barley may be substituted for corn, and in some parts of the United States, is more commonly used in swine rations. Barley has a higher fiber content than corn and, therefore, slightly less digestible energy. It has a higher protein content than corn, but is lacking in some of the essential amino acids. Barley must be supplemented with protein, minerals, and vitamins when used in swine rations.

Barley is ground medium fine for swine. It may also be rolled or pelleted. Barley is not as palatable as corn and is more successfully used in complete rations.

Relative to corn, barley has a feeding value of 90 to 95 percent. The feeding value of barley also varies with the weight per bushel. When up to one-third of the grain in the ration is barley, hogs will gain as fast as on an all-corn ration. When 100 percent of the grain in the ration is barley, gains are slightly slower.

Barley is sometimes infested with *scab*, a disease that attacks barley and may make it poisonous to hogs. Scabby barley should never be used in swine rations.

Milo (grain sorghum) is grown widely in the southwestern part of the United States. It has a higher protein content than corn and can replace all the corn in swine rations. Milo must be supplemented with protein, minerals, and vitamins in swine rations. It has a relative feeding value of 90 to 95 percent, compared to corn. Milo should be ground when used for swine feeding. Some varieties are unpalatable to hogs and should be mixed with protein in complete rations for feeding.

Wheat is equal to or slightly higher in feeding value than corn, with a relative value of 100 percent. Wheat is slightly higher in protein than corn, but because of its relative cost, wheat is seldom used in swine feeding. Low-quality or damaged wheat may be cheap enough to use in swine rations, and if used should be coarsely ground for hogs. If it is ground too finely, it forms a pasty mass in the pig's mouth. Wheat is sometimes used in pig-starter rations.

Oats have a higher protein content than corn, but the quality of the protein is lower. A protein supplement must be used when oats are included in swine rations. Oats have a high fiber content and a relative feeding value of 70-80 percent compared to corn. Oats should not be used for more than 25-30 percent of the grain in the growing-finishing ration. Higher levels in the ration will reduce the rate of gain. Up to 50 percent or more of the ration for pregnant sows may be composed of oats.

Oats for hog rations should be medium to finely ground. Dehulled, rolled oats are an excellent feed as starter rations for baby pigs.

Rye is not a good feed for swine because it is relatively unpalatable. It has a relative feeding value of 75-80 percent, compared to corn. Rye should not be substituted for more than 20 percent of the grain in the ration. Rye is harder than corn and should be ground when fed to hogs.

Rye is sometimes infested with a fungus called *ergot*. Ergot will cause abortions in pregnant sows, and therefore, ergot-infested rye should never be used in sow rations. Ergot-infested rye will slow down the rate of gain when fed to growing-finishing hogs.

Triticale is a hybrid cereal grain, the result of a cross between wheat and rye. Triticale has more lysine than corn, but is not as palatable as corn in swine rations. No more than 50 percent of the ration should be made up of triticale. Some varieties of triticale may be infested with ergot and should not be fed to pregnant sows.

Potatoes may be used in swine rations but, because they are mainly carbohydrate, must be supplemented with protein. Heavier hogs make better use of potatoes in the ration than do younger hogs. It takes about 400 pounds (181 kg) of potatoes to equal the feeding value of 100 pounds (45 kg) of corn. Feed potatoes at the rate of 1 part potatoes to 3 parts of grain. Cook potatoes before feeding them to hogs.

Bakery wastes such as stale bread, bread crumbs, cookies, and crackers may be fed as part of the swine ration. The average protein content of these foodstuffs is about 10 percent. Little is known about their amino acid content. Bakery wastes may substitute for about 50 percent of the grain in a swine ration and a good protein supplement must be included.

Fats, tallow, and greases provide a high-energy source in swine rations. These substances usually make up less than 5 percent of the ration, depending on the price of fat. They are used to improve the binding qualities of pelleted feeds. *Binding quality* refers to how well the feed particles stick together in the pellet. Carcass quality is decreased if too much fat, tallow, or grease is included in the ration. Because these substances contain no protein, minerals, or vitamins, appropriate supplements are essential when fat, tallow, or grease is used in swine rations.

Molasses provides carbohydrates in the ration and may be substituted for part of the grain. Limit molasses to no more than 5 percent of the ration. Scours may result when too much molasses is fed.

Protein

Plant proteins. Soybean meal ranges from 42.6 to 48.5 percent total protein, depending on the process used. Soybean meal containing 48.5 percent total protein is often used in formulating prestarter and starter rations for pigs. The protein quality of soybean meal is excellent with a good amino acid balance. Soybean meal is the most widely used protein supplement for swine feeding. The basic ration for growing-finishing hogs is corn-soybean meal.

Soybean meal is highly palatable and hogs will overeat this feed if it is fed free choice.

Prevent overeating of soybean meal by mixing it in complete rations or with a less palatable protein supplement.

Cottonseed meal is 40-45 percent protein, but the quality of the protein is poor. It is low in lysine and must be fed with other protein sources in swine rations. Ten to 15 percent of the protein in swine rations may be made up of cottonseed meal. Some cottonseed meal contains gossypol, which is toxic to hogs. If the gossypol has been removed, it may replace up to 50 percent of the soybean meal in the ration. Cottonseed meal is low in minerals and fair in vitamin B content. It is not highly palatable to swine.

Linseed meal contains about 35 percent protein, which is of poor quality. Therefore, linseed meal must be limited to about 10 percent of the total protein in the swine ration. Linseed meal contains more calcium than cottonseed meal or soybean meal and has about the same level of vitamin B. It is recommended that linseed meal be used in combination with animal protein sources in swine rations. In large quantities, linseed meal is laxative.

Peanut meal is 45-47 percent protein, but is low in several amino acids. Therefore, peanut meal must be fed in combination with other protein sources. It becomes rancid if stored for more than a few weeks and is low in minerals and vitamins. Peanut meal affects carcass composition, causing soft pork.

Whole cooked soybeans contain 37 percent protein and may be used to replace soybean meal in swine rations. Rate of gain is about the same as when soybean meal is used, but feed efficiency is improved by 5-10 percent. Whole cooked soybeans have a higher fat content than soybean meal, which provides more energy in the ration. Because of their lower protein content, more whole cooked soybeans must be used in the ration to balance for amino acid needs. The protein level in the diet must be 1-2 percent higher to maintain the correct protein to energy ratio.

Animal proteins. Tankage and meat scraps range from 50-60 percent protein, but they are low in the amino acid tryptophan and, therefore, must be used with other protein sources in swine rations. Tankage and meat scraps are not as palatable to hogs as soybean meal, but can provide up to one-half the protein in the ration.

Meat and bone scraps contain 45-50 percent protein, but are low in lysine. Another protein source needs to be used to provide enough lysine in the diet.

Fish meal is 60-70 percent protein, which is of excellent quality. It is palatable to swine and high in minerals and vitamins. While fish meal is a good protein source, it is generally too expensive to use except in creep rations.

Skim milk and buttermilk contain about 33 percent protein when dried. In liquid form, they are worth only about one-tenth as much as dried milk, because the liquid form contains about 90 percent water. The protein quality of skim milk and buttermilk is good, and they are good sources of B vitamins. These milk products are often used in creep rations in dried form. Young pigs cannot consume enough in liquid form to meet their protein needs.

Whey, in liquid form, contains only about 1 percent protein. Dried whey contains 13-14 percent protein, which is of excellent quality. However, liquid whey has a high water content, and hogs cannot consume enough to meet their protein needs.

ROUGHAGES

Growing-finishing hogs can make only limited use of roughage in the diet. Generally, it must be processed for inclusion in a complete ration.

Alfalfa meal contains 13-17 percent protein and has large amounts of vitamins A and B, as well as minerals. It may be used in growing-finishing complete rations when limited to no more than 5 percent of the total ration. It may be as much as 35 percent of the ration for brood sows where, because of its fiber content, it helps keep them from getting too fat.

Alfalfa hay and other hays are generally not used in swine rations except for feeding the breeding herd. Hays must be ground and mixed in the ration for self-feeding sows and gilts. Good quality hay can make up as much as one-third of the ration for breeding animals.

Silage is most valuable in rations for the breeding herd. Ten to 12 pounds (4.5-5.4 kg) of corn or grass-legume silage can be fed per day to sows and gilts during the gestation period. This must be supplemented with protein and minerals. Never feed moldy silage to sows and gilts. (Generally, moldy feed should never be fed to any animal.)

Pasture is of most value to the breeding herd. Good-quality pasture supplies the same nutrients as alfalfa meal and hay. Growing-finishing hogs on pasture will not gain as rapidly as when fed in confinement. Good-quality pasture may supply enough nutrients to permit as much as a 40 percent reduction in the amount of concentrates needed in the ration of sows and gilts. Care must be taken to properly balance the ration when feeding sows and gilts on pasture. Pasture is also of value when feeding the herd boar.

MINERALS

Ground limestone is a common source of calcium used in swine rations. Dicalcium phosphate, steamed bone meal, and defluorinated rock phosphate all supply both calcium and phosphate.

Trace minerals needed in the diet are generally supplied through the use of trace-mineralized salt or a trace-mineral supplement incorporated in a complete ration.

Iron and copper are especially important to baby pigs to prevent anemia. In addition to the iron in the ration, baby pigs should be given iron shots when they are two to four days old. Zinc is required to prevent para-keratosis. Early weaned pigs have a higher zinc requirement than pigs weaned at an older age.

Do not overfeed minerals in the diet or the rate of gain will be reduced. Do not add minerals to a complete ration containing commercial protein supplements unless the feed tag indicates that they are required. A mineral mix can be fed free choice, as hogs will not overeat minerals if they are receiving enough minerals in the ration.

VITAMINS

Vitamins may be added to the ration as part of the protein supplement, in mineral-vitamin premixes, or as vitamin premixes. The major differences among these sources of vitamins is the amount of vitamins they contain and their cost. It is difficult to determine the exact amount of vitamins that sources contain, because the feed tag usually does not list the amount. Past experience with a particular mix is the best guide to follow in selecting a vitamin source.

Complete supplements and mineral-vitamin premixes usually cost more than vitamin premixes. However, if the producer does not have the mixing equipment capable of incorporating small amounts of ingredients thoroughly in the complete ration, it may be best to use complete supplements or mineral-vitamin premixes. Vitamin premixes are used in such small amounts per ton that it is difficult to mix them into the ration properly, Figure 13-4.

ADDITIVES

Feed additives are commonly used in swine feeding because of the increased efficiency that results. Pigs gain faster, feed conversion is improved, and disease stress is reduced when additives are used. The additives most com-monly found in swine rations are: anthel-mintics, antibiotics, arsenicals, nitrofurans, and sulfa compounds. (See Chapter 6 for descriptions of these additives.)

Additive premixes must be carefully mixed into the ration to secure an even distribution. Additives are sometimes included in protein supplements, Figure 13-5.

Feed tag instructions for use and with-drawal times must be followed when using additives in rations. Federal regulations govern the use of additives, and there are penalties for improper use or not withdrawing them from use at the proper time prior to marketing animals.

FEEDING THE BREEDING HERD

Boars

Young boars, up to 125 pounds (57 kg), need a ration containing 18 percent total protein. The protein level in the ration may be reduced to 16 percent as the boar grows from 125 pounds (57 kg) to 200 pounds (91 kg).

Mature boars must be in good condition for the breeding season, but not too fat. They may be fed the same ration as gilts prior to breeding. About two weeks prior to the breeding season, increase the amount of feed by about 50 percent and continue feeding at a level that will maintain body weight during the breeding season without allowing the boar to become fat.

Weight gain is controlled by adjusting the energy level of the diet. Increase the amount of fiber in the ration by adding alfalfa meal or hay to the ration or using pasture. Replacing one-half of the corn in the ration with oats will also increase the fiber level and reduce the energy level of the ration.

Gilts—Before Breeding

After selection of gilts for the breeding herd, they should be fed a ration that limits weight gains to approximately 1 pound (0.45 kg) per

Figure 13-4. Premixes are used to include small amounts of trace minerals and vitamins in swine rations.

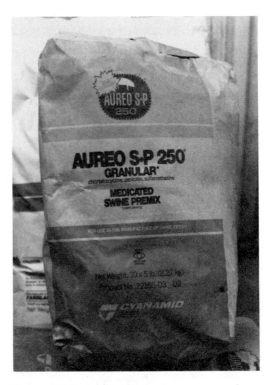

Figure 13-5. Additive premixes are used to provide antibiotics in swine rations.

day. For breeding purposes, the gilt should weigh about 250 pounds (113 kg) at eight months of age. During the summer, a complete ration that contains 12-16 percent total protein may be fed at the rate of 4-6 pounds (1.8-2.7 kg) per day. This ration is composed of 3-5 pounds (1.4-2.3 kg) of corn or other grain and 0.8-1.0 pound (0.36-0.45 kg) of protein supplement. During the winter a ration containing 11-14 percent total protein may be fed at the rate of 5-7 pounds (2.3-3.2 kg) per day. This ration is composed of 4-6 pounds (1.8-2.7 kg) of corn or other grain and 0.8-1.0 pound (0.36-0.45 kg) of protein supplement.

Gilts and Sows

Flushing and Breeding. *Flushing* is the practice of increasing the amount of feed fed, for a short period of time prior to breeding, to improve the condition of the gilt. Bigger litters are farrowed if this practice is followed. Flushing does not appear to be necessary for gilts that are in good condition at breeding time. Flushing has not been shown to be beneficial for mature sows.

Feed a ration containing 11-14 percent total protein at the rate of 6-9 pounds (2.7-4.1 kg) per day for approximately three weeks prior to breeding. This ration may contain 5-8 pounds (2.3-3.6 kg) of corn or other grain and 1.0-1.2 pounds (0.45-0.54 kg) of protein supplement. Do not continue feeding this ration after breeding.

Sows will usually come in heat within one week after pigs are weaned and should be bred back at that time. The same rations as used during the gestation period may be fed to sows during the breeding season. Adding

200 grams of antibiotics to the ration for one week prior to breeding will increase the number of pigs born alive.

Gestation Period. Weight gain must be controlled during the gestation period. This is done by controlling the intake of energy. A gain of 30-60 pounds (13.6-27.2 kg) is about right for sows and 70-100 pounds (31.8-45.4 kg) for gilts.

Several methods of restricting energy intake may be used when feeding sows and gilts during the gestation period. Four to 5 pounds (1.8-2.3 kg) of feed may be hand-fed during the first two-thirds of gestation. Increase this to 6 pounds (2.7 kg) during the final one-third of gestation.

Feeding on pasture will also restrict energy intake, Figure 13-6. Alfalfa, ladino, and red clover are good legume pastures to use. Legume pasture may increase estrogen activity, which could impair reproduction. Orchard grass and Kentucky bluegrass are good non-legume pastures to use. Concentrate intake for gilts should be limited to 2-3 pounds (0.9-1.4 kg) of a 14 percent protein ration per day during the first ten weeks on pasture. Increase this to 4-5 pounds (1.8-2.3 kg) from ten weeks until farrowing. Sows usually do not need any additional concentrate for the first ten weeks of gestation when they are on good pasture. After ten weeks, feed 2-3 pounds (0.9-1.4 kg) per day.

Self-feeding may be used during gestation if bulky rations are used. Adding a good-quality ground alfalfa hay, dehydrated alfalfa meal, or ground corn cobs to the ration provides the needed bulk. Ground ear corn and oats may also be used to add bulk to the ration, Figure 13-7.

Sows and gilts may be self-fed a high-energy ration if access to the feeders is limited to one day out of each three. Be sure adequate feeder space is available with this method of limited feeding.

Feeding corn silage at the rate of 10-15 pounds (4.5-6.8 kg) per day to sows or 8-12 pounds (3.6-5.4 kg) per day to gilts will restrict energy intake. Supplement silage with 0.5-1.0 pound (0.22-0.45 kg) of protein per day. Do not feed moldy silage, as it may cause abortion.

Provide a salt-mineral mix free choice during gestation. Also be sure a good supply

Figure 13-6. Feeding the breeding herd on pasture will help control energy intake and provide needed exercise.

Figure 13-7. The breeding herd may be self-fed on pasture if bulky rations are used to limit energy intake.

Lactation

Feed a ration containing 13-16 percent total protein at the rate of 10-14 pounds (4.5-6.4 kg) per day to gilts and 11-15 pounds (5-6.8 kg) per day to sows. A full feed of a high-energy, low fiber ration is necessary for maximum milk production during lactation.

FEEDING BABY PIGS

About one-fourth of the baby pigs lost before weaning die because of poor nutrition or starvation. Each pig needs to nurse shortly after birth to receive colostrum milk, which helps protect the pig from disease. The sow reaches maximum milk production in three to four weeks and then production falls off. Pigs must be eating well by then, in order to get the nutrients they need.

Baby pigs will start to nibble on a creep feed within a week after being born. Provide fresh feed daily in pans. Commercial pre-starters containing 20-22 percent total protein are usually used. These are fed to early weaned pigs or to pigs when the sow is not providing an adequate amount of milk. Pre-starters usually contain corn, soybean meal, dried skim milk, minerals, vitamins, and antibiotics and are fed in pelleted or crumbled form. Sugar (sucrose) is sometimes added to increase palatability. Use a prestarter that has the sugar incorporated in the pellet, rather than sugar-coated pellets. Provide plenty of clean, fresh water to baby pigs.

of clean, fresh water is available. If gilts or sows appear to be too thin during the last part of the gestation period, increase the concentrate by 1-2 pounds (0.45-0.9 kg) per day for the last 3-5 weeks before farrowing.

Farrowing

Additional bulk may be added to the ration 3-5 days before farrowing by substituting wheat bran, oats, dried beet pulp, ground legume hay, or dehydrated alfalfa meal for part of the corn. At farrowing, about one-third of the ration may be made up of these bulky feeds. Adding bulk to the ration at farrowing may help prevent constipation and reduce problems with mastitis-metritis-agalactia (MMA). Make sure there is a good supply of fresh water at farrowing time, Figure 13-8.

When weaning pigs that are more than three weeks of age, creep-feed a starter feed containing 18-20 percent total protein. Starter rations typically contain corn, soybean meal, dried skim milk, minerals, vitamins, and anti-biotics. Sugar is sometimes added to starter rations to increase palatability. One to 3 percent fat may be included, which improves the binding characteristics of the pellets, reduces dustiness, improves palatability, and increases feed efficiency. Commercial pig starters are available.

Figure 13-8. Feed and water are always available to the sow in this farrowing crate.

FEEDING MARKET PIGS

The nutritional requirements of young pigs are greater than those of older pigs. Protein content of the ration can be reduced as pigs get older. A 16 percent total protein pig grower can be used from weaning to about 75 pounds (34 kg). A complete ration containing corn and protein supplement with antibiotics is typical.

Scours (diarrhea) is sometimes a problem during the first two or three weeks after weaning, especially with early weaned pigs. Replacing 10-15 percent of the corn in the ration with ground oats for two or three weeks may help prevent scouring.

Finishing rations usually contain 12-15 percent total protein. From 75-125 pounds (34-56.7 kg), the pig should receive a ration containing about 14-15 percent total protein. From 125 pounds (56.7 kg) to market, the protein level can be reduced to 12-14 percent.

Complete rations contain corn, soybean meal, minerals, vitamins, and antibiotics. After the pigs reach 125 pounds (56.7 kg), the use of antibiotics in the ration is optional.

While protein-feeding recommendations for swine usually are stated in terms of crude protein, it is important to remember that the critical need is for the essential amino acids. Corn-soybean meal diets fed at the recommended crude protein levels will provide the ten essential amino acids in sufficient quantity to meet nutritional needs. However, as soybean meal is replaced with corn in the ration, lysine is the first amino acid to become inadequate. It is possible to add synthetic lysine, which is relatively inexpensive, to the diet when replacing soybean meal with corn to make a more economical ration.

If too much soybean meal is replaced with corn in the diet, tryptophan is the next amino acid to become limiting. Synthetic tryptophan

Figure 13-9. Feed storage bins, feed augers, and a grinding/mixing facility are used by this swine feeder to provide complete growing/finishing rations.

is currently too expensive to be used as a replacement in swine rations.

High-lysine corn may be used in the swine ration; however, lysine content appears to be quite variable. It is important to have high-lysine corn analyzed for lysine content to assure an adequate level of this essential amino acid in the swine ration.

While rations may be fed free choice, it is a more common practice to feed complete rations, Figure 13-9. More uniform growth results from feeding complete rations. Pigs tend to overeat protein (especially soybean meal) if it is fed free choice. On pasture, pigs gain faster when fed complete rations. Growing-finishing pigs may be fed on pasture and if they are on good legume pasture, the protein content of the ration can be reduced by about 2 percent. However, using pasture reduces the rate of gain and is not as common a practice among swine producers as it once was.

SUMMARY

Carbohydrates and lipids (fats and oils) are the primary sources of energy in swine rations. The energy intake of the breeding herd must be limited or the animals will get too fat.

Energy intake may be limited by increasing the level of fiber in the ration.

Swine are simple-stomached animals and must have the ten essential amino acids included in the diet. Lysine is the first limiting amino acid in swine growth.

Necessary minerals and vitamins are provided in the ration. A good supply of fresh, clean water must be available for maximum production.

A corn-soybean meal diet is the basic ration used for feeding swine. Other energy and protein sources may be used, but are evaluated against this basic diet. Growing-finishing swine make only limited use of roughage in the diet. Roughage is used in rations for the breeding herd to limit energy intake.

Feed additives are commonly used in swine rations and improve efficiency of production.

Creep-feeding of baby pigs is recommended. Prestarters should be used for early weaned pigs.

As growing-finishing pigs get older, the level of protein in the diet may be reduced. Pasture may be used for growing-finishing pigs, but the rate of gain is slower.

REVIEW

1. What nutrients in feedstuffs provide energy in the ration?
2. Why can mature hogs use more fiber in the diet than immature hogs?
3. What is the recommended source of carbohydrate in prestarter rations for baby pigs?
4. What effect does the addition of lipids in the diet have on the requirements for other nutrients?
5. List the ten essential amino acids that must be provided in swine diets.
6. List the major and trace minerals needed in swine diets.
7. List the vitamins that need to be included in swine diets.
8. How much water is needed for market hogs and how much for sows with litters?
9. What are the characteristics of corn that make it the grain of choice when feeding swine?
10. Corn is low in what essential amino acids?
11. Discuss the use of high-moisture corn in swine feeding.
12. Why are self-fed, complete rations more commonly used in swine feeding than rations fed free choice?
13. Compare other grains to corn as an energy source for swine.
14. Why is soybean meal the protein source most commonly used in swine rations?
15. Compare other protein sources to soybean meal as protein sources in swine rations.
16. Discuss the use of roughages in feeding growing-finishing hogs and in feeding the breeding herd.
17. What sources are commonly used to provide minerals in swine diets?
18. How are vitamins usually included in the swine ration?
19. Discuss the use of additives in swine feeding.
20. Describe recommended diets for herd boars.
21. Describe recommended rations for feeding gilts before breeding.
22. What is flushing, and why is it used when breeding gilts?
23. Describe a flushing ration.
24. Describe rations that might be fed to gilts and sows during the gestation period.
25. Discuss how to limit energy intake of gilts and sows during gestation.
26. Describe recommended feeding at farrowing time.
27. Describe recommended rations during lactation.
28. Discuss the feeding of baby pigs.

29. Describe the typical content of prestarter and starter rations.
30. Describe finishing rations for various weight groups of growing-finishing hogs.

14

FEEDING SHEEP

OBJECTIVES

After completing this chapter you will be able to:

- Describe nutrient requirements for sheep.
- Feed sheep during gestation and lactation.
- Feed the ram.
- Feed lambs to weaning and from weaning to market.
- Feed replacement ewes.

NUTRITIONAL NEEDS

Sheep are ruminants, as are cattle, but have a much smaller digestive system. As much as 89 percent of the total feed used for sheep in the United States comes from roughages. Roughages are more economical than concentrates and, in order to keep production costs as low as possible, it is important to balance the use of roughages and concentrates to meet the nutritional needs of sheep. The nutrient requirements for sheep are given in tables in the Appendix.

Energy

A deficiency of energy in the diet has a greater limiting effect on production than does any other nutrient. Not providing enough feed or using poor quality feeds may result in energy shortages.

Age, size, stage of lactation, pregnancy, and growth are factors that affect the energy requirements of sheep. Additionally, stress of

any kind, environment, and shearing all may increase the energy requirement.

Sheep on pasture or range have maintenance requirements from 10-100 percent greater than those fed in confinement. The availability of feed and water, kind of terrain, and distance sheep must travel daily for feed all affect the maintenance requirement, Figure 14-1.

Energy requirements are less for dry ewes and for pregnant ewes during the first fifteen

Figure 14-1. Maintenance requirements vary depending on several factors when sheep are on pasture.

weeks of pregnancy. During the last six weeks of pregnancy, more energy is needed to provide for fetal growth and the development of high milk producing potential. Too much energy intake may result in excessive fat deposits, leading to increased lambing difficulty. A shortage of energy may result in higher death loss of lambs, low birth weights, and pregnancy toxemia in the ewe.

Ewes nursing twins have higher energy requirements than those nursing one lamb. Energy requirements of all ewes are reduced during the last 8 weeks of lactation.

Breeding, finishing, and early weaned lambs all have different energy requirements. These are listed in the tables in the Appendix.

Protein

Although sheep are ruminants, research has shown that the quality of protein does influence its utilization. Protein produced by microorganisms in the rumen does not supply all the essential amino acids in the quantity needed for maximum production. Methionine is the first limiting amino acid for weight gain and wool growth, followed by lysine and threonine. Cystine can replace methionine for wool growth. Attention to these essential amino acids when balancing rations for sheep will improve production.

Urea or other nonprotein nitrogen sources may be used for all the supplemental nitrogen needed in high-energy, grain-based diets. For proper utilization, diets must be properly balanced and the NPN source fed continuously. While urea can be used in high roughage diets, it is not as satisfactory as other protein supplements as a nitrogen source. Biuret, which releases ammonia more slowly than urea and is less soluble, is more satisfactory as a NPN source under range conditions.

When NPN sources are used in the diet, time must be allowed for the rumen microorganisms to become adapted, which may take as much as three to five weeks. Generally, urea should be limited to not more than 1 percent of the dry matter in the diet and a readily available source of carbohydrate, such as grain, should be included in the ration.

Urea may be toxic, especially if there is a shortage of available energy in the diet. Toxicity will vary with the form in which urea is fed and the degree to which the rumen has adapted to its use.

Minerals

There are 15 mineral elements that are essential for sheep. Major minerals needed are sodium, chlorine, calcium, phosphorus, magnesium, potassium, and sulfur. Trace minerals needed are iodine, iron, copper, molybdenum, cobalt, manganese, zinc, and selenium.

Fluorine is toxic to sheep when the intake level is too high. Levels above 150 parts per million (ppm) on a dry-matter basis are toxic to finishing lambs. Acute toxicity occurs when the level reaches 200 ppm. Fluorine levels in the diet of breeding sheep should be limited to not more than 60 ppm on a dry-matter basis.

Minerals are provided in the diet through the use of salt and mineral supplements. Salt is usually added to mixed feeds at the rate of 0.5 percent of the complete diet or 1 percent of the concentrate mix.

Vitamins

Sheep require vitamins A, D, E, and the B complex vitamins. Green roughages are a good source of vitamin A. Vitamin D usually does not need to be added to the diet if the sheep are receiving enough sunshine. Sheep fed in confinement may need vitamin D added to the ration.

Young nursing lambs need vitamin E, but no demonstrated need for supplementing the diets of older sheep with this vitamin has been shown. A vitamin E deficiency in nursing lambs can cause white muscle disease (nutritional muscular dystrophy or stiff lamb disease).

Figure 14-2. Many kinds of pasture may be used for the ewe flock.

The B complex vitamins are synthesized in the rumen of sheep. Early weaned young lambs (less than two months old) may need some supplementation of B complex vitamins before the rumen becomes functional. Other vitamins needed by sheep are usually present in the diet in sufficient amounts so they do not have to be added to the ration.

Water

Water needs of sheep vary with air temperature, rainfall, snow cover, amount of dew on pasture, age, breed, stage of production, wool covering, amount of feed intake, frequency of watering, composition of feed, and amount of exercise. During late pregnancy and lactation, ewes need more water than when dry. Generally, a clean, fresh supply of water should be maintained where possible. Sheep require from 1 quart (0.9 litre) to more than 1 gallon (3.8 litres) of water daily, depending on the factors mentioned above.

FEEDING THE BREEDING FLOCK

Breeding Time (Flushing)

Flushing is feeding the ewe a ration for ten days to two weeks before breeding and two weeks after breeding, which causes it to gain weight rapidly. This practice may increase the lambing percentage by 10-20 percent. If the ewes are already in good condition, flushing will not increase the lamb crop by as high a percentage. Ewes that are already fat should not be fed a flushing ration because fat ewes may not breed at all.

One method of flushing is to put the ewes on a better quality pasture. Another method is to feed corn or oats or a mixture of the two at the rate of 0.5-0.75 pounds (0.2-0.3 kg) per head per day. Use caution when putting ewes on lush legume pastures when the rainfall has been heavy. Legume forages under these conditions may contain a high level of coumestrol, a plant estrogen, which can delay conception.

Gestation Feeding

Use roughage as the basic ration during the early part of the gestation period. High quality hay or haylage from legumes or grasses or a mixture of the two may be used. Corn, grass, or legume silage may be used in the ration. Chop silage finer than for beef or dairy cattle. Ewes may be put on stubble or stalk fields for pasture. Native range pastures, permanent pastures, rotation, or temporary pastures may all be used as a roughage source during this period, Figure 14-2.

Rotating pastures increases the amount of feed available to the ewes. Rotation also helps to break the internal parasite cycle. Pastures should be rotated every two to three weeks, depending on their kind and quality. Do not overstock the pasture, because sheep graze very close to the ground and may kill the vegetation. The stocking rate for pasture depends on the kind and quality of pasture used.

Provide salt, mineral mix, water, and shade on pasture. More even grazing of the pasture will result when the salt and mineral feeders are moved occasionally from place to place. It may be necessary to use supplemental pastures when native range pasture is used.

Poor quality pastures may require the feeding of some additional hay. One to 2 pounds (0.45-0.9 kg) of legume hay will generally meet the ewes' needs. Silage may be substituted for hay at the rate of 2-3 three parts of silage for each 1 part of hay. Corn silage is low in protein and calcium and requires additional supplementation to provide these nutrients unless one-half the roughage is from legume hay.

Finely chopped corn silage containing 32-38 percent dry matter may be fed as a complete diet to ewes during gestation, if the following materials have been added per ton of silage at ensiling time:

- 15 pounds (6.8 kg) of urea
- 10 pounds (4.6 kg) of limestone
- 4 pounds (1.8 kg) of dicalcium phosphate
- Equivalent of 1 pound (0.45 kg) of pure sulfur

This diet has also proved satisfactory for ewes nursing single lambs. Ewes nursing twins must have the ration supplemented with ⅔ pound (0.3 kg) of soybean meal per head per day for the first four weeks after lambing. Another protein source that supplies the same amount of protein per day may be substituted for the soybean meal.

Some concentrate mixture needs to be fed during the last four to six weeks of the gestation period to meet the energy needs for the rapidly developing fetus. The fetus gains about two-thirds of its birth weight during the last six weeks of pregnancy. Feed a ration that will allow the ewe to gain about 20-30 pounds (9-14 kg) during the gestation period.

Corn, grain, sorghum, oats, barley, and bran are often used in concentrate mixtures for ewes. A protein supplement may be needed, especially if the pasture or hay is of poor quality. Soybean meal, linseed meal, urea, or commercial protein supplements are used. Limit the use of urea in sheep rations to no more than one-third of the protein source.

Use of Antibiotics

Studies done in South Dakota and Wyoming suggest that the feeding of an antibiotic may reduce the mortality rate of lambs. The South Dakota research showed that feeding Aureomycin to ewes at the rate of 60 milligrams per head daily for 80 days, beginning six weeks before lambing, lowered the lamb mortality rate. The average mortality rate of the lambs in the group fed Aureomycin was 3.9 percent over a three year period. Lamb mortality rate in the control group averaged 14.5 percent. There was no effect on weight changes of the ewes or in the rate of gain in the lambs from birth to weaning.

The Wyoming research produced similar results when 65 milligrams of Aureomycin was fed per head per day to the ewes. Feeding Aureomycin began six weeks before lambing and ended six weeks after lambing began. Lamb mortality between birth and 14 days of age in the group fed Aureomycin was 4.8 percent in the first year of the study and 10.9 percent in the second. This compared to mortality rates of 13.8 percent and 15.6 percent in the control group. The general health of the lambs also appeared to be better in the group fed Aureomycin.

Table 14-1. Rations that may be self-fed to ewes.

Feed	Early gestation		Late gestation		Lactation	
	(lb)	(kg)	(lb)	(kg)	(lb)	(kg)
Corncobs, ground	70.00	31.75	65.00	29.48	60.00	27.22
Alfalfa meal	5.00	2.27	5.00	2.27	5.00	2.27
Corn, ground	10.00	4.54	15.00	6.80	20.00	9.07
Soybean meal	15.00	6.80	15.00	6.80	15.00	6.80
Steamed bone meal	1.00	.45	1.00	.45	1.00	.45
Salt, trace-mineralized	1.00	.45	.50	.23	.50	.23
Oat hay, ground	80.00	36.29	75.00	34.02	70.00	31.75
Alfalfa meal	5.00	2.27	5.00	2.27	5.00	2.27
Corn, ground	10.00	4.54	15.00	6.80	20.00	9.07
Soybean meal	5.00	2.27	5.00	2.27	5.00	2.27
Steamed bone meal	1.00	.45	1.00	.45	1.00	.45
Salt, trace-mineralized	1.00	.45	.50	.23	.50	.23
Straw			37.50	17.01	29.00	13.15
Legume hay (high quality)			37.50	17.01	29.00	13.15
Shelled corn			25.00	11.34	35.00	15.88
Soybean meal					7.00	3.18

Self-Feeding Ewes

Ewes may be self-fed complete ground mixed rations during gestation and lactation. This practice reduces labor requirements and increases the use of lower quality roughage; however, care must be taken to prevent the ewes from becoming too fat. Total feed intake can be controlled by limiting the amount of time the ewes have access to the self-feeders. Rations designed for self-feeding must contain a high percentage of roughage. Ewes nursing twins require a higher percentage of grain and protein in the ration than do ewes nursing single lambs. Self-fed rations that have been used successfully at the University of Illinois are shown in Table 14-1.

A ration made up of 34 percent ground ear corn and 66 percent ground grass-legume hay may be self-fed starting thirty days before lambing and continuing until sixty days after lambing. If there is a high percentage of multiple births in the flock, substitute ground shelled corn for ground ear corn. During late pregnancy, a ration of 80-85 percent high quality legume hay and 15-20 percent grain may be self-fed. Increase the grain to 20-25 percent of the ration after lambing.

Lactation Feeding

Nutrient requirements for ewes nursing lambs are higher than during the gestation period. The National Research Council recommends approximately 0.13-0.14 pounds (0.06 kg) more protein per head per day during the early part of the lactation period for ewes nursing twins as compared to ewes nursing single lambs. Current research indicates that the protein requirement may be as much as 0.25-0.30 pounds (0.11-0.14 kg) higher per head per day for ewes nursing twins compared to ewes nursing singles.

During the first few days after lambing, reduce the amount of grain in the ration and provide plenty of fresh water. A mix of equal parts of oats and bran with hay fed free choice may be used.

Ewes should be eating a normal ration by the third day after lambing, for increasing milk production to meet the needs of the nursing lambs. Some suggested rations are:

a. 1.5-2.25 lb (0.68-1.02 kg) shelled corn; 4-5 lb (1.8-2.3 kg) alfalfa hay;

b. 1.5-2.25 lb (0.68-1.02 kg) shelled corn; 1.5 lb (0.68 kg) alfalfa hay; 7.5-10 lb (3.4-4.5 kg) alfalfa silage;

c. 1.25-1.75 lb (0.57-0.79 kg) shelled corn; 0.25-0.30 lb (0.11-0.14 kg) soybean meal; 1-1.5 lb (0.45-0.68 kg) alfalfa hay; 7.5-9.5 lb (3.4-4.3 kg) corn silage.

The lower amounts in the above rations are fed to lighter weight ewes nursing singles and the higher amounts are fed to heavier ewes nursing twins.

A salt-mineral mix is fed free choice with the above rations to assure adequate mineral intake in the diet. Because current research indicates a possible need for more protein during the first four weeks of lactation, it may be desirable to increase the soybean meal allowance by 0.3 pounds (0.14 kg) during that part of the lactation period. If low quality roughage is used in the ration, then the grain and protein allowance will need to be increased to meet the nutrient requirements of the ewes. Other rations may be developed, using locally available feeds, which will meet the nutrient needs of the lactating ewes.

Ewes need additional nutrients for about eight weeks after lambing to allow for maximum milk production to nurse the lambs. Pasture can be used as the roughage source for ewes that lamb in late April and May. Additional grain may not be needed if the pasture is high quality and not overstocked. Ewes lambing in September and October can receive most of their roughage needs for the first four to six weeks of the lactation period from pasture. Supplemental feeding of grain and protein should begin if the ewes are losing weight or the lambs are not growing as fast as they should be.

After two months, the ration may be reduced to the amounts fed during the last six weeks of the gestation period. Milk production is dropping at this time, and lambs should be eating dry rations. About a week before weaning lambs, reduce the ewes' intake of feed. Eliminate all concentrate (grain and protein supplement) from the diet and feed only roughage. Reduce water consumption during the last two days before weaning. This helps to decrease milk production so there will be fewer udder problems at weaning time. This also forces the lambs to eat more creep feed.

Ewes that are nursing triplets need more nutrients than those nursing singles or twins. During the first four weeks of lactation, the total diet should include 16-17 percent total protein. Roughage intake should be limited to 2-2.5 pounds (0.90-1.1 kg) of high quality legume hay. Feed 3-3.75 pounds (1.4-1.7 kg) of concentrate (grain and protein supplement) per head per day to ewes weighing 150 pounds (68 kg) with more being fed to heavier ewes. Ewes nursing triplets should also be fed twice a day rather than once a day.

Feeding the Ram

Rams that are not adequately fed have lowered fertility or may even become infertile. They do not have the strength and vigor necessary for the breeding season. Rams must not be too thin or too fat at breeding time.

Before the breeding season, the ram may receive all the nutrients needed from pasture. If the ram is losing weight or is thin, add 1-1.5 pounds (0.45-0.68 kg) of shelled corn or other grain mix to the diet per day. Ram lambs require more grain than mature rams.

During the winter, 3.5-5 pounds (1.6-2.3 kg) of hay and 1.5 pounds (0.68 kg) of concentrate mix should be fed. Heavier rams will need the higher level of hay in the diet. Silage may be used as the roughage in the diet and will substitute at the rate of about 3 parts silage for each 1 part of hay. The ram should gain some weight but not become too fat.

FEEDING LAMBS

Birth to Weaning

Colostrum. *Colostrum* is the first milk produced by the ewe after lambing. It provides energy, protein, vitamins, minerals, and antibodies needed by the lamb to prevent disease. Lambs are born with a low level of vitamin A in their system. Colostrum is especially rich in vitamin A, which helps overcome this deficiency. Ewes that produce twins have about 46 percent more vitamin A in the colostrum milk than ewes that give birth to single lambs.

Antibodies for disease prevention are not transferred through the umbilical cord in sheep as they are in many other species. This makes it especially important that the lamb get the colostrum milk to prevent infections.

The lamb must receive the colostrum milk as soon as possible after birth, preferably within thirty minutes and no later than one hour. If the lamb does not nurse during this time, the chances of saving it are minimal.

Colostrum may be frozen for use with orphan lambs or with lambs whose mothers do not produce milk. Colostrum from ewes whose lambs are born dead or die shortly after birth may be frozen for later use.

Creep-Feeding. *Creep-feeding* is setting up an area that lambs may enter to eat, but that ewes cannot enter. Creep-feeding should begin about seven to ten days after lambing starts. Lambs will start to eat grain at about ten days to two weeks after birth. Feed small amounts of grain and clean out the trough daily. Crack, crimp, or roll the grain until the lambs are six weeks of age. Complete pelleted rations may also be used.

Creep-feeding must be used with an early weaning program. It gives greater benefits under conventional feeding programs with an early lambing system than with a late lambing system.

Advantages of creep feeding:

1. Lambs gain faster, especially those from multiple births.
2. Supplemental feed is utilized more efficiently early in the feeding period as compared to after weaning.
3. Lambs may be marketed at an earlier age.
4. Earlier marketed lambs usually bring higher prices.
5. Early lambs can be sold without being fed on pasture, which frees the pasture for ewes and reduces internal parasite problems.

Corn, oats, grain sorghum, and barley are grains that may be used in a creep ration. Barley is not as palatable as the other grains and probably should not be used during the first two weeks of creep-feeding. A high quality legume hay from second or third cutting should be available to the lambs at all times. The creep ration should contain from 15-16 percent total protein. Early weaned lambs need 18-19 percent total protein in the ration. Adding molasses to the concentrate mix increases its palatability and reduces the dustiness of the ration. Limit molasses to 5-10 percent of the concentrate mix. Ten to 15 percent of the ration may be bran, which is highly palatable and somewhat laxative. Soybean meal, cottonseed meal, linseed meal, or a commercial protein supplement may be used to provide the necessary protein in the ration. Do not use urea in creep rations. Antibiotics (Aureomycin or Terramycin are most commonly used) may be included in the ration at the rate of 20-25 grams per ton. Provide clean fresh water in the creep feeding area. Creep rations may be hand-fed or self-fed. It is a common practice to hand-feed until the lambs are eating well and then put them on self-feeders.

Utilize home grown grains as much as possible in creep rations to hold down the cost. Simple creep-feeding rations work as

Figure 14-3. Self-feeding lambs require less labor. Water may be provided using automatic waterers.

well as more complex ones. Some simple mixes (100 lb; 45 kg) for creep-feeding include (hay and salt-mineral mix fed free choice):

a. 79.5 lb (36 kg) shelled corn; 20.5 lb (9 kg) soybean meal;
b. 41.5 lb (18.7 kg) shelled corn; 41 lb (18.5 kg) oats; 17.5 (7.8 kg) soybean meal.

Other rations may be formulated to meet the needs of the lambs using locally available feed sources.

Weaning to Market

Lambs may be weaned from two to four months of age depending on when they were born, how many multiple births there were, whether or not they were creep fed, how much grain or pasture is available, parasite problems, type of sheep, predators, market price, and market outlook. The most common time of weaning is from two to three months of age. Lambs born later in the season are more commonly weaned at the older age, with early lambs that were creep fed being weaned at the younger age. Early weaning at 6-10 weeks of age is being practiced by a number of producers. If the lambs have been eating creep

feeds, the rumen should be functioning properly at this time. Lambs should weigh 40-50 pounds (18-22.7 kg) for early weaning.

Use high quality feeds and do not make abrupt changes in the ration when feeding lambs. Vaccinate lambs for enterotoxemia (overeating disease) when starting on a full feed to protect lambs on high concentrate rations.

Market lambs may be fed out in drylot, utilizing self-feeding and complete pelleted rations, Figure 14-3. Lambs may also be hand-fed or pasture fed. Lambs that are eating well and weigh over 40 pounds (18 kg) should not be put on pasture, as this slows the rate of gain. Later lambs weaned after July 1 should be drenched before being put on pasture. Drenching is done to control internal parasites. Phenothiazine, thiabendazole, haloxon, and tetramizole are commonly used for drenches.

Rations for early weaned lambs need to be high in energy and total protein. The ration may contain from 10-25 percent roughage if self-fed. All concentrate rations may be self-fed successfully, but require careful management to prevent enterotoxemia (overeating disease).

Pasture may be used for the early part of the feeding period and then the lambs may be moved to drylot for finishing. Legume or legume-grass mixtures make the best pastures, but gains are slower when lambs are fed on pasture. Lightweight lambs make better use of pasture than do heavier lambs.

Lambs fed on pasture are grazed until about the first of December, weather permitting. A supplemental ration is then fed at the rate of 1 pound (0.45 kg) per head per day. This ration is made up of grain and hay to balance the nutrient needs of the animals. Increase the grain portion and reduce the hay beginning about the first of January. When the lambs are fed out in drylot, a protein supplement is added to the ration. Lambs fed out on high quality legume or legume-grass pastures do not need protein supplement in their diet.

Drylot rations use more grain and less roughage. Faster gains are achieved when the concentrate makes up about 65 percent of the ration and the roughage about 35 percent. Lightweight lambs can use more roughage during the early part of the feeding period, when the ration should be about 15 percent protein.

Corn, grain sorghum, barley, wheat, and oats are all popular grain feeds for lambs. Legume or legume-grass hays are commonly used for the roughage portion, with cottonseed hulls, peanut hulls, or peanut hay being used in areas where they are available.

Soybean meal, cottonseed meal, linseed meal, peanut meal, or urea may be used for the protein supplement. The ration should contain about 16 percent protein. Lambs will gain faster at protein levels above 16 percent, but the additional cost is not offset by the faster gains. Urea must not be used for more than one-third of the total protein in the ration. It is better to limit the use of urea to the last 25 pounds (11 kg) of feeding before market.

When lambs reach 85-90 pounds (38-41 kg) the protein level may be reduced to 13-14 percent of the ration. Lambs close to market weight require less protein than do younger lambs.

Feed salt and mineral free choice and provide a good supply of fresh clean water. Calcium and phosphorus are essential for proper bone development, but must be in the proper ratio to prevent the formation of urinary calculi (water belly). The ratio of calcium to phosphorus should be about 2:1. Copper is essential but must not be at too high a level in the diet, because it is toxic. If copper toxicity is a problem, then select a trace-mineral mix that does not contain copper.

It is a common practice to add a vitamin supplement to the ration of early lambs. Vitamins A, D, and E are usually included in the supplement.

The concentrate-roughage ratio in the ration should be changed every seven to ten days as the lambs become heavier. By the end of the feeding period, the lambs should be receiving 90 percent concentrate and 10 percent roughage.

Feeding Orphan Lambs

There are usually orphan lambs in the flock because some ewes will not accept their lambs and some ewes die at lambing. Also, in some cases of multiple births, the ewe cannot feed all of her lambs. These orphan lambs can be saved, if proper feeding and management practices are followed.

Sometimes a ewe whose lamb has died can be made to accept an orphan lamb. If this cannot be done, then the orphan lamb must be bottle-fed.

It is essential that the lamb receive colostrum milk, if it is to have much of a chance to live. Frozen colostrum milk may be used or fresh colostrum may be milked from another ewe that has just lambed. If no sheep colostrum is available, it is possible to use cow's milk by adding one tablespoon of fat, such as corn oil, per pint of cow's milk. The milk must be warmed to body temperature, but do not boil it, and be sure to use clean utensils. Feed

4-6 ounces (113-170 g) every 4 hours for the first 12 to 18 hours. By the end of the first week, feed as much as the lamb will drink in five minutes, four times per day. This is about 0.5-0.75 pints (0.22-0.35 litre).

A number of good commercial milk replacers are available for use in feeding orphan lambs. Milk replacers need to contain 30-32 percent fat, 22-24 percent total protein, and 22-25 percent lactose (dry-matter basis). Follow the directions on the package for mixing. Calf milk replacers are not recommended for lambs except in an emergency and, if used, must be fed at a more concentrated mix than that used for calves.

Keep orphan lambs in dry, well-bedded pens and free from drafts and cold. Use heat lamps if necessary, to provide enough warmth during cold weather.

Start the lambs on creep feed as soon as possible. If the lambs are eating well, they can be weaned from the milk or milk replacer at four to six weeks of age. At the latest, they should be ready to be weaned at eight weeks of age.

FEEDING REPLACEMENT EWES

Many producers are breeding ewe lambs at seven to eight months of age for lambing when they are one year old. These lambs need to be fed differently from those being finished for market and, therefore, should be kept in a separate group. Do not feed a high energy finishing ration because this will cause them to become too fat and will cause the udder to develop too many fat cells. These lambs should be fed in drylot until about September, because this permits better control of the diet. A ration of 1 pound (0.45 kg) of shelled corn and 2 pounds (0.90 kg) of high quality legume or legume-grass hay will meet the nutrient requirements for growth. An increase in the ration to 1.5-2 pounds (0.68-0.90 kg) of shelled corn will increase the rate of growth. A

ration of 1.5 pounds (0.68 kg) of shelled corn and 2.5 pounds (1.1 kg) of hay will also increase the rate of growth. During the fall and winter months, these lambs may be fed a ration of 1-1.5 pounds (0.45-0.68 kg) of grain and pasture or hay free choice. In all cases, care must be taken to adjust the ration before the lambs become too fat. This requires careful management on the part of the producer.

Lambs being kept to breed as yearlings and lamb at fifteen months of age may be fed a ration that allows for slower growth. More than one-half the ration may be roughage. Care must be taken not to allow the developing ewes to become too fat. Adjust the rations to the growth and condition of the individual flock.

SUMMARY

Because sheep are ruminants, a high percentage of their diet comes from roughage. Energy is the most limiting nutrient for production in sheep rations. Energy requirements for nursing ewes are higher than for dry ewes.

Although sheep are ruminants, the quality of protein in the diet does affect its utilization. Methionine is the first limiting amino acid for weight gain and wool growth. Urea can be used in sheep diets, but its use must be limited.

Ewes need to be in good condition at breeding time. Feeding a diet that causes rapid weight gains at breeding time is called *flushing* and is recommended unless the ewes are too fat.

Roughage is the basic ration to use during the gestation period. Some grain and protein supplementation may be necessary if the roughage is of poor quality.

Research has shown that feeding antibiotics such as Aureomycin can reduce lamb mortality.

Ewes may be self-fed but care must be taken to not allow the ewes to become too fat. Control access to the self-feeders.

Nutrient requirements for nursing ewes are higher than for dry ewes. Protein requirements may be higher than those recommended by the National Research Council. Pasture can be used as the roughage source for ewes that lamb in April and May.

The ram must be adequately fed or fertility problems may result. Rams must not be too fat or too thin at breeding time.

Lambs must have colostrum milk within one hour after birth or they probably will not survive. Colostrum contains antibodies that are essential for disease prevention.

Begin creep feeding lambs seven to ten days after lambing begins. Simple rations of cracked, crimped, or rolled grains with some protein and roughage will meet the needs of the lambs.

Market lambs may be fed out in drylot or feeding may start on pasture and be completed in drylot. Early weaned lambs need rations high in energy and total protein. Lightweight lambs can use more roughage early in the feeding period.

Orphan lambs can be saved by using good management practices. The best way is to get another ewe to accept the lamb. If this is not possible, begin feeding with colostrum milk, and then use a good sheep milk replacer. Begin creep feeding as soon as possible.

Replacement ewes need a different ration than market lambs and so should be fed in a separate group. Use grain and roughage but do not let them become too fat.

REVIEW

1. Why is energy so important in sheep nutrition and what factors affect the energy requirements of sheep?

2. Compare the maintenance requirements of sheep on pasture to those raised in confinement.

3. Compare the energy requirements of sheep nursing lambs with those of sheep that are dry.

4. What are the limiting amino acids for sheep weight gain and wool growth?

5. Discuss the use of urea in feeding sheep.

6. What are the major mineral elements needed by sheep?

7. How are minerals usually provided in the diet of sheep?

8. Discuss the use of vitamins in sheep nutrition.

9. Discuss water requirements of sheep.

10. What is flushing and why is it done?

11. What kind of ration is needed for flushing?

12. Discuss the use of roughage in sheep gestation rations.

13. How should silage be prepared if it is to be used as a complete diet for sheep during the gestation period?

14. How should the increased nutrient needs of the last four to six weeks of the gestation period be met?
15. Discuss the use of antibiotics in rations for sheep during the gestation period.
16. Describe a ration that might be used for self-feeding ewes.
17. Describe a ration that will meet the nutrient needs of sheep during the early part of the lactation period.
18. How should the ration for ewes be changed at weaning time?
19. Describe how the ram should be fed for best breeding performance.
20. Why is colostrum milk vital for the newborn lamb?
21. What are the advantages of creep feeding lambs?
22. Describe a ration that might be used for creep feeding lambs.
23. How should pasture be utilized when feeding market lambs?
24. Describe a ration that might be used for feeding market lambs in drylot.
25. How may orphan lambs be saved?
26. Describe the nutrient content of a good milk replacer for lambs.
27. How should replacement ewes that are bred at seven to eight months of age be fed?
28. How should replacement ewes that are bred at one year of age be fed?

15

FEEDING GOATS

OBJECTIVES

After completing this chapter you will be able to:

- Describe nutrient requirements of goats.
- Feed dairy goats.
- Feed Angora goats.

NUTRIENT REQUIREMENTS

Energy

Energy is a vital component of goat diets affecting the utilization of other nutrients and overall productivity. The basic maintenance requirement for energy in goat diets is similar to the requirements for sheep. Additional energy is needed in the diet for increased activity, type of terrain, amount of vegetation on range, and distance traveled to get feed. Stable-fed goats with minimum activity need a basic maintenance level in the diet. Light activity requires about 25 percent more energy. Goats on hilly, semiarid range land need an increase of about 50 percent above basic maintenance requirements. When vegetation is sparse and goats must travel long distances to graze, the energy requirement is about 75 percent above the basic maintenance requirement. Angora and Cashmere goats require more energy in the diet after shearing, especially during cold weather.

Energy requirements can be met by good quality roughage in the diet, except for early

weaned kids, for does during the last two months of gestation, and for lactating dairy goats. Concentrate needs to be added to the diet to meet the energy requirements of these animals. Angora goats will respond to supplemental feeding of concentrates with higher production of mohair. Goats will also gain weight faster if more energy is provided in the diet.

Goats probably require a minimum level of fats in the diet under some conditions, but currently, these requirements have not been defined. More research is needed in this area of goat nutrition.

Protein

The basic requirement for protein in goat diets is similar to that needed in sheep and dairy cattle diets. A minimum level of 6 percent total protein needs to be provided or feed intake will be reduced. This leads to deficiencies in both energy and protein, which results in reduced rumen activity and lowers the efficiency of feed utilization.

Additional protein is required in the diet for growth, pregnancy, lactation, and mohair production. Goats on range need higher levels of protein in the diet than do stable-fed goats, because of the increased activity required to get feed. Adding concentrate to the ration will provide the additional protein needed. An excessive amount of protein in the diet of goats with light activity is undesirable.

Urea may be used to replace part of the protein needed in goat diets. An excessive amount of urea in the diet can be toxic; therefore, the amount in the diet must be limited. In forage diets, replace no more than one-third of the total crude protein with urea. Up to one-half of the protein in the concentrate part of the diet may be replaced with urea. It requires about three weeks for the rumen to adapt to the use of urea in the diet.

Minerals

While mineral needs for goats have not been determined through experimental studies, it is believed that they need the same basic minerals required by other ruminants. Major mineral needs include calcium, chlorine, magnesium, phosphorus, potassium, sodium, and sulfur. Trace-mineral needs include cobalt, copper, fluorine, iodine, iron, manganese, molybdenum, selenium, and zinc. Other trace minerals may be needed in small amounts. Generally, feeds used in goat nutrition provide adequate quantities of the necessary minerals. In some instances, deficiencies may occur, especially of the major minerals.

Calcium may need to be added to the diets of lactating dairy goats. A shortage of calcium in the diet can lead to reduced milk production. Bone meal, dicalcium phosphate, ground limestone, and oyster shell added to the ration will provide calcium. Angora or meat-type goats usually get enough calcium from the forage they eat.

Phosphorus deficiency may occur with goats grazing on range land if the forage is deficient in this mineral. Supplementation of the diet with phosphorus is more likely to be needed with high-producing dairy goats. A calcium to phosphorus ratio of 1.2:1 is recommended in goat diets.

Salt provides the sodium and chlorine needed. The recommended level is 0.5 percent of the ration. Salt may be fed free choice.

Sulfur shortages in the diet may occur when the diet includes forages grown on some types of soils. Goats grazing on browse, which includes plants containing tannic acid, may show a sulfur deficiency. If the diet includes a high level of nonprotein nitrogen, a sulfur deficiency may result.

Iron deficiency is not common in goats grazing on range land. Young goat kids may show an iron deficiency, which can be corrected by injections of iron-dextran (150 mg) at two-to-three week intervals.

Iodine deficiency may occur in some parts of the United States, but can be corrected by using iodized salt. Do not force-feed iodized salt, as this will result in an excessive intake of iodine.

Vitamins

Goats on range or pasture will usually get enough of the necessary vitamins in the diet. A vitamin supplement may be necessary for goats not on range or pasture and high-producing lactating dairy goats. A vitamin B complex supplement may be needed for young, nursing kids, sick goats, or those with poorly functioning rumens.

Water

While goats are less dependent on available water sources than other domestic animals, it is recommended that a clean, fresh supply of water be available to meet their needs. Goats are less likely to drink water that is foul-tasting than are other animals.

Goats are more efficient users of water than many other animal species. There is some evidence to indicate that they can store water in the body for later use. They are not affected as much by high temperature stress and require less water to control body temperature than other species of domestic animals. Under some conditions, the daily intake of water from forage is higher than that of other domestic animals. The daily range of free water intake is from zero to several quarts.

Factors affecting free water needs include lactation, air temperature, water content of forage, exercise, and the salt and mineral content of the water. A lack of sufficient water in the diet results in a reduction in feed intake, with reduced production. Starvation can result if the water deficiency continues for too long a time.

HERBAGE AND BROWSE UTILIZATION

Compared to other domestic species, goats are unique in their ability to forage on herbage and browse. *Herbage* refers to vegetation that does not develop woody tissues and is commonly used for pasture, such as the grasses. *Browse* refers to the leaves and twigs of trees and shrubs.

Goats eat a wider variety of vegetation, especially woody plants, than other species. Goats particularly like the leaves of woody plants, such as small trees and shrubs, both deciduous and evergreen.

The crude protein and phosphorus content of browse is higher than that of grasses during the growing season. However, some browse species contain inhibitors that limit the availability of the nutrients that they contain. For example, some browse plants contain high levels of tannins, some contain excessive lignification of leaves and twigs, and others contain oils that affect the digestibility of the nutrients in the plants. Experience has

shown, however, that generally the browse plants selected by goats provide a higher level of nutrient value than might be expected from their chemical analysis.

The utilization of browse plants are an important aspect of goat nutrition. Some studies indicate that goats may be more efficient in their digestion of some forages than other domestic species of ruminants. However, for purposes of developing rations, consider that goats have approximately the same forage digestive efficiency as sheep. Further research may suggest ways of expanding the utilization of otherwise low value herbage, browse, plant refuse, and byproducts in goat nutrition.

DAIRY GOATS

Good nutrition is essential for efficient milk production by dairy goats. Roughages form the basis for dairy goat diets, with concentrates added as necessary to balance the ration for needed nutrients. Roughages are used because they provide the lowest cost source of nutrients, and goats, being ruminants, can effectively utilize them. Concentrates are needed during lactation and to supplement low quality roughages. Concentrates may be needed for efficient growth of young goats.

Energy feeds commonly used in goat diets include corn, oats, barley, bran, beet pulp, milo, wheat, and triticale. Cereal grains must be supplemented with protein, minerals, and sometimes vitamins. When corn is used as the main ingredient in the concentrate mix, it may cause digestive disturbances. Adding oats, wheat bran, brewer's dried grains, beet pulp, or other cereal grain to the ration will increase bulkiness and reduce the chances of digestive problems. A common concentrate mix is one-half corn and one-half oats.

Protein sources often used include soybean meal, cottonseed meal, linseed meal, beans, and peas. Urea may be used in the ration, but should be limited as outlined earlier. Dairy cow supplements, mineral mixes, vitamin mixes, calf starters, and milk replacers (calf or lamb) are examples of commercial feeds that may be used in dairy goat rations.

Common roughages used in dairy goat diets include legume and grass pasture, hay, and silage. Corn silage may be used in goat rations. Garden refuse and roots, such as sugarbeets, may also be used as part of the roughage in the ration. A good quality hay, which has been cured to retain the leaves and color, will provide more nutrients than low quality hay.

Gestation Feeding

High quality roughages provide the basic nutrients needed during the last six to eight weeks of gestation. If the does are thin, add some grain to the ration. A maximum of 2 pounds (0.90 kg) will usually provide enough energy to keep does in good condition. The grain supplement should contain 14 to 16 percent protein. Do not allow does to become too fat as this will increase health problems during kidding and early lactation. Reduce the grain in the diet if does are becoming too fat.

Provide pasture, if available, and make sure does get plenty of exercise. Locating the drinking water away from the feeding area will help ensure goats get more exercise.

During the winter, provide hay free choice and supplement with up to 1 pound (0.45 kg) of grain, if needed to keep them in good condition. Always keep the feed clean and provide fresh, clean water. It is a good practice to keep a block of trace-mineralized salt available at all times.

Several days before the does freshen, reduce the grain ration to one-half and add bran to provide more bulk. After kidding, feed a bran mash for a few days, gradually bringing the doe to full feed for milk production.

Lactation Feeding

Nutrient requirements are higher during lactation. The ration for milking does should contain 12-14 percent crude protein. When low quality roughages are fed, a 14-16 percent protein supplement is needed. High quality alfalfa hay or pasture will meet the protein requirements. Grain fed at the rate of 1 pound (0.45 kg) for each 3-4 pounds (1.4-1.8 kg) of milk produced will provide the needed energy for maximum milk production and prevent weight loss during lactation.

Keep a clean, fresh supply of water available at all times. Low quality roughages require additional minerals in the diet. High quality alfalfa provides enough minerals in the diet. Keep a trace-mineralized block of salt available.

Commercial dairy mixes may be used to supplement the roughage in the diet. Increase the amount of concentrate in the diet by 20-25 percent if fair-to-good legume-grass hay is used. Reduce the amount of concentrate mix by one-half if does are on good pasture.

Table 15-1 shows some suggested rations and concentrate mixtures for lactation.

Grains used in goat rations should be coarsely ground, rolled, crimped, or pelleted. Add 1 percent trace-mineralized salt and 1 percent calcium-phosphorus mineral mix to concentrate rations. Molasses (5-7 percent of concentrate mix) may be used to increase palatability and reduce dustiness of the ration.

Feeding the Buck

During the nonbreeding season, the buck does not need additional grain if he is on good pasture. During the breeding season, the same concentrate mix fed to the does may be used at the rate of 1-2 pounds (0.45-0.90 kg) daily. Provide roughage free choice along with clean, fresh water and minerals. Care must be taken not to allow the buck to get too fat. Reduce the intake of energy feeds as needed to prevent this. Make sure the buck gets plenty of exercise.

Table 15-1. Rations and concentrate mixtures for lactating dairy goats.

Ration	Amount per day	
	(lb)	(kg)
Concentrate mixture	1-2	0.45-0.90
Alfalfa or clover hay	Feed free choice	
Concentrate mixture	1-2	0.45-0.90
Corn silage or roots	1.5	0.68
Alfalfa or clover hay	Feed free choice	
CONCENTRATE MIXTURES	Percent of mix	
Crushed oats	37.5	
Cracked corn	37.5	
Wheat bran	12.0	
Soybean meal	12.0	
Salt	1.0	
(Provides 13.5% digestible protein. Use with alfalfa or other legume forage.)		
Crushed oats	32	
Cracked corn	32	
Wheat bran	10	
Soybean meal	25	
Salt	1	
(Provides 16% digestible protein. Use with high producing does or when forage is only grass or mixed fair-quality legume-grass.)		
Corn meal	40	
Ground oats	20	
Wheat bran	25	
Linseed meal	10	
Cottonseed meal	5	
(Provides 16% crude protein. Use with alfalfa forage.)		
Hominy	35	
Wheat bran	25	
Ground oats	20	
Gluten feed	10	
Linseed meal	10	
(Provides 16% crude protein. Use with alfalfa forage.)		
Corn meal	17.5	
Wheat bran	35	
Ground oats	10	
Gluten feed	37.5	
(Provides 18% crude protein. Use with mixed and legume hay.)		
Corn and cob meal	20	
Wheat bran	30	
Ground oats	12.5	
Brewer's grains	30	
Cottonseed meal	7.5	
(Provides 18% crude protein. Use with mixed and legume hay.)		

Feeding Kids

Kids must receive colostrum milk from the doe within a few hours after birth. Kids that do not receive colostrum milk will probably not survive. They may nurse the doe or be hand-fed colostrum from a nipple bottle. Feed colostrum milk for 2 to 3 days and then change to whole milk or milk replacer, if they are not nursing. Begin by feeding one-half pint (0.24 litre) 3 to 4 times daily and gradually increase the amount to 5 pints (2.4 litres) daily by the time kids are six weeks old. Teach kids to drink from a flat-bottomed pan or dish. They may be fed twice a day after the first few days. Be sure all feeding utensils are kept clean and sterilize them before use each day.

To prevent digestive upset, gradually switch from whole milk to milk replacer by blending the whole milk and milk replacer for the first week. Cow's milk may be used in place of goat's milk after the kid is a few days old. After the kids have adjusted, milk replacer may be fed without causing digestive problems. Milk replacers will provide the needed nutrients for growth during the first few weeks.

Provide a good legume hay (or fresh, green grass) and calf starter or rolled grain, along with fresh water at three to four weeks of age. Equal parts of cracked corn, crushed oats, wheat bran, and 10 percent linseed meal may be fed as the concentrate mix. Rumen activity will develop quicker and kids will start chewing their cud by the time they are three to four weeks of age. Provide fresh feed daily at the rate of 0.25-0.5 pound (0.11-0.23 kg) daily.

Milk replacer may be fed until kids are four months old. During this time, feed roughages and grains. Kids may be weaned as late as six months of age. Commercial calf starters may be used in the ration.

From four months to breeding, kids may be fed roughages that will provide enough nutrients for normal growth. If low quality roughages (grass pasture, grass hay, or corn silage) are fed, supplement the ration with a 12-14 percent dairy calf growing mix at the rate of 0.75-1 pound (0.34-0.45 kg) daily. Do not allow growing dairy goats to become too fat. Reduce the intake of energy feeds as necessary to prevent this.

Always provide clean, fresh water and minerals to kids as they grow. Commercial mineral mixes may be used.

ANGORA GOATS

Range Feeding

Angora goats are able to utilize a wide variety of browse. When year-around browse is available on range, additional feeding is generally not necessary. Supplemental feeding may be necessary during the winter when range is not available.

The quality of fleece is affected by the type of feed available on range. The fleece is light-weight and lacks natural oil when forage consists mainly of dry grass and weeds. The fleece is heavier and contains more natural oil when goats are on range that provides green browse or green grass and weeds. When goats are grazing on small grains, the fleece becomes coarse.

Guajillo and live oak are rated as excellent browse for Angora goats. Excessive amounts of guajillo may produce a condition called limber leg. Live oak provides year-round browse.

Brush rated as good includes briar, cat-claw, elm, ill-scented sumac, post oak, shin oak, small-leaved sumac, Spanish oak, and yaupon. Young goats should not be grazed on briar because they may become entangled in it and not have the strength to free themselves. Mature goats like briar and are sometimes used to clear this brush from range land. Cat-claw provides browse during the spring and summer for mature goats, but do not graze kids and young goats on catclaw, because of the danger of entanglement. Elm is available as browse during the spring and summer. Shin oak does not provide browse during the winter, and during early spring, care must be taken to prevent oak bud poisoning, which may occur if grazing is limited only to this one type of browse.

Cedar, coral bean, and wild plum are rated fair as browse for Angora goats. Cedar provides browse during the fall and winter. Second growth cedar is preferred by goats.

Black persimmon, mesquite, and white brush are rated poor browse. Mesquite is eaten by goats in the fall after frost. While they like mesquite beans, excessive consumption can cause compaction in the digestive tract.

Supplemental Feeding

Supplemental feeding may be necessary during dry periods and when winter range is not available. Heavier fleece weights are produced when supplemental feeding is used. The amount of supplemental feed needed depends on the condition of the range and the goats. Supplemental feeding may include pelleted feeds, 20 percent protein range cubes, shelled yellow corn, or other feeds locally available.

Pregnant does require more feed than dry does. Supplemental feeding may prevent problems with abortion.

Protein and energy may be provided by feeding 0.25 to 0.5 pound (0.11-0.23 kg) of cottonseed cake along with yellow corn at the rate of 0.5 to 1.0 pound (0.23-0.45 kg) per head per day. Goat cubes may be used in place of the cottonseed cake and corn at the rate of 0.5 to 0.75 pound (0.23-0.34 kg) per head per day.

Goats may be self-fed on range, with salt added to the mixture to limit feed intake. Mix 7 parts of concentrate to 1 part of salt, if the feeders are located a mile or more from water. If the feeders are placed near water, mix 3 parts of concentrate to 1 part of salt, to limit feed intake.

A common concentrate mix is 1 part cottonseed meal, 3 parts ground grain, and 1 part salt. Alfalfa or other ground roughages may be used in the self-fed mixture. If ground roughage is used, lower the amount of salt in the mix. Moving the feeders from time to time results in better utilization of the pasture. Salt-controlled feeding should be used only if all other methods of feeding are impractical.

Roughages, such as alfalfa, sorghum, peanut, Sudan, and Johnson grass hay, may be fed. Provide a minimum of 1 pound (0.45 kg) of hay per head per day.

Singed pricklypear and tasajillo may be fed free choice along with 0.25 pound (0.11 kg) of cottonseed cake per head per day. Eating raw pear or tasajillo causes sore mouths and makes it difficult for goats to graze.

Live oak brush may be cut and fed to goats. If it is cut at about 3 feet in height, the stump will sprout and provide browse.

Feeding Angora Kids

As with other animals, Angora kids need to receive colostrum milk within a few hours after birth. Orphaned kids should be allowed to nurse another doe that has just given birth. Cow's milk may be used to raise orphaned kids after they have received goat colostrum milk. Milk may be fed four times per day for the first two weeks and then reduce the feedings to three times per day. Start kids on grain when they are about two weeks of age.

Providing does with protein blocks during kidding season helps prevent kids from becoming lost from the does. For best growth, kids should be fed during the winter months. Use roughages with some supplemental grain feeding.

SUMMARY

Goats require energy in the diet for effective utilization of nutrients and overall productivity. The amount of energy needed increases with activity, rough terrain, sparse vegetation, and distance to feed. Roughage provides the basic ration for goats, but some supplemental feeding of grain and protein may be needed, especially for lactating dairy goats.

While mineral and vitamin needs are usually met by the forage and grain in the diet, it may be necessary under some conditions to add these nutrients to the ration. Additional calcium may be needed in the diet of lactating dairy goats. Salt should generally be fed free choice.

Goats are efficient users of water but a clean, fresh supply should be readily available at all times.

Goats have the ability to utilize browse on range land, an important aspect of goat nutrition.

REVIEW

1. Under what conditions do goats need additional energy above basic maintenance requirements in their diet?
2. What groups of goats may need additional energy above that provided by the roughage in the diet?
3. Under what conditions is additional protein required in goat diets?
4. Discuss the use of urea in goat rations.
5. How may the mineral needs of goats be met?
6. Discuss the use of vitamins in goat nutrition.
7. Discuss the water needs of goats.
8. Define herbage and browse.
9. Discuss the importance of browse in goat nutrition.
10. Why are other cereal grains often mixed with corn in dairy goat rations?
11. What protein sources are often used in goat rations?
12. Describe a ration recommended for feeding dairy goats during gestation.
13. Discuss the feeding of dairy goats during lactation.
14. Describe how the buck should be fed.
15. Describe feeding kids during the first four weeks after they are born.
16. Describe feeding kids from four months to breeding.
17. How does the quality of forage effect the quality of the fleece of Angora goats?
18. Rank the various kinds of browse as to their value for Angora goats.
19. Describe methods of providing supplemental feeding to Angora goats.
20. What kinds of rations may be used for supplemental feeding of Angora goats?
21. How should Angora kids be fed?

16

FEEDING HORSES

OBJECTIVES

After completing this chapter you will be able to:

- Describe nutrient requirements for horses.
- Select feeds for horses.
- Balance rations for horses.

NUTRIENT REQUIREMENTS— ENERGY

Horses with free access to feeds will tend to select feeds with higher energy content. They will eat higher quality forage before poorer quality forage and grain rather than hay. Energy intake can be controlled by feeding a combination of roughage and concentrate to meet the horse's requirements.

A deficiency of energy in the diet results in a loss of weight, poor condition, and delayed estrus in mares. Young horses must have sufficient energy to grow. Too much energy in the diet causes horses to become too fat. Fat horses are more susceptible to stress and founder, have more problems in reproduction, and do not live as long as horses that are not overweight.

Maintenance

Energy for maintenance is that amount of energy needed to maintain weight of the horse and normal activity of the nonworking horse. The energy requirements for horses are given in terms of digestible energy (DE) and total digestible nutrients (TDN) in tables in the Appendix. Energy needed for maintenance varies with the weight of the horse and its temperament.

Work

Additional energy above the maintenance requirement is needed for horses at work. The kind of activity is a major factor in determining how much additional energy is needed. Other factors include the length of time at work, the condition of the horse, its training, the ability and weight of the rider or driver, how tired the horse is, and the air temperature. The more intense the work, the higher the energy requirement. The DE energy above maintenance requirement needed per hour ranges from 0.2 kcal/lb. (0.5 kcal/kg) of body weight for a horse that is walking to 17.7 kcal/lbs. (39 kcal/kg) of body weight for a horse racing at full speed.

Pregnancy

Most of the fetal development takes place during the last 90 days of gestation. During this time, the energy requirement is estimated to be about 12 percent above maintenance requirement. As the fetus grows, the intake of forage decreases; therefore, the amount of energy from concentrates needs to be increased during the last 90 days of gestation.

Lactation

The amount of milk produced during lactation governs the amount of additional energy needed. The maximum amount of milk production occurs about six to eight weeks after foaling. Energy requirements are higher during the first three months of lactation and then decrease. The size of the horse also influences the energy needed for lactation, with ponies requiring less than larger horses. Horses require approximately 0.18 pound (0.36 Mcal) of DE per pound (0.45 kg) of milk produced.

NUTRIENT REQUIREMENTS— PROTEIN

A protein deficiency in the diet of horses results in reduced feed intake, which in turn leads to a deficiency of energy. The result is a loss of weight in mature horses and slower growth of young horses. Fertility and milk production may also be lowered by protein deficiency in the diet.

Mature horses have less need for protein than young horses. The protein content of feeds used in mature horse diets generally will meet their needs without additional protein supplementation.

The maintenance requirement of protein for mature horses averages approximately 1.3 pounds (0.59 kg) of crude protein per 1,000 pounds (453 kg) of body weight. Heavier horses require less protein and lighter horses require more protein in the diet for maintenance.

Quality

Some synthesis of amino acid occurs in the cecum and colon of the horse; however, the amount that is digested and absorbed is not known. It is recommended that the essential amino acids be included in the diet of horses, just as they are in the diets of other non-ruminant animals. Studies have shown that increasing the lysine content of the diet increases growth rate and feed efficiency of young horses.

Nonprotein Nitrogen

Nonprotein nitrogen sources such as urea may be used in horse diets. Horses are not as susceptible to urea toxicity as are ruminants. Up to 5 percent of the horse's diet may be made up of nonprotein nitrogen sources.

Maintenance

Diets composed of high quality hay and grain usually provide enough protein and sufficient balance of amino acids for maintenance of horses. The exact requirements of amino acids for maintenance of horses is not known.

Growth

Additional protein is required for efficient growth of young horses. They are less able to synthesize amino acids in the cecum and colon, as compared to mature horses. The diet of weanlings requires 0.6-0.7 percent lysine and yearlings need 0.4 percent lysine. If the quality of the feed is poor, additional protein may be needed in the diet.

Work

The level of activity of the horse has little effect on the protein requirement above the maintenance level. The amount of feed fed to meet energy needs will meet the protein needs of working horses. Little additional protein above maintenance level is needed for working horses.

While some nitrogen is lost by sweating, the amount is insignificant and does not require additional supplementation of the diet. Exercise does not result in muscle breakdown, so no additional protein is needed for muscle replacement during work.

Too much protein in the diet of working horses causes increased heart and respiration rates and sweating. These effects reduce performance; therefore, care must be taken not to overfeed protein.

Pregnancy

Maintenance levels of feeding will meet the protein needs during the early part of gestation. Protein requirements increase significantly during the last 90 days of gestation, to meet the needs of the developing fetus.

An average of 0.26 pound (0.11 kg) of total protein is needed per day, above the maintenance requirement, during the last 90 days of gestation. Lighter horses need less and heavier horses require more total protein for the developing fetus. The requirements for several different mature weights of horses are given in the tables in the Appendix.

Lactation

Additional protein above maintenance requirements is needed during lactation. The first three months of lactation require approximately 0.048 pound (0.022 kg) of total protein per pound (0.45 kg) of milk produced. The remainder of the lactation period requires approximately 0.046 pound (0.021 kg) of total protein per pound (0.45 kg) of milk produced.

NUTRIENT REQUIREMENTS— MINERALS

Horses require calcium, phosphorus, salt, and iodine in the diet. The need for salt is critical in the diet of working horses because of loss through sweating.

The need for calcium and phosphorus is especially high for growing, lactating, and pregnant horses. The skeleton and teeth make up about 5 percent of the body weight and are composed mainly of calcium and phosphorus. Milk is also high in these two minerals. The level of milk production during lactation

determines the need for additional calcium and phosphorus in the diet. The need for calcium and phosphorus is higher during the last 90 days of gestation, when the fetus makes most of its growth.

The ratio of calcium to phosphorus in the diet should be no less than 1:1. Do not feed more phosphorus than calcium, as this may interfere with the absorption of calcium in the body resulting in a calcium deficiency. Mature horses can tolerate ratios of 6:1 with adequate phosphorus intake; however, the recommended ratio is about 2:1. Mature horse diets should contain about 0.6 percent calcium and 0.5 percent phosphorus. The recommended ratio for foals and yearlings is about 1.5:1 and should not exceed 3:1. Diets for young horses should include about 0.8 percent calcium and 0.6 percent phosphorus.

Feed salt at the rate of 0.5 to 1 percent of the diet. It may be provided free choice as loose salt along with a readily available, adequate supply of fresh water. If horses are salt-starved, do not allow them free access to salt because the excess intake can be toxic. Bring them to a full feed of salt over a period of about one week, gradually increasing the amount available each day.

Little is known about the trace-mineral requirements of horses. Generally, an adequate supply, with the exception of iodine, is available in feeds normally used in horse diets. Iodine and other trace minerals may be supplied by feeding trace-mineralized salt free choice. Do not force-feed trace minerals, as this may be toxic to horses.

NUTRIENT REQUIREMENTS— VITAMINS

Little is known about the vitamin requirements of horses, but it is accepted that a need exists in the diet. Horses undergoing stress have a higher requirement for vitamins. Generally, vitamin needs can be met by high quality pasture or hay, especially when at least one-half is legume. Horses in confinement for long periods of time or those receiving poor quality forage may show signs of vitamin deficiency.

Vitamin requirements of horses can be easily met by including a vitamin supplement in the ration. Overfeeding vitamins increases the cost of the ration.

Horses with access to green forages generally do not need additional vitamin A in the diet. When the forage is low in carotene, additional vitamin A supplementation is recommended. Supplement at the rate of 5,000-15,000 IU of vitamin A per day, depending on the weight of the horse. Higher levels are recommended during gestation and lactation. Tables in the Appendix show the recommended levels for different mature weights of horses. Feeding excessive amounts of vitamin A over a period of time may cause bone fragility, hyperostosis, and exfoliated epithelium.

Horses on pasture or outside in the sunlight part of the day do not need additional amounts of vitamin D. When horses are kept inside much of the time, additional vitamin D should be supplied in the diet. The recommended level is 6.6 IU of vitamin D per kilogram of body weight. Excessive intake of vitamin D can cause calcification of the blood vessels, heart, and other soft tissues. It can also cause bone abnormalities and swollen joints. A level of vitamin D intake about 50 times the requirement may be toxic to horses. A high calcium intake increases vitamin D toxicity.

Vitamin E is usually found in adequate amounts in horse rations. There is no evidence that additional amounts of vitamin E has any effect on fertility in horses.

Good quality forage contains adequate amounts of the B-complex vitamins. B-complex vitamins are also synthesized in the cecum and colon of the horse. Deficiencies of B-complex vitamins rarely occur; however, horses under severe stress may need some supplementation in the diet.

NUTRIENT REQUIREMENTS— WATER

It is important that horses have an adequate supply of fresh, clean drinking water at all times. A lack of water can lead to digestive trouble such as colic. The average intake of water is 10-12 gallons (37.8-45.4 litres) daily. The requirement for water is increased by 20-300 percent for working horses. Lactation requirements are 70-80 percent above maintenance.

The amount of dry matter in the diet greatly affects the intake of water. Studies show that a horse needs about 0.2-0.4 gallon (0.8-1.5 litres) of water per 1 pound (0.45 kg) of dry-matter intake.

Higher air temperatures also increase the need for water. A rise in air temperature from 55 to 77°F (13-25°C) increases the water requirement by 15-20 percent.

Water horses at work at frequent intervals to reduce the chances of heat exhaustion. Horses that are hot should be cooled out or given only small amounts of water before being allowed to drink all they want. Idle horses should be watered at least twice a day when water is not available free choice. Infrequent drinking can result in impaction in the digestive tract.

ROUGHAGE

Roughages are the major source of energy in the diet of horses. The horse's stomach can only hold approximately 2 to 4 gallons (7.6-15 litres), a much smaller capacity than a cow's stomach. Roughages are digested mainly in the large intestine, rather than in the stomach. When horses are fed only once or twice a day, they tend to eat rapidly and some of the feed, especially grain, passes through the digestive tract without being completely digested. When horses are grazing, they spend about three-fourths of their time eating and the feed passes through more slowly, allowing more complete digestion.

Mature, idle horses can usually get enough energy from roughage alone. Working horses, pregnant mares, lactating mares, and growing foals often need some concentrate added to the ration to provide needed energy.

Pastures

Except as noted above, most adult horses can get the nutrients they need from a good legume-grass pasture. The amount of pasture needed varies with the use of the pasture, type of forage, moisture, and management of the pasture, Figure 16-1.

If the primary use of the area is to be an exercise lot, then one acre or less per horse is enough, but this will not provide enough feed for the horse. Tall fescue should be used for small areas such as this because it is more resistant to trampling. The horse must be fed as though it were in confinement.

Two acres per horse during the grazing season is the minimum amount of space needed to provide adequate feed, exercise space, and to reduce the problem of internal parasites. In western range areas with native grasses, 15 to 50 acres per horse are needed to provide adequate feed from the pasture. An improved, irrigated pasture in western range areas can provide enough feed to support one horse per acre.

Rotating horses from one part of the pasture to another will increase the carrying capacity of the pasture. This practice will also help to control internal parasites. Using an electric fence to divide pastures is a simple method of rotating pasture areas. Horses should be moved from one pasture area to another every three to four weeks. Periodic clipping of the pasture will help maintain fresh forage growth if rotating pastures is not practical.

Better use is made of the pasture if cattle are grazed with horses. Horses tend to undergraze some areas and overgraze others. Cattle will graze the pasture more evenly, thus eliminating the need for clipping the pasture

Figure 16-1. Adult horses can get most of the nutrients they need from good pastures.

periodically to maintain fresh growth. Because of different grazing habits, a better balance of legumes and grasses is maintained in the pasture when cattle are grazed with horses.

Horses and cattle tend not to graze around their own feces droppings but will graze around each other's droppings. By grazing both on the pasture, better utilization of the forage results with more uniform grazing.

Internal parasite eggs are deposited in the feces. Rotating pastures or grazing with cattle helps to control this problem. Cattle feces tend to be more scattered while horse feces tend to be more concentrated in one area. Cattle are not susceptible to the internal parasites of horses nor are horses susceptible to those of cattle. Therefore, each can graze around the feces of the other without danger of parasite infestation.

Generally horses do not bloat on pasture, but may founder when first put on lush spring pasture. For several days, provide a full feed of hay before putting them on pasture and allow grazing for only a few hours. This will allow horses to become used to the succulent feed, while preventing founder. Fat horses are more susceptible to this problem than others.

Overgrazing of pasture is a problem when pastures are continuously grazed. When pastures are overgrazed, the legume tends to first increase and then die out as overgrazing continues. This results in many bare areas or the growth of undesirable plants such as weeds in the pasture, reducing its productivity. Large bare areas will increase erosion problems. In those parts of the pasture that are undergrazed, the grasses tend to smother out the legume. Frequent clipping and rotating the pasture helps to prevent problems with overgrazing, Figure 16-2.

Horse pastures must have adequate fertilization to maintain productivity. Top-dress

Figure 16-2. Horses tend to graze pasture unevenly, resulting in overgrazed and undergrazed areas. Bare ground in the pasture may increase erosion problems.

grass-legume pastures annually with a minimum of 30 pounds (13.5 kg) of phosphorus and 60 pounds (27 kg) of potash per acre. On grass pastures with no legume, add 90 pounds (41 kg) of nitrogen to the above recommendation for top-dressing. For best results, apply the fertilizer during the winter or early spring.

Kinds of Pastures

In the western United States pastures may be irrigated, dryland, or native grass. In some parts of the United States summer pastures may contain permanent or annual forages. The management of the pasture varies with the type.

Irrigated pastures.   Graze horses on irrigated pastures for six to seven days and then move them to a different pasture area. Allow the grazed pasture to regrow for two to three weeks before moving horses back on it.

Care must be taken to leave enough stubble to allow regrowth of the pasture. Begin grazing when the grass is 6-8 inches (15-20 cm) high and take the horses off the pasture when it is grazed down to 2-4 inches (5-10 cm) in length.

Do not irrigate the pasture while horses are grazing on it. To do so results in compacted soil, restricted water penetration, and reduced plant growth, thus lowering the carrying capacity of the pasture.

Dryland pastures.   Dryland pastures are usually seeded to crested wheatgrass, intermediate wheatgrass, or Russian wildrye. Carrying capacity is increased when about 1 pound (0.45 kg) of alfalfa is seeded with the grass. Use nitrogen fertilizer in the spring to increase grass growth.

Crested wheatgrass produces good forage during the spring and fall. Early summer grazing is provided by intermediate wheatgrass. Russian wildrye provides forage throughout the grazing season. Clipping crested and

intermediate wheatgrass or grazing with cattle will reduce the coarse material and produce more uniform grazing with better utilization of the pasture.

Native ranges.   Native ranges have a much lower carrying capacity than other pastures. Overgrazing can be a serious problem on native ranges, as horses tend to graze much closer to the ground than cattle. Do not allow horses to remove more than one-half of the desirable forage when grazing native ranges. Rotation grazing helps to maintain the productivity of the pasture. To maintain the species growing on the native range, keep livestock off every third year. Range lands are fragile and overgrazing can cause damage that may take many years to repair.

Permanent pastures.   Permanent pastures which contain a mixture of legumes and grasses give the highest yields of forage for horses. Kentucky bluegrass is preferred by horses over other taller grasses. Horses also prefer clovers over alfalfa and birdsfoot trefoil. However, studies show that horses get the nutrients they need on any type of good legume-grass mixture.

When horses are grazed continuously on permanent pastures, there may be a problem with them tearing up the sod with their sharp hoofs. Kentucky bluegrass produces a tough turf that is not as easily destroyed as the turf of other grasses and legumes. Bluegrass is highly palatable and withstands close grazing. In many areas, Kentucky bluegrass is the primary grass used in permanent pastures for horses.

Orchard grass and brome grass are more productive during the cool growing season than Kentucky bluegrass. Horses readily eat these grasses unless they are allowed to get too mature. Clipping the pasture or stocking heavily enough to prevent excess growth helps prevent this problem. Neither orchard grass nor brome grass can tolerate overgrazing. Including a legume in the pasture mix provides nitrogen needed by the grasses and increases productivity of the pasture during the hot summer.

Timothy is a palatable forage for horses, but does not stand up well under close grazing. It is slower to recover from grazing than many other pasture forages. While timothy has long been considered the forage of choice for horses, it is no more nutritious than other grasses, and studies have shown that horses do just as well on other grasses and grass-legume combinations.

Tall fescue is a hardy grass and is recommended for exercise and other high-traffic areas because it develops a tough turf. However, it is not as palatable to horses as some other forages, and rations need to be fortified with grain. A more valuable pasture for horses will result if fescue is seeded with a legume. It is recommended that pregnant mares carrying extremely valuable colts not be grazed on tall fescue during the last four months of pregnancy because of the possibility of abortions. While the danger of abortion does not appear to be high, there have been some reports of this occuring on tall fescue pastures even though the ration appeared to be adequate in all respects. A fungus infestation on the fescue is the cause of the abortions.

Birdsfoot trefoil is deeper rooted than some other legumes and thus is more drought resistant. It also does better than some other legumes on poorly drained soils. It may be difficult to establish a stand of trefoil. A common mixture to use when renovating permanent pastures is 6 pounds (2.7 kg) birdsfoot trefoil mixed with 2 pounds (0.9 kg) Kentucky bluegrass or 2-3 pounds (0.9-1.4 kg) timothy per acre. Birdsfoot trefoil pastures should not be grazed closer than 3-4 inches (7.6-10 cm) in order to maintain the stand. Once established, birdsfoot trefoil will persist for many years, with proper management, in permanent pastures.

Alfalfa seeded with grasses such as brome grass or orchard grass makes a good pasture for horses. This combination is often used in rotation pastures used for other species of livestock. It is also used for hay.

Alsike and ladino clovers seeded with timothy or brome grass make a good permanent pasture on poorly drained soils. In some parts of the United States, the clover may not persist more than one to three years.

Other grasses and legumes that are locally adapted may be used for horse pastures. Because horses do not bloat, there is no danger in using any of the legumes that are adapted to local soil and moisture conditions. Legumes should be limited to no more than 35-40 percent of the mixture to minimize slobbering, which may occur with an excessive intake of legumes in the diet. Simple mixtures of one grass and one or two legumes are best to use for horse pastures. Differences in palatability and maturity dates make it undesirable to use more than one grass in a pasture mix.

Annual pastures. Annual grasses planted alone can provide summer grazing for horses. Typically, several different grasses planted in separate fields are used to provide grazing for the entire season. Oats or barley can be used for spring grazing up until about June. Sudan grass might be used from July until the first frost. Wheat and rye could be used in late fall and early spring.

Sudan grass should not be grazed after a frost or when regrowing after being stunted by drought. The prussic acid content is high in the new growth and, while horses are not as susceptible to prussic acid poisoning as other species, it is better not to graze Sudan grass at these times. There have been reports from the southwestern United States of problems with cystitis in horses grazed on Sudan grass and sorghum-Sudan hybrids that are in the active growing stage. The problem is more prevalent in mares than in stallions or geldings. It is recommended that these forages not be grazed by horses during the active growing stage, but hay harvested from these crops does not appear to cause cystitis.

Figure 16-3. Alfalfa-brome hay yields more digestible energy per acre than grass hay. Proper harvesting and curing is essential to prevent mold.

Hay

Use a high quality hay that is free from dust and mold for winter feeding of horses. Idle, mature horses and nonlactating brood mares can get enough nutrients from high quality hay to meet their needs. Low quality hay requires additional supplementation of the diet. Growing horses, working horses, and pregnant mares need additional feed to supplement even the nutrients found in high quality hay.

Hay may be a grass-legume mixture, or legume or grass alone. Harvesting at an early stage of maturity improves the quality of hay. Alfalfa-brome grass or alfalfa-timothy mixtures generally yield more digestible energy per acre than grasses alone, Figure 16-3.

Legume hays. Pure legume hays may be used in horse rations. They are higher in protein and minerals and more palatable than grass hays. Legume hays are particularly recommended for growing horses, pregnant brood mares, and working horses. Green legume hays may be slightly laxative to horses.

Alfalfa hay is highly recommended for horses when properly cured. It contains from 50-56 percent TDN and 15-18 percent crude protein and also has a high calcium and vitamin content.

Clovers may be used for horse hay but care must be taken that red clover and alsike clover are properly cured or they may contain mold. Do not feed moldy hay to horses or colic may result, as well as abortion in pregnant mares. When clovers are properly cured, they can be substituted for alfalfa but are not as high in nutrients.

Lespedeza cut at an early stage of growth makes an excellent hay for horses. It is higher in protein than red clover, but the calcium content is about one-half that of alfalfa hay. If lespedeza is cut when mature it is not as nutritious and has a lower digestibility.

Grass hays. Grass hays are less likely to be moldy than legume hays but they are not as high in nutrients and yield less per acre. Pure grass hays are often cut at too late a stage of maturity to provide maximum nutrients. Grass forages grown in combination with legumes make an excellent hay for horses.

Timothy has long been the hay of choice for horses. It is adapted to a wide range of climatic conditions, is easy to cure, and is generally free from mold and dust. Timothy is low in protein and must be supplemented with a high protein grain, such as oats, which increases the cost of the ration. Timothy is better for mature work horses than for growing horses, stallions, or mares. When it is harvested at too mature a stage, it is not a good hay to use for horses.

Brome grass is more palatable when harvested in the bloom stage. It makes a good horse hay and contains 46-52 percent TDN. Orchard grass hay is a little lower than brome grass hay in feeding quality. Prairie hay is not as palatable and is lower in protein than timothy hay. Oat, barley, wheat, and rye hays are often used in the Pacific Coast region of the United States. They are cut in the soft to stiff dough stage for highest quality.

Silage. Several kinds of silage may be used to replace up to one-half the hay in horse rations. While corn silage is preferred, grain sorghum and grass silages can be used. Use only high quality silage that is finely chopped and free from mold, spoilage, or frozen material.

Silage is generally too bulky for working horses and foals. It may be fed to mature idle horses, growing horses, brood mares, and stallions. If used in these rations, begin feeding it slowly to allow the horse to adapt to the change in diet. In horse rations, legume haylage is just as good or better than silage.

GRAIN

Additional energy is provided for horses by including grain in the ration. Hard working horses generally need grain in the ration, while idle horses may get fat on grass alone.

Grain is highly palatable to horses and they eat too much if it is fed free choice. It is best to feed small amounts several times a day rather than all the grain for the day at one feeding. When working horses are idle, the grain ration should be reduced to one-half and the amount of hay fed should be increased. Feeding too much grain on days

when working horses are idle can cause Azoturia ("Monday morning" disease).

When changing from one grain to another in the diet, do so gradually to allow the horse to become adjusted to the change in diet. Grains may be ground or rolled for feeding to horses, but do not grind any grain too fine.

Oats

Oats have been the traditional grain of choice to feed horses. Oats are bulky and palatable and do not cause digestive problems. Crimping or crushing oats will increase their palatability for horses. Oats are higher in protein than corn, but lower in energy. They are especially valuable as a supplement to low-quality grass hay. When fed with a legume-grass hay mixture, they usually will provide all the additional energy needed without feeding other grain.

Oats are variable in quality and may not be the most economical grain to use on a cost-of-TDN basis. Satisfactory horse rations can be developed using grains other than oats.

Corn

Corn is often a better buy on an energy basis than oats. About 15 percent less corn will provide the same amount of energy as a given weight of oats. Corn is especially good for improving the condition of thin horses and for feeding hard working horses. Coarse cracked shelled corn is better utilized by horses than whole shelled corn. Ear corn or ground ear corn can be fed in horse rations to reduce the tendency to eat the grain too fast.

Horses fed corn are more susceptible to colic than when oats are fed. Corn and oats mixed half and half make an excellent grain ration for horses. Because of its higher energy and lower fiber content, care must be taken when feeding corn to prevent horses becoming too fat.

Grain sorghum (milo). Grain sorghum can be substituted for corn in horse diets, and is best utilized in grain mixtures. Some varieties of grain sorghum are unpalatable to horses; it is low in vitamin A; and it is highly variable in protein content, ranging from 6-12 percent. Grain sorghum needs to be cracked or rolled when fed to horses or it may pass through undigested.

Barley

Ground barley is a satisfactory grain for horses, and it may be substituted for corn in the ration. When fed with wheat bran or oats, there is little risk of colic.

Wheat

Because of its high cost, wheat is rarely used in horse rations, except in the Pacific Northwest. Feed wheat with a bulky feed and use no more than one-third in the grain mix. Wheat needs to be rolled or coarsely ground for feeding to horses. When wheat is moist is gets doughy, which reduces its palatability.

Wheat Bran. Wheat bran is a bulky, palatable, slightly laxative feed for horses. It is often used in rations for horses that are under stress such as fatigue, foaling, or sickness. Wheat has a high cost per unit of TDN and is generally limited to no more than 5-15 percent of the ration.

Rye

Rye is unpalatable for horses and needs to be cracked or crushed when included in horse rations. It tends to gum when chewed and should be limited in horse rations.

Cane Molasses

Cane molasses may be added to the ration at the rate of 5-10 percent to reduce dustiness and increase palatability. At higher levels, it is laxative. Cane molasses is low in protein and expensive per unit of TDN. Dried molasses is sometimes added to the ration to increase consumption.

PROTEIN SUPPLEMENTS

Usually, little protein supplement is needed in horse rations. If at least one-half the ration is legume hay, the protein needs of the horse, except for mares in lactation, will be met. Young growing horses may need additional protein to make sure they grow adequately and to stimulate appetite. Show horses may be fed about 1 pound (0.45 kg) of linseed meal per day to make their hair coat bloom. Larger amounts are laxative. Add protein supplements to horse rations when the quality of the forage is low, as in the case of late-harvested grass hays.

Soybean Meal

Soybean meal is an excellent protein supplement for horses. It is high in protein and has a good balance of amino acids. Generally, soybean meal is the most economical protein supplement to use in many parts of the United States, Figure 16-4.

Cottonseed Meal

In the southwestern United States, cottonseed meal is used extensively when protein supplement is needed in horse rations. Cottonseed meal is not as palatable to horses as some other protein supplements and where other, more palatable protein supplements are economically competitive, its use is not widespread.

Linseed Meal

Linseed meal may be too laxative when fed with legume hay. Expeller type linseed meal contains the fatty acid *linoleic*, which is often lacking in horse diets. Expeller type linseed meal was used to put a bloom on the haircoat of show horses. Linoleic is removed in solvent-process linseed meal. Linseed meal does not contain a good balance of amino acids; therefore, there is little reason to use solvent-process linseed meal in horse rations.

Figure 16-4. Harvesting soybeans on an Illinois farm. Soybean meal is an excellent protein supplement for horses.

Other Protein Supplements

Alfalfa meal, corn gluten meal, and meat meals, as well as other protein supplements, may be used in horse rations.

Commercial Protein Supplements

Horse owners who do not want to mix their own rations find commercial protein supplements convenient to use. Most commercial supplements are developed for a particular feeding program; therefore, the feeding directions on the label must be closely followed.

PELLETING RATIONS

The use of complete pelleted feeds is gaining popularity among horse owners, Figure 16-5. These rations are carefully balanced by the manufacturer and are convenient to use. There is often less waste when pelleted rations are fed to horses; therefore, less pelleted feed is required as compared to conventional feeds. Pelleted feeds are not as dusty and less storage space is required. They may be easily transported and fed to horses on range. However, using complete pelleted feeds is

usually more expensive than providing a ration of hay and grain.

Horses being fed complete pelleted rations are more likely to chew on wood and eat bedding. The vice called *cribbing* is more likely to

Figure 16-5. Complete horse feeds are gaining in popularity among horse owners.

develop when pellets are fed, unless horses are allowed to run in exercise lots.

Horses may require an adjustment period to adapt to the use of complete pelleted rations when they have been accustomed to roughage and grain rations. Continue to feed hay free choice and gradually replace the grain portion of the diet with pellets. Feed 1 to 2 pounds (0.45-0.9 kg) more pellets each day and gradually reduce the amount of hay fed. After several days, horses will generally eat only the pellets.

Hay is sometimes pelleted for feeding. The long hay pellets are digested with about the same efficiency as unpelleted hay. However, the pellets are eaten faster and may pass through the digestive tract more quickly. When hay pellets are fed free choice, the horse may tend to overeat, as compared to eating unpelleted hay, and get too fat.

ANTIBIOTICS

As a general rule, antibiotics are not included in rations for horses. Little research has been done on the effect of including antibiotics in horse rations. Current Food and Drug Administration regulations allow the use of 85 mg chlortetracycline per head per day in the rations of horses under one year of age for the purpose of stimulating growth and improving feed efficiency.

FEEDING HORSES—GENERAL

Feeding Guidelines

Horses are fed according to their size, stage of growth, condition, function, and amount of work they are doing. Some suggested guidelines (amount per 100 pounds (45 kg) of body weight) for determining rations for horses are listed below. Amounts given are on an air-dry (as-fed) basis. To calculate a ration on a 100 percent dry-matter basis, divide these amounts by 0.9.

Foals, before weaning

> 0.5-0.75 pound (0.22-0.34 kg) concentrate
> 0.5-0.75 pound (0.22-0.34 kg) hay
> 16-20 percent crude protein

Foals at weaning

> 1-1.25 pounds (0.45-0.56 kg) concentrate
> 1-1.5 pounds (0.45-0.68 kg) hay
> 16-18 percent crude protein

Foals, weaning to one year

> 1-1.5 pounds (0.45-0.68 kg) concentrate
> 1-1.5 pounds (0.45-0.68 kg) hay
> 16-17 percent crude protein

Yearlings

> 0.5-1 pound (0.22-0.45 kg) concentrate
> 1-1.5 pounds (0.45-0.68 kg) hay
> 13-15 percent crude protein

Mature, idle horses

> 1.5-2 pounds (0.68-0.9 kg) hay
> 7.7-11 percent crude protein

Mature horses, light work (1-3 hours per day)

> 0.25-0.5 pound (0.11-0.22 kg) concentrate
> 1.25-1.5 pounds (0.56-0.68 kg) hay
> 7.7 percent crude protein

Mature horses, medium work (3-5 hours per day)

> 0.66-1 pound (0.29-0.45 kg) concentrate
> 1-1.25 pounds (0.45-0.56 kg) hay
> 7.7 percent crude protein

Mature horses, heavy work (5-8 hours per day)

> 1.25-1.5 pounds (0.56-0.68 kg) concentrate
> 1 pound (0.45 kg) hay
> 7.7 percent crude protein

Pregnant mares (last 90 days gestation)

> 0.5-1 pounds (0.22-0.45 kg) concentrate
> 1-1.5 pounds (0.45-0.68 kg) hay
> 10 percent crude protein

Lactating mares, idle, first 3 months

> 1 pound (0.45 kg) concentrate
> 1.2 pounds (0.54 kg) hay
> 12.5 percent crude protein

Lactating mares, idle, 3 months to weaning

> 0.6 pound (0.27 kg) concentrate
> 1.5 pounds (0.68 kg) hay
> 11 percent crude protein

Stallions, breeding season

> 0.75-1.5 pounds (0.34-0.68 kg) concentrate
> 0.75-1.5 pounds (0.34-0.68 kg) hay
> 14 percent crude protein

Feeding Management

Horses are creatures of habit and should be fed at about the same time each day. Failure to do so can lead to horses going off feed. This is especially true of young growing horses. Feed horses at least twice each day and more frequently if it is practical to do so. The capacity of their stomachs is limited and cannot hold the total amount of feed needed per day at one time. This is especially important for horses needing a high feed intake, such as growing horses, mares during lactation, performance, and working horses. Horses that are mature and not working can be fed once a day. Feed working horses at least one or two hours before work so feed has enough time to digest before work begins.

Do not make sudden changes in the ration. Allow 7-10 days for the horse to adjust to changes to prevent digestive upset. Keep feed troughs and waterers clean; remove any uneaten grain from one feeding to the next; and remove all uneaten forage at the end of each day.

The use of roughages is the basis for formulating horse rations. Use as much grass forage as is practical in the ration, because it is economical and an excellent horse feed from the nutritional viewpoint. Except in the case of hard working horses receiving a large amount of grain, legumes may be used for up to one-half the roughage in the ration. Using legumes in the rations of hard working horses may result in colic or loose feces. Alfalfa is the recommended legume forage; however, other legumes may be used. Only use forages that are free from mold, dust, and other damage.

The minimum amount of roughage to use in the ration is 1 pound (0.45 kg) per 100 pounds (45 kg) of body weight. The general range is 1-2 pounds (0.45-0.9 kg) per 100 pounds (45 kg) of body weight. Silage may be substituted for roughage up to one-third to one-half of the total dry matter in the roughage.

Total feed consumption varies as a function of body weight. Horses weighing under 600 pounds (272 kg) consume approximately 2.5 percent of their body weight each day. Those weighing 600-1,000 pounds (272-454 kg) consume about 2.25 percent of their body weight daily. Horses weighing over 1,000 pounds (454 kg) consume about 2 percent of their body weight daily.

Use grains when additional energy is needed in the diet, Figure 16-6. Palatability and feeding value of grains are improved by coarse grinding, cracking, rolling, or crushing. Do not grind grains too finely for horse rations. It is recommended that some hay be fed prior to feeding the grain. Molasses may be used for 5-10 percent of the ration to reduce dustiness, increase palatability, and provide energy.

Protein supplements may be included in the ration when additional protein is required to meet the needs of the horse. Feeding low-quality grass hay may require the use of a protein supplement. Generally, 0.5-1 pound (0.22-0.45 kg) per day will meet the additional protein requirements. In some cases, as much as 2 pounds (0.9 kg) may be required.

Complete pelleted diets should be made up of 60-70 percent coarsely ground hay and 30-40 percent concentrate. The recommended pellet size for mature horses is one-half inch (1.3 cm), while one-fourth inch (0.6 cm) pellets are better for weanlings and yearlings. Avoid extremely hard pellets, because horses will not eat them.

Cool out working horses before feeding. A tired horse should not receive a full feed of grain. Feed one-half the grain, allow the horse

Figure 16-6. During the winter horses may need additional energy in the diet. Feeding grain with the hay provides additional energy.

to rest for an hour, then feed the rest of the grain. Do not feed a full feed of grain just before working a horse.

Always provide plenty of clean fresh water. If water is not available free choice, water the horse before feeding rather than after. Cool out the horse before allowing it to drink water. Do not allow a heated horse to drink excessive amounts of water.

Horses are highly individual in their eating habits, temperament, and nutritional needs. Rations must be balanced to the individual needs of the horse and fed according to the horse's condition. Horses are fed for efficiency of service over a period of many years. Temporary economy in formulating rations may not meet this goal. Care must be taken to prevent horses from becoming too fat.

FEEDING BREEDING HORSES

Stallions

During the nonbreeding season, stallions may be fed the same kind of rations as mature, idle horses, with the amount increased as appropriate for the weight of the stallion. For stallions weighing 900-1,400 pounds (408-635 kg), a daily allowance of 1.5-1.75 pounds (0.68-0.79 kg) of hay per 100 pounds (45 kg) of body weight will usually meet their needs, if legume-grass forage is fed. When grass hay is fed, add 0.5-0.75 pound (0.22-0.34 kg) of 30-35 percent crude protein supplement to the ration. Stallions may be maintained on good quality pasture without additional feed. Maintain the stallion in good condition without allowing him to become too thin or too fat. Table 16-1 shows some sample concentrate mixes that may be fed to stallions during the breeding season.

Additional feed may need to be fed to stallions during the breeding season, depending upon the number of mares to be bred, quantity and quality of pasture, and his condition. For stallions weighing 900-1,400 pounds (408-635 kg), a daily allowance of 0.75-1.25 pounds (0.34-0.56 kg) of concentrate per 100 pounds (45 kg) of body weight may be fed along with an equal amount of hay, if the stallion is not on pasture. A mature stallion in good condition will not need additional feed when pasture-breeding a limited number of mares, if the pasture is of good quality.

Table 16-1. Stallion (breeding season)—sample concentrate mixes.

Ingredient	Percent
(1,300 lbs. (590 kg) moderate service, legume-grass hay)	
Oats	57
Cracked corn	29
Wheat bran	7
Linseed meal	7
(grass hay)	
Oats	55
Cracked corn	28
Wheat bran	7
Linseed meal	10
(1,200 lbs. (544 kg) heavy service, legume-grass hay)	
Oats	64
Wheat bran	22
Molasses	7
Soybean meal	7
(900-1,400 lbs. (408-635 kg) general breeding season legume-grass hay)	
Oats	55
Wheat	20
Wheat bran	20
Linseed meal	5

Table 16-2. Pregnant mares—sample concentrate mixes.

Ingredient	Percent
Oats	80
Wheat bran	20
Oats	45
Barley	45
Wheat bran	10
Oats	45
Corn	40
Supplement (30-35% crude protein)	15

Pregnant Mares

At breeding and during the early part of gestation, mares may be fed a maintenance ration if it contains adequate minerals and vitamins. If the mare is nursing a foal, she should be fed a lactation ration plus an additional amount of feed to provide for some weight gain.

During the last three months of gestation, the mare requires additional energy, protein, and other nutrients to provide for the rapidly developing fetus. The fetus makes about 60 percent of its weight gain during this period of time.

Nonlactating pregnant mares weighing 900-1,400 pounds (408-635 kg) should be fed 1.5-2 pounds (0.68-0.9 kg) of hay per 100 pounds (45 kg) of body weight daily during the first eight months of gestation. During the last three months of gestation, feed 0.5-1 pound (0.22-0.45 kg) of concentrate along with 1-1.5 pounds (0.45-0.68 kg) of hay per 100 pounds (45 kg) of body weight daily. Table 16-2 shows some sample concentrate mixes that may be fed to pregnant mares.

A few days before foaling, the ration should be changed to contain more bulk. This will reduce constipation at foaling time. Bring the mare to full feed over a period of 7-10 days after foaling.

Lactating Mares

Lactating mares should be fed 1-1.5 pounds (0.45-0.68 kg) of concentrate per 100 pounds (45 kg) of body weight along with an equal amount of hay per day. Lower the level of concentrate feeding if it appears the mare is gaining too much weight. Lactating mares on good pasture combined with water, salt, and minerals may need little additional feed to meet their needs. Mares that have foundered or had colic in the past may need to be fed a lower level of concentrate. Table 16-3 shows some sample concentrate mixes that may be fed to lactating mares.

FEEDING GROWING HORSES

Foals

From a nutrition standpoint, the first year of a horse's life is critical. Foals attain 55-60 percent of their adult weight by the time they are

Table 16-3. Lactating mares—sample concentrate mixes.

Ingredient	Percent
(on pasture)	
Oats	66
Cracked corn	17
Bran	17
(grass hay)	
Oats	61
Cracked corn	26
Soybean meal	13
(legume-grass hay)	
Oats	64
Cracked corn	27
Soybean meal	9
(legume-grass hay)	
Oats	90
Supplement (30-35% crude protein)	10

Table 16-4. Foals, creep feeding—sample concentrate mixes.

Ingredient	Percent
Crushed oats	50
Wheat bran	40
Linseed meal	10
Crimped oats	56
Cracked corn	33
Linseed pellets	11
Crushed oats	62
Cracked corn	31
Soybean meal	7
Crushed oats	64
Wheat bran	21
Molasses	10
Linseed meal	5
Crimped oats	40
Cracked corn	30
Soybean meal	20
Dehydrated alfalfa meal	4
Dried molasses	4
Dicalcium phosphate	1
Trace-mineralized salt	0.5
Vitamin mix (5,000 IU vitamin A plus B vitamins)	0.5
Aurofac (40 mg antibiotic/lb ration)	

one year old. They use feed two to three times more efficiently to make weight gains than do older horses.

Make sure the foal nurses within two hours after it is born. As with other species, it is critical that the foal receive colostrum milk. Colostrum milk is higher in protein and vitamin A, contains antibodies that protect against infection, and is a natural purgative, which removes fecal matter accumulated in the digestive tract. Do not milk out the mare shortly before foaling, as this removes colostrum milk needed by the foal. Assist the foal to nurse if necessary.

Foals receive sufficient nutrients when nursing well-fed mares for the first three to four weeks. They begin to eat grass and other solid feeds when about ten days to three weeks of age. Begin creep-feeding at four to six weeks of age, unless they are on excellent pasture. Rolled oats and wheat bran with a little brown sugar added is a highly palatable feed to use when starting creep-feeding. After the foal is eating, use a mixture of crushed or ground oats, cracked or ground corn, wheat bran, and linseed meal. Commercial creep feeds may be used.

Creep-feed a concentrate mix of grain and protein supplement containing a minimum of 12 percent crude protein and 5 percent fiber. A ratio of 9 parts grain to 1 part oil meal is recommended. One-half of the grain may be corn, milo, or wheat, with the other half consisting of oats or barley. Feed mineral supplement free choice or mix it with the concentrate. Table 16-4 shows some sample creep-feeding concentrate mixtures that may be fed to foals.

Foals are weaned at five to seven months of age. By the time foals are weaned, they should be consuming 1 to 1.5 pounds (0.45-0.68 kg) of good legume hay and about the same amount of concentrate mix per 100 pounds (45 kg) of body weight. The ration should contain 16-17 percent crude protein. If a high quality roughage is fed, use 65 percent con-

Table 16-5. Foals, six months to one year—sample concentrate mixes.

Ingredient	Percent
(alfalfa hay)	
Crushed oats	50
Cracked corn	50
(timothy-clover hay)	
Crushed oats	50
Cracked corn	17
Wheat bran	17
Molasses	8
Linseed meal	8
(legume-grass hay)	
Oats	70
Wheat bran	15
Linseed meal	15
(legume-grass hay)	
Oats	92
Linseed meal	8
(grass hay)	
Oats	43
Cracked corn	43
Soybean meal	14

Table 16-6. Yearlings and two-year olds—sample concentrate mixes.

Ingredient	Percent
(legume-grass hay)	
Oats	63
Cracked corn	32
Soybean meal	5
(legume-grass hay)	
Oats	80
Wheat bran	20
(legume-grass hay)	
Oats	35
Barley	35
Bran	15
Linseed meal	15
(legume-grass hay)	
Crushed oats	33
Cracked corn	67
(grass hay)	
Oats	60
Cracked corn	30
Soybean meal	10
(grass hay)	
Crushed oats	62
Molasses	31
Soybean meal	7
(grass hay)	
Crimped oats	47.5
Cracked corn	30
Wheat bran	5
Pelleted supplement (33-35% crude protein)	17.5
Limestone	0.5
add:	
Vitamin A - 4 million IU/ton	

centrate and 35 percent roughage in the ration. When feeding a lower quality roughage, increase the proportion of concentrate to 70 percent of the ration. Gradually increase the proportion of hay and reduce the concentrate in the ration as the foals grow older. By the time foals are one year of age they should be consuming about 1 pound (0.45 kg) of concentrate to 2 pounds (0.9 kg) of hay per 100 pounds (45 kg) of body weight. The amount of grain in the ration may be increased if foals are being fitted for show or sale. Table 16-5 shows some sample concentrate mixtures that may be fed to foals from six months to one year of age.

Yearlings and Two-year-olds

The nutrient needs of yearlings can be met on good pasture without feeding concentrate. Salt and minerals should be provided on a free choice basis. When yearlings are fed in drylot, they will need some additional concentrate in the ration, depending on the quality of the roughage being fed. Additional supplementation to increase amino acid intake will increase growth and body weight.

Two-year-olds do not require extra feed unless they are in training. Feed additional concentrate when they are in training to meet the needs for additional stress and growth.

Close observation of the condition of the individual horse is necessary for proper feeding. Adjust the ration as necessary to prevent

Table 16-7. Mature working horses—sample concentrate mixes.

Ingredient	Percent
(light work; legume-grass hay)	
Crimped oats	50
Cracked corn	50
(light work; legume-grass hay)	
Crimped oats	46
Cracked corn	31
Crimped barley	15
Linseed meal	8
(light work; grass hay)	
Oats	70
Corn	30
(medium work; legume-grass hay)	
Oats	67
Cracked corn	33
(medium work; legume-grass hay)	
Cracked corn	56
Barley	44
(medium work; grass hay)	
Oats	63
Cracked corn	32
Soybean meal	5
(heavy work; grass hay)	
Oats	53
Corn	40
Molasses	7
(heavy work; grass hay)	
Crimped oats	47
Cracked corn	47
Wheat bran	6

horses from becoming too thin or too fat. Table 16-6 shows some sample concentrate mixtures that may be fed to yearlings and two-year-olds.

Dry Mares and Geldings

Hay or pasture can provide the necessary nutrients for dry mares and for geldings that are idle or not being worked hard. Additional feed is necessary to meet the needs of working horses, depending on the amount of work being done. The guidelines given earlier may be used to formulate rations for different levels of work. Table 16-7 shows some sample concentrate mixtures that may be fed to mature working horses.

Good pasture will meet the nutrient needs of mature idle horses. In drylot, feed 1.5-2 pounds (0.68-0.9 kg) of legume-grass hay per 100 pounds (45 kg) body weight. When grass hay is fed in drylot, add 0.5-0.75 pound (0.22-0.34 kg) of protein supplement per day.

SUMMARY

Nutrient requirements of horses vary with their size, condition, stage of growth, and amount of work required. Horse diets are built around forages with grain, protein supplement, and minerals added as needed to provide additional nutrients.

Good legume-grass pastures or hay provide the needed nutrients with additional supplement needed when low quality or grass forages are fed. Never feed moldy, dusty, or damaged hay to horses. Horses need additional nutrients for growth, lactation, work, and the latter part of gestation.

Oats and corn are commonly used as grains for horse diets. Other grains may be substituted, but amounts must be adjusted to maintain a balanced ration.

Soybean meal is often the most economical protein supplement to use in horse diets. Other protein supplements may be used when they are economically competitive, but some are not as palatable for horses. Part of the nitrogen needed may come from non-protein nitrogen sources, such as urea.

Salt, calcium, and phosphorus are important minerals in horse diets. Provide salt and minerals in the concentrate mix or free choice. Trace minerals are generally provided in trace-mineralized salt.

Provide clean, fresh water for horses. Cool down hard working horses before allowing them to drink.

Horses tend to graze pastures unevenly. Clipping pastures, rotating pastures, and grazing horses with cattle will provide better utilization of the forage.

Complete feeds, such as pelleted rations, are gaining popularity with horse owners. Pelleted feeds have less waste and are easy to transport and store, but must not be too hard or horses will not eat them.

Feed horses at the same time each day. It is better to divide the ration into several feedings because of the limited capacity of the horse's digestive tract. Make changes in the ration over a period of several days to prevent digestive upset.

Foals attain 55-60 percent of their growth during the first year of life. They make more efficient use of feed during this time than later in life. Make sure foals receive colostrum milk shortly after birth. Begin creep feeding when the foal is four to six weeks of age.

REVIEW

1. What is the effect of a deficiency of energy in horse diets?
2. How much should the energy level of the diet be increased above maintenance requirements during the last 90 days of gestation?
3. During lactation, how much DE is required per pound of milk produced?
4. What are the effects of protein deficiency in a horse's diet?
5. Why do young horses have a greater need for protein than mature horses?
6. What effect does the amount of work a horse is doing have on its protein requirements?
7. How much should the protein level of the diet be increased above maintenance requirements during the last 90 days of gestation?
8. Describe the protein requirements during lactation.
9. Describe mineral requirements for horses.
10. What is the recommended calcium-to-phosphorus ratios for mature horses, foals, and yearlings?
11. How should salt and minerals be provided in horse diets?
12. How may the horse owner be sure the vitamin requirements of the horse are being met?
13. How much water does a horse need and how is this affected by the amount of work the horse is doing?
14. Describe recommendations for watering horses.
15. What forage is recommended for use in exercise and high traffic areas? Why?
16. What is the minimum amount of pasture needed to provide adequate feed for one horse?
17. How much pasture is needed per horse in western range areas to provide adequate feed?
18. Why should horse pastures be rotated?

19. Describe other methods for better pasture utilization.
20. How may pasture overgrazing be controlled?
21. Discuss the management of irrigated pastures.
22. Discuss the management of dryland pastures.
23. Discuss the management of native ranges.
24. What forages are recommended for permanent pastures?
25. What precaution should be followed when grazing pregnant mares on tall fescue?
26. What proportion of the pasture should be made up of legumes?
27. What annual grasses may be used for pasture?
28. What precautions should be followed when using Sudan grass and sorghum-Sudan hybrids for forage?
29. Compare legume-grass hay to grass hay in terms of nutrient yield per acre.
30. Compare legume hays to grass hays in terms of nutrient value.
31. What disadvantage does legume hay have as compared to grass hay for horses?
32. What grass hays are commonly used in horse rations?
33. Discuss the use of silage in horse rations.
34. Why have oats been the traditional grain of choice for feeding horses?
35. Discuss the use of corn in horse rations.
36. List and briefly discuss other grains that may be used in horse rations.
37. Why is soybean meal used more often in horse rations than most other protein supplements?
38. Discuss the use of complete pelleted rations for horses.
39. List guidelines for feeding foals at several stages of growth.
40. What are the guidelines for feeding yearlings and mature, idle horses?
41. Describe the guidelines for feeding horses at light, medium, and hard work.
42. List guidelines for feeding pregnant mares during the last 90 days of gestation.
43. What are the guidelines for feeding idle lactating mares during the first three months and during the period of three months to weaning?
44. What are the guidelines for feeding stallions during the breeding season?
45. Discuss some general rules for managing the feeding of horses.
46. What is the general range of roughage to be fed per 100 pounds (45 kg) of body weight?
47. How may the palatability and feeding value of grains be improved for horses?
48. Describe the feeding of stallions during the breeding and nonbreeding seasons.
49. Describe the feeding of pregnant mares during the early part and the last three months of gestation.

50. Describe the feeding of lactating mares.
51. Why is the proper feeding of foals so critical?
52. How important is colostrum milk for foals?
53. When should creep feeding begin for foals?
54. Describe a good creep-feeding ration for foals.
55. Discuss the feeding of foals from weaning to one year of age.
56. Discuss the feeding of yearlings and two-year-olds.
57. Discuss the feeding of dry mares and geldings.
58. Discuss the feeding of mature idle horses.

17

FEEDING POULTRY

OBJECTIVES

After completing this chapter you will be able to:

- Describe nutrient requirements for poultry.
- Provide proper feeding and watering space for poultry.
- Select correct feeds for poultry.
- Feed young poultry.
- Feed the laying flock.
- Feed broilers.
- Feed the breeding flock.
- Feed turkeys, ducks, and geese.

NUTRIENT REQUIREMENTS OF POULTRY

Tables in the Appendix give the nutrient requirements for various classes and species of poultry. Values are given for both percentages or amounts per pound and per kilogram of feed, as well as daily amounts per head. The amounts shown per day per head are based on normal feed consumption. As there are a number of factors that can affect daily feed intake, these amounts should serve only as general guides in poultry nutrition. Usually, rations for poultry are formulated on the basis of units per unit-weight of feed, because feeding is done on a group basis, *ad libitum*.

Energy

The energy values of rations for poultry may be calculated on the basis of apparent metabolizable energy (ME), true metabolizable energy (TME), or net energy (NE). The ME value is the one most commonly used by feed manufacturers. This is defined as the gross energy from feed intake minus fecal and urinary energy. In poultry, little energy is produced from the gaseous products of digestion and can be ignored when calculating ME. Correcting for nitrogen retained in the body yields an ME_n value, which is the most common measure of available energy for poultry. Most of the ME values for poultry have been determined from research with young chicks. Little data are available for chickens and turkeys of different ages.

Because poultry tend to eat to satisfy their energy needs when fed free choice, it is generally possible to control the intake of all nutrients by including them in the ration in a definite proportion to the available energy level. Water intake is not controlled by the energy level of the diet because water is made available separately from the rest of the ration.

Diets with a high energy level will reduce feed consumption; therefore, the proportion of other needed nutrients must be increased to ensure meeting the nutritional needs of the birds. Conversely, lowering the energy level of the diet will increase feed consumption, permitting a lower proportion of other needed nutrients in the ration. While some variation in the level of energy in the diet can be made, it is generally not practical to deviate too widely from recommended levels. For example, increasing the energy level with a concomitant increase in protein and other nutrients will generally increase the cost of the ration, because of the higher price of the protein supplements. Increasing the nutrient concentration in the rations of laying hens may result in obesity (overweight), higher mortality, cannibalism, and feather picking.

Feed efficiency in producing eggs is higher when a high energy diet is fed. The amount of feed intake must be controlled under these conditions to offset the potential negative results of feeding a high energy diet.

Feed intake is affected by a number of factors:

a. daily energy requirement
b. environmental temperature
c. health of the bird
d. genetics
e. form of the feed
f. nutritional balance of the diet
g. stress on the bird
h. body size
i. rate of egg production or growth
j. amount of glucose in the blood

Because of these factors affecting daily feed intake, it is not possible to establish an energy requirement in terms of kilocalories per pound (kcal/lb) or kilocalories per kilogram (kcal/kg) of diet.

Environmental temperature has a major influence on daily feed intake. Intake is lower when the temperature is high and increases with lower temperatures. The percentage of all nutrients needed in the diet is highly dependent on environmental temperature. Values given in the tables of nutrient requirements are based on the assumption that the environmental temperature is as near to optimum (60-75°F <16-24°C>) as possible for efficient production. Dietary adjustments for prolonged variation of temperature from the optimum level may need to be made to maximize production.

The ratio of energy to protein in the diet affects the amount of fat deposition in the body of a bird. When the protein level is low in relation to the energy contained in the diet, fat deposition is substantially increased. The addition of more protein to the diet reduces the amount of fat deposition in the body.

Protein and Amino Acids

The protein and amino acid requirements of poultry are listed in tables in the Appendix. Rate of growth and level of egg production are both affected by the level of amino acids in the diet. Both deficiencies and excesses of amino acids in the diet have undesirable effects upon growth and egg production. The requirements for the 14 essential amino acids for poultry must be met by the diet, for efficient reproduction, growth, and egg production.

As discussed above, there is a relationship between the energy level of the diet and the intake of feed. This means that the amount of protein required in the ration is proportional to the energy level of the ration. Reducing the protein level of the diet may increase feed intake slightly.

Environmental temperature affects the level of protein required in the diet. Recommended protein levels are based on the assumption of air temperatures in the range of 60-75°F (16-24°C). When temperatures rise above this range the percentage of protein in the diet needs to be increased, and it should be decreased when temperatures are below this range. This adjustment is necessary because of the effect of temperature upon feed intake and the necessity of maintaining an adequate level of amino acid intake.

Lowering the protein level of the diet results in a slower rate of gain in broilers. The decision to use least-cost rations as compared to least-time rations is dependent, at least in part, upon the anticipated trend in broiler prices. Delaying marketing when prices are declining may offset any economic gain realized by lowering the protein content of the diet.

The need for some amino acids can be met by other amino acids because of the ability of the metabolic system of poultry to convert one to another. The cystine requirement can be met by methionine; however, the methionine requirement cannot be met by cystine. The requirement for tyrosine can be met by phenylalanine; however, the phenylalanine requirement cannot be met by tyrosine. Glycine and serine can be interchanged in poultry diets; that is, each can meet the requirement for the other.

Some amino acids can compensate for deficiencies of some vitamins in the diet. However, it is not practical to substitute high priced amino acids for lower priced vitamins in the ration.

In some cases, the availability of protein in a particular feedstuff is relatively low, because of the poor digestibility of the feed. Overheating or underheating during processing can affect the availability of some amino acids. For example, some animal byproducts, such as blood meal and feathers, while relatively high in protein, may be low in digestibility and, therefore, not good protein sources. Meat scraps and soybean meal, when underheated during processing, have a lower availability of some amino acids.

Minerals

Mineral requirements for poultry are listed in tables in the Appendix. These are expressed as percentages or amounts per kilogram or pound of feed.

The major minerals needed in poultry diets are calcium, phosphorus, sodium, and chloride. Trace minerals may need to be added to the diet if feeds used are grown on soils that have become deficient in trace minerals.

The recommended ratio of phosphorus to calcium in the diet of poultry varies with the age of the birds and is more critical with young poultry than with older birds. The ideal ratio for young poultry is approximately 1:1.2 (P:Ca); however, a range of 1:1 to 1:1.5 is satisfactory. Poultry laying eggs need a ratio of phosphorus to calcium of about 1:4.

Calcium is important for bone formation in growing poultry and for shell formation in laying poultry. Too much calcium in the diet will interfere with the utilization of other minerals such as magnesium, manganese, and zinc. An excess of calcium will also lower the palatability of the ration. Under optimum conditions, laying hens require about 3.25 percent calcium in the diet. When subjected to extended periods of temperatures around 90°F (33°C), the calcium level in the diet should range from 3.25-3.75 percent. When growing pullets have too little or too much calcium in the diet, they have increased difficulty with calcium metabolism when they begin egg production.

Finely pulverized limestone may reduce feed intake if used to provide high calcium levels in the diet. Dolomitic limestone has a lower level of biologicaly available calcium than other calcium sources. It contains a high level of magnesium.

Inorganic phosphorus sources have a higher availability than do organic sources. It is important to maintain the minimum recommended level of available inorganic phosphorus in the diet of young poultry. Older birds can utilize most of the organic phosphorus in the diet.

Salt provides sodium and chloride in the diet. The recommended level of salt in the ration is 0.5-1 percent of the total ration. When water is readily available, the level of salt in the diet can exceed 1 percent without adversely affecting production.

Vitamins

Vitamin requirements of poultry are given in the Appendix tables. These are expressed as amounts per kilogram or pound of feed. The requirements for vitamins A and E are given in International Units (IU), while the vitamin D requirement is given as International Chick Units (ICU) per kilogram or pound of feed. The use of units to express vitamin A, D, and E requirements is made necessary because of the different levels of activity of the various forms of these vitamins.

Because vitamin C is synthesized by poultry, it normally is not included in diet formulation. However, there is some evidence that birds under stress from high temperatures benefit from the addition of vitamin C to the diet.

Vitamin premixes are generally used when mixing poultry feeds to assure an adequate amount of the required vitamins. Feedstuffs normally used in poultry rations usually contain sufficient amounts of the water-soluble vitamins, such as the B-complex vitamins. If substitutions in normal feedstuffs are made when formulating poultry diets, it may be necessary to supplement the ration with water-soluble vitamins.

The dietary requirement for several vitamins is related to the type of carbohydrate used, the protein level, and the amino acid balance in the ration. Also, some vitamins can replace others in the diet. For example, betaine and choline are interchangeable in the diet of growing chicks to provide methylating agents; however, betaine cannot be used to replace choline in the prevention of perosis. Vitamin B_{12} reduces the need for choline in the diet.

Xanthophylls

The yellow color in egg yolks and the yellow skin color in broiler and roaster carcasses is produced by xanthophylls. Xanthophylls are crystalline, unsaturated compounds ($C_{40}H_{46}O_2$) found in plants and are yellow in a dilute solution. Some of the feedstuffs used in poultry diets contain xanthophylls but most do not contain significant amounts. Alfalfa meal and corn gluten meal are two possible natural sources of xanthophylls for poultry rations. However, both corn and alfalfa lose xanthophylls fairly rapidly when stored.

Antibiotics

Antibiotics are added to poultry rations to stimulate growth, improve feed efficiency, and increase egg production. Generally, they are used at low levels ranging from 1 to 10 mg/kg of diet. For young birds, they may be used at levels as high as 50 mg/kg of diet.

There is increasing concern that the use of low levels of antibiotics in animal nutrition may be a hazard to human health. Current regulations governing the use of antibiotics must be followed at all times. Sources of information on current regulations for use are listed in Chapter 6.

Unidentified Nutrients

There is a great deal of evidence of a number of unidentified nutrient factors that affect poultry nutrition. These appear to stimulate growth, increase reproduction, improve the quality of poultry meat, or reduce the toxicity of mineral elements in poultry.

Some of the materials that appear to contain these factors include egg yolk, whey, yeast, marine-fish and packing house by-products, soybeans, distiller's solubles, corn, green forages, fermentation residues, and some other natural feedstuffs. Marine-fish products are often used in formulating poultry diets to provide some of these unidentified factors. Young birds appear to have a greater response to these unidentified factors than do older birds.

Water

As with all animals, water is essential for poultry. Water consumption of poultry is shown in a table in the Appendix. A general guide to poultry water consumption is that a bird will drink about twice as much water by weight as the amount of feed it consumes. Water consumption as shown in the Appendix is based on air temperatures of about 70°F (21°C). As temperatures increase or decrease from this level, water consumption will increase or decrease. If poultry cannot consume enough water under extremely hot temperatures, they cannot survive.

Some medications are administered in the drinking water of poultry. The pH level and salt content of the water sometimes influences the acceptability of the medicated water. Turkeys are especially sensitive to minor variations in water content.

FEEDS FOR POULTRY

Energy Feeds

Grains, grain byproducts, and animal and vegetable fats and oils supply most of the energy in poultry diets. High-fiber feeds are of little value for poultry because of the relative inability of birds to digest the fiber.

Corn is the most common grain used in formulating poultry feeds. Other grains such as milo, wheat, oats, and proso millet may be substituted for part of the corn in the diet. The nutrient content of each grain must be considered when making substitutions, in order to maintain the desired level of nutrient intake. Grains, other than corn, should be used in limited quantities in poultry rations because of their lower energy level and higher fiber content. For example, limit oats or barley to not more than 10-15 percent of the ration.

Animal and vegetable fats are added in limited amounts (5-10 percent of diet) to poultry rations to supply energy. As discussed earlier in this chapter, the addition of fat to the diet affects the level of feed intake. Fats also increase the palatability of the diet, decrease dustiness, and improve the texture of the feed. Fats are used more often in broiler rations to increase the energy level. In hot weather, feed containing added fat may become rancid, unless it has been properly stabilized.

Molasses will supply energy but should be limited to not more than 2 percent of the ration. Excessive amounts of molasses will have a laxative effect on birds.

Protein Supplements

Protein supplements are added to poultry rations to provide the essential amino acids. Balancing poultry rations on amino acid requirements rather than on crude protein assures that nutritional needs are more closely met. Because the various sources of protein vary in their content of amino acids, it is recommended that several protein sources be used to achieve a better balance of the needed amino acids. Both plant and animal protein sources are generally used in formulating poultry rations. Animal protein sources are usually more variable in their amino acid content than are plant protein sources.

Plant proteins. The plant protein source most commonly used in poultry rations is soybean meal. Other common plant protein sources include cottonseed meal, corn gluten meal, alfalfa meal, and linseed meal.

Soybean meal has a better balance of amino acids than other plant protein sources. It is highly palatable, easily digested, and generally free from toxic substances.

Corn gluten meal and alfalfa meal provide amino acids. They also contain xanthophylls, which add the yellow color to egg yolks and skin. Normally, the use of these protein sources is limited to not more than 10 percent of the ration.

Cottonseed meal is usually not used in rations for laying hens, because it contains gossypol and cyclopropenoic fatty acids.

Gossypol causes egg yolks to have a mottled, greenish appearance. Cyclopropenoic fatty acid causes egg whites to have a pink color. Cottonseed meal may be used in grower poultry rations to replace up to 50 percent of the soybean meal.

Linseed meal in excessive amounts can slow growth and cause diarrhea. For this reason, its use is generally limited to not more than 3-5 percent of the diet.

Animal proteins. Commonly used animal protein sources in poultry diets include fish meals, meat byproducts, milk byproducts, blood meal, hydrolyzed poultry feathers, and poultry by-product meal.

Fish meals provide a source of unidentified nutrients in poultry diets. Because they are high in fat and may cause a fishy flavor in eggs and poultry meat when used in large quantities, their use is generally limited to not more than 2-5 percent of the diet. Fish meals contain a fairly good balance of amino acids.

The other listed animal protein sources may be used to replace up to 10 percent of the soybean meal in the ration. In addition to amino acids, they also provide calcium and phosphorus.

Mineral Supplements

Common sources of calcium in poultry rations include crushed oystershells, oystershell flour, ground limestone, steamed bone meal, and dicalcium phosphate.

Inorganic phosphorus is supplied in poultry rations by adding steamed bone meal, dicalcium phosphate, defluorinated rock phosphate, rock phosphate, and monosodium phosphate.

Sodium and chlorine are added to poultry rations by including salt at levels ranging from 0.5-1 percent of the ration. The use of iodized salt will add needed iodine to the ration, but adding too much salt increases water intake, resulting in wet droppings.

Manganese and zinc may need to be added to the diet in small amounts. Manganese

sulfate may be used to add manganese and zinc chloride or zinc sulfate may be used to add zinc to the ration. Other trace minerals may be needed in some areas and can be added by using a trace-mineral premix.

Vitamin Supplements

Natural feedstuffs provide some vitamins for poultry. However, vitamin premixes are commonly used to provide the required vitamins in poultry diets. This represents the most economical way to ensure that vitamin needs of birds are met.

FEED PREPARATION

Commercial feeds for poultry are available as mash, pellets, or crumbles. Any of these physical forms of feed are acceptable for feeding poultry. There is less waste when using pelleted or crumbled feeds and poultry grow faster.

Mash feeds are ground medium-fine. Pellets are mash feeds that have been pelleted and crumbles are pellets that have been processed by rolling. The manufacturer's directions for use must be closely followed, with special attention given to feed-tag instructions regarding use when the feed contains additives.

Pellets and crumbles are more expensive than mash feeds but are easier to use in mechanized feeding equipment. They are more commonly used for broiler and turkey rations than for laying hen rations. Laying hens fed pellets or crumbles tend to become too fat, unless they are on a restricted feeding program.

PREPARING RATIONS

Four systems which may be used for preparing rations are:

1. Use a complete commercial feed.
2. Use a commercial protein concentrate and mix with local or homegrown grains.
3. Use a commercial vitamin-mineral premix

and soybean meal or other protein supplement and mix with local or homegrown grain.

4. Buy all of the individual ingredients and mix the ration.

The choice of system depends on the age of the poultry being fed, the relative costs, the size of the enterprise, and the equipment available for grinding and mixing feeds. Farmers with small flocks usually find it more economical to use a complete commercial feed. Large commercial poultry enterprises generally use some system of mixing, either at a local mill or on the farm. For large operations, mixing the feeds usually results in a lower feed cost per ton.

FEEDING MANAGEMENT

Use only fresh feed, and avoid waste by not overfilling feeders. Hanging feeders should be filled only three-fourths full and trough feeders only two-thirds full. Fill feeders in the early morning and refill during the day as needed. When refilling feeders, remove any feed that has become dirty and unpalatable. Keep the feed clean and never use moldy or dirty feed. Clean feeders periodically.

Feed should be stored in a dry place where rats or mice cannot contaminate it. If practical, do not store feed more than two weeks in order to keep it as fresh as possible. Feed for large laying flocks (over 2,000 birds) should be handled in the bulk.

A good manager watches the feed consumption of the flock closely. A drop in feed intake is often the first sign of trouble. Stress, disease outbreak, molting, or other management problems are often first indicated by a drop in feed consumption.

GRIT

Grit helps the bird grind coarse feeds in the gizzard. Flocks being fed all mash rations have less need for grit, but will still get some benefit from its use. The use of grit when feeding whole grains improves feed efficiency. Grit is generally made from granite and is insoluble in the digestive acid. It is not a substitute for oyster shell or ground limestone, as these materials are soluble and provide a source of calcium for the birds.

Grit comes in small (chick), medium (growing), and large (hen) sizes. Do not use small or medium sizes for the laying flock, as these will pass through the digestive system too quickly. Feed grit in separate feeders or mix it with the complete feed. It may also be sprinkled on top of the feed about once a week.

FEEDING REPLACEMENT PULLETS

The recommended protein level for starter rations used for replacement pullets for egg-type chickens is 18-20 percent during the first six weeks of feeding. This recommendation is based on an ME level of 1250-1400 kcal/lb (2756-3086 kcal/kg). The ration during this time should consist only of mash, pellets, or crumbles. Do not include grain in the diet.

From 6-14 weeks of age, feed a grower diet containing 14-16 percent crude protein at an ME level of 1250-1400 kcal/lb (2756-3086 kcal/kg). From 14-20 weeks of age, a developer ration containing 12-14 percent crude protein at an ME level of 1200-1400 kcal/lb (2645-3086 kcal/kg) is recommended. Suggested rations for replacement pullets of the egg laying breeds are shown in Table 17-1.

Proper physical development at sexual maturity is important, if maximum potential egg production is to be reached. When birds come into production without attaining proper body size and at too early an age, egg size is reduced. Light has a major effect on the age of sexual maturity and egg production. Sexual maturity in pullets is reached quicker when the day is lengthened; therefore, do not

Table 17-1. Replacement pullets, egg laying breeds—example feed formulations.

Ingredients	Starter 0-6 wks		Grower 6-14 wks		Developer 14-20 wks	
	(%)	(%)	(%)	(%)	(%)	(%)
Sorghum grain or corn	60.6		69.1		74.6	
Corn, yellow, finely ground		54.2				
Corn, yellow, medium ground				54.7		42.2
Oats, ground		5.0		5.0		5.0
Alfalfa meal, dehydrated (17% protein)	5.0	2.5	5.0	2.0	5.0	2.0
Soybean meal (44% protein)	26.5		18.0		12.5	
Soybean meal, solvent (45% protein)		25.5		16.0		4.5
Poultry by-product meal (60% protein)	5.0		5.0		5.0	
Wheat middlings, standard		5.0		15.0		20.0
Meat and bone scraps (50% protein)		2.5		2.5		2.5
Whey, dried		1.0		1.0		
Fish meal (60% protein)		1.5		1.0		1.5
Salt	.2	.3	.2	.3	.2	.3
Defluorinated rock phosphate (18% P, 32% Ca)	1.5		1.5		1.5	
Dicalcium phosphate		.5		.5		.5
Calcium carbonate oyster shell flour (38% Ca)	.5		.5		.5	
Limestone, ground		1.5		1.5		1.0
Manganese sulphate	.025		.025		.025	
Zinc sulphate	.025		.025		.025	
DL-methionine	.15		.15		.15	
Vitamin-trace mineral premix	.5	.5	.5	.5	.5	.5
Coccidiostat	(Follow manufacturers directions for use)					
Protein (%)	21.00	20.00	18.05	17.00	16.00	14.00
Metobolizable energy (ME) (kcal/lb)	1275	1265	1320	1249	1350	1208
Metabolizable energy (ME) (kcal/kg)	2810	2789	2910	2754	2976	2663
Lysine (%)	1.10	1.10	.85	.32	.70	.26
Methionine (%)	.48	.34	.42	.29	.40	.25
Cystine (%)	.30	.31	.28	.27	.26	.23
Phosphorus—total (%)	.72	.70	.72	.60	.72	.60
Calcium (%)	1.00	1.2	1.00	1.1	1.00	.9

increase the day length during the last eight weeks of growth. A suggested lighting program for replacement pullets is given in Table 17-2.

Combining light management and feed intake control will help bring pullets to the proper physical size and sexual maturity at the appropriate time, as will restricting the intake of feed after the pullets are eight weeks of age. Other methods are restricting the amount of time feed is available each day, feeding every other day, or feeding a high-

Table 17-2. Suggested lighting program for replacement pullets—egg laying breeds.

Age	Hours of light per day
Chicks, first 3-7 days	23-24
Pullets, 1-20 weeks	6-9
Pullets, at 20 weeks	12
Laying hens	Increment 15 minutes/ week till reach 14 hours

Table 17-3. Suggested feeder and waterer space for replacement layers.

Age in weeks	Linear space per 100 chicks			
	Feeder		Waterer	
	Inches	Cm	Inches	Cm
0-2	100	254	25	63.5
3-6	200	508	50	127
7-12	250	635	50	127
13-20	300	762	100	254

fiber diet. The simplest method is providing a well-balanced ration on a controlled-intake basis. Starting at eight weeks of age, limit feed intake to 10 pounds (4.5 kg) per 100 birds per day (lightweight breeds). With heavy-breed birds, the amount must be proportionately increased. When feed intake is limited, ample feeder space must be provided so all birds can have access to the feed. Research shows that restricted feeding programs are of greater value with heavy-breed layers than with light-breed layers.

An all-mash ration should be fed the first six weeks. After this, an all-mash ration may be continued or a mash and grain ration may be used.

When grains are included in the ration, the use of insoluble grit is recommended. There is less value in using grit when all-mash rations are fed. Poultry do not have teeth and the grit helps break up the feed in the gizzard of the bird.

Provide the proper feed and waterer space for replacement pullets. Table 17-3 gives recommended space allowances for replacement pullets at various ages.

When pullets begin laying eggs, the ration is changed to a laying formula. Make this change gradually over a two-week period.

FEEDING LAYING HENS

To get the maximum economic return from the laying flock (light-weight breeds) a feed efficiency of 3.5-4.0 pounds (1.6-1.8 kg) of feed per dozen eggs produced is needed.

Heavier breeds will require proportionately more feed per dozen eggs produced. The choice of ration should be based on the cost per dozen eggs produced and not on the cost per pound (kilogram) of feed. Records must be kept of the amount of feed used and the egg production, in order to calculate feed efficiency. Divide the total pounds (kilograms) of feed fed by the number of dozen eggs produced to calculate how many pounds (kilograms) of feed it took to produce one dozen eggs. Wasted feed, low rate of lay, health problems, or other management problems lower feed efficiency.

A laying ration should contain about 15 percent protein, based on 1315 kcal/lb (2,900 kcal/kg) of ME. Egg production, body size, health, and air temperature all affect the amount of feed a laying hen eats. Light breeds need to eat about 24 pounds (10.9 kg) (per 100 hens per day) of a 15 percent ration to get the recommended amount of protein in the diet. A hen of the light-weight breeds eats a total of 85-90 pounds (38.6-40.8 kg) of feed per year. A heavy-breed hen eats 95-115 pounds (40.8-52.2 kg) per year. About 2-5 pounds (0.9-2.3 kg) of oyster shell and one pound (0.45 kg) of grit per year per hen are required.

Provide adequate feeder and waterer space for the laying flock. Table 17-4 gives recommendations for feeder and waterer space.

Sudden changes in the diet of the laying flock should be avoided. This creates stress in hens, which may result in a drop in egg production or may force a molt with a complete stop in egg production. Even changing from a

Table 17-4. Suggested feeder and waterer space for laying chickens.

| Age in weeks | Linear space per 100 chicks | | | |
| | Feeder | | Waterer | |
	Inches	Cm	Inches	Cm
Light Breeds (Leghorn type)	300	762	50	127
General Purpose Breeds (Heavy type)	400	1016	100	254

coarse-ground feed to a fine-ground feed may create stress and a drop in egg production.

Three systems of feeding are commonly used for laying hens: (1) all mash, (2) mash and grain, and (3) cafeteria.

The all-mash system is the use of complete ground feed. This system is well adapted to use with mechanical feeding systems and is often used by commercial egg producers.

Using the grain and mash system, feed grain and mash separately. A 20-26 percent protein supplement is fed with a light grain feeding in the morning, and corn is fed in the evening. Feeding some of the grain in the litter causes the hens to stir up the litter by scratching in it and helps to keep the litter dry.

The cafeteria system allows the birds to balance their own rations. Grain is fed in one feeder and a 26-32 percent protein supplement is fed in another feeder. Feed is kept in the feeders at all times. When this system is used, older hens may tend to eat too much grain and not enough protein supplement. About one-fourth of the feeders should contain protein supplement and three-fourths of the feeders should contain grain.

Phase feeding is a system of making specific feeds to be used to meet the changing nutritional requirements of chickens. A number of factors such as rate of lay, egg size, body maintenance needs, and air temperature are used to determine the content of the ration. The purpose of phase feeding is to lower feed cost. When making up a ration, the procedure for taking into account all of the phase-feeding

factors is a complicated one. When more than one age group of chickens is being fed, more feed storage bins and feeders are needed. It requires a high level of management ability on the part of the producer to successfully use phase feeding. Phase feeding is more practical for large commercial operations than it is for small farm flocks.

Table 17-5 gives some suggested feed formulations for laying flocks. When diets are based on corn-soybean meal formulations, methionine is the first limiting amino acid. Adding feeds high in methionine may be necessary to assure an adequate level of this amino acid.

Laying hens need additional amounts of calcium for egg shell formation, and rations must be formulated to meet this requirement at an adequate level. Limit the level of total phosphorus in the diet to 0.5-0.6 percent. An excess of phosphorus will interfere with the absorption of calcium in the bird's body.

FEEDING BROILERS

Breeds of chickens used in the broiler industry have been developed for rapid growth and weight gain. Feeding programs for broilers usually follow a three-step sequence: (1) starter; (2) grower; and (3) finisher. The protein levels are different for each phase of the feeding period.

Starter rations are fed for the first three weeks and contain approximately 23-24 percent protein. Do not use starter rations designed for egg-production-type chickens. Broiler rations may contain 3-5 percent added fat and a coccidiostat. Feed a grower ration containing approximately 20-22 percent protein from three to five weeks of age. Use a finisher ration containing approximately 18-20 percent protein from five weeks to market.

The energy level of the ration varies by stage of feeding and season. Unless housing is well-insulated, more energy is required during the winter because of higher maintenance requirements. In warm climates, there is less

Table 17-5. Example mash formulation for laying hens.

Ingredients	1 (%)	2 (%)	3 (%)	4 (%)	5 (%)	Free Choice[c] (%)
Corn	57.0		69.15	46.95	47.95	17.95
Wheat		79.15				
Oats				25.00		20.00
Barley[a]					25.00	
Wheat middlings, standard	9.0					
Soybean meal, solvent (45% protein)	19.5					
Soybean meal (48% protein)		8.00	18.00	15.20	14.20	39.00
Meat and bone meal (50% protein)	2.0	2.00	2.00	2.00	2.00	6.00
Alfalfa meal, dehydrated (17% protein)	2.5	2.00	2.00	2.00	2.00	4.00
Fish meal (60% protein)	1.0					
Salt	.5					
Salt, iodized[b]		.30	.30	.30	.30	.90
Dicalcium phosphate	1.5	2.00	2.00	2.00	2.00	4.50
Limestone, ground (or Marine shell)	6.5	6.00	6.00	6.00	6.00	6.00
Vitamin-trace mineral premix	.5	.50	.50	.50	.50	1.50
Methionine		.05	.05	.05	.05	.15
Protein (%)	17.0	15.48	16.20	15.61	15.48	26.22
Metabolizable energy (ME) (kcal/lb)	1204	1268	1450	1218	933	1020
Metabolizable energy (ME) (kcal/kg)	2654	2795	3197	2685	2057	2249

[a]Millet may be substituted for barley on a pound for pound basis.

[b]Iodized salt should contain approximately 0.7% manganese, 0.5% iron, and 0.25% zinc.

[c]This supplement is fed free choice with a grain or grain mixture. Oyster shell or large particle crushed limestone and granite grit should be available at all times.

variation needed in the energy level of the diet, based on the season. Suggested energy levels of starter rations are 1475-1500 kcal/lb (3252-3307 kcal/kg). Energy levels of grower rations are 1475-1525 kcal/lb (3252-3362 kcal/kg). Finisher rations contain 1500-1550 kcal/lb (3307-3417 kcal/kg). In all cases, the required level of essential amino acids must be met. Because feed intake varies with the energy level of the diet, amino acid levels must be adjusted as energy levels change.

Adequate feeder and waterer space is essential for efficient production. Table 17-6 shows suggested feeder and waterer space for broiler type chickens.

Broiler rations are fed as complete feeds in mash, pellet, or crumbles form. Separate grain and mash rations are not used for broiler production. Broiler rations normally contain higher levels of vitamins than do rations for egg-type chickens. Antibiotics may also be added to broiler rations. Table 17-7 shows some suggested rations for feeding broilers.

Proper mixing and quality control are essential in the preparation of broiler rations. Small amounts of some ingredients such as vitamins and antibiotics are used in the rations and without proper mixing and quality control feed efficiency is reduced. Good inventory control of ingredients at the mixing mill is one way of maintaining quality control. Regular assaying of the mixed feed is another method of assuring quality control in the final product. Accurate information concerning the ingredients used in mixing broiler rations is also essential to maintain quality. Proper formulation of the ration can only be done when accurate information is available concerning the ingredients used in the ration.

Table 17-6. Suggested feeder and waterer space for broilers.

| Age in weeks | Linear space per 100 chicks | | | |
| | Feeder | | Waterer | |
	Inches	Cm	Inches	Cm
0-2	100	254	25	63.5
3-6	300	762	50	127
7-12	350	889	75	190.5
13-20	400	1016	100	254

Table 17-7. Suggested feed formulations for broilers.

Ingredient	Starter 0-3 wks (%)	Grower 3-6 wks (%)	Finisher 6-9 wks (%)
Corn, ground yellow	58.35	66.85	66.61
Soybean meal, 48% protein	34.00	27.00	
Soybean meal, 44% protein			16.50
Meat and bone meal, 50% protein	2.50	2.00	
Dehydrated alfalfa, 17% protein	2.00	1.00	2.00
Fish meal, 60% protein			5.00
Poultry byproduct meal			2.50
Dried whey			1.50
Dried distiller's solubles			2.00
Ground limestone	0.50	0.50	
Calcium carbonate			1.00
Dicalcium phosphate	1.80	1.80	
Steamed bonemeal			1.50
Salt, iodized	0.30	0.30	0.30
Manganese sulfate			0.05
Vitamin premix	0.5	0.5	0.76
DL-Methionine	0.05	0.05	0.18
Antibiotic supplement			0.05
Arsonic acid			0.05
Protein (%)	23.24	20.16	18.7

Feed Conversion

Feed conversion refers to the amount of feed necessary to produce one pound of live weight. Feed conversion in the broiler industry is around 2.2 pounds to produce 1 pound of live weight. Feed conversion may be found by dividing the total pounds of feed fed by the total weight of the birds produced. This method is known as *feed conversion before condemnations*. Feed conversion after condemnations is found by dividing the total amount of feed used by the pounds of birds marketed after the pounds of condemned meat have been subtracted from the weight of the total birds produced.

Feed conversion is affected by the type of feed fed, genetics of the strain of birds fed, air temperature, disease and condemnations, age and weight of the birds, rodent and flying bird control in the feeding area, antibiotics and medications used, debeaking, size of baby chicks, feed wastage, form of feed, and overall management.

High energy diets give a better feed conversion than low energy diets. Diets high in fiber give poorer feed conversion. Well-balanced diets improve feed conversion.

Higher temperatures tend to improve feed conversion up to the point of heat stress, after which feed conversion is poorer. In cold climates, part of the feed must be used for maintenance of body heat; therefore, feed conversion is better in the summer. Insulating broiler houses in cold climates improves feed conversion.

A high incidence of sickness or a high level of condemnations causes poorer feed conversions. Sick birds have lower rates of gain.

As birds get older and gain more weight, it takes more feed to put on a pound of gain. Feed conversion of older, heavier birds is poorer as compared to younger, lighter birds. The break-even point is approximately an increase of 5 points in weight for each 1 point increase in feed conversion. Each point is 0.05 pounds of weight or feed conversion.

Rodents and flying birds that get into the broiler house consume feed. This causes a poorer calculated rate of feed conversion when the broilers are marketed.

The use of antibiotics and other medications generally improves feed conversion, and broiler feeds generally contain these additives.

Broilers that are not debeaked waste more feed, causing poorer feed conversion. Improperly debeaked birds may not grow as fast or may also waste feed—both of which causes poorer feed conversions.

Baby chicks hatched from small eggs do not grow as fast as those hatched from normal size eggs. Using baby chicks hatched from small eggs will cause poorer feed conversion.

Improper managing of feed causes waste, resulting in poor feed conversion. Filling feeders too full can waste as much as 75 percent of the feed. Tube feeders must be properly adjusted to prevent feed wastage. Proper ventilation and proper heat adjustment on the brooder are good management practices that improve feed conversion. Watching the broilers for early signs of disease and taking preventative measures in a timely manner also improves feed conversion.

Experimental evidence shows that pelleted feeds give better feed conversion than do all-mash rations. Feed conversion as much as 5 to 9 points better has been demonstrated in experimental work using pelleted rations as compared to all-mash rations.

Feeding Broiler Type Breeding Pullets

Chickens bred for meat production grow rapidly and reach sexual maturity at an early age. This results in too many small eggs, which are not good for hatching purposes, and in lower egg production at peak production. When birds are kept for breeding purposes, it is necessary to slow down their rate of growth and development of sexual maturity. Several methods are used to accomplish this.

Restricting feed intake to approximately 70 percent (of the amount which would be consumed if fed free choice) is one method. This program is usually started when the pullets are seven to nine weeks old and continued until they are approximately 23 weeks of age. At this time, they are fed a laying mash for production of hatching eggs.

A skip-a-day program involves full feeding every other day. On alternate days, only 2 pounds (0.9 kg) of grain is fed per 100 birds. This feeding program is begun at seven to nine weeks of age and continued to the 23rd week.

Another method is feeding a diet containing 10 percent protein. This program is begun at seven to nine weeks of age and continued to the 23rd week.

A fourth method is feeding a low-lysine diet, which slows down the rate of growth and development of sexual maturity, because of the imbalance in the amino acids in the ration. The diet contains 0.4-0.45 percent lysine and 0.6-0.7 percent arginine, with a protein level of 12.5-13 percent. This diet is begun at seven to nine weeks of age and continued to the 23rd week.

FEEDING TURKEYS

The general principles of feeding turkeys are similar to those for feeding broilers. Major differences are in the protein levels required and the importance of the vitamins biotin and pyridoxine in the turkey diet. The nutrient requirements of turkeys are shown in tables in the appendix.

Grit

An insoluble grit may be used when feeding turkey poults but there is little evidence to indicate that it is necessary when all-mash rations are fed. If used, do not feed grit free choice as poults will consume too much of the grit and not enough of the ration. Sprinkle grit at the rate of 1 percent of the ration on top of the feed once a week for the first four weeks. After eight weeks, supply turkey-size insoluble grit to all turkeys.

Poults

Poults must be fed and watered as soon as possible after hatching. If feeding is delayed beyond 36 hours after hatching, poults have difficulty learning to eat and drink. Force-feeding may be necessary to save these birds. A thin mixture of starting mash and water may be made and forced into the gullet of the poult with a pipette or a large eye dropper with the tip removed. Be sure the pipette or eye dropper is inserted far enough into the gullet to be below the entrance to the windpipe. Fill the crop by forcing the feed out of the pipette or eye dropper. Usually one feeding is enough to get the poult started to eat and drink. If dehydration is severe, two or three force-feedings may be necessary.

Poults are fed starting mash or crushed pellets for the first four weeks of age. Colored glass marbles or rolled oats may be sprinkled on the feed and placed in the water to attract the birds' attention and get them started to eat and drink. Pellets may be fed starting at about four weeks of age.

Prestarters may be fed for the first three weeks. The prestarter contains 30-33 percent protein with higher levels of amino acids, energy, vitamins, and antibiotics. Growers who have experienced high mortality rates in poults may find it desirable to use a prestarter ration. A prestarter may also be desirable if unusual stress conditions such as disease, chilling, or crowding are present.

Starter rations contain 25-29 percent protein, 1.5 percent calcium, and 1.1 percent phosphorus. The energy level is 1330-1360 kcal/lb (2930-3000 kcal/kg). A coccidiostat is often included in starter rations, but is seldom needed in rations beyond 8-12 weeks of age. Water should always be provided free choice.

Growing and Finishing Turkeys

Poults are fed a growing diet starting at eight weeks of age. As turkeys grow older, the protein level of the ration is reduced and the energy level is increased. From 8-12 weeks of age the protein level of the ration is 20-22 percent and the energy level is approximately 1400 kcal/lb (3100 kcal/kg). From 13-16 weeks of age, the protein level is approximately 19 percent and the energy level is about 1450 kcal/lb (3200 kcal/kg). Change to a protein level of about 16 percent and an energy level of 1485 kcal/lb (3275 kcal/kg) at 17-20 weeks of age. At 21 weeks to market, feed a ration containing a protein level of 13-14 percent protein and an energy level of 1520 kcal/lb (3350 kcal/kg).

While growing turkeys can be fed on mash and grain diets, it is recommended that complete pelleted feeds be used. Turkeys do not do a good job of balancing their diet when fed mash and grain separately, and more economical gains are achieved when complete rations are fed.

At about 10-12 weeks of age, separate hens from toms. The protein and energy requirements are different as the birds get older, and more efficient gains are made when hens and toms are fed separately. Overall feed efficiency for growing and finishing turkeys is about 3.5 pounds of feed per pound of gain. Toms reach a market weight of 30-33 pounds (13.6-15 kg) at about 24 weeks of age and hens reach a market weight of 18-20 pounds (8-9 kg) at 20-22 weeks of age.

Turkeys may be raised in confinement or on range after they are 8-10 weeks of age. Confinement raising is becoming more common among large commercial turkey growers. When a good forage or grain crop is available on the range, savings of up to 10 percent in feed costs may be realized. Alfalfa, ladino clover, bluegrass, brome grass, and orchard grass are often used when permanent range areas are used for turkeys. Annual crops often used when range areas are rotated include soybeans, rape, kudzu, kale, sunflowers, reed canary grass, and Sudan grass.

Feeding Breeding Turkeys

Breeding turkeys are fed the same diets as market turkeys until they are about 18 weeks of age. Turkey breeder hens are fed a holding diet beginning at about 18 weeks of age. The holding diet slows down sexual maturity in the hens. Restricted feeding is not recommended for turkeys. The holding ration contains 13-15 percent protein and approximately 1340 kcal/lb (2950 kcal/kg) of energy. This is fed until about 3 weeks before egg laying begins.

Provide 15 hours of light when egg laying begins and a ration containing 15-18 percent protein and 1330 kcal/lb (2925 kcal/kg) of energy. When peak production of eggs is reached after about two weeks of laying, increase the protein level to 19 percent with 1250 kcal/lb (2755 kcal/kg) of energy. As the weather becomes warmer, increase the protein level to 21 percent and decrease the energy level to 1210 kcal/lb (2667 kcal/kg). This change in protein and energy level is necessary because the turkeys tend to eat less as the weather gets warmer.

Fortification of the diet with vitamins and minerals is essential for good hatchability of turkey eggs. Vitamin and mineral requirements are given in tables in the appendix.

FEEDING DUCKS AND GEESE

Commercial feeds in mash, pelleted, or crumble form are available for ducks and geese. Pelleted or crumble forms are recommended. If a commercial feed for ducks or geese is not available, a chicken feed may be used if it does not contain a coccidiostat.

For the first two weeks of brooding, feed a starter feed free choice. Feed an insoluble grit in addition to the starter feed. Ducks are changed to a grower ration at two weeks of age while geese are put on a grower ration at three to four weeks of age. The grower ration is fed until the birds are ready for market. Ducks are ready for market in seven to eight weeks while geese are marketed at 24-30 weeks. Tables in the appendix show the nutritional requirements for ducks and geese.

Provide a good supply of clean, fresh water for ducks and geese. Use waterers which prevent the ducks and geese from getting into them to swim.

Geese will start to eat pasture when they are only a few days old. They can live entirely on pasture after they are five to six weeks of age. It is recommended that a growing ration be used for more efficient gains. Timothy, bluegrass, ladino clover, white clover, and brome grass make good pastures for geese. Barley, wheat, and rye make good fall pastures. Geese do not eat alfalfa, sweet clover, or lespedeza.

Feed additional grain if the pasture is not of good quality. Cracked corn, wheat, or milo can be used. Pasture rotation is recommended, with 1 acre of pasture feeding about 20 birds.

Ducks will eat some green feed, but are not as good at foraging as geese are. Pasture is not necessary for feeding ducks.

Feed a breeder-developer ration to ducks being kept as breeders, and use a breeder diet when they are older. Breeder diets contain less energy than grower diets and should be fed starting about one month before eggs are to be kept for hatching. Make sure the diet contains enough calcium, which can be provided by feeding oyster shell.

SUMMARY

Poultry are usually fed on a group basis; therefore, rations are formulated on the basis of units per unit weight of feed. The most commonly used measure of energy for poultry diets is apparent metabolizable energy (ME). Poultry generally eat to meet their energy needs; therefore, the intake of nutrients is controlled by including them in the proper proportion to the energy content of the ration. High energy levels tend to reduce feed intake.

The 13 essential amino acids must be provided in poultry diets for efficient reproduction, growth, and egg production. Proper balance of amino acids is necessary.

Poultry need calcium, phosphorus, sodium, and chloride provided in the diet. Trace minerals may need to be added if feedstuffs come from soils deficient in trace minerals.

Vitamins A, D, and E need to be provided in the ration, so the use of a vitamin premix is recommended.

Grains, especially corn, provide energy in the diet. Animal and vegetable fats are sometimes added to rations to provide additional energy.

Soybean meal is the most commonly used protein supplement in poultry rations. Some animal proteins may be used, and fish meals provide a source of unidentified nutrients in the diet.

Oyster shells, steamed bonemeal, and salt provide the major minerals needed. Trace-mineral premixes may be used if needed.

Poultry feeds may be in the form of mash, pellets, or crumbles. There is less waste when pellets are used, and pellets and crumbles are easier to use with mechanized feeding equipment.

Poultry may be fed with a complete feed or mash and grains may be used. Large-scale producers may find it more economical to buy the various ingredients and have rations custom mixed.

The appropriate protein and energy levels need to be used when formulating rations for different purposes. Broilers have different requirements than do laying flocks or breeding flocks. Turkeys, ducks, and geese must have rations formulated to meet their unique needs as well.

REVIEW

1. Why are poultry rations generally formulated on the basis of units per unit weight of feed?
2. What is the most common method of measuring energy when formulating poultry rations?
3. What effect does the energy level of the ration have on feed intake?
4. List the factors that affect feed intake in poultry.
5. What effect does environmental temperature have on feed intake?
6. What effect on fat deposition in poultry does the ratio of energy to protein in the ration have?
7. Why is the proportion of energy to other nutrients in the ration of importance in formulating poultry diets?
8. What are the major mineral requirements of poultry and how can these requirements be met?
9. What are recommended ratios of phosphorus to calcium for young chickens and for laying hens?
10. What vitamins need to be included in poultry rations and how is this done?
11. Discuss unidentified nutrients in poultry nutrition.
12. List and discuss some common energy feeds used in poultry diets.
13. List and discuss some common protein sources used in poultry diets.
14. Discuss the preparation of feeds for poultry.

15. What are the four systems of preparing rations that may be used with poultry?

16. List and discuss some recommended feed management practices.

17. Discuss the use of grit in feeding poultry.

18. What are the recommended protein and energy levels for feeding replacement pullets of the egg-laying breeds?

19. How may the rate of growth and development of sexual maturity be controlled in replacement pullets?

20. What is the recommended protein and energy level of a laying ration?

21. List and describe the three systems of feeding that may be used for laying flocks.

22. Describe phase feeding and tell why it might be used.

23. What are the recommended protein and energy levels in broiler feeds for starting, growing, and finishing rations?

24. List and discuss the factors that affect feed conversion in broilers.

25. Discuss methods that may be used to control rate of growth and development of sexual maturity in broiler-strain breeding flocks.

26. Why is it important that turkey poults eat and drink within 36 hours after hatching?

27. Under what conditions is the use of a prestarter feed recommended for turkey poults?

28. What is the recommended level of protein and energy in turkey starter rations?

29. At what age are turkey poults put on a growing ration?

30. What are the recommended protein and energy levels in turkey rations during the different periods of growth and finishing?

31. Why are complete rations recommended for turkeys?

32. At what age are the hens separated from the toms? Why?

33. Discuss the feeding of breeding turkeys.

34. Discuss the feeding of ducks.

35. Discuss the feeding of geese.

18

MYCOTOXINS

OBJECTIVES

After completing this chapter you will be able to:

- Define mycotoxins.
- Describe the general effects of mycotoxins.
- Describe the effects of specific mycotoxins.
- Describe methods of using mycotoxin contaminated feeds.

MYCOTOXINS—DEFINITION

Fungi are a group of thallophytic plants, which are composed of molds, mildews, rusts, smuts, and mushrooms. As a group, they do not contain chlorophyll and generally reproduce by means of asexual spores. Metabolites are products of metabolism. Some metabolites produced by fungi are harmful to animals. These are called *mycotoxins* (*myco* meaning *fungi* and *toxin* meaning *poison*). The literal meaning of mycotoxins is *poisons from fungi*.

Mycotoxins are sometimes named on the basis of the fungi that produce them. For example, *aflatoxin* is named by using *a* for the fungi *Aspergillus* and *fla* for the species *flavus* and appending the word *toxin*. *Deoxynivenol* is an example of the actual chemical name of a toxin. In other cases, the name is based on the effect of the toxin, for example, *vomitoxin* because it causes a vomiting reaction in the animal that ingests it.

GENERAL EFFECTS OF MYCOTOXINS

The effects of mycotoxins have been known throughout recorded history. However, most of the research on mycotoxins has been done since the discovery of aflatoxins in Great Britain in 1960.

Mycotoxicosis in animals is generally the result of exposure to a group of mycotoxins, rather than to a specific mycotoxin. However, one or two of the mycotoxins in the group usually have the major impact in creating a problem.

Nonlethal toxic effects of mycotoxins include:

Hepatotoxins—toxic to the liver
Nephrotoxins—toxic to the kidneys
Neurotoxins—toxic to the nerves
Genitotoxins—toxic to the reproductive system
Dermatoxins—toxic to the skin

Mycotoxins may cause mutagenic effects (causing biological mutations) and teratogenic effects (causing fetal malformations). In larger doses, mycotoxins can cause death. Some mycotoxins also exhibit carcinogenic (cancer causing) effects when ingested over a period of time.

The acute effects of mycotoxins are of greater concern in livestock production because animals are generally kept for relatively short periods of time, except in the case of breeding stock, where the chronic effects are also of concern. The acute effects reduce overall animal productivity. The carcinogenic effect is of greater concern in human health because of the danger of humans consuming animal products (especially milk) that may contain mycotoxins.

Mycotoxins appear to affect basic metabolic processes in the animal. These include carbohydrate metabolism, mitochondrial functions (conversion of food to energy), lipid metabolism, and the biosynthesis of proteins and nucleic acids.

The specific mechanisms believed to be involved in the action of mycotoxins include:

1. inhibiting key enzymes required in metabolic processes;
2. hindering the synthesis of proteins containing DNA and RNA;
3. interfering with molecular transport in the cell and releasing hydrolytic enzymes to decompose vital molecules;
4. reducing enzyme activities by reacting with enzyme cofactors.

Mycotoxins lower resistance to diseases and adversely affect immunization. A level of mycotoxin intake below that needed to produce visable chronic effects can affect resistance and immunity. Aflatoxin effects on immunization have been demonstrated in poultry with fowl cholera and in swine with erysipelas. Aflatoxin has also been shown to increase poultry susceptibility to salmonellosis, coccidiosis, and candidosis (fungus infection caused by members of the genus *Candida*). It has also increased the susceptibility of calves to fascioliasis (infestation with a trematode worm of the genus *Fasciola*). Further research is needed on the effects of mycotoxins on resistance to disease and immunization.

Table 18-1. Mycotoxicoses commonly found in the United States.

Disease caused by mycotoxin	Organism producing mycotoxin	Name of disease
aflatoxicosis	*Aspergillus flavus, Aspergillus parasiticus*	yellow mold
ochratoxicosis	*Aspergillus ochraceus, Penicillium* spp.	blue eye
paspalum staggers	*Claviceps paspali*	ergot
ergotism	*Claviceps purpurea*	ergot
T-2 poisoning, fusariotoxicoses	*Fusarium tricinctum*	
swine refusal, hyperestrogenism	*Gibberella zeae*	scab, Gibberella ear rot
slobbers	*Rhizoctonia leguminicola*	black patch

INCIDENCE OF MYCOTOXICOSES IN THE UNITED STATES

At one time it was commonly believed that livestock could be safely fed feeds containing molds, because the animals had the ability to avoid that portion of the feed that was harmful to them. It is now known that this is not true. Animals will eat corn, other grains, forages, or processed feeds that contain harmful levels of mycotoxins.

While aflatoxin is the major nutritional problem related to mycotoxins in the United States, there are several other mycotoxins that cause problems with livestock feeding. Table 18-1 shows the major mycotoxin problems found in the United States.

OCHRATOXIN

Poultry

A number of species of the fungi *Aspergillus* and *Penicillium* can produce ochratoxin A and ochratoxin B in poultry. The structural formula for ochratoxin A is shown in Figure 18-1. While this mycotoxin has generally been of low incidence in the United States, there have been some large outbreaks in poultry in several states.

Ochratoxin is comparatively more toxic to poultry than either aflatoxin or T-2 toxin. Broiler chickens exhibit severe dehydration and emaciation, proventricular hemorrhages, and visceral gout. The kidneys appear to be the major body organ affected by ochratoxin A; however, the liver is also affected, though to a lesser degree. Kidneys in affected birds are enlarged. Livers show a tan color and have increased glycogen content.

Bone strength is reduced in young broiler chicks affected by ochratoxin. Laying hens show lowered egg production, slower growth, and lowered feed consumption. The liver, kidneys, and red and white muscle show residues of ochratoxin A; however, no residues have been found in eggs, fat, or skin. Residues

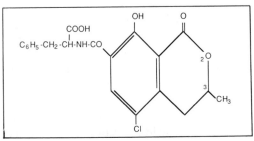

Figure 18-1. Structural formula of Ochratoxin A.

disappear from the white muscle within 24 hours after feeding of contaminated feed ends; however, residues persist in the kidneys and liver more than 48 hours after feed containing ochratoxin A is withdrawn from the diet. Chick embryos also appear to be highly sensitive to the presence of ochratoxin A in the diet, with a high mortality rate found with low levels of feed contamination.

Swine

There is little evidence of a problem with ochratoxin in swine in the United States. In some parts of Europe, this mycotoxin has been a major problem in swine feeding. Ochratoxin causes liver damage in swine.

Ruminants

Naturally occurring ochrotoxicosis in ruminants is rare to the point of being almost non-existent. Experimental evidence indicates that rumen microorganisms have the ability to detoxify ochratoxin.

PASTURE GRASS MYCOTOXINS

Several fungi attack pasture grasses such as tall fescue, Bermuda grass, Dallis grass, rye grass, wheat grass, smooth brome grass, orchard grass, redtop, bluegrass, and Reed's canary grass. Among these the fungi *Claviceps purpurea* and *Claviceps paspali* probably have the greatest impact, from an animal nutrition point of view. The effects of these mycotoxins on livestock include ergotism, fescue foot, Bermuda grass staggers, and Dallis grass tremors or poisoning.

Ergot

While the fungus *Claviceps purpurea* attacks a large number of grasses, it is most important as a mycotoxin on the range grasses of the Western plains. It also attacks rye, triticale, wheat, barley, and oats. Hybrid wheat and barley are less susceptible, but naturally open-pollinated varieties of grain such as rye and triticale are highly susceptible. This fungus is rarely found on crops in the southeastern United States.

Claviceps purpurea causes a disease on plants called *ergot*. The disease may be recognized on plants because of the blue-black sclerotia that take the place of the grain or grass seeds. The sclerotia are generally larger than the seeds that they replace and may be harvested with the grain or may fall to the ground to overwinter, forming a source of continued infestation of the fungus in the field.

In the spring the sclerotia germinate, producing spores carried by wind currents, rain, and insects to the flowers of the grain or grass plants. Fungus is formed from the spores and the threads of the fungus form the sclerotia on the grain or grass plants.

The flowering heads of the plant are susceptible to infestation for a short period during and after pollination. The incidence of ergotism is increased during moist weather, which promotes the germination and growth of the spores from the sclerotia. A high level of rainfall may reduce the incidence of infection.

An early symptom of ergotism is lameness in the hind legs. Vitality of newborn animals is reduced and milk flow is reduced or stopped. Acute ergotism (caused by ingesting large amounts of the ergot-infested feed in a short period of time) results in nervousness and increased sensitivity of the skin, muscular trembling, uncoordinated gait, painful contractions, and convulsions. Animals may die when they reach the convulsive stage. Chronic ergotism (caused by ingesting small amounts of the ergot-infested feed over a longer period of time) results in animals going off feed and becoming dull and depressed. The lower legs become stiff and sore and gangrene may become evident in areas of poor blood circulation. Poultry generally show symptoms of gangrene in the comb, wattles, and beak as a result of ingesting ergot-infested feed.

Dallis Grass Poisoning

Claviceps paspali is the fungus that infects Dallis grass. Cattle that ingest Dallis grass infected with this fungus develop a disease called *paspalum staggers*, or Dallis grass poisoning. The symptoms of this disease include tremors, hyperexcitability, ataxia (inability to coordinate voluntary muscular movements), and exaggerated movement.

This disease is confined mainly to the southern United States, where Dallis grass has been an important forage crop. Because the acreage of Dallis grass is declining, the importance of this mycotoxin is less than it was at one time.

The life cycle of the fungus is similar to that of *Claviceps purpurea*, with wet weather producing conditions conducive to the growth of the fungus. There is some indication that dampness from heavy dew may be more important than rainfall in promoting the growth of the fungus.

Fescue Foot

The fungus *Balansia epichloe* is a clavicipitaceous fungus that produces ergot-like symptoms in cattle, including Bermuda grass tremors and fescue foot. The alkaloids produced by this fungus are toxic to duck and chick embryos. Another fungus belonging to the Clavicipitaceae family is *Epichloe typhina*, which is a parasitic fungi of pasture grasses. It has been found on tall fescue, and experiments with toxic grass extracts being injected into cattle have produced symptoms of fescue foot.

The early symptoms of fescue foot in cattle include a slight arching of the back, roughened hair coat, and soreness in one or both hind legs. As the disease progresses, the lameness becomes more severe and the coronary band and the area between the dewclaws and hoofs reddens and becomes swollen. The end of the tail also shows redness and swelling. As the tissue dies, it changes to a purple-black color. After several weeks, the tail tip or the entire switch may drop off. Sloughing of the ear tips is also common. Tremors beginning in the rear leg muscles and advancing forward to the head may occur, especially at lower ambient temperatures (below $16°F$ ($-8.9°C$)). As the disease becomes more severe, swelling above the hoof increases, sometimes extending to the hock. A dead tissue line develops just below the dewclaw with the resultant loss of the hoof. If the case is especially severe, with swelling occurring above the dewclaws, the fleshy part of the leg midway between the hock and the dewclaws may be lost, leaving only the long bone intact. Feed intake is reduced in affected animals and emaciation is common.

A number of other conditions such as foot rot, frozen feet, and mechanical injury may produce symptoms similar to fescue foot. It is important to consult a veterinarian concerning treatment when any of the symptoms of fescue foot appear in the herd. Remove animals from the fescue as soon as symptoms appear. Mild to moderate cases may recover slowly. Animals with a severe case of fescue foot will continue to get worse and should be salvaged by marketing.

FUSARIUM TOXINS

A number of species of *Fusarium* are found on plants. Among the cereal grains, these fungi cause a variety of diseases such as scab in wheat and barley and stalk rot and cob rot in corn. In some cases, toxins harmful to animals may be produced.

Gibberella

Gibberella zeae is the name of the sexual stage of a fungus causing Gibberella ear rot in corn. It

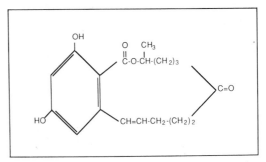

Figure 18-2. Structural formula of Zearalenone.

also causes scab on small grains, where it produces toxic compounds. Scab is a blight of the flowering heads and kernels of small grains. From a mycotoxin point of view, *Gibberella zeae* is a more serious problem on corn than it is on small grain. It is easier to remove the infected kernels from small grain than it is from corn.

The imperfect stage of this fungus is called *Fusarium roseum* or *Fusarium graminearum*. This fungus produces deoxynivalenol and other substances when it grows on grain in the field, which sometimes cause swine to refuse the feed and also induces vomiting when they ingest contaminated feed. Other symptoms of mold poisoning include scours, bloody feces and urine, skin hemorrhage, and poor growth. Other species of *Fusarium* have also been shown to produce symptoms of mold poisoning.

When *Gibberella zeae* grows on grain in storage, it produces *zearalenone* (F-2 toxin), which is a female estrogen. The structural formula for zearalenone is shown in Figure 18-2. Grain contaminated with zearalenone may cause hyperestrogenism when consumed by swine. Hyperestrogenism in female hogs can cause swollen vulvas and mammary glands, irritated teats, descending vaginas or rectums, increased size and weight of the uterus, false heat, and infertility. Pregnant gilts or sows may abort. Young males may show atrophy of the testes and enlarged mammary glands. Other livestock appear to be less sensitive to zearalenone than are swine. How-

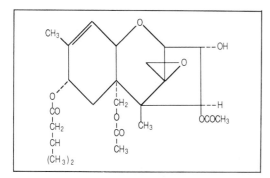

Figure 18-3. Structural formula of T-2 toxin.

ever, there is some indication that zearalenone and other related compounds may cause infertility problems in ruminants.

The *fungus Gibberella zeae* attacks corn during silking, when the weather is wet and cool. Cool, wet weather in the fall also increases the incidence of the fungus. Corn varieties with especially tight husks are more susceptible to Gibberella ear rot. The problem also appears to be intensified in fields with a high level of foliage blight. *Gibberella zeae* can also stalk rot in corn.

Infected ears have a pink to white colored mold growth at the tip, which may extend several inches along the ear. The kernels also show the pink to white color. In some cases it is difficult to remove the husks from the ears. A shiny black coating may appear on the tips of the husks.

The Gibberella fungus continues to grow in stored ear corn when the ambient temperature is above 35°F (1.7°C) at moisture contents above 25 percent. The fungus also grows in the cob, which may make ground corn and cob meal more dangerous to feed than shelled corn from infected grain.

It is more difficult to identify the fungus in shelled corn, especially after the corn is dried. In heavy infestations, kernels are shrunken, with blotches of red or pink color. In some cases, the kernel may show discoloration only at the tip where it was attached to the cob. It is easy to mistake heat-damaged kernels for Gibberella-infested grain.

Experimental work done at Purdue University indicates that swine feeding problems are likely to occur if 5 percent or more of the kernels show the pink to white mold. At infestation levels of 2-3 percent, there is generally no problem in feeding the grain to swine. It is important to note that the level of visible damage is not necessarily related to the amount of toxin present in the corn. The actual level of toxin present can only be detected by laboratory analysis of the grain.

Trichothecenes (T-2 Toxins)

Numerous fungi of the species *Fusarium* produce a group of 20 to 30 toxins called trichothecenes. *Fusarium tricinctum* is found on many plants, but the toxin it produces (T-2) is usually found only on corn. The structural formula of T-2 toxin is shown in Figure 18-3. The fungus usually develops at moisture contents above 23 percent and requires a period of low or alternating moderate and low temperature for growth in stored grain. It will occur on frost-damaged ears of corn in the northern part of the Corn Belt in the United States. However, *Fusarium tricinctum* is not as serious a problem as *Gibberella zeae* in corn.

T-2 toxins have been associated with oral lesions, impairment of the nervous system, reduced feed intake, body weight loss, lowered egg production, thinner eggshells, and abnormal feathering in poultry. The hatchability of fertile eggs may also be reduced.

In cattle, the reported effects of T-2 toxin have included feed refusal, rumen ulcers, congestion in the small intestine, depression, listlessness, anorexia, ataxia in the hindquarters (inability to control voluntary muscular movement), decreased milk production, hair loss, bloody diarrhea, and hemorrhage of the mucous membranes. It is believed that T-2 toxin has little effect upon rumen function. However, the evidence indicates that the toxin is not destroyed or detoxified completely by rumen microorganisms.

T-2 toxin has been associated with hemorrhagic effects and hair loss in swine.

It has been difficult to determine the exact causes of these problems because of the difficulty of detecting the trichothecenes involved. It is believed that these toxins act together as a group, rather than alone, in producing the observed effects.

SLOBBER

Forage infected by the fungus *Rhizoctonia leguminicola* and ingested by animals can cause excessive salivation, or slobber. Forages that are sometimes attacked by this fungus include red clover, white clover, soybean, Kudzu, cow pea, blue lupine, alsike clover, alfalfa, Korean lespedeza, and black medic. The disease on the crop is called *black patch* and more commonly develops during periods of wet weather and high humidity. The fungus grows on the surface of the leaf, showing a large black mycelium. Infected seed can spread the fungus. Outbreaks of this mycotoxin are not common in the United States.

AFLATOXIN

Field and Storage Infestation

The fungi *Aspergillus flavus* and *Aspergillus parasiticus* produce aflatoxin in a variety of feeds. The structural formulas for several *Aspergillus* toxins are shown in Figure 18-4. Corn, peanuts, and cottonseed may be infested both before and after harvest. Soybeans and small grains usually are not attacked prior to harvest, but contamination may occur in storage.

Aflatoxin contamination of corn generally occurs in the southeastern part of the United States, with little contamination occurring in other corn growing areas. This appears to be related to more favorable climatic conditions (warm and humid) for the growth of the fungus and for the presence of insects that damage the kernels. Insect damage to the kernels is a major factor in the infestation of corn by the fungus. However, general stress of

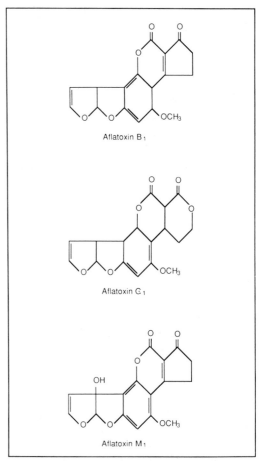

Figure 18.4 Structural formulas of several *Aspergillus* toxins.

the corn plant appears to also be a factor in field contamination.

The major geographic areas of infestation of the cotton plant are in the Imperial Valley of California and in Arizona. High ambient temperatures appear to favor field infestation of the cotton plant. When cotton harvest occurs with dew on the plant, storage infestation is more likely to occur.

Factors conducive to infestation of peanuts by the fungus are overmature pods, drought stress, pod growth cracks, and fungal and insect damage to the pod. Contamination may occur in storage if the kernels are damaged and dried slowly in the windrow.

Relative humidity above 85 percent and lower than 94 percent is favorable to infestation by the fungus in storage. Ambient temperatures between 75°F (24°C) and 95°F (35°C) appear to permit the maximum production of aflatoxin. The growth of the fungus is stimulated at a moisture content of 14 percent or above. When moisture can get into the storage area by leaking, condensation, exposure to rain, or other sources, the growth of the fungus is enhanced. Poor aeration of storage areas and stored insect damage can also improve conditions for the growth of the fungus. The growth of the fungus is favored by the presence of fines such as broken kernels, dust, and weed seeds. Contamination by aflatoxin is increased by poor sanitary practices in feed storage and feeding areas.

Aflatoxin Effects—General

The early symptoms of aflatoxin poisoning in livestock include lowered feed intake, loss of weight, fatty livers, and rapid death. The positive identification of an aflatoxin problem is made by an analysis of feed, urine, or milk samples, which reveals the presence of aflatoxin in the sample.

Factors that affect the susceptibility of animals to aflatoxin poisoning include species, age, form of the toxin, and nutritional status of the animal. There are several forms of aflatoxin, with B_1 being the most common and the most toxic. Other forms include B_{2a}, G_1, G_2, M_1, and M_2. The body organ most generally affected in animals is the liver. Aflatoxins have also shown carcinogenic properties.

Aflatoxin Effects in Swine

Younger swine appear to be more susceptible to aflatoxin poisoning than are older swine. Effects of aflatoxin in swine include liver damage, jaundice, and hemorrhaging in many parts of the body. Hemorrhaging is especially pronounced in the ham area, with the increased pressure on the muscles of the buttocks resulting in an inability to control voluntary muscle movement in the area. The animals exhibit a doglike sitting position with panting. Aflatoxin can also produce cancer in swine.

At higher levels of aflatoxin intake, rate of gain and feed efficiency are reduced. Depression, with shivering, muscle tremors, weakness, and internal hemorrhaging occur with acute aflatoxicoses in young pigs. Newborn pigs nursing sows that were consuming feeds containing aflatoxin are stunted in growth.

A number of factors influence the susceptibility of swine to aflatoxin B_1. These include age and protein and vitamin K levels in the diet. The presence of aflatoxin M_1 in sows' milk of nursing pigs is also a factor. In addition, the extent and level of exposure to aflatoxin influences its effect on swine. The presence of other mycotoxins in the feed may increase the severity of the effect of aflatoxin B_1 in swine. The effects of aflatoxin poisoning are quickly overcome when swine are removed from the contaminated feed and placed on good rations.

Aflatoxin Effects in Ruminants

Young calves affected by aflatoxin poisoning show poor feed utilization. Dairy cows show the following symptoms of aflatoxicoses: reduced feed intake, unthriftiness, lethargy, decreased milk production, dry, peeling skin on the nose, liver damage, prolapse of the rectum, and higher blood levels of cholesterol.

Experimental work suggests that the presence of aflatoxin B_1 affects the growth and function of microorganisms in the rumen. Volatile fatty acid production, ammonia formation, and cellulolysis is reduced by the presence of aflatoxin. The proportions of the volatile fatty acids is also altered, with acetic acid decreased and propionic acid and butyric acid increased. The amount of valeric acid is not altered.

Ruminants have a greater resistance to the effects of aflatoxin poisoning than do non-ruminants. There is an indication that some rumen microorganisms may partially detoxify aflatoxin by transforming it to less toxic metabolites. More research is needed on the effects of aflatoxin on rumen function.

Aflatoxin is secreted into the milk of dairy cows, resulting in a health hazard for humans consuming the milk because of the carcinogenic effect of aflatoxin B_1. High-producing cows may excrete higher levels of aflatoxin in the milk than low producers. Experimental evidence suggests that cows excrete an average of 0.91 percent of ingested aflatoxin in the milk.

DETECTING MYCOTOXINS IN FEED

Because of the increased awareness of the importance of mycotoxins in animal and human health, more research has been done in recent years regarding the detection of mycotoxins in feedstuffs. This is an area in which continued research is needed.

One of the problems associated with the detection of mycotoxins in feedstuffs is the fact that a small infestation of fungi can contaminate fairly large quantities of feed. Therefore, it is important to get representative samples of feedstuffs when attempting to detect the presence of mycotoxins.

Another problem has been the difficulty of establishing a clear cause and effect relationship between feeds suspected of containing mycotoxins and symptoms of mycotoxicoses in animals. This is especially true in cases of low-level ingestion of mycotoxins causing chronic effects.

Mycotoxins also interact with other feed components, making their detection more difficult. For example, aflatoxin has been shown to bind to molecules of nucleic acids, proteins, polysaccharides, and many other smaller molecules found in feeds.

Problems associated with the detection of mycotoxins in feedstuffs include:

1. sporadic occurrence of fungi in fields in a geographic area;

2. year-to-year weather variations that influence the growth of various fungi;
3. variations in plant stress such as those caused by insect infestation;
4. soil differences affecting plant susceptibility to insects and fungi;
5. variation in the occurrence of toxins on individual kernels of grain;
6. toxin distribution not uniform within contaminated kernels;
7. variation of toxin distribution in initial sample and in subsamples used for analysis in the laboratory;
8. difficulty of extracting toxin from contaminated feedstuffs.

A qualitative test for the presence of aflatoxin in feedstuffs involves the use of ultraviolet light. Suspected samples of feed are exposed to ultraviolet light and if a bright greenish-yellow (BGY) fluorescence occurs, it is an indication of the possible presence of aflatoxin. When the BGY emission under ultraviolet light occurs, the feed is subjected to further analysis in the laboratory to confirm the presence of aflatoxin. Laboratory testing is necessary because the presence of BGY emission does not always indicate the presence of aflatoxin.

MYCOTOXIN RESIDUES

Based on data from research, it appears that there is little danger of human consumption of mycotoxins from residues in animal tissues. A mature, nonlactating animal effectively modifies and eliminates mycotoxins with little or no residue remaining in animal tissues. Toxins and/or their metabolites are excreted from the animal's body within 24 hours of ingestion with no significant quantity being retained in edible tissues.

The only significant residue of mycotoxins has been aflatoxin M_1, found in milk, and ochratoxin A, found in swine kidneys. Fluid milk is considered contaminated if a level of 0.5 ng/ml of aflatoxin M_1 is found in a milk sample. In swine, it is recommended that a period of at least four weeks of feeding non-contaminated feed occur to reduce the level of ochratoxin A to less than 2 ng/g of animal tissue.

USING FEEDS CONTAINING MYCOTOXINS

Prevention and Decontamination

The best method of controlling mycotoxin contamination of feedstuffs is by preventing the initial contamination. Good farm management practices help to prevent the infestation of feed crops with fungi.

Ergot

Crop rotation and deep plowing help to prevent ergot infestation in crops. Use ergot-free seed for planting and if that is not available, clean the seed to remove the ergot that is present. Controlling grasses and weeds near grain fields helps to reduce the availability of host plants for the fungi causing ergot. Insect control helps to reduce ergot infestation in crops.

Grain contaminated with ergot may be cleaned for feeding by physically separating the ergot from the grain. This can be done by screening the grain to remove the ergot bodies, which are generally larger than either rye or wheat kernels. The use of a gravity separator may be necessary when the ergot bodies are broken and pass through a screen. Because the ergot is lower in density than the grain, it can be removed in a gravity separator.

Aflatoxin

Good farm management practices help to prevent aflatoxin contamination of feeds. Use fungus-free seed and control insects and diseases in the field. Harvest promptly at maturity, using properly adjusted equipment to avoid damage to the grain. Avoid getting excessive amounts of dirt and leaves in the grain during harvest.

The use of propionic acid and ammonia has been demonstrated to reduce significantly the formation of mold in stored corn and subsequently to reduce the infestation of the grain by mycotoxins.

The early detection of the presence of contamination in feed and the diversion of these batches to prevent contamination of larger lots of feed also is a possible method of preventing mycotoxin infestation. The use of BGY fluorescence on corn and cottonseed and detecting the presence of molds on peanuts during grading, represent methods of early screening and detection that may have practical application.

It is possible to extract aflatoxin from oils by using solvents. Solvents that can be used include 95 percent ethanol, 90 percent aqueous acetone, 80 percent isopropanol, hexane-ethanol, hexane-methanol, hexane-acetone-water, and hexane-ethanol-water. Using these solvents will remove essentially all of the aflatoxins present in the oils.

Experimental work indicates that it may be possible to decontaminate aflatoxin-contaminated corn by treating it with aqueous ammonia, followed by roasting.

The use of chemicals to detoxify aflatoxin has been extensively studied. Ammonia, methyl amine, sodium hydroxide, and formaldehyde show the greatest promise as detoxifying agents. Further testing and feeding trials are necessary before chemical detoxification of aflatoxins can be considered to be an accepted procedure.

Little work has been done on methods of detoxifying other mycotoxins. This is an area that needs more research.

Blending

The Food and Drug Administration (FDA) sets a limit of 20 parts per billion (ppb) of aflatoxin in corn sold in interstate commerce. Section 402(a)(1) of the Federal Food, Drug and Cosmetic Act, 21 USC 342(a)(1) specifically prohibits the blending of aflatoxin-contaminated feed with uncontaminated feed to achieve a mixture within the legal tolerance level for mycotoxins.

A one-time special blending policy was established by the FDA for corn contaminated with aflatoxin in the southeastern United States in 1978 (1977 crop) to alleviate a hardship situation. The contamination of that crop was so widespread that it was difficult for farmers to find sufficient supplies of noncontaminated corn to feed livestock. This policy applied only to the 1977 corn crop.

While it is possible to blend feeds to achieve lower levels of mycotoxin contamination, the practical problem exists of finding enough uncontaminated feed to reduce the level of mycotoxin contamination to an acceptable level. If the corn contained 1000 ppb of aflatoxin B_1, it would require 49,000 bushels of uncontaminated corn to mix with 1,000 bushels of contaminated corn in order to achieve a legal level of aflatoxin in the resulting mix.

The mixing of feeds to lower the mycotoxin level is not currently permitted for feeds being sold in interstate commerce. However, it may be practical for an individual farmer to mix noncontaminated feed with contaminated feed to achieve an overall level of mycotoxin intake in the ration for on-farm feeding below the level considered harmful to animals.

Young pigs and colts and lactating dairy cows should not have aflatoxin levels above 20 ppb in their rations. An aflatoxin level of 100 ppb is acceptable in rations of young calves and lambs and animals nursing their young. Dry sows, boars, dry mares, geldings, stallions, and growing-finishing pigs can tolerate aflatoxin levels of 200 ppb in their rations. Dry cows, bulls, dry ewes, and rams can tolerate aflatoxin levels of 400 ppb in their rations. Feedlot cattle and lambs can tolerate levels up to 500 ppb in the ration.

General Formula for Blending with Examples

A general formula may be used to determine the mix of contaminated and noncon-taminated feeds that may be used in the ration for a selected class of livestock. The information given for maximum levels of aflatoxin apply only to that mycotoxin and should not be construed to apply to other mycotoxin contaminated feeds.

In general, if livestock show symptoms of mycotoxicoses, the suspected feed source should be analyzed for mycotoxin content. If it is found to be contaminated, it is usually best to remove it from the ration.

The following designations are used in the general formula:

a = Percent of contaminated feed component(s) in ration.
b = Percent of noncontaminated feed component(s) in ration.
c = Aflatoxin level in each feed (ppb).
x = Total aflatoxin in ration (ppb).

The general formula is then stated:

$$x = \{[(a_1)(c_1)] + [(a_2)(c_2)] + \ldots [(a_n)(c_n)] + [(b_1)(c_1)] + \ldots [(b_n)(c_n)]\}/100$$

Example 1: Calculate the overall aflatoxin level in a ration made up of 80 percent corn which has an aflatoxin level of 500 ppb and 20 percent other feed components that have an aflatoxin level of 0 percent. Insert the given numbers in the general formula and solve:

$$x = \{[(80)(500)] + [(20)(0)]\}/100$$
$$x = (40000 + 0)/100$$
$$x = 40000/100$$
$$x = 400 \text{ ppb (Total aflatoxin level in ration)}$$

Example 2: Calculate the overall aflatoxin level in a ration made up of 65 percent corn that has an aflatoxin level of 400 ppb; 10 percent soybean meal that has an aflatoxin level of 600 ppb; and 25 percent other feeds that have an aflatoxin level of 0 percent. Insert the given numbers in the general formula and solve:

$$x = \{[(65)(400)] + [(10)(600)] + [(25)(0)]\}/100$$
$$x = (26000 + 6000 + 0)/100$$
$$x = 32000/100$$
$$x = 320 \text{ ppb (Total aflatoxin level in ration)}$$

Example 3: Calculate the quantity of aflatoxin contaminated corn that can be used in a ration for lactating dairy cattle. Laboratory tests show the aflatoxin content of the corn to be 150 ppb. The other feeds to be used in the ration contain no aflatoxin. As "a" represents the percent of aflatoxin contaminated feed (in this example, corn) in the ration, then "100 − a" represents the balance of the ration that may include corn not containing aflatoxin. The maximum tolerance level of lactating dairy cattle for aflatoxin in the ration is 20 ppb. Insert the given information in the general formula and solve:

$$20 = \{[(a)(150)] + [100-a)(0)]\}/100$$
$$20 = (150a + 0)/100$$
$$20 = 150a/100$$
$$100(20) = 100(150a)/100$$
$$150a = 2000$$
$$a = 2000/150$$
$$a = 13.3\% \text{ (percent of the ration which can be composed of the aflatoxin contaminated corn)}$$
$$100 - a = 86.7\% \text{ (percent of the ration which contains feeds with no aflatoxin content)}$$

Example 4: Calculate the amount of aflatoxin-contaminated corn and aflatoxin-contaminated soybean meal that may be used in a ration for growing-finishing swine. The balance of the ration contains no aflatoxin. The maximum level of aflatoxin in a ration for growing-finishing swine is 200 ppb. The feeder must first decide what the ratio of corn to soybean meal is to be in the ration. A ratio of 4 parts (80%) corn to 1 part (20%) soybean meal will be used for this example. Laboratory tests show that the aflatoxin content of the corn is 540 ppb and the aflatoxin content of the soybean meal is 950 ppb. The average aflatoxin level in the corn/soybean meal mix is then calculated by using the general formula.

$$x = \{[(80)(540)] + [(20)(950)]\}/100$$
$$x = (43200 + 19000)/100$$
$$x = 62200/100$$
$$x = 622 \text{ ppb (aflatoxin content of the corn/soybean meal mix)}$$

This average aflatoxin content of the corn/soybean meal mix is then used when calculating the amount of this mix to be used in the total ration. The percent of this mix is represented by "a" and the balance of the ration is represented by "100−a". Again, substituting in the general formula:

$$200 = \{[(a)(622)] + [(100-a)(0)]\}/100$$
$$200 = (622a + 0)/100$$
$$200 = 622a/100$$
$$100(200) = 100(622a)/100$$
$$622a = 20000$$
$$a = 20000/622$$
$$a = 32.2\% \text{ (percent of aflatoxin contaminated corn/soybean meal mix which can be used in the ration)}$$
$$100 - a = 67.8\% \text{ (percent of noncontaminated feeds to be used in the ration)}$$

SUMMARY

Fungi, such as molds that grow on plants, produce metabolites, some of which are harmful to animals that ingest them. These harmful metabolites are called *mycotoxins*, the literal meaning of which is *poisons from fungi.*

Mycotoxins affect the liver, kidneys, nerves, reproductive system, and skin of animals. In large amounts, some mycotoxins can cause death. Over a period of time, some mycotoxins are carcinogenic. They also lower resistance to disease and adversely affect immunization to some diseases. The overall effect of mycotoxins on animals is a lowering of productivity. Mycotoxins appear to act by affecting basic metabolic processes in the body.

The major mycotoxin problem in the United States is aflatoxin. Other mycotoxins that have affected livestock in the United States include ochratoxin, pasture grass mycotoxins, fusarium toxins, and trichothecenes.

It is sometimes difficult to identify a mycotoxin problem. Extremely small quantities of mycotoxin can contaminate a fairly large quantity of feed. The incidence of mycotoxins is varied, with plant stress and weather conditions being two of the factors affecting the growth of the fungi that cause the mycotoxins.

There is little evidence of any significant mycotoxin residue in edible animal tissues, with the exception of swine kidneys and cows' milk.

The best way to prevent mycotoxin contamination of feeds is through good farm management practices. Crop rotation, using clean seeds, reducing insect infestations, and proper drying and storage of feed are practices that help reduce the incidence of mycotoxins in feedstuffs.

Some experimental work has been done on detoxifying feeds contaminated with mycotoxins. More research is needed in this area.

The blending of feeds to reduce mycotoxin levels to those permitted in interstate commerce is prohibited. However, a farmer may wish to blend aflatoxin-contaminated feed with noncontaminated feed to lower the overall level of aflatoxin intake in animal rations.

REVIEW

1. Define the term *mycotoxin*.
2. List five nonlethal effects of mycotoxins.
3. What is the major chronic danger of mycotoxins?
4. List the four basic metabolic processes in animals that mycotoxins appear to affect.
5. List the four specific mechanisms believed to be involved in the action of mycotoxins.
6. List five mycotoxins that affect livestock in the United States.
7. Name the two fungi responsible for producing ochratoxin.
8. What is the effect of ochratoxin on poultry?
9. Name two fungi that produce pasture grass mycotoxins.
10. List four pasture grasses affected by fungi that produce mycotoxins.
11. What is *ergot*, and what fungus causes it?
12. Describe the appearance of ergot on plants.
13. What are the symptoms of ergotism in livestock?
14. What fungus causes Dallis grass poisoning?
15. What are the symptoms of Dallis grass poisoning?
16. What are the symptoms of fescue foot in cattle?
17. What is the name of the fungus which causes Gibberella ear rot in corn?
18. What mycotoxin does Gibberella produce on grain?
19. What are the symptoms of hyperestrogenism in swine?
20. What weather conditions favor the growth of the fungus that causes Gibberella ear rot in corn?
21. Describe the appearance of Gibberella ear rot on corn.

22. What is the name of the fungus that produces T-2 toxin?

23. What are some of the effects on livestock that are associated with T-2 toxin?

24. Name four feeds that may be infested with aflatoxin.

25. What field and weather conditions appear to be conducive to infestation of corn with aflatoxin?

26. What storage conditions favor the growth of fungi that cause aflatoxin?

27. Name two fungi that cause aflatoxin.

28. Which is the most toxic form of aflatoxin?

29. What are some of the general symptoms of aflatoxin poisoning in livestock?

30. What are some of the effects of aflatoxin poisoning in swine?

31. What are some of the indications of aflatoxin poisoning in dairy cattle?

32. What appears to be some of the effects of aflatoxin on rumen function?

33. How may aflatoxin be a health hazard to humans?

34. List five problems associated with the detection of mycotoxins in feedstuffs.

35. Describe a qualitative test for the presence of aflatoxin in feedstuffs.

36. Describe the possible dangers of mycotoxin residues in animal tissues or products.

37. What are some of the methods of controlling mycotoxin contamination of feedstuffs?

38. Name two solvents that have been successfully used to remove aflatoxin from oils.

39. Discuss the use of chemicals to detoxify aflatoxin.

40. Is it permitted to blend feeds to reduce their aflatoxin content to the legal level for interstate commerce?

41. List the acceptable levels of aflatoxin in the rations of (a) young pigs; (b) animals nursing their young; (c) growing-finishing pigs; and (d) feedlot cattle.

BIBLIOGRAPHY

Albert, W. W. 1972. *Feeding suggestions for horses*. University of Illinois Circular 1034(rev).

Anderson, R. H. and Steevens, B. J. 1975. *Complete rations for dairy cows*. University of Wyoming Bulletin 549R.

Ankerman, D., and Petersen, L., eds. 1980. *Feed analysis and livestock feeding guide*. Fort Wayne, Indiana: A & L Agricultural Laboratories.

Appleman, R. D., Otterby, D. E., and Foley, J. A. 1977. *Using colostrum to raise dairy calves*. University of Minnesota Dairy Husbandry Fact Sheet No. 9.

Barnes, W. J. 1977. *Using urea in dairy rations*. Auburn University Circular A-25.

Blakely, J., and Bade, D. H. 1982. *The science of animal husbandry*. 3rd ed. Reston, Virginia: Reston Publishing.

Bradley, C. M., Browning, C. W., Case, A. A., et. al. 1970. *Missouri confinement swine manual*. University of Missouri Manual 74.

Bradley, M., and Pfander, W. H. 1977. *Feeds for light horses*. University of Missouri Science and Technology Guide 2806.

Bradley, M. and Pfander, W. H. 1975. *Rations for light horses*. University of Missouri Science and Technology Guide 2808.

Brannon, W. F., 1975. *The dairy goat*. Cornell University Information Bulletin 78.

Buckner, R. C., and LaBore, D. E. 1971. *Fescue foot in cattle*. USDA Leaflet No. 546.

Carlson, C. W., and Bonzer, B. J. 1975. *Feeding chickens*. South Dakota State University FS 502(rev).

Carlson, C. W., and Bonzer, B. J. 1973. *Feeding turkeys*. South Dakota State University FS 389.

Carlson, C. W., Marion, W. W., Miller, B. F., and Goodwin, T. L. 1975. *Factors affecting poultry meat yields*. South Dakota State University Agricultural Experiment Station Bulletin 630.

Carpenter, J. C., Jr., Klett, R. H., and Hembry, F. G. 1971. *Producing slaughter steers with grain self-fed on pasture*. Louisiana State University Bulletin No. 659.

Chance, C. M. and Cason, J. L. 1968. *Feeding corn silage to dairy cattle*. University of Maryland Fact Sheet 197.

Chance, C. M., Davis, R. F., and Vandersall, J. H. 1974. *Economical grain feeding for dairy cows*. University of Maryland Bulletin 202.

Colby, B. E., Evans, D. A., Lyford, S. J., Jr., Nutting, W. B., and Stern, D. N. 1972. *Dairy goats breeding/feeding/management*. Spindale, North Carolina: American Dairy Goat Association.

Cornell, C. N. and Garner, G. B. 1976. *Fescue foot*. University of Missouri Science & Technology Guide 2100.

Council for Agricultural Science and Technology. 1984. *Energy use and production in agriculture*. Ames, Iowa: CAST Report No. 99.

Creger, C. R., Cawley, W. O., and Krueger, W. F. 1976. *Feeding turkey breeders*. Texas A&M University Fact Sheet L-596.

Creger, C. R., Couch, J. R., and Quisenberry, J. H. 1964. *Feeding growing turkeys*. Texas A&M University L-597.

Cullison, A. E. 1979. *Feeds and feeding*. 2nd ed. Reston, Virginia: Reston Publishing.

Cunningham, M. and Hill, D. L. 1970. *Concentrate mixtures for dairy cows*. Purdue University Animal Science Dept. Mimeo DH-103.

Dean, B. T., Cromwell, G. L., Edwards, R. L., Liptrap, D. O., and Whiteker, M. D. 1975. *Profitable pig and pork production in Kentucky*. University of Kentucky Bulletin ASC-26.

Deeb, B. S., and Sloan, K. W. 1975. *Nitrates, nitrites, and health*. Illinois Agricultural Experiment Station Bulletin 750.

Ellis, B., and Hutto, D. C. 1973. *Poultry feed formulas*. University of Wyoming Bulletin 473 R.

Ensminger, M. E. 1977. *Animal science*, 7th ed. Danville, Illinois: The Interstate.

Fiez, E. A. 1975. *Feeding dairy goats*. University of Idaho Current Information Series No. 296.

Fiez, E. A., and Cleveland, G. W. 1972. *Manage corn silage for maximum value*. Idaho Agricultural Experiment Station Current Information Series No. 188.

Fisher, G. 1970. *Urea in dairy rations*. North Dakota State University Circular A-536.

Fox, D. G., and Newman, L. E. 1976. *Michigan beef production: New feeder cattle: nutrition, health and treatments*. Michicgan State University Extension Bulletin E-992.

Frischknecht, W. D. 1973. *Some feeding alternatives for wintering beef cattle*. Oregon State University FS 200.

Gay, Nelson, Herbst, G. R., and Boling, J. A. 1974. *Beef: Implants for improving growth rate and feed efficiency*. University of Kentucky ASC-25.

Giesler, F. J., ed. 1974. *More profit from pigs*. University of Wisconsin Bulletin A1992.

Gillespie, J. R. 1983. *Modern livestock and poultry production*, 2nd ed. Albany, New York: Delmar Publishers.

Gray, J. A., and Groff, J. L. 1970. *Texas angora goat production*. Texas A&M University B-926.

Greathouse, T. R. 1967. *Principles of ruminant digestion* Illinois Agricultural Experiment Station Circular 894.

Guenthner, H. R. 1974. *Horse pastures*. University of Nevada C151.

Hall, R. E., and Nichols, R. E. 1970. *Moldy corn management*. University of Wisconsin Fact Sheet 60.

Harrow, B., and Apfelbaum, P. M. 1946. *An introduction to organic chemistry*. 6th ed. New York: John Wiley & Sons.

Hartman, D. A., Bibb, T. L., and Murley, W. R. 1979. *Tube feeding weak calves*. Virginia Polytechnic Institute and State University Series 294.

Hartman, D. A., and Murley, W. R. 1979. *Feeding the calf*. Virginia Polytechnic Institute and State University Series 292.

Herrick, J. 1978. *Proper use of colostrum*. Iowa State University Pm-844f.

Hays, V. W. 1985. Antibiotics for animals. *Science of food and agriculture* V.3, No.2, April 1985, pp. 17-26. Council for Agricultural Science and Technology, Ames, Iowa.

Heineman, W. W., and Dyer, I. A. 1975. *High crude fiber rations for finishing cattle*. Washington State University Bulletin 816.

Hillman, D., Huber, J. T., Emery, R. S., Thomas, J. W., and Cook, R. M. 1979. *Basic dairy cattle nutrition*. Michigan State University Bulletin E-702.

Hollis, G. R. 1984. Alternate protein and energy feed sources for swine. In *1984 Illinois Swine Seminar*, pp. 93-100. University of Illinois.

Hollis, G. R. 1980. New developments in swine nutrition and management. In *1980 Illinois Swine Seminar*, pp. I-3 - I-14. University of Illinois.

Hollis, G. R. 1985. Nutritional recommendations for swine. In *1985 Illinois Swine Seminar*, pp. 65-79. University of Illinois.

Hollis, G. R. 1984. Nutrition and management of the weanling pig: a review. In *1984 Illinois Swine Seminar*, pp. 59-80. University of Illinois.

Hudson, L. W. 1972. *Feeding light horses*. Clemson University Livestock Leaflet 14.

Hutjens, M. F., Otterby, R. D., Appleman, R. D., and Linn, J. G. 1981. *Feeding the dairy herd*. Illinois Extension Circular M-1183.

Hutjens, M. F., Otterby, D. E., Martin, N. P., and Linn, J. G. 1977. *Corn silage in dairy cattle rations*. University of Minnesota Dairy Husbandry Fact Sheet No. 7.

Jefferies, N. W. 1970. *Better pastures for horses and ponies*. Montana State University Circular 294.

Jones, G. M. 1977. *Grouping cows for complete ration feeding*. Virginia Polytechnic Institute and State University Series 113.

Jurgens, M. H. 1982. *Animal feeding and nutrition*. Dubuque, Iowa: Kendall/Hunt Pub. Co.

Jurgens, M. H. 1976. *Rations for horses*. Iowa State University AS-387(rev).

Keeton, W. T. 1972. *Biological science*. 2d ed. New York: W. W. Norton & Co.

Kohler, P. 1970. *Wheat as a feed for horses*. South Dakota State University FS 511.

Kortan, L. J., and Wahlstrom, R. C. 1977. *Feeding brood sows*. South Daktoa State University FS 367(rev).

Lazarus, S. S., Hill, L. D., and Thompson, S. R. 1980. *Grain production and consumption for feed in the north central and southern states with projections for 1985, 1990, and 2000*. Illinois Agricultural Experiment Station Bulletin 763.

Linn, J. G. 1979. *Silage for dairy cattle*. Iowa State University Pm-417f(rev).

Linn, J. G., and Otterby, D. E. 1980. *Vitamins for dairy cattle*. University of Minnesota Dairy Husbandry Fact Sheet No. 12.

Linn, J. G., Otterby, D. E., and Hutjens, M. F. 1978. *NPN in the dairy ration*. Iowa State University Pm-867.

Lucas, L. E. 1970. *Reproductive efficiency of swine: The effect of mycotoxins on reproduction.* Des Moines, Iowa: National Pork Producers Council.

McCarty, R. E. 1984. ATP: The energetic compound. *Science of food and agriculture* V.2, No.4, Nov. 1984. pp. 7-11.

Marsden, S. J. 1971. *Turkey production.* USDA Agriculture Handbook No. 393.

Marshall, C. W. 1984, Vitamin A: Marvel and menace. *Science of food and agriculture* V.2, No.4, Nov. 1984. pp. 2-6.

Mayfield, W., Watson, H., and Donald, J. 1977. *Harvesting, storing, and feeding silage.* Auburn University Circular R-45.

Meekma, A. M., and Groff, J. L. 1975. *Dairy goat care and management.* Texas A&M University MP-1178.

Meinershagen, F. H. 1980. *Silage from grasses and legumes.* University of Missouri Science and Technology Guide 3248.

Miller, E. C., Miller, E. R., and Ullrey, D. E. 1975. *Swine technology nutrition: introductory animal nutrition.* Michigan State University Extension Bulletin 536.

Miller, E. C., Miller, E. R., and Ullrey, D. E. 1975. *Swine technology nutrition: swine feeds and feeding.* Michigan State University Extension Bulletin 537.

Miller, L. E. ca1969. *Advanced livestock production: a course of study.* Agriculture Education, Purdue University and Department of Public Instruction.

Mills, J. M., ed. 1974. *Beef cattle backgrounding handbook.* Colorado State University.

Mork, I. J., Sell, J. L., and Johnson, R. L. 1974. *Feeding chickens.* North Dakota State University Circular AS-587.

Morter, R. L. 1980. *Feeding colostrum to calves.* Purdue University VY-55.

Moseley, B. L., and Sewell, H. B. 1974. *Care of newly-purchased feeder cattle.* University of Missouri Science and Technology Guide 2102.

Murley, W. R. 1978. *Feeding guidelines for dairy cows.* Virginia Polytechnic Institute and State University Pub. 630.

Murley, W. R. 1980. *Feeding guidelines for dairy replacements.* Virginia Polytechnic Institute and State University Series 102.

Murley, W. R. 1979. *A guide for feeding dairy cows.* Virginia Polytechnic Institute and State University Series 105.

Murley, W. R., and Jones, G. M. 1976. *Complete rations for dairy herds.* Virginia Polytechnic Institute and State University Series 111.

National Research Council. 1981. *Effect of environment on nutrient requirements of domestic animals.* Washington: National Academy Press.

National Research Council. 1980. *The effects on human health of subtherapeutic use of antimicrobials in animal feeds.* Washington: National Academy Press.

National Research Council. 1981. *Feeding value of ethanol production by-products.* Washington: National Academy Press.

National Research Council. 1979. *Interactions of mycotoxins in animal production.* Washington: National Academy Press.

National Research Council. 1980. *Mineral tolerance of domestic animals*. Washington: National Academy Press.

National Research Council. 1976. *Nutrient requirements of domestic animals, No. 4. Nutrient requirements of beef cattle*. 5th ed. Washington: National Academy Press.

National Research Council. 1978. *Nutrient requirements of domestic animals, No. 3. Nutrient requirements of dairy cattle*. 5th ed. Washington: National Academy Press.

National Research Council. 1981. *Nutrient requirements of domestic animals, No. 15. Nutrient requirements of goats*. Washington: National Academy Press.

National Research Council. 1978. *Nutrient requirements of domestic animals, No. 6. Nutrient requirements of horses*. Washington: National Academy Press.

National Research Council. 1977. *Nutrient requirements of domestic animals, No. 1. Nutrient requirements of poultry*. Washington: National Academy Press.

National Research Council. 1975. *Nutrient requirements of domestic animals, No. 5. Nutrient requirements of sheep*. Washington: National Academy Press.

National Research Council. 1978. *Nutrient requirements of domestic animals, No. 2. Nutrient requirements of swine*. Washington: National Academy Press.

National Research Council. 1981. *Nutritional energetics of domestic animals*. Washington: National Academy Press.

National Research Council. 1978. *Plant and animal products in the U.S. food system*. Washington: National Academy Press.

National Research Council. 1983. *Underutilized resources as animal feedstuffs*. Washington: National Academy Press.

National Research Council. 1982. *United States-Canadian tables of feed composition*. Washington: National Academy Press.

Nelson, D. K. 1977. *The middle-13 method for balancing grain mix protein*. Iowa State University Pm-652.

O'Dell, G. D., Jenny, B. F., and Brannon, C. C. 1976. *Lactational response of dairy cows fed an ensiled complete ration*. South Carolina Agricultural Experiment Station TB 1059.

Oldfield, J. E. 1984. Ruminating on the ruminants. *Science of food and agriculture* V.2, No.3, Sept. 1984, pp. 2-8.

Otterby, D. E., and Linn, J. G. 1980. *Calcium and phosphorus for dairy cattle*. University of Minnesota Dairy Husbandry Fact Sheet No. 8.

Otterby, D. E., and Linn, J. G. 1980. *Use of high-moisture corn for dairy cattle*. University of Minnesota Dairy Husbandry Fact Sheet No. 19.

Oxford University Press. 1968. *Dictionary of organic compounds*. 4th ed. 4th supplement. New York: Oxford University Press.

Pardue, F. 1968. *Feeding dairy cows in South Carolina*. Clemson University Dairy Science Extension Leaflet 1.

Parrott, C., and Koers, W. C. 1976. *Implant growth stimulants for growing and finishing cattle*. Texas A&M University L-1402.

Petersen, P. W. 1962. *Electrically operated feed handling equipment*. Utah Agricultural Experiment Station Leaflet 26.

Pfander, W. H., and Bradley, M. 1975. *Feeding light horses.* University of Missouri Science and Technology Guide 2807.

Purchase, G. H. 1977. *Farm poultry management.* USDA Farmer's Bulletin No. 2197.

Purdue University. 1983. Purdue Research Facts. October 1983.

Putnam, P. A., and Warwick, E. J. 1980. *The farm beef herd.* USDA Farmer's Bulletin No. 2126.

Rea, J. C., 1976. *Evaluating vitamin premixes for swine.* University of Missouri Science and Technology Guide 2351.

Rea, J. C., 1976. *Grain-protein supplement ratios in swine rations.* University of Missouri Science and Technology Guide 2350.

Rea, J. C., and Veum, T. L. 1974. *Antibiotics and other additives for swine.* University of Missouri Science and Technology Guide 2353.

Rea, J. C., and Veum, T. L. 1975. *By-products, damaged feeds, and nontraditional feed sources for swine.* University of Missouri Science and Technology Guide 2355.

Rea, J. C., and Veum, T. L. 1974. *Evaluating additives for swine rations.* University of Missouri Science and Technology Guide 2354.

Rea, J. C., and Veum, T. L. 1975. *Feeding the sow and gilt.* University of Missouri Science and Technology Guide 2356.

Rea, J. C., and Veum, T. L. 1975. *Mineral needs for swine rations.* University of Missouri Science and Technology Guide 2322.

Rea, J. C., and Veum, T. L. 1975. *Vitamin E and Selenium in swine rations.* University of Missouri Science and Technology Guide 2357.

Rea, J. C., and Veum, R. L. 1976. *Vitamin requirements of swine.* University of Missouri Science & Technology Guide 2321.

Ricketts, R. E. 1980. *Minerals for dairy cattle.* University of Missouri Science and Technology Guide 3110.

Ricketts, R. E. 1979, *Ration work sheet with example.* University of Missouri Science and Technology Guide 3108.

Ricketts, R., and Behan, S. 1980. *Calculating rations for dairy cattle, part 1.* University of Missouri Science and Technology Guide 3104.

Ricketts, G. E., Scoggins, R. D., and Thomas, D. L. 1983. *Recommendations for a sheep management program.* University of Illinois Circular 1221.

Rohweder, D. A., and Antoniewicz, R. J. 1976. *Forages for horses.* University of Wisconsin Bulletin A2460.

Ross, C. V. 1971. *Missouri sheep producer's manual.* University of Missouri Manual 69.

Roylance, H. B., and Hall, R. F. 1974 *Ergot - a loser for grain growers and livestock owners.* Idaho Agricultural Experiment Station Current Information Series No. 145.

Schryver, H. F., and Hintz, H. F. 1975. *Feeding horses.* New York State College of Agriculture Information Bulletin 94.

Sell, J. L., Rose, R. J., Johnson, R. J., and Mark, I. J. 1970. *Nutrition of the growing turkey.* North Dakota State University Extension Bulletin No. 11.

Sewell, H. B. 1975. *Feeding grain to beef cattle on pasture.* University of Missouri Science and Technology Guide 2072.

Sewell, H. B. 1976. *Minerals for beef cattle-needs.* University of Missouri Science and Technology Guide 2080.

Sewell, H. B. 1975. *Minerals for beef cattle-supplements.* University of Missouri Science and Technology Guide 2081.

Sewell, H. B. 1976. *Rations for growing and finishing beef cattle.* University of Missouri Science and Technology Guide 2066.

Sewell, H. B. 1974. *Sorghum grain for beef cattle rations.* University of Missouri Science and Technology Guide 2055.

Sewell, H. B. 1975. *Urea and mineral additives for corn silage.* University of Missouri Science and Technology Guide 2073.

Sewell, H. B. 1977. *Urea supplements for beef cattle.* University of Missouri Science and Technology Guide 2071.

Sewell, H. B. 1977. *Vitamins for beef cattle.* University of Missouri Science and Technology Guide 2058.

Sewell, H. B. 1974. *Whole shelled-corn rations for beef cattle.* University of Missouri Science and Technology Guide 2054.

Sewell, H. B., and Dyer, A. J. 1975. *Corn silage for beef cattle.* University of Missouri Science and Technology Guide 2061.

Sewell, H. B., and Thompson, B. B. 1976. *High-moisture grain for beef cattle.* University of Missouri Science and Technology Guide 2056.

Sheppard, C. C., Flegal, C. J., Wolford, J. H., and Thorburn, T. 1974. *The small poultry flock.* Michigan State University Extension Bulletin 773.

Shumway, C. R., and Jones, J. R. 1975. *Least cost swine rations.* North Carolina State University Circular 567(rev).

Skinner, J. L., Brevik, T. J., Sunde, M. L., and Hall, R. E. 1973. *The laying flock: feeding - managing - housing.* University of Wisconsin A1774.

Smith, S. E., and Loosli, J. K. 1970. *The mineral and vitamin requirements of livestock.* New York State College of Agriculture Extension Bulletin 1149.

Spooner, A. E. and Ray, M. L. 1972. *Finishing yearling steers on pasture with grain.* University of Arkansas Bulletin 772.

Stallings, C. C., Murley, W. R., and Carr, S. B. 1980. *Harvesting and storing of legume and grass forages.* Virginia Polytechnic Institute and State University Series 159.

Steevens, B. J. 1975. *Guidelines for feeding dairy cattle.* University of Wyoming B-627.

Steevens, B. J. 1975. *Swine nutrition guide.* University of Wyoming B-553R.

Strand, O. E., Hutjens, M. F., and Martin, N. P. 1975. *Interpreting forage test results.* University of Minnesota Extension Folder 297.

Strohbehn, D., Zmolek, B., and Geasler, M. 1975. *Developing the replacement heifer.* Iowa State University Pm-629.

Strum, G. E., and Schipper, I. A. 1973. *New calves for your feedlot*. North Dakota State University Circular AS 302.

Texas A&M University. 1971. *Keys to profitable turkey production*. Texas A&M Fact Sheet L-963.

Thacker, D. L. 1970. *Corn silage for dairy cattle*. Idaho Agricultural Experiment Station Current Information Series Leaflet No. 50.

Thomberry, F. D., and Creger, C. R. 1977. *Managing and feeding egg strain replacement pullets*. Texas A&M University L-593.

Thompson, G. R., George, R. M., Finley, R. M., Sewell, H. B., Gwin, P., Dyer, A. J., Grimes, G. A., and Ricketts, R. L. 1973. *Missouri cattle feeder's manual*. University of Missouri Manual 65.

Thompson, U. D. 1970. *Keys to feeding cattle for slaughter*. Texas A&M University L-906.

Thompson, U. D., and Allen, W. S. 1971. *Beef cattle feedlot facilities*. Texas A&M University MP-680.

United States Department of Agriculture 1981. *Fact book of U.S. agriculture*. USDA Misc. Pub. 1063.

United States Department of Agriculture. 1973. *Farm poultry management*. USDA Farmer's Bulletin No. 2197.

United States Department of Agriculture. 1973. *Finishing beef cattle*. USDA Farmer's Bulletin No. 2196.

University of Illinois. Agronomy News No. 523. April 1981.

University of Illinois. Agronomy News No. 524. May 1981.

University of Illinois. 1980. *Animals in world agriculture*. VAS 1058.

University of Illinois. 1984. *1984 Beef cattle research report*. Illinois Agricultural Experiment Station.

University of Illinois. 1985. *Beef handbook*. University of Illinois College of Agriculture Department of Animal Science.

University of Illinois. 1978. *Beef report*. No. 45. Dec. 1978.

University of Illinois. 1978. *Beef report*. No. 44. Oct. 1978.

University of Illinois. 1980. *Beef report*. No. 53. Dec. 1980.

University of Illinois. 1981. *Beef report*. No. 54. Apr. 1981.

University of Illinois. 1950. *Digestion in animals*. VAS 1026.

University of Illinois. 1965. *Feed additives*. VAS 1038.

University of Illinois. 1981. *Feeding and managing sheep and lambs*. VAS 1060.

University of Illinois 1968. *Horse science: 4-H horse program*. Cooperative Extension Service.

University of Illinois. 1984. *Pork industry handbook*. University of Illinois College of Agriculture.

University of Illinois. 1960. *Silage as an animal feed*. VAS 1043.

University of Illinois. 1978. *Swine feeds and feeding*. VAS 1036a.

University of Illinois. 1979. *Swine report*. No. 52. Dec. 1979.

University of Illinois. 1980. *Swine report*. No. 53. Mar. 1980.

University of Illinois. 1980. *Swine report*. No. 54. June 1980.

University of Illinois. 1980. *Swine report*. No. 56. Dec. 1980.

University of Illinois. 1981. *Swine report*. No. 58. June 1981.

University of Illinois. 1981. *Swine report*. No. 59. Sept. 1981.

University of Illinois. 1981. *Swine report*. No. 60. Dec. 1981.

University of Illinois. 1982. *Swine report*. No. 63. Sept. 1982.

University of Illinois. 1969. *Urea as a ruminant supplement*. VAS 1042.

University of Missouri. 1980. *Calculating rations for dairy cattle, part 2*. University of Missouri Science and Technology Guide 3105.

University of Wisconsin. 1973. *Gibberella ear rot in wisconsin corn*. University of Wisconsin Fact Sheet A2468.

Vander Noot, G. W., and Kniffen, D. M. 1972. *Dairy goat management*. Rutgers University Extension Bulletin 334.

Veum, T. L., and Rea, J. C. 1976. *Swine feeds and their composition*. University of Missouri Science and Technology Guide 2352.

Waldern, D. E., Blosser, T. H., Murdock, F. R., and Colenbrander, V. F. 1968. *Nutritive value of alfalfa hay for dairy cows*. Washington Agricultural Experiment Station Bulletin 699.

Waldroup, P. W., Maxey, J. F., Luther, L. W., Jones, B. D., and Meshew, M. L. 1976. *Factors affecting the response of turkeys to biotin and pyridoxine supplementation*. University of Arkansas Bulletin 805.

Wallenius, R. W., Whitchurch, R. E., and Bryant, J. M. 1974. *Alfalfa hay cubes for lactating dairy cows*. Washington State University Bulletin 797.

Ward, J. B. 1970. *Feeding broilers*. North Carolina State University PS&T Guide No. 7.

Weast, R. C., ed. 1985. *CRC handbook of chemistry and physics*. 66th ed. Boca Raton: CRC Press.

Weichenthal, B. A., and Russell, H. G. 1970. *Beef cattle feeding suggestions*. University of Illinois Circular 1025.

Wheaton, H. N., and Bradley, M. 1973. *Horse pastures*. University of Missouri Science and Technology Guide 4695.

Williams, J. N. 1972. *Urea for beef cattle feeding*. Clemson University Livestock Leaflet 16.

Willson, F. S., Thomas, O. O., and Jacobsen, N. A. 1972. *Beefing up your cattle profits through: nutrition and management*. Montana State University Bulletin 339.

Wilson, E. B. 1963. *Your feedlot: build it—mechanize it*. University of Idaho PNW Bulletin 53.

GLOSSARY

Abomasum—the fourth compartment, or true stomach, of the ruminant animal.

Absorption—the process by which digested nutrients are taken into the bloodstream.

Additive—a drug or druglike material added to a basic feed mix.

ADF (acid detergent fiber)—mainly insoluble residues such as cellulose and lignin, which remain after treating a feed sample with an acid detergent solution.

ADG (average daily gain)—the average of the daily increase in live weight of farm animals.

Adipose—refers to animal fat, such as fat in the connective tissue in the animal's body.

Ad Libitum—providing feed on a free-choice basis.

Aerial part—that part of a plant that is above ground.

Aerobic—living or growing where free oxygen is present.

Air-dry—See *as-fed basis*.

Alanine—a nonessential amino acid.

Alimentary tract—the passage in the body through which food passes; same as digestive tract.

Ambient temperature—the temperature of the environmental air in a locality.

Amino acid—a compound that contains carbon, hydrogen, oxygen, and nitrogen; certain amino acids are essential for growth and maintenance of cells.

Amylase—an enzyme instrumental in changing starch to sugar.

Anabolism—the process of changing food into living cells; constructive metabolism.

Anaerobic—living or growing where free oxygen is not present.

Animal protein—a protein supplement that comes from animals or animal byproducts.

Antibiotic—a chemical agent that prevents the growth of a germ or bacteria.

Anus—the opening at the end of the large intestine, which is the termination of the digestive system and through which feces pass out of the body.

Arsenical—a drug containing arsenic.

As-fed basis—data on feed consumption or nutrient requirements calculated on the basis of the average amount of moisture found in the feed as it is used on the farm; sometimes referred to as *air-dry*.

Ash—the mineral part of the feed left after burning off the organic matter.

Average daily gain—see ADG.

Backgrounding—the growing and feeding of calves from weaning until they are ready to enter the feedlot.

Bacteria—one-celled microorganisms.

Balanced ration—feed allowance, for an animal during a 24-hour period, which has all the nutrients the animal needs in the right proportions and amounts.

Basal metabolism—amount of energy used for the functioning of the heart, for breathing, and for other vital body processes while the organism is at rest; measured by the rate at which heat is given off and expressed in calories per hour per square meter of skin surface.

Bile—a fluid produced by the liver, which aids in the digestion of fats and fatty acids.

Biochemistry—chemistry that deals with the life processes of plants and animals.

Biological value—the extent to which a protein feed provides the essential amino acids in the necessary proportions and amounts. High biological value of a protein feed is referred to as *good quality*.

Biotin—one of the B vitamins.

Blending—mixing several feed ingredients together.

Bomb calorimeter—a device for measuring the gross energy content of feed.

Bran—the seed coat (pericarp) or husk of grain removed during processing and used as animal feed.

Brix—a measure of the sugar (sucrose) content of molasses; each degree brix equals one percent sucrose.

Browse—the shoots, twigs, and leaves of brush plants found growing on rangeland.

Bushel—dry measure equal to a volume of 2150.42 cubic inches (approximately 1.25 cubic feet).

Butyric acid—a volatile fatty acid generally found in the rumen and in low quality silage.

Caecum—see *cecum*.

Calorie—the amount of heat energy needed to raise 1 gram of water 1 degree centigrade.

Calorimeter—a device for measuring heat.

Carbohydrate—organic compound containing carbon, hydrogen, and oxygen.

Carcinogenic—causing cancer.

Cardia—a valve, located at the end of the esophagus, that prevents food in the stomach from coming back into the esophagus.

Carotene—yellow- or orange-colored compound found in plants, which is the precursor of vitamin A.

Carrier—an edible material to which micronutrients are added to make it easier to mix them in feed.

Carrying capacity—the number of animals that can be grazed on a pasture during the grazing season.

Casein—protein precipitated from milk by acid and/or rennin.

Catabolism—process in a plant or animal that changes complex substances into simpler substances; destructive metabolism.

Catalyst—a substance that causes or speeds up a chemical reaction but is itself not changed by the reaction.

Cecum (caecum)—a blind pouch located at the point where the small intestine joins the large intestine.

Cellulose—substance that is low in digestibility, found in the cell walls or woody parts of plants.

Celsius—scale for measuring temperature in which water freezes at 0° and boils at 100°; also called *Centigrade*.

Centigrade—see *Celsius*.

Choline—one of the B vitamins.

Chopped—reducing a feed material in size by cutting.

Chyme—partially digested feed, which moves from the stomach to the small intestine.

Citrulline—one of the nonessential amino acids.

Coenzyme—a substance that usually contains a vitamin and is needed by some enzymes to cause their action.

Colostrum—the milk produced the first few days after parturition.

Complete ration—a feed mix which contains all the necessary ingredients, except water, required for any specific class of livestock.

Concentrate—feed containing less than 18-20 percent crude fiber when dry; grains and protein supplements are concentrates.

Cracked—reducing a feed material in size by breaking and crushing.

Creep—a feeding area for the young that keeps older animals out.

Crimping—passing a feed crop through a set of corrugated rollers set close together.

Crop—in poultry, an enlargement of the gullet that serves as a storage area for feed.

Crop residue—material remaining after harvesting grain crops; i.e., straw, husklage, stover, stalks, etc.

Crude fat—the portion of a feed that is soluble in ether; also called *ether extract*.

Crude fiber—the part of a feed, containing mainly cellulose and lignin, which is of low digestibility.

Crude protein—the amount of ammoniacal nitrogen in a feed multiplied by 6.25.

Crumbles—pelleted feed broken into a granular form.

Cubing—processing a feed by grinding and then forming it into a hard form called a cube; cubes are larger than *pellets*.

Cud—in ruminants, a ball-like mass of feed brought up from the stomach to be rechewed.

Curd—material formed when acid or rennin is mixed with milk; contains the protein *casein*.

Cystine—a nonessential amino acid containing sulphur; can replace part of the methionine required in a diet.

Deficiency disease—disease resulting from insufficient dietary intake of a necessary nutrient.

Defluorinated—fluorine removed to a non-toxic level.

Degermed—the embryo or germ of the seed is removed from the starchy endosperm.

Dehulled—the outer covering or hull of the seed is removed.

Dehydrated—moisture removed by an artificial process.

Digestible energy (DE)—gross energy of a feed minus the energy remaining in the feces of the animal after the feed is digested.

Digestible protein (DP)—that portion of the crude protein in a feed that can be utilized by an animal.

Digestion—the process of breaking down feed into simple substances that can be absorbed by the body.

Digestive system (digestive tract)—the passage beginning at the mouth and ending at the anus through which feed passes as it is subjected to the digestive process.

Disaccharide—compound sugar such as sucrose, maltose, and lactose, which can yield two monosaccharide molecules upon hydrolysis.

Dormant—stage of maturity of a plant at which the plant is no longer growing and is cured on the stem.

Dough stage—stage of maturity of a plant at which the seeds are soft and immature.

Drench—medicine in a liquid form administered to the animal through the mouth.

Drylot—an area in which animals are confined for feeding as opposed to being on pasture.

Dry matter—the portion of a feed that is not water.

Dry rolling—pressing feed between rollers without adding moisture.

Duodenum—the upper part of the small intestine from the stomach to the jejunum.

Early bloom—a stage of forage plant maturity between the time the plants start to bloom and the time one-tenth of the plants are in bloom.

Early leaf—a stage of plant maturity at which the plant has reached one-third of its growth before blooming.

Electrolyte—a solution containing salts and energy sources used to feed young animals suffering from scours (diarrhea).

Emaciated—body is excessively thin.

Endosperm—that part of the seed composed of carbohydrate (starch).

Endotoxin—poisonous substances produced by certain bacteria.

Energy feed—livestock feed containing less than 20 percent crude protein; most grains are energy feeds.

Ensilage—see *silage*.

Ensiled—preserved by chopping, being put into a silo, and being subjected to anaerobic fermentation.

Essential amino acid—any one of several amino acids needed by animals, which cannot be synthesized in the animal's body and must, therefore, be included in the diet.

Environment—the total of the external conditions and influences which affect the life and development of living organisms.

Enzyme—an organic catalyst that speeds up the digestive process without being used up in the process.

Epithelium—in poultry, a thick, horny membranelike material that lines the muscular stomach or gizzard.

Ergot—disease of rye and some other grain crops caused by a fungus.

Esophagus—the tubelike passage from the mouth to the stomach; sometimes called the *gullet*.

Essential Fatty acid—fatty acid necessary for animal nutrition but one that cannot be synthesized in the animal's body in sufficient quantities to meet its needs.

Excreta—material excreted from the body after feed is digested; mainly feces and urine.

Expanded—feed subjected to moisture, pressure, and temperature and then increased in volume by sudden reduction of pressure.

Expeller process—mechanical extraction of oil from seeds by using a screw press.

Extruded—forcing feed through a die at high pressure.

Factor—any chemical substance found in feed.

Fahrenheit—scale for measuring temperature in which water freezes at 32° and boils at 212°.

Fat—organic compound composed of carbon, hydrogen, and oxygen; fats contain more carbon and hydrogen than do carbohydrates and are mainly glyceryl esters of certain acids soluble in ether but not in water.

Fat soluble—material soluble in fats and fat solvents but usually not in water.

Fattening—depositing unused energy in the form of fat in the tissues of the body.

Fatty acid—any of a number of organic compounds that contain carbon, hydrogen, and oxygen and combine with glycerol to form fat.

Feces—undigested material that is passed out of the digestive system through the anus.

Feed (feedstuff)—any material fed to animals as part of the diet to provide necessary nutrients.

Feed additive—any of a number of non-nutritive materials added to a feed to reduce the incidence of disease, improve the rate of gain, improve feed efficiency, or preserve feed.

Feed Additive Compendium—a publication that lists feed additives in current use and the regulations for their use.

Feed composition table—a table showing the nutrients found in feeds.

Feed conversion—see *feed efficiency*.

Feed efficiency—the ratio of units of feed needed per one unit of production.

Feed grade—material suited for animal but not human consumption.

Feed grain—seeds from cereal plants usually used for animal feeding.

Feeding standard—a table of nutrient requirements for an animal.

Fermentation—changes in feed produced by the action of enzymes produced by some microorganisms.

Fiber—complex carbohydrates such as cellulose and lignin.

Finish—fattening a slaughter animal; also refers to the degree of fatness in a slaughter animal.

Fibrous—contains a high level of cellulose and/or lignin.

Fistula—an artificial opening or passage from one part of the body to another or to the body surface.

Flaking—see *rolling*.

Flushing—increasing the amount of feed fed to an animal for a short period of time, usually just prior to breeding.

Fodder—the above ground part of the plant in fresh or cured form used for feed; especially corn or sorghum.

Folacin—one of the B vitamins; also called *folic acid*.

Folic acid—see *folacin*.

Forage—plants fed as fresh, dried, or silage to animals; examples: pasture, hay, silage, green chop, or haylage.

Formula feed—feed containing two or more ingredients that are mixed in specific proportions.

Fortify—adding one or more nutrients to a feed.

Founder—indigestion caused by overeating; may also refer to laminitis (a crippled condition of an animal).

Free-choice—making two or more feeds available to the animal at all times.

Fresh—green or wet form of a feed.

Fructose—fruit sugar; a hexose monosaccharide found in ripe fruits and honey.

Full bloom—stage of maturity of a plant at which two-thirds or more of the plants are in bloom.

Full feed—giving an animal all it wants to eat.

Gastric juice—a fluid secreted by glands in the wall of the stomach, containing hydrochloric acid and the enzymes gastric lipase, pepsin, and rennin.

Germ—the embryo of a seed.

Gizzard—in poultry, the muscular stomach, which crushes and grinds the feed and mixes it with digestive juices.

Glutamic acid—one of the nonessential amino acids.

Glycine—one of the nonessential amino acids.

Gossypol—a material found in some cottonseed meal that is toxic to swine and certain other simple-stomached animals.

Grain—the seed of cereal plants.

Green chop—harvesting fresh forage and feeding it green.

Grinding—processing a feed by breaking it up into smaller particles.

Grit—small particles of granite used in poultry diets to help in grinding the feed in the gizzard.

Groat—grain after the hull is removed.

Gross energy—total amount of heat released by completely burning a feed in a bomb calorimeter.

Growth—increase in size of bones, muscles, vital organs, and connective tissue by means other than fat deposition.

Gullet—see *esophagus*.

Hay—aerial part of forage crop that is cut, dried, and stored for animal feed.

Haylage—forage crop (legume or grass) stored as low-moisture (40-60 percent) silage.

Heat increment—heat produced by an animal by the processes of digestion and utilization of the products of digestion.

Heat processed—processing a feed material with high temperatures either with or without pressure.

Heat rendered—melting, extracting, or clarifiying feed material with heat; process usually removes water and fat.

High-lysine corn (opaque 2)—a variety of corn with a higher lysine and tryptophan content than other corn varieties; provides better balance of amino acids for simple-stomached animals such as swine.

High-moisture storage—harvesting a feed crop at a high moisture content and storing it in a silo.

Histidine—one of the essential amino acids.

Hull—the outer protective coating on a seed.

Husk—leaves surrounding ear of corn; also dry, membranous outer covering of kernels or seeds.

Hydraulic process—mechanical extraction of oil from seeds using a hydraulic press; now seldom used and sometimes called *old process*.

Hydroxyproline—one of the nonessential amino acids.

Ileum—that part of the small intestine extending from the jejunum to the cecum.

Impermeable—cannot be penetrated.

Ingest—to take into the body through the mouth.

Inositol—one of the B vitamins.

Insulin—a protein hormone secreted by the pancreas, which regulates metabolism of sugar in the body.

Intestinal juice—fluid produced by the walls of the small intestine which contains peptidase, sucrase, maltase, and lactase.

Intestinal tract—the small and large intestine.

Intestine, large—the tube from the small intestine to the anus; shorter and larger in diameter than the small intestine.

Intestine, small—the long, folded tube attached to the lower end of the stomach which leads to the large intestine.

Isoleucine—one of the essential amino acids.

Jejunum—the middle part of the small intestine between the duodenum and the ileum.

Kernel—in a cereal, the whole grain; dehulled seed in other species.

Kilocalorie—1,000 calories.

Lactase—enzyme in the intestinal juice that acts on lactose producing glucose and galactose.

Late bloom—stage of maturity in plants at which blossoms begin to dry and fall from the plant.

Legume—a plant with nitrogen fixing nodules on its roots, that can take nitrogen from the air; alfalfa, clovers, and soybeans are examples of legumes.

Leucine—one of the essential amino acids.

Lignin—compound of low digestibility found in the cell wall of some plant materials.

Limited-feeding—see *restricted feeding*.

Limit-fed—a method of feeding in which the amount of feed given the animal is controlled or limited to less than the animal would eat if given free access to the feed.

Limiting amino acid—the essential amino acid in a protein feed that has the highest percentage deficit in relation to the animal's needs.

Linolein—an unsaturated fat formed when linoleic acid reacts with glycerol.

Lipase—an enzyme which splits fat to produce fatty acids and glycerol.

Lipids—fats and fatlike substances.

Lysine—one of the essential amino acids.

Maltase—enzyme found in the mouth and intestine that acts on maltose to produce glucose.

Maltose—compound sugar.

Meal—a ground feed ingredient that has a particle size a little larger than flour.

Mechanically extracted—removal of fat with heat and mechanical pressure; may be either hydraulic or expeller process.

Megacalorie—1,000,000 calories.

Metabolism—total of all the physical and chemical processes that occur in any living organism.

Metabolizable energy (ME)—for ruminants, the gross energy in the feed eaten minus the energy found in the feces, the energy in the gaseous products of digestion, and the energy in the urine; for poultry and simple-stomached animals, the energy in the gaseous products of digestion is not considered when determining metabolizable energy.

Methionine—one of the essential amino acids; contains sulfur and part of the methionine requirement may be met by *cystine*.

Micronutrients—feed ingredients, such as minerals and vitamins, used in small amounts in the ration.

Mid-bloom—stage of plant maturity at which one-tenth to two-thirds of the plants are in bloom.

Middlings—byproduct of flour milling; contains bran, endosperm, and germ.

Milk stage—stage of plant maturity at which plants have bloomed and the seeds are beginning to form.

Mineral—inorganic substance needed in small amounts for proper nutrition.

Mouth—that part of the digestive system through which feed enters the animal's body.

Mycotoxin—bacterial or fungus toxin sometimes formed by molds in feed.

NE_g—net energy used for animal growth.

NE_l—net energy used for milk production (lactation).

NE_m—net energy used for animal maintenance.

Net energy (NE)—metabolizable energy minus the heat increment; used for maintenance, for production, or both.

Niacin—one of the B vitamins; also called *nicotinic acid*.

Nicotinic acid—see *niacin*.

Nitrogen-free extract—simple carbohydrates, such as sugar and starches.

Nonruminant—an animal that has a simple, one-compartment stomach; for example: pigs, horses, poultry.

Nonessential amino acid—an amino acid required in animal nutrition but one that can be synthesized by the animal in sufficient quantities to meet its needs.

Nonprotein nitrogen (NPN)—nitrogen from a nonamino acid source that can be used by the rumen bacteria to synthesize protein; a common example is urea.

Nutrient—a chemical element or compound that aids in the support of life.

Oil—fat that is soluble at room temperature.

Omasum—the third compartment of the ruminant stomach.

One hundred percent (100%) dry-matter basis—data on feed composition or nutrient requirement calculated on the basis of all the moisture being removed from the feed.

Orts—feed that an animal refuses to eat.

Palatable—good tasting.

Pancreatic juice—fluid secreted by the pancreas, which contains the enzymes trypsin, pancreatic amylase, pancreatic lipase, and the hormone insulin.

Pantothenic acid—one of the B vitamins.

Papillae—in ruminants, the small projections on the inside of the rumen wall.

Para-aminobenzoic acid—one of the B vitamins.

Pelleting—grinding a feed into small particles and then forming it into a small, hard form called a pellet.

Pepsin—enzyme in the gastric juice that acts on protein to form proteoses, peptones, and peptides.

Phenylalanine—one of the essential amino acids.

Phospholipids—substances that are like fats and contain phosphorus, nitrogen, fatty acids, and cholesterol.

Prebloom—stage of plant maturity during which the plant makes the last one-third of its growth before bloom.

Preconditioning—the process of preparing calves for the stress of being moved into the feedlot.

Precursor—compound used by the body to form another compound.

Premix—a carrier that has micronutrients mixed in it in a uniform manner, which is used to mix the micronutrient in a larger batch of feed.

Proline—one of the nonessential amino acids.

Propionic acid—a volatile fatty acid often found in the rumen.

Protein—an organic compound made up of amino acids and containing carbon, hydrogen, oxygen, and nitrogen.

Protein supplement—livestock feed that contains 20 percent or more protein.

Provitamin—substance that an animal can use to form a vitamin in its body.

Provitamin A—carotene.

Pulp—solid residue left after the juice has been removed from a plant material.

Pyridoxine—vitamin B_6.

Ration—the total amount of feed that an animal is allowed during a 24-hour period.

Rectum—the last part of the large intestine.

Restricted feeding—feeding 75 to 80 percent of full feed; also referred to as *limited feeding*.

Reticulum—the second compartment of the ruminant stomach.

Riboflavin—vitamin B_2.

Rolling—processing grain through a set of smooth rollers that are close together; sometimes called *flaking*.

Roughage—a feed containing more than 18-20 percent crude fiber when dry; examples: hay, silage, and pasture.

Rumen—the first and largest compartment of the ruminant stomach.

Rumen organisms—bacteria and protozoa found in the rumen of cattle and other ruminant animals.

Ruminant—an animal that has a stomach divided into several compartments; for example, cattle, sheep, goats.

Rumination—in ruminants, the process of chewing the cud.

Saliva—fluid secreted into the mouth by glands and containing enzymes that aid in digestion.

Self-fed—a method of feeding in which the animal is given free access to all the feed it will eat.

Serine—one of the nonessential amino acids.

Shorts—flour mill byproduct that contains bran, germ, the aleurone layer, and coarse flour.

Silage—feed produced by storing wet or green crops in anaerobic conditions under which fermentation occurs.

Solvent extracted—removing oil from seeds using an organic solvent; also called *solvent process*.

Starch—the main part of the seed endosperm; a major energy source in livestock feeding.

Steam flaking—heating grain with steam and putting it through a set of smooth rollers that are set close together.

Stomach—the organ in the digestive system that receives the feed and adds chemicals which help in the digestive process.

Stover—residue, such as stalks and leaves, left after the ears of corn or the heads of sorghum are harvested.

Straw—the part of the plant remaining after the grain has been harvested; usually used in reference to small grains.

Sucrase—intestinal juice enzyme that acts on sucrose to produce glucose and fructose.

Sucrose—a compound sugar that can be broken down to glucose and fructose; commonly called *cane, beet,* or *table sugar.*

Sun-cure—dry by exposure to the sun.

Supplement—a concentrated source of one or more nutrients that is added to an animal diet to provide needed nutrients; generally used to add protein, minerals, vitamins, or antibiotics to the ration.

Tankage—animal tissues and bones from animal slaughter houses and rendering plants that are cooked, dried, and ground and used as a protein supplement.

Thiamine—vitamin B_{12}; also called thiamin or thiamine hydrochloride.

Threonine—one of the essential amino acids.

Toasted—browning, drying, or parching a feed by exposure to heat.

Tocopherol—vitamin E.

Total digestible nutrients (TDN)—the total of the digestible protein, digestible nitrogen-free extract, digestible crude fiber, and 2.25 times the digestible fat.

Toxin—a poisonous substance produced by the metabolism of plant or animal organisms.

Trace mineral—a mineral element needed by animals in small quantities; micromineral.

Trachea—the passage (windpipe) through which air passes to and from the lungs.

True protein—nitrogen compound that will hydrolize completely to amino acid.

Tryptophan—one of the essential amino acids.

Tyrosine—one of the nonessential amino acids.

Urea—a synthetic nitrogen source that is manufactured from air, water, and carbon.

Valine—one of the essential amino acids.

Vegetable protein—protein supplement that comes from plant sources.

Vent—the external opening of the lower end of the digestive system in poultry.

Villi—small, fingerlike projections that line the walls of the small intestine to increase the absorption area.

Viscera—the internal organs of an animal.

Vitamin—an organic compound needed in small amounts for nutrition.

Withdrawal period—the length of time a feed additive must not be fed to an animal prior to slaughter.

PROBLEM SECTION

PROXIMATE ANALYSIS PROBLEMS

1. The following results are obtained in laboratory analysis of feed samples:

	Sample #		
	1	2	3
Original sample weight	2.00	1.00	2.00
Original sample dried weight:	1.78	0.92	1.78
Nitrogen content:	0.03072	0.02768	0.14272
Crude fat weight:	0.076	0.027	0.028
Sample weight containing crude fiber and mineral:	0.078	0.337	0.254
Ash weight:	0.026	0.097	0.13

From this information determine the following for each sample:
a. Percent water
b. Percent dry matter
c. Percent protein
d. Percent crude fat (ether extract)
e. Percent crude fiber
f. Percent ash
g. Percent NFE

CONVERSION PROBLEMS

CONVERT FROM 100% DRY-MATTER BASIS TO AS-FED BASIS:

FEED #:	1	2	3	4	5
Dry matter (%):	90.0	89.0	89.0	30.0	93.0
Amount fed (100% dry matter):	14.0	1,602	1.513	6.0	80.0

CONVERT FROM AS-FED BASIS TO 100% DRY-MATTER BASIS:

FEED #:	1	2	3	4	5
Dry matter (%):	90.0	89.0	89.0	30.0	93.0
Amount fed (as-fed basis):	15.0	150	500	32.0	5.0

RATION BALANCING PROBLEMS

1. Formulate 1 ton of feed containing 14 percent crude protein using ground shelled corn (IFN 4-02-935) and 44 percent soybean meal (IFN 5-20-637).

2. Prepare 1,000 pounds of concentrate mix containing 16 percent crude protein using corn (IFN 4-02-935) with a mixture of soybean meal (IFN 5-20-637) (2 parts) and tankage (IFN 5-00-387) (1 part).

3. Prepare 1,000 pounds of feed containing 18 percent crude protein using corn (IFN 4-02-935) and soybean meal (IFN 5-20-637). Fixed ingredients in the feed mix make up 10 percent of the total and do not provide any protein.

4. Formulate 1,000 pounds of a swine finishing ration containing 12 percent crude protein using grain sorghum (IFN 4-04-383) and 44 percent soybean meal (IFN 5-20-637).

5. Formulate one ton of a swine finishing ration containing 15 percent crude protein using a grain mix of 80 percent ground shelled corn (IFN 4-02-935) and 20 percent ground oats (IFN 4-03-309) plus 44 percent soybean meal (IFN 5-20-637).

6. Formulate 1 ton of a swine growing ration containing 15 percent crude protein using ground corn (IFN 4-02-309) and a mixture of 70 percent soybean meal (IFN 5-20-637) and 30 percent meat meal (IFN 5-00-386).

7. Balance a ration for 45 pound growing pigs using corn (IFN 4-03-309) and soybean meal (IFN 5-04-604) as the basic feedstuffs. Fixed ingredients in the ration include salt at 0.5 percent, trace-mineral premix at 0.10 percent, vitamin premix at 1.0 percent, dicalcium phosphate at 1.0 percent, and limestone at 1.0 percent. Balance the ration on the basis of lysine content.

8. Utilizing the balanced ration formulated in problem 7, assume a farmer is feeding 200 head of pigs of this weight.
   a. Using the average expected daily gain from the Nutrient Requirements table, calculate the number of days of feeding until the pigs reach 77 pounds.
   b. Calculate the total amount of each feed required for this feeding period.
   c. Using current local prices, calculate the total cost of the feed for this feeding period.

9. Balance a ration for a 200 kg growing-finishing steer calf with a daily gain of 0.9 kg. Use corn silage (IFN 3-09-912), corn (IFN 4-02-935), and soybean meal (IFN 5-04-600). Add mineral supplement if needed.

10. Balance a ration for a pregnant yearling beef heifer in the last four months of pregnancy weighing 425 kg with an average daily gain of 0.6 kg. Use corn silage (IFN 3-09-912) and alfalfa hay (IFN 1-20-681) to provide the roughage. The concentrate mix is corn (IFN 4-02-935) and oats (IFN 4-03-309) if needed to balance the ration. Add mineral supplement if needed.

11. Balance a ration for a dry, pregnant beef cow weighing 500 kg in the last third of the gestation period. Use alfalfa (IFN 1-20-681) -brome grass (IFN 1-05-633) hay mixed half and half to provide the roughage. Balance with ground ear corn (IFN 4-28-238) and soybean meal (IFN 5-20-637) if needed to meet the nutrient requirements. Add mineral supplement if needed.

12. Balance a ration for a 70 kg ewe during the first 8 weeks of lactation suckling twins. Use red clover hay (IFN 1-01-415) to provide roughage plus shelled corn (IFN 4-02-935). Add mineral supplement if needed.

13. Balance a ration for a 110 pound finishing lamb gaining 0.48 pounds per day. Use orchard grass hay (IFN 1-03-425) and shelled corn (IFN 4-02-935). Add mineral supplement if needed.

14. Balance a ration for a mare (mature weight 500 kg) during the last 90 days of gestation. Use timothy hay (IFN 1-04-883) and oats (IFN 4-03-309). Add mineral supplement if needed.

15. Balance a ration for a mature working horse (mature weight 500 kg) doing moderate work. Use orchard grass hay (IFN 1-03-425), oats (IFN 4-03-309), and soybean meal (IFN 5-20-637) if needed. Add mineral supplement if needed.

16. Balance a ration for a growing dairy heifer of the small breeds weighing 300 pounds and gaining 1.2 pounds per day. Use alfalfa (IFN 4-03-309) -brome grass (IFN 1-05-633) hay (50:50 mix) with ground shelled corn (IFN 4-02-935) and ground oats (IFN 4-03-309) (3:1 ratio). Add mineral supplement if needed.

17. Balance a ration for a 1,200 pound dairy cow, producing 60 pounds of 4.0 percent milk per day. Use alfalfa hay (IFN 1-20-681), corn silage (IFN 3-02-912) (50:50 on dry matter basis), corn and cob meal (IFN 4-28-238) and oats (IFN 4-03-309) (2:1 ratio), and soybean meal (IFN 5-20-637). Add mineral supplement if needed.

APPENDIX

Table 1 Nutrient Requirements for Growing-Finishing Steer Calves and Yearlings (Daily Nutrients per Animal)

Weight[a] (kg)	(lb)	Daily Gain (kg)	(lb)	Minimum Dry Matter Consumption[b] (kg)	(lb)	Roughage[b] (%)	Total Protein (kg)	(lb)	Digestible Protein (kg)	(lb)
100	220	0	0	2.1	4.6	100	0.18	0.40	0.10	0.22
		0.5	1.1	2.9	6.4	70-80	0.36	0.79	0.24	0.58
		0.7	1.5	2.7	6.0	50-60	0.40	0.88	0.28	0.62
		0.9	2.0	2.8	6.2	25-30	0.46	1.01	0.33	0.73
		1.1	2.4	2.7	6.0	15	0.49	1.08	0.36	0.79
150	331	0	0	2.8	6.2	100	0.23	0.51	0.13	0.29
		0.5	1.1	4.0	8.8	70-80	0.44	0.97	0.28	0.62
		0.7	1.5	3.9	8.6	50-60	0.49	1.08	0.33	0.73
		0.9	2.0	3.8	8.4	25-30	0.54	1.19	0.37	0.82
		1.1	2.4	3.7	8.2	15	0.58	1.28	0.41	0.90
200	441	0	0	3.5	7.7	100	0.30	0.66	0.17	0.37
		0.5	1.1	5.8	12.8	80-90	0.57	1.26	0.35	0.77
		0.7	1.5	5.7	12.6	70-80	0.61	1.34	0.39	0.86
		0.9	2.0	4.9	10.8	35-45	0.61	1.34	0.40	0.88
		1.1	2.4	4.6	10.1	15	0.63	1.39	0.43	0.95
250	551	0	0	4.4	9.7	100	0.35	0.77	0.20	0.44
		0.7	1.5	5.8	12.8	55-65	0.62	1.37	0.39	0.86
		0.9	2.0	6.2	13.7	45-50	0.69	1.52	0.44	0.97
		1.1	2.4	6.0	13.2	20-25	0.73	1.61	0.48	1.06
		1.3	2.9	6.0	13.2	15	0.76	1.68	0.51	1.12
300	661	0	0	4.7	10.4	100	0.40	0.88	0.23	0.51
		0.9	2.0	8.1	17.9	55-65	0.81	1.79	0.50	1.10
		1.1	2.4	7.6	16.8	20-25	0.82	1.81	0.52	1.15
		1.3[d]	2.9	7.1	15.6	15	0.83	1.83	0.54	1.19
		1.4[d]	3.1	7.3	16.1	15	0.87	1.92	0.57	1.26
350	772	0	0	5.3	11.7	100	0.46	1.01	0.26	0.57
		0.9	2.0	8.0	17.6	45-55	0.80	1.76	0.49	1.08
		1.1	2.4	8.0	17.6	20-25	0.83	1.83	0.52	1.15
		1.3	2.9	8.0	17.6	15	0.87	1.92	0.55	1.21
		1.4[d]	3.1	8.2	18.1	15	0.90	1.98	0.57	1.26
400	882	0	0	5.9	13.0	100	0.51	1.12	0.29	0.64
		1.0	2.2	9.4	20.7	45-55	0.87	1.92	0.54	1.19
		1.2	2.6	8.5	18.7	20-25	0.87	1.92	0.54	1.19
		1.3	2.9	8.6	19.0	15	0.90	1.98	0.56	1.23
		1.4[d]	3.1	9.0	19.8	15	0.94	2.07	0.59	1.30
450	992	0	0	6.4	14.1	100	0.54	1.19	0.31	0.68
		1.0	2.2	10.3	22.7	45-55	0.96	2.12	0.57	1.26
		1.2	2.6	10.2	22.5	20-25	0.97	2.14	0.58	1.28
		1.3	2.9	9.3	20.5	15	0.97	2.14	0.59	1.30
		1.4[d]	3.1	9.8	21.6	15	0.98	2.16	0.60	1.32
500	1,102	0	0	7.0	15.4	100	0.60	1.32	0.34	0.75
		0.9	2.0	10.5	23.1	45-55	0.95	2.09	0.56	1.23
		1.1	2.4	10.4	22.9	20-25	0.96	2.12	0.57	1.26
		1.2[d]	2.6	9.6	21.2	15	0.96	2.12	0.58	1.28
		1.3[d]	2.9	10.0	22.0	15	0.97	2.14	0.60	1.32

Source: *Nutrient Requirements of Beef Cattle*, fifth revised edition, National Research Council, National Academy of Sciences, Washington, D.C., 1976.

[a] Average weight for a feeding period.

[b] Dry matter consumption, ME and TDN allowances are based on NE requirements and the general types of diets indicated in the roughage column. Most roughages will contain 1.9-2.2 Mcal of ME/kg dry matter and 90-100% concentrate diets are expected to contain 3.1-3.3 Mcal of ME/kg.

NE$_m$ (Mcal)	NE$_g$ (Mcal)	ME[b] (Mcal)	TDN[b, c] (kg)	(lb)	Calcium (g)	Phosphorus (g)	Vitamin A (1000 IU)
2.43	0	4.2	1.2	2.6	4	4	5
2.43	0.89	6.6	1.8	4.0	14	11	6
2.43	1.27	7.1	2.0	4.4	19	13	6
2.43	1.68	7.7	2.1	4.6	24	16	7
2.43	2.10	8.4	2.3	5.1	28	19	7
3.30	0	5.6	1.6	3.5	5	5	6
3.30	1.20	9.0	2.5	5.5	14	12	9
3.30	1.73	9.6	2.7	6.0	18	14	9
3.30	2.27	10.7	3.0	6.6	23	17	9
3.30	2.84	11.3	3.1	6.8	28	20	9
4.10	0	7.0	1.9	4.2	6	6	8
4.10	1.49	12.1	3.4	7.5	14	13	12
4.10	2.14	13.0	3.6	7.9	18	16	13
4.10	2.82	13.3	3.7	8.2	23	18	13
4.10	3.52	14.1	3.9	8.6	27	20	13
4.84	0	8.2	2.3	5.1	8	8	9
4.84	2.53	14.4	4.0	8.8	18	16	14
4.84	3.33	16.2	4.5	9.9	22	19	14
4.84	4.17	17.0	4.7	10.4	26	21	14
4.84	5.04	18.6	5.2	11.5	30	23	14
5.55	0	9.4	2.6	5.7	9	9	10
5.55	3.82	19.5	5.4	11.9	22	19	16
5.55	4.78	20.4	5.6	12.3	25	22	16
5.55	5.77	21.6	6.0	13.2	29	23	16
5.55	6.29	22.5	6.2	13.7	31	25	16
6.24	0	10.6	2.9	6.4	10	10	12
6.24	4.29	20.8	5.8	12.8	20	18	18
6.24	5.36	22.4	6.2	13.7	23	20	18
6.24	6.48	24.2	6.8	15.0	26	22	18
6.24	7.06	25.3	7.0	15.4	28	24	18
6.89	0	11.8	3.3	7.3	11	11	13
6.89	5.33	24.5	6.8	15.0	21	20	19
6.89	6.54	25.4	7.0	15.4	23	21	19
6.89	7.16	26.5	7.3	16.1	25	22	19
6.89	7.80	28.0	7.7	17.0	26	23	19
7.52	0	12.8	3.6	7.9	12	12	14
7.52	5.82	26.7	7.4	16.3	20	20	20
7.52	7.14	28.6	7.9	17.4	23	22	20
7.52	7.83	29.0	8.0	17.6	24	23	20
7.52	8.52	30.5	8.4	18.5	25	23	20
8.14	0	13.9	3.8	8.4	13	13	15
8.14	5.60	27.1	7.5	16.5	19	19	23
8.14	7.01	29.2	8.1	17.8	20	20	23
8.14	7.73	29.7	8.2	18.1	21	21	23
8.14	8.47	31.4	8.7	19.2	22	22	23

[c] TDN was calculated by assuming 3.6155 Mcal of ME per kg of TDN.

[d] Most steers of the weight indicated, and not exhibiting compensatory growth, will fail to sustain the energy intake necessary to maintain the rate of gain for an extended period.

Table 2 Nutrient Requirements for Growing-Finishing Heifer Calves and Yearlings (Daily Nutrients per Animal)

Weight[a] (kg)	(lb)	Daily Gain (kg)	(lb)	Minimum Dry Matter Consumption[b] (kg)	(lb)	Roughage[b] (%)	Total Protein (kg)	(lb)	Digestible Protein (kg)	(lb)
100	220	0	0	2.1	4.6	100	0.18	0.40	0.10	0.22
		0.5	1.1	3.0	6.6	70-80	0.37	0.82	0.25	0.55
		0.7	1.5	2.9	6.4	50-60	0.42	0.93	0.29	0.64
		0.9	2.0	3.0	6.6	25-30	0.48	1.06	0.34	0.75
		1.1	2.4	3.0	6.6	15	0.53	1.17	0.39	0.86
150	331	0	0	2.8	6.2	100	0.24	0.53	0.14	0.31
		0.5	1.1	4.1	9.0	70-80	0.45	0.99	0.29	0.64
		0.7	1.5	4.0	8.8	50-60	0.50	1.10	0.33	0.73
		0.9	2.0	4.0	8.8	25-30	0.54	1.19	0.37	0.82
		1.1	2.4	4.0	8.8	15	0.60	1.32	0.42	0.93
200	441	0	0	3.5	7.7	100	0.30	0.66	0.17	0.37
		0.3	0.7	5.4	11.9	100	0.49	1.08	0.29	0.64
		0.5	1.1	6.0	13.2	80-90	0.58	1.28	0.35	0.77
		0.7	1.5	6.0	13.2	70-80	0.61	1.34	0.39	0.86
		0.9	2.0	5.3	11.7	35-45	0.62	1.37	0.40	0.88
		1.1	2.4	5.0	11.0	15	0.64	1.41	0.43	0.95
250	551	0	0	4.1	9.0	100	0.35	0.77	0.20	0.44
		0.3	0.7	6.4	14.1	100	0.57	1.26	0.33	0.73
		0.5	1.1	6.5	14.3	80-90	0.62	1.37	0.37	0.82
		0.7	1.5	5.8	12.8	55-65	0.62	1.37	0.38	0.84
		0.9	2.0	5.9	13.0	35-45	0.65	1.43	0.42	0.93
		1.1	2.4	6.5	14.3	20-25	0.74	1.63	0.48	1.06
		1.2	2.6	6.3	13.9	15	0.75	1.65	0.49	1.08
300	661	0	0	4.7	10.4	100	0.40	0.88	0.23	0.51
		0.3	0.7	7.4	16.3	100	0.63	1.39	0.36	0.79
		0.5	1.1	7.4	16.3	80-90	0.67	1.48	0.40	0.88
		0.7	1.5	6.6	14.6	55-65	0.67	1.48	0.40	0.88
		0.9	2.0	6.8	15.0	35-45	0.70	1.54	0.44	0.97
		1.1	2.4	7.5	16.5	20-25	0.78	1.72	0.49	1.08
		1.2	2.6	7.2	15.9	15	0.79	1.74	0.50	1.10
350	772	0	0	5.3	11.7	100	0.46	1.01	0.26	0.57
		0.3	0.7	8.2	18.1	100	0.69	1.52	0.39	0.86
		0.5	1.1	8.3	18.3	80-90	0.73	1.61	0.42	0.93
		0.7	1.5	7.9	17.4	55-65	0.73	1.61	0.43	0.95
		0.9	2.0	8.1	17.9	35-45	0.77	1.70	0.46	1.01
		1.1	2.4	8.3	18.3	20-25	0.81	1.79	0.50	1.10
		1.2[d]	2.6	8.1	17.9	15	0.81	1.79	0.50	1.10
400	882	0	0	5.9	13.0	100	0.51	1.12	0.29	0.64
		0.3	0.7	9.1	20.0	100	0.76	1.68	0.43	0.95
		0.5	1.1	8.5	18.7	70-80	0.78	1.72	0.43	0.95
		0.7	1.5	8.7	17.2	55-65	0.79	1.74	0.46	1.01
		0.9	2.0	8.4	18.5	20-25	0.79	1.74	0.47	1.04
		1.1	2.4	8.3	18.3	15	0.81	1.79	0.49	1.08
450	992	0	0	6.4	14.1	100	0.55	1.21	0.31	0.68
		0.2	0.4	8.7	19.2	100	0.74	1.63	0.41	0.90
		0.5	1.1	9.3	20.5	70-80	0.80	1.76	0.46	1.01
		0.8	1.8	9.1	20.1	35-45	0.82	1.81	0.48	1.06
		1.0[d]	2.2	8.5	18.7	15	0.83	1.83	0.48	1.06

Source: *Nutrient Requirements of Beef Cattle*, fifth revised edition, National Research Council, National Academy of Sciences, Washington, D.C., 1976.

[a] Average weight for a feeding period.

[b] Dry Matter Consumption, ME and TDN allowances are based on NE requirements and the general type of diet indicated in the roughage column. Most roughages will contain 1.9 to 2.2 Mcal of ME/kg dry matter and 90-100% concentrate diets are expected to have 3.1 to 3.3 Mcal of ME/kg.

NE_m (Mcal)	NE_g (Mcal)	ME^b (Mcal)	$TDN^{b,c}$ (kg)	$TDN^{b,c}$ (lb)	Calcium (g)	Phosphorus (g)	Vitamin A (1000 IU)
2.43	0	4.2	1.2	2.6	4	4	5
2.43	0.99	6.9	1.9	4.2	14	11	6
2.43	1.44	7.5	2.1	4.6	19	14	6
2.43	1.92	8.3	2.3	5.1	24	17	7
2.43	2.43	9.2	2.5	5.5	29	19	7
3.30	0	5.6	1.6	3.5	5	5	6
3.30	1.34	9.4	2.6	5.7	14	12	9
3.30	1.95	10.4	2.8	6.2	18	14	9
3.30	2.60	11.3	3.1	6.8	23	17	9
3.30	3.30	12.4	3.4	7.5	28	20	9
4.10	0	7.0	1.9	4.2	6	6	8
4.10	0.95	10.8	3.0	6.6	10	10	12
4.10	1.66	12.7	3.5	7.7	14	13	13
4.10	2.42	13.8	3.8	8.4	18	16	13
4.10	3.23	14.3	4.0	8.8	22	17	13
4.10	4.09	15.4	4.3	9.5	25	19	13
4.84	0	8.3	2.3	5.1	7	7	9
4.84	1.13	12.8	3.5	7.8	12	12	14
4.84	1.96	14.2	3.9	8.6	13	13	14
4.84	2.86	15.0	4.1	9.1	17	15	14
4.84	3.81	16.5	4.6	10.1	21	17	14
4.84	4.84	18.7	5.2	11.5	25	20	14
4.84	5.37	19.4	5.4	11.9	27	21	14
5.55	0	9.5	2.6	5.7	9	9	10
5.55	1.29	14.5	4.0	8.4	13	13	16
5.55	2.25	16.3	4.5	9.9	14	14	16
5.55	3.37	17.1	4.7	10.4	16	15	16
5.55	4.37	19.0	5.2	11.5	19	17	16
5.55	5.55	21.5	6.0	13.2	23	20	16
5.55	6.16	22.3	6.2	13.7	24	20	16
6.24	0	10.6	2.9	6.4	10	10	12
6.24	1.45	16.5	4.6	10.0	15	15	18
6.24	2.52	18.3	5.1	11.2	15	15	18
6.24	3.68	19.7	5.4	16.9	15	15	18
6.24	4.91	21.8	6.0	13.2	17	17	18
6.24	6.23	24.0	6.6	14.5	20	19	18
6.24	6.91	25.0	6.9	15.2	21	20	18
6.89	0	11.8	3.3	7.3	11	11	13
6.89	1.61	18.2	5.0	11.1	16	16	19
6.89	2.79	19.5	5.4	11.9	15	15	19
6.89	4.06	21.7	6.0	13.2	16	16	19
6.89	5.43	23.5	6.5	14.3	17	17	19
6.89	6.88	25.9	7.2	15.9	19	18	19
7.52	0	12.9	3.6	7.9	12	12	14
7.52	1.14	17.4	4.8	10.6	16	16	19
7.52	3.05	21.3	5.9	13.0	17	17	20
7.52	5.17	24.5	6.8	15.0	16	16	20
7.52	6.71	26.8	7.4	16.3	19	19	20

c TDN was calculated by assuming 3.6155 Kcal of ME per gram of TDN.

d Most heifers of the weight indicated, and not exhibiting compensatory growth, will fail to sustain the energy intake necessary to maintain this rate of gain for an extended period.

Table 3 Nutrient Requirements for Beef Cattle Breeding Herd
(Daily Nutrients per Animal)

Weight[a] (kg)	(lb)	Daily Gain (kg)	(lb)	Minimum Dry Matter Consumption[b] (kg)	(lb)	Roughage[b] (%)	Total Protein (kg)	(lb)	Digestible Protein (kg)	(lb)
Pregnant Yearling Heifers — Last 3-4 months of pregnancy										
325	716	0.4[c]	0.9	6.6	14.5	100[d]	0.58	1.28	0.34	0.756
		0.6	1.3	8.5	18.7	100	0.75	1.65	0.42	0.93
		0.8	1.8	9.4	20.7	85-100	0.85	1.87	0.50	1.10
350	772	0.4[c]	0.9	6.9	15.2	100	0.61	1.34	0.35	0.772
		0.6	1.3	8.9	17.6	100	0.78	1.72	0.45	0.99
		0.8	1.8	10.0	22.0	85-100	0.88	1.94	0.51	1.12
375	827	0.4[c]	0.9	7.2	15.9	100	0.63	1.39	0.36	0.797
		0.6	1.3	9.3	20.5	100	0.81	1.79	0.46	1.01
		0.8	1.8	11.0	24.2	85-100	0.96	2.12	0.55	1.21
400	882	0.4[c]	0.9	7.5	16.5	100	0.65	1.43	0.38	0.84
		0.6	1.3	9.7	21.4	100	0.84	1.85	0.48	1.06
		0.8	1.8	11.6	25.6	85-100	1.01	2.23	0.57	1.26
425	937	0.4[c]	0.9	7.8	17.2	100	0.69	1.52	0.40	0.88
		0.6	1.3	10.1	22.3	100	0.88	1.94	0.50	1.10
		0.8	1.8	12.1	26.7	85-100	1.05	2.31	0.60	1.32
Dry Pregnant Mature Cows — Middle third of pregnancy										
350	772			5.5	12.2	100[d]	0.32	0.70	0.15	0.33
400	882			6.1	13.4	100	0.36	0.79	0.17	0.37
450	992			6.7	14.8	100	0.39	0.86	0.19	0.42
500	1,102			7.2	15.9	100	0.42	0.93	0.20	0.44
550	1,213			7.7	17.0	100	0.45	0.99	0.22	0.48
600	1,323			8.3	18.3	100	0.49	1.08	0.23	0.51
650	1,433			8.8	19.4	100	0.52	1.15	0.25	0.55
Dry Pregnant Mature Cows — Last third of pregnancy										
350	772	0.4[c]	0.9	6.9	13.9	100[d]	0.41	0.90	0.19	0.42
400	882	0.4	0.9	7.5	15.4	100	0.44	0.97	0.21	0.46
450	992	0.4	0.9	8.1	16.5	100	0.48	1.06	0.23	0.51
500	1,102	0.4	0.9	8.6	17.9	100	0.51	1.12	0.24	0.52
550	1,213	0.4	0.9	9.1	19.0	100	0.54	1.19	0.25	0.55
600	1,323	0.4	0.9	9.7	20.3	100	0.57	1.26	0.27	0.60
650	1,433	0.4	0.9	10.2	22.4	100	0.60	1.32	0.29	0.64
Cows nursing calves — Average milking ability[e] — First 3-4 months postpartum										
350	772			8.2	18.1	100[d]	0.75	1.65	0.44	0.97
400	882			8.8	19.4	100	0.81	1.79	0.48	1.06
450	992			9.3	20.5	100	0.86	1.90	0.50	1.10
500	1,102			9.8	21.6	100	0.90	1.98	0.53	1.17
550	1,213			10.5	23.1	100	0.97	2.14	0.57	1.26
600	1,323			11.0	24.2	100	1.01	2.23	0.59	1.30
650	1,433			11.4	25.1	100	1.05	2.31	0.62	1.37
Cows nursing calves — Superior milking ability[f] — First 3-4 months postpartum										
350	772			10.2	22.4	100[g]	1.11	2.45	0.65	1.43
400	882			10.8	23.8	100	1.17	2.58	0.69	1.52
450	992			11.3	24.9	100	1.23	2.71	0.72	1.59
500	1,102			11.8	26.0	100	1.29	2.84	0.76	1.68
550	1,213			12.4	27.3	100	1.35	2.98	0.79	1.74
600	1,328			12.9	28.4	100	1.41	3.11	0.83	1.83
650	1,433			13.4	29.5	100	1.46	3.22	0.86	1.90
Bulls, growth and maintenance (moderate activity)										
300	661	1.00	2.2	8.8	19.4	70-75	0.90	1.98	0.55	1.21
400	882	0.90	2.0	11.0	24.2	70-75	1.03	2.27	0.62	1.37
500	1,102	0.70	1.5	12.2	26.9	80-85	1.07	2.36	0.62	1.37
600	1,323	0.50	1.1	12.0	26.4	80-85	1.02	2.25	0.60	1.32
700	1,543	0.30	0.7	12.9	28.4	90-100[g]	1.08	2.38	0.60	1.32
800	1,764	0	0	10.5	23.1	100[g]	0.89	1.96	0.50	1.10
900	1,984	0	0	11.4	25.1	100[g]	0.99	2.18	0.55	1.21
1000	2,205	0	0	12.4	27.3	100[g]	1.05	2.31	0.60	1.32

Source: *Nutrient Requirements of Beef Cattle*, fifth revised edition, National Research Council, National Academy of Sciences, Washington, D.C., 1976.

[a] Average weight for a feeding period.

[b] Dry Matter Consumption, ME and TDN requirements are based on the general type of diet indicated in the roughage column.

[c] Approximately 0.4 (±) 0.1 kg of weight gain/day over the last third of pregnancy is accounted for by the products of conception. These nutrients and energy requirements include the quantities estimated as necessary for conceptus development.

NE_m (Mcal)	NE_g (Mcal)	ME^b (Mcal)	TDN[b, c] (kg)	TDN[b, c] (lb)	Calcium (g)	Phosphorus (g)	Vitamin A (1000 IU)
5.89	0.62	12.6	3.5	7.7	15	15	19
5.89	1.52	16.2	4.5	9.9	18	18	23
5.89	2.49	20.1	5.6	12.3	22	20	26
6.23	0.65	13.2	3.7	8.1	15	15	19
6.23	1.60	16.9	4.7	10.3	19	19	25
6.24	2.63	21.1	5.8	12.9	22	21	28
6.56	0.68	13.7	3.8	8.4	15	15	20
6.56	1.68	17.7	4.9	10.8	19	19	26
6.56	2.76	22.1	6.1	13.5	22	22	31
6.89	0.71	14.2	3.9	8.7	16	16	21
6.89	1.76	18.5	5.1	11.3	19	19	27
6.89	2.90	23.0	6.4	14.0	22	22	33
7.21	0.74	14.8	4.1	9.0	16	16	22
7.21	1.84	19.2	5.3	11.7	19	19	28
7.21	3.03	24.0	6.6	14.6	22	22	34
6.23		10.8	3.0	6.6	10	10	15
6.89		11.9	3.3	7.3	11	11	17
7.52		13.0	3.6	7.9	12	12	19
8.14		14.1	3.9	8.6	13	13	20
8.75		15.1	4.2	9.2	14	14	22
9.33		16.1	4.4	9.8	15	15	23
9.91		17.1	4.7	10.4	16	16	25
7.8		13.2	3.6	8.0	12	12	19
8.4		14.3	4.0	8.7	14	14	21
9.1		15.4	4.2	9.4	15	15	23
9.7		16.4	4.5	10.0	15	15	24
10.3		17.5	4.8	10.7	16	16	26
10.9		18.5	5.1	11.2	17	17	27
11.5		19.6	5.4	11.9	18	18	29
9.2		15.9	4.4	9.7	24	24	19
9.9		17.0	4.7	10.4	25	25	21
10.5		18.1	5.0	11.0	26	26	23
11.1		19.2	5.3	11.7	27	27	24
11.9		20.3	5.6	12.3	28	28	26
12.3		21.3	5.9	13.0	28	28	27
12.9		22.3	6.2	13.7	29	29	29
12.3		21.0	5.8	12.8	45	40	32
13.0		22.1	6.1	13.5	45	41	34
13.6		23.2	6.4	14.1	45	42	36
14.2		24.3	6.7	14.8	46	43	38
14.9		25.3	7.0	15.4	46	44	41
15.5		26.4	7.3	16.1	46	44	43
16.2		27.5	7.6	16.8	47	45	45
5.6	3.8	20.4	5.6	12.3	27	23	34
6.9	4.1	25.2	7.0	15.4	23	23	43
8.5	3.7	27.0	7.5	16.5	22	22	48
9.8	3.0	26.4	7.3	16.1	22	22	48
11.0	2.0	27.7	7.7	17.0	23	23	50
12.2	0	21.0	5.8	12.8	19	19	41
13.3	0	22.8	6.3	13.9	21	21	44
14.4	0	24.8	6.9	15.2	22	22	48

[d] Average quality roughage containing about 1.9–2.0 Mcal ME/kg dry matter.

[e] 5.0 ($\pm$) 0.5 kg of milk/day. Nutrients and energy for maintenance of the cow and for milk production are included in these requirements.

[f] 10 ($\pm$) 0.5 kg of milk/day. Nutrients and energy for maintenance of the cow and for milk production are included in these requirements.

[g] Good quality roughage containing at least 2.0 Mcal ME/kg dry matter.

Table 4 Nutrient Requirements for Growing—Finishing Steer Calves and Yearlings (Nutrient Concentration in Diet Dry Matter)[a]

Weight[b]		Daily gain[c]		Minimum Dry Matter Consumption[c]		Roughage[c] (%)	Total Protein (%)	Digestible Protein (%)
(kg)	(lb)	(kg)	(lb)	(kg)	(lb)			
100	220	0	0	2.1	4.6	100	8.7	5.0
		0.5	1.1	2.9	6.4	70-80	12.4	8.3
		0.7	1.5	2.7	6.0	50-60	14.8	10.7
		0.9	2.0	2.8	6.2	25-30	16.4	11.8
		1.1	2.4	2.7	6.0	<15	18.2	13.3
150	331	0	0	2.8	6.2	100	8.7	5.0
		0.5	1.1	4.0	8.8	70-80	11.0	7.0
		0.7	1.5	3.9	8.6	50-60	12.6	8.5
		0.9	2.0	3.8	8.4	25-30	14.1	9.7
		1.1	2.4	3.7	8.2	<15	15.6	11.1
200	441	0	0	3.5	7.7	100	8.5	4.8
		0.5	1.1	5.8	12.8	80-90	9.9	6.0
		0.7	1.5	5.7	12.6	70-80	10.8	6.8
		0.9	2.0	4.9	10.8	35-45	12.3	8.2
		1.1	2.4	4.6	10.1	<15	13.6	9.3
250	551	0	0	4.1	9.7	100	8.5	4.8
		0.7	1.5	5.8	12.8	55-65	10.7	6.7
		0.9	2.0	6.2	13.7	45-50	11.1	7.1
		1.1	2.4	6.0	13.2	20-25	12.1	8.0
		1.3	2.9	6.0	13.2	<15	12.7	8.5
300	661	0	0	4.7	10.4	100	8.6	4.8
		0.9	2.0	8.1	17.9	55-65	10.0	6.2
		1.1	2.4	7.6	16.8	20-25	10.8	6.8
		1.3	2.9	7.1	15.6	<15	11.7	7.6
		1.4	3.1	7.3	16.1	<15	11.9	7.8
350	772	0	0	5.3	11.7	100	8.5	4.8
		0.9	2.0	8.0	17.6	45-55	10.0	6.1
		1.1	2.4	8.0	17.6	20-25	10.4	6.5
		1.3[e]	2.9	8.0	17.6	<15	10.8	6.9
		1.4[e]	3.1	8.2	18.1	<15	10.9	7.0
400	882	0	0	5.9	13.0	100	8.5	4.8
		1.0	2.2	9.4	20.7	45-55	9.4	5.7
		1.2	2.6	8.5	18.7	20-25	10.2	6.3
		1.3[e]	2.9	8.6	19.0	<15	10.4	6.5
		1.4[e]	3.1	9.0	19.8	<15	10.5	6.6
450	992	0	0	6.4	14.1	100	8.5	4.8
		1.0	2.2	10.3	22.7	45-55	9.3	5.5
		1.2	2.6	10.2	22.5	20-25	9.5	5.7
		1.3[e]	2.9	9.3	20.5	<15	10.4	6.3
		1.4[e]	3.1	9.8	21.6	<15	10.0	6.1
500	1,102	0	0	7.0	15.4	100	8.5	4.8
		0.9	2.0	10.5	23.1	45-55	9.1	5.3
		1.1	2.4	10.4	22.9	20-25	9.2	5.5
		1.2[d]	2.6	9.6	21.2	<15	10.0	6.0
		1.3[d]	2.9	10.0	22.0	<15	9.7	6.0

Source: *Nutrient Requirements of Beef Cattle*, fifth revised edition, National Research Council, National Academy of Sciences, Washington, D.C., 1976.

[a] The concentration of vitamin A in all diets for finishing steers is 2,200 IU/kg of dry diet.

[b] Average weight for a feeding period.

[c] Dry matter consumption, ME and TDN allowances are based on NE requirements and the general types of diet indicated in the roughage column. Most roughages will contain 1.9-2.2 Mcal of ME/kg dry matter and 90-100 percent concentrate diets are expected to contain 3.1-3.3 Mcal of ME/kg.

NE$_m$[f] (Mcal/kg)	(Mcal/lb)	NE$_g$[f] (Mcal/kg)	(Mcal/lb)	ME[f] (Mcal/kg)	(Mcal/lb)	TDN[d,f] (%)	Ca (%)	P (%)
1.17	0.53	—	—	2.0	0.91	55	0.18	0.18
1.32	0.60	.70	.32	2.2	1.00	62	0.48	0.38
1.56	0.71	.95	.43	2.5	1.13	70	0.70	0.48
1.81	0.82	1.18	0.54	2.8	1.27	77	0.86	0.57
2.07	0.94	1.37	0.62	3.1	1.41	86	1.04	0.70
1.17	0.53	—	—	2.0	0.91	55	0.18	0.18
1.32	0.60	.70	.32	2.2	1.00	62	0.35	0.32
1.56	0.71	.95	0.43	2.5	1.13	70	0.46	0.36
1.81	0.82	1.18	0.54	2.8	1.27	77	0.61	0.45
2.07	0.94	1.37	0.62	3.1	1.41	86	0.76	0.54
1.17	0.53	—	—	2.0	0.91	55	0.18	0.18
1.25	0.56	0.60	0.27	2.1	0.95	58	0.24	0.22
1.40	0.64	0.78	0.35	2.3	1.04	64	0.32	0.28
1.70	0.78	1.10	0.50	2.7	1.22	75	0.47	0.37
2.07	0.94	1.37	0.62	3.1	1.41	86	0.59	0.43
1.17	0.53	—	—	2.0	0.91	55	0.18	0.18
1.56	0.71	0.95	0.43	2.5	1.13	70	0.31	0.28
1.64	0.74	1.02	0.46	2.6	1.18	72	0.35	0.31
1.81	0.82	1.18	0.54	2.8	1.27	77	0.43	0.35
2.07	0.94	1.37	0.62	3.1	1.41	86	0.50	0.38
1.17	0.53	—	—	2.0	0.91	55	0.18	0.18
1.56	0.71	0.95	0.43	2.5	1.18	70	0.27	0.23
1.81	0.82	1.18	0.54	2.8	1.27	77	0.33	0.29
1.98	0.90	1.31	0.59	3.0	1.36	83	0.41	0.32
2.07	0.94	1.37	0.62	3.1	1.41	86	0.42	0.34
1.17	0.53	—	—	2.0	0.91	55	0.18	0.18
1.64	0.74	1.02	0.46	2.6	1.18	72	0.25	0.22
1.81	0.82	1.18	0.54	2.8	1.27	80	0.29	0.25
1.98	0.90	1.31	0.59	3.0	1.36	83	0.32	0.28
2.07	0.98	1.37	0.62	3.1	1.41	86	0.34	0.29
1.17	0.53	—	—	2.0	0.91	55	0.18	0.18
1.64	0.74	1.02	0.46	2.6	1.18	72	0.22	0.21
1.81	0.82	1.18	0.54	2.8	1.27	80	0.27	0.25
2.07	0.98	1.37	0.62	3.1	1.41	86	0.29	0.26
2.07	0.98	1.37	0.62	3.1	1.41	86	0.29	0.26
1.17	0.53	—	—	2.0	0.91	55	0.18	0.18
1.64	0.74	1.02	0.46	2.6	1.18	72	0.19	0.19
1.81	0.82	1.18	0.54	2.8	1.27	80	0.23	0.22
2.07	0.98	1.31	0.62	3.1	1.41	86	0.26	0.25
2.07	0.98	1.37	0.62	3.1	1.41	86	0.26	0.23
1.17	0.53	—	—	2.0	0.91	55	0.18	0.18
1.64	0.74	1.02	0.46	2.6	1.18	72	0.18	0.18
1.81	0.82	1.18	0.54	2.8	1.27	80	0.19	0.19
2.07	0.98	1.31	0.62	3.1	1.41	86	0.22	0.22
2.07	0.98	1.37	0.62	3.1	1.41	86	0.22	0.22

[d] TDN was calculated by assuming 3.6155 Mcal of ME per kg of TDN.

[e] Most steers of the weight indicated, and not exhibiting compensatory growth, will fail to sustain an energy intake necessary to maintain this rate of gain for an extended period.

[f] Due to conversion and rounding variation, the figures in these columns may not be in exact agreement with a similar energy concentration figure calculated from the data of Table 1.

Table 5 Nutrient Requirements for Growing—Finishing Heifer Calves and Yearlings (Nutrient Concentration in Diet Dry Matter)[a]

Weight[b]		Daily Gain		Minimum Dry Matter Consumption[c]		Roughage[c]	Total Protein	Digestible Protein
(kg)	(lb)	(kg)	(lb)	(kg)	(lb)	(%)	(%)	(%)
100	220	0	0	2.1	4.6	100	8.7	5.0
		0.5	1.1	3.0	6.6	70-80	12.4	8.3
		0.7	1.5	2.9	6.4	50-60	14.4	10.0
		0.9	2.0	3.0	6.6	25-30	15.9	11.3
		1.1	2.4	3.0	6.6	<15	17.8	13.0
150	331	0	0	2.8	6.2	100	8.7	5.0
		0.5	1.1	4.1	9.0	70-80	11.0	7.1
		0.7	1.5	4.0	8.8	50-60	12.4	8.2
		0.9	2.0	4.0	8.8	25-30	13.5	9.2
		1.1	2.4	4.0	8.8	<15	15.0	10.5
200	441	0	0	3.5	7.7	100	8.5	4.9
		0.3	0.7	5.4	11.9	100	9.1	5.4
		0.5	1.1	6.0	13.2	80-90	9.6	5.8
		0.7	1.5	6.0	13.2	70-80	10.2	6.5
		0.9	2.0	5.3	11.7	35-45	11.7	7.5
		1.1	2.4	5.0	11.0	<15	12.8	8.6
250	551	0	0	4.1	9.0	100	8.5	4.9
		0.3	0.7	6.4	14.1	100	8.9	5.2
		0.5	1.1	6.5	14.3	80-90	9.5	5.7
		0.7	1.5	5.8	12.8	55-65	10.5	6.5
		0.9	2.0	5.9	13.0	35-45	11.1	7.1
		1.1	2.4	6.5	14.3	20-25	11.4	7.4
		1.2	2.6	6.3	13.9	<15	11.9	7.8
300	661	0	0	4.7	10.4	100	8.6	4.9
		0.3	0.7	7.4	16.3	100	8.5	4.9
		0.5	1.1	7.4	16.3	80-90	9.2	5.4
		0.7	1.5	6.6	14.6	55-65	10.1	6.1
		0.9	2.0	6.8	15.0	35-45	10.4	6.5
		1.1	2.4	7.5	16.5	20-25	10.4	6.5
		1.2	2.6	7.2	15.9	<15	10.9	6.9
350	772	0	0	5.3	11.7	100	8.5	4.8
		0.3	0.7	8.2	18.1	100	8.5	4.8
		0.5	1.1	8.3	18.3	80-90	8.7	5.1
		0.7	1.5	7.9	17.4	55-65	9.2	5.4
		0.9	2.0	8.1	17.9	35-45	9.5	5.7
		1.1	2.4	8.3	18.3	20-25	9.9	6.0
		1.2[e]	2.6	8.1	17.9	<15	10.0	6.2
400	882	0	0	5.9	13.0	100	8.5	4.8
		0.3	0.7	9.1	20.0	100	8.5	4.8
		0.5	1.1	8.5	18.7	70-80	8.8	5.1
		0.7	1.5	8.7	19.2	55-65	9.0	5.3
		0.9	2.0	8.4	18.5	20-25	9.4	5.6
		1.1[e]	2.4	8.3	18.3	<15	9.7	5.9
450	992	0	0	6.4	14.1	100	8.5	4.8
		0.2	0.4	8.7	19.2	100	8.5	4.7
		0.5	1.1	9.3	20.5	70-80	8.6	4.9
		0.8	1.8	9.1	20.1	35-45	9.0	5.3
		1.0[e]	2.2	8.5	18.7	<15	9.5	5.6

Source: *Nutrient Requirements of Beef Cattle*, fifth revised edition, National Research Council, National Academy of Sciences, Washington, D.C., 1976.

[a] The concentration of vitamin A in all diets for finishing heifers is 2,200 IU/kg of dry diet.

[b] Average weight for a feeding period.

[c] Dry matter consumption, ME and TDN allowances are based on NE requirements and the general type of diet indicated in the roughage column. Most roughages will contain 1.9-2.2 Mcal of ME/kg dry matter and 90-100% concentrate diets are expected to have 3.1 to 3.3 Mcal of ME/kg.

NE$_m^f$ (Mcal/kg)	(Mcal/lb)	NE$_g^f$ (Mcal/kg)	(Mcal/lb)	ME^f (Mcal/kg)	(Mcal/lb)	TDN^{d,f} (%)	Ca (%)	P (%)
1.17	0.53	—	—	2.0	0.91	55	0.18	0.18
1.32	0.60	0.70	0.32	2.2	1.00	61	0.47	0.37
1.56	0.71	0.95	0.43	2.5	1.13	69	0.66	0.48
1.81	0.82	1.18	0.54	2.8	1.27	77	0.80	0.57
2.07	0.94	1.37	0.62	3.1	1.41	86	0.97	0.63
1.17	0.53	—	—	2.0	0.91	55	0.18	0.18
1.32	0.60	0.70	0.32	2.2	1.00	61	0.34	0.29
1.56	0.71	0.95	0.43	2.5	1.13	69	0.45	0.35
1.81	0.82	1.18	0.54	2.8	1.27	77	0.57	0.42
2.07	0.94	1.37	0.62	3.1	1.41	86	0.70	0.50
1.17	0.53	—	—	2.0	0.91	55	0.18	0.18
1.17	0.53	0.50	0.23	2.0	0.91	55	0.18	0.18
1.24	0.56	0.60	0.27	2.1	0.95	58	0.23	0.22
1.40	0.64	0.87	0.39	2.3	1.04	64	0.30	0.27
1.72	0.78	1.10	0.50	2.7	1.22	75	0.41	0.32
2.07	0.94	1.37	0.62	3.1	1.41	86	0.50	0.38
1.17	0.53	—	—	2.0	0.91	55	0.18	0.18
1.17	0.53	0.50	0.23	2.0	0.91	55	0.18	0.18
1.24	0.56	0.60	0.27	2.1	0.95	58	0.20	0.20
1.64	0.74	1.02	0.46	2.6	1.18	72	0.29	0.26
1.81	0.82	1.18	0.54	2.8	1.27	77	0.36	0.29
1.89	0.86	1.25	0.57	2.9	1.31	80	0.38	0.31
2.07	0.94	1.37	0.62	3.1	1.41	86	0.43	0.33
1.17	0.53	—	—	2.0	0.91	55	0.18	0.18
1.17	0.53	0.50	0.23	2.0	0.91	55	0.18	0.18
1.32	0.60	0.70	0.32	2.2	1.00	61	0.19	0.19
1.64	0.74	1.02	0.46	2.6	1.18	72	0.24	0.23
1.81	0.82	1.18	0.54	2.8	1.27	77	0.28	0.25
1.89	0.86	1.25	0.57	2.9	1.31	80	0.31	0.27
2.07	0.94	1.37	0.62	3.1	1.41	86	0.33	0.28
1.17	0.53	—	—	2.0	0.91	55	0.18	0.18
1.17	0.53	0.50	0.23	2.0	0.91	55	0.18	0.18
1.32	0.60	0.70	0.32	2.2	1.00	61	0.18	0.18
1.56	0.71	0.95	0.43	2.5	1.13	69	0.19	0.19
1.72	0.78	1.10	0.50	2.7	1.22	75	0.21	0.21
1.89	0.86	1.25	0.57	2.9	1.31	80	0.24	0.23
2.07	0.94	1.37	0.62	3.1	1.41	86	0.26	0.25
1.17	0.53	—	—	2.0	0.91	55	0.18	0.18
1.17	0.53	0.50	0.23	2.0	0.91	55	0.18	0.18
1.40	0.64	0.78	0.35	2.3	1.04	64	0.18	0.18
1.56	0.71	0.95	0.43	2.5	1.09	66	0.18	0.18
1.81	0.82	1.18	0.54	2.8	1.27	77	0.20	0.20
2.07	0.94	1.37	0.62	3.1	1.41	86	0.23	0.22
1.17	0.53	—	—	2.0	0.91	55	0.18	0.18
1.17	0.53	0.50	0.23	2.0	0.91	55	0.18	0.18
1.40	0.64	0.78	0.35	2.3	1.04	64	0.18	0.18
1.72	0.78	1.10	0.50	2.7	1.22	75	0.18	0.18
2.07	0.94	1.37	0.62	3.1	1.41	86	0.22	0.22

[d] TDN was calculated by assuming 3.6155 kcal of ME per g of TDN.

[e] Most heifers of the weight indicated, and not exhibiting compensatory growth, will fail to sustain the energy intake necessary to maintain this rate of gain for an extended period.

[f] Due to conversion and rounding variation, the figures in these columns may not be in exact agreement with a similar energy concentration figure calculated from the data of Table 2.

Table 6 Nutrient Requirements for Beef Cattle Breeding Herd (Nutrient Concentration in Diet Dry Matter)[a]

Weight[b] (kg)	(lb)	Daily Gain (kg)	(lb)	Dry Matter[c] Consumption (kg)	(lb)	Roughage[c] (%)	Total Protein (%)	Digestible Protein (%)
Pregnant yearling heifers—Last third of pregnancy								
325	716	0.4[d]	0.9	6.6	14.5	100[e]	8.8	5.1
		0.6	1.3	8.5	18.7	100	8.8	5.1
		0.8	1.8	9.4	20.7	85-100	9.0	5.3
350	772	0.4[d]	0.9	6.9	15.2	100	8.8	5.1
		0.6	1.3	8.9	19.6	100	8.8	5.1
		0.8	1.8	10.0	22.0	85-100	8.8	5.1
375	827	0.4[d]	0.9	7.2	15.9	100	8.7	5.0
		0.6	1.3	9.3	20.5	100	8.7	5.0
		0.8	1.8	11.0	24.2	85-100	8.7	5.0
400	882	0.4[d]	0.9	7.5	16.5	100	8.7	5.0
		0.6	1.3	9.7	21.4	100	8.7	5.0
		0.8	1.8	11.6	25.6	85-100	8.7	5.0
425	937	0.4[d]	0.9	7.8	17.2	100	8.8	5.1
		0.6	1.3	10.1	22.3	100	8.7	5.0
		0.8	1.8	12.1	26.7	85-100	8.7	5.0
Dry pregnant mature cows—Middle third of pregnancy								
350	772	—	—	5.5	12.2	100[e]	5.9	2.8
400	882	—	—	6.1	13.4	100	5.9	2.8
450	992	—	—	6.7	14.8	100	5.9	2.8
500	1,102	—	—	7.2	15.9	100	5.9	2.8
550	1,213	—	—	7.7	17.0	100	5.9	2.8
600	1,323	—	—	8.3	18.3	100	5.9	2.8
650	1,433	—	—	8.8	19.4	100	5.9	2.8
Dry pregnant mature cows—Last third of pregnancy								
350	772	0.4[d]	0.9	6.9	13.9	100[e]	5.9	2.8
400	882	0.4	0.9	7.5	15.4	100	5.9	2.8
450	992	0.4	0.9	8.1	16.5	100	5.9	2.8
500	1,102	0.4	0.9	8.6	17.9	100	5.9	2.8
550	1,213	0.4	0.9	9.1	19.0	100	5.9	2.8
600	1,323	0.4	0.9	9.7	20.3	100	5.9	2.8
650	1,433	0.4	0.9	10.2	22.4	100	5.9	2.8
Cows nursing calves—Average milking ability[f]—First 3-4 months postpartum								
350	772	—	—	8.2	18.1	100[e]	9.2	5.4
400	882	—	—	8.8	19.4	100	9.2	5.4
450	992	—	—	9.3	20.5	100	9.2	5.4
500	1,102	—	—	9.8	21.6	100	9.2	5.4
550	1,213	—	—	10.5	23.1	100	9.2	5.4
600	1,323	—	—	11.0	24.2	100	9.2	5.4
650	1,433	—	—	11.4	25.1	100	9.2	5.4
Cows nursing calves—Superior milking ability[g]—First 3-4 months postpartum								
350	772	—	—	10.2	22.4	100[h]	10.9	6.4
400	882	—	—	10.8	23.8	100	10.9	6.4
450	992	—	—	11.3	24.9	100	10.9	6.4
500	1,102	—	—	11.8	26.0	100	10.9	6.4
550	1,213	—	—	12.4	27.3	100	10.9	6.4
600	1,323	—	—	12.9	28.4	100	10.9	6.4
650	1,433	—	—	13.4	29.5	100	10.9	6.4
Bulls, growth and maintenance (moderate activity)								
300	661	1.00	2.2	8.8	19.4	70-75	10.2	6.3
400	882	0.90	2.0	11.0	24.2	70-75	9.4	5.6
500	1,102	0.70	1.5	12.2	26.9	80-85	8.8	5.1
600	1,323	0.50	1.1	12.0	26.4	80-85	8.8	5.0
700	1,543	0.30	0.7	12.9	28.4	90-100[h]	8.5	4.8
800	1,764	0.00	0.0	10.5	23.1	100[h]	8.5	4.8
900	1,984	0.00	0.0	11.4	25.1	100[h]	8.5	4.8
1,000	2,205	0.00	0.0	12.4	27.3	100[h]	8.5	4.8

Source: *Nutrient Requirements of Beef Cattle*, fifth revised edition, National Research Council, National Academy of Sciences, Washington, D.C., 1976.

[a] The concentration of vitamin A in all diets for pregnant heifers and cows is 2,800 IU/kg dry diet; for lactating cows and breeding bulls, 3,900 IU/kg.

[b] Average weight for a feeding period.

[c] Dry matter consumption, ME, and TDN requirements are based on the general type of diet indicated in the roughage column.

[d] Approximately 0.4 ± 0.1 kg of weight gain/day over the last third of pregnancy is accounted for by the products of conception.

NE_m[i] (Mcal/kg)	(Mcal/lb)	NE_g[i] (Mcal/kg)	(Mcal/lb)	ME[i] (Mcal/kg)	(Mcal/lb)	TDN[i] (%)	Ca (%)	P (%)
1.09	0.49	0.38	0.17	1.9	0.86	52	0.23	0.23
1.09	0.49	0.38	0.17	1.9	0.86	52	0.21	0.21
1.24	0.56	0.60	0.27	2.1	0.95	58	0.23	0.21
1.09	0.49	0.38	0.17	1.9	0.86	52	0.22	0.22
1.09	0.49	0.38	0.17	1.9	0.86	52	0.21	0.21
1.24	0.56	0.60	0.27	2.1	0.95	58	0.22	0.21
1.09	0.49	0.38	0.17	1.9	0.86	52	0.21	0.21
1.09	0.49	0.38	0.17	1.9	0.86	52	0.20	0.20
1.17	0.53	0.50	0.23	2.0	0.91	55	0.20	0.20
1.09	0.49	0.38	0.17	1.9	0.86	52	0.21	0.21
1.09	0.49	0.38	0.17	1.9	0.86	52	0.20	0.20
1.17	0.53	0.50	0.23	2.0	0.91	55	0.19	0.19
1.09	0.49	0.38	0.17	1.9	0.86	52	0.20	0.20
1.09	0.49	0.38	0.17	1.9	0.86	52	0.19	0.19
1.17	0.63	0.50	0.23	2.0	0.91	55	0.18	0.18
1.09	0.49	—	—	1.9	0.86	52	0.18	0.18
1.09	0.49	—	—	1.9	0.86	52	0.18	0.18
1.09	0.49	—	—	1.9	0.86	52	0.18	0.18
1.09	0.49	—	—	1.9	0.86	52	0.18	0.18
1.09	0.49	—	—	1.9	0.86	52	0.18	0.18
1.09	0.49	—	—	1.9	0.86	52	0.18	0.18
1.09	0.49	—	—	1.9	0.86	52	0.18	0.18
1.09	0.49	—	—	1.9	0.86	52	0.18	0.18
1.09	0.49	—	—	1.9	0.86	52	0.18	0.18
1.09	0.49	—	—	1.9	0.86	52	0.18	0.18
1.09	0.49	—	—	1.9	0.86	52	0.18	0.18
1.09	0.49	—	—	1.9	0.86	52	0.18	0.18
1.09	0.49	—	—	1.9	0.86	52	0.18	0.18
1.09	0.49	—	—	1.9	0.86	52	0.18	0.18
1.09	0.49	—	—	1.9	0.86	52	0.29	0.29
1.09	0.49	—	—	1.9	0.86	52	0.28	0.28
1.09	0.49	—	—	1.9	0.86	52	0.28	0.28
1.09	0.49	—	—	1.9	0.86	52	0.28	0.28
1.09	0.49	—	—	1.9	0.86	52	0.27	0.27
1.09	0.49	—	—	1.9	0.86	52	0.25	0.25
1.09	0.49	—	—	1.9	0.86	52	0.25	0.25
1.17	0.53	—	—	2.0	0.91	55	0.44	0.39
1.17	0.53	—	—	2.0	0.91	55	0.42	0.38
1.17	0.53	—	—	2.0	0.91	55	0.40	0.37
1.17	0.53	—	—	2.0	0.91	55	0.39	0.36
1.17	0.53	—	—	2.0	0.91	55	0.37	0.35
1.17	0.53	—	—	2.0	0.91	55	0.36	0.34
1.17	0.53	—	—	2.0	0.91	55	0.35	0.33
1.40	0.64	0.78	0.35	2.3	1.04	64	0.31	0.26
1.40	0.64	0.78	0.35	2.3	1.04	64	0.21	0.21
1.32	0.60	0.70	0.32	2.2	1.00	61	0.18	0.18
1.32	0.60	0.70	0.32	2.2	1.00	61	0.18	0.18
1.17	0.53	0.50	0.17	2.0	0.91	55	0.18	0.18
1.17	0.53	—	—	2.0	0.91	55	0.18	0.18
1.17	0.53	—	—	2.0	0.91	55	0.18	0.18
1.17	0.53	—	—	2.0	0.91	55	0.18	0.18

[e] Average quality roughage containing about 1.9-2.0 Mcal ME/kg dry matter.

[f] 5.0 ± 0.5 kg of milk/day.

[g] 10 ± 1 kg of milk/day.

[h] Good quality roughage containing 2.0 Mcal ME/kg dry matter.

[i] Due to conversion and rounding variation, the figures in these columns may not be in exact agreement with a similar figure calculated from the data in Table 3.

Table 7 Daily Nutrient Requirements of Sheep
(100% Dry Matter Basis)

Body Weight (kg)	(lb)	Gain or Loss (kg)	(lb)	Dry Matter[a] Per Animal (kg)	(lb)	% Live-weight	Energy TDN (kg)	(lb)	DE[b] (Mcal)	ME (Mcal)
EWES[b]										
Maintenance										
50	110	0.01	0.022	1.0	2.2	2.0	0.55	1.21	2.42	1.98
60	132	0.01	0.022	1.1	2.4	1.8	0.61	1.34	2.68	2.20
70	154	0.01	0.022	1.2	2.6	1.7	0.66	1.46	2.90	2.38
80	176	0.01	0.022	1.3	2.9	1.6	0.72	1.59	3.17	2.60
Nonlactating and first 15 weeks of gestation										
50	110	0.03	0.07	1.1	2.4	2.2	0.60	1.32	2.64	2.16
60	132	0.03	0.07	1.3	2.9	2.1	0.72	1.59	3.17	2.60
70	154	0.03	0.07	1.4	3.1	2.0	0.77	1.70	3.39	2.78
80	176	0.03	0.07	1.5	3.3	1.9	0.82	1.81	3.61	2.96
Last 6 weeks of gestation or last 8 weeks of lactation suckling singles[e]										
50	110	0.175(+0.045)	0.39(+0.10)	1.7	3.7	3.3	0.99	2.18	4.36	3.58
60	132	0.180(+0.045)	0.40(+0.10)	1.9	4.2	3.2	1.10	2.42	4.84	3.97
70	154	0.185(+0.045)	0.41(+0.10)	2.1	4.6	3.0	1.22	2.69	5.37	4.40
80	176	0.190(+0.045)	0.42(+0.10)	2.2	4.8	2.8	1.28	2.82	5.63	4.62
First 8 weeks of lactation suckling singles or last 8 weeks of lactation suckling twins[f]										
50	110	-0.025(+0.08)	-0.06(+0.18)	2.1	4.6	4.2	1.36	3.00	5.98	4.90
60	132	-0.025(+0.08)	-0.06(+0.18)	2.3	5.1	3.9	1.50	3.31	6.60	5.41
70	154	-0.025(+0.08)	-0.06(+0.18)	2.5	5.5	3.6	1.63	3.59	7.17	5.88
80	176	-0.025(+0.08)	-0.06(+0.18)	2.6	5.7	3.2	1.69	3.73	7.44	6.10
First 8 weeks of lactation suckling twins										
50	110	-0.06	-0.13	2.4	5.3	4.8	1.56	3.44	6.86	5.63
60	132	-0.06	-0.13	2.6	5.7	4.3	1.69	3.73	7.44	6.10
70	154	-0.06	-0.13	2.8	6.2	4.0	1.82	4.01	8.01	6.57
80	176	-0.06	-0.13	3.0	6.6	3.7	1.95	4.30	8.58	7.04
Replacement lambs and yearlings[g]										
30	66	0.18	0.40	1.3	2.9	4.3	0.81	1.79	3.56	2.92
40	88	0.12	0.26	1.4	3.1	3.5	0.82	1.81	3.61	2.96
50	110	0.08	0.18	1.5	3.3	3.0	0.83	1.83	3.65	2.99
60	132	0.04	0.09	1.5	3.3	2.5	0.82	1.81	3.61	2.96
RAMS										
Replacement lambs and yearlings[g]										
40	88	0.25	0.55	1.8	4.0	4.5	1.17	2.58	5.15	4.22
60	132	0.20	0.44	2.3	5.1	3.8	1.38	3.04	6.07	4.98
80	176	0.15	0.33	2.8	6.2	3.5	1.54	3.40	6.78	5.56
100	220	0.10	0.22	2.8	6.2	2.8	1.54	3.40	6.78	5.56
120	265	0.05	0.11	2.6	5.7	2.2	1.43	3.15	6.29	5.16
LAMBS										
Finishing[h]										
30	66	0.20	0.44	1.3	2.9	4.3	0.83	1.83	3.65	2.99
35	77	0.22	0.48	1.4	3.1	4.0	0.94	2.07	4.14	3.39
40	88	0.25	0.55	1.6	3.5	4.0	1.12	2.47	4.93	4.04
45	99	0.25	0.55	1.7	3.7	3.8	1.19	2.62	5.24	4.30
50	110	0.22	0.48	1.8	4.0	3.6	1.26	2.78	5.54	4.54
55	121	0.20	0.44	1.9	4.2	3.5	1.33	2.93	5.85	4.80
Early weaned[i]										
10	22	0.25	0.55	0.6	1.3	6.0	0.44	0.97	1.94	1.59
20	44	0.275	0.60	1.0	2.2	5.0	0.73	1.61	3.21	2.63
30	66	0.30	0.66	1.4	3.1	4.7	1.02	2.25	4.49	3.68

Source: *Nutrient Requirements of Sheep*, fifth revised edition, National Research Council National Academy of Sciences, Washington, D.C., 1975.

[a] To convert dry matter to an as-fed basis, divide dry matter by percent of dry matter.

[b] 1 kg TDN=4.4 Mcal DE (digestible energy). DE may be converted to ME (Metabolizable energy) by multiplying by 82%.

[c] DP= Digestible Protein

[d] Values are for ewes in moderate condition, not excessively fat or thin. Fat ewes should be fed at the next lower weight, thin ewes at the next higher weight. Once maintenance weight is established, such weight would follow through all production phases.

Total Protein (kg)	(lb)	DP[c] (kg)	(lb)	Grams DP per Mcal DE	Calcium (g)	Phosphorus (g)	Carotene (mg)	Vitamin A IU	Vitamin D IU
0.089	0.20	0.048	0.11	20	3.0	2.8	1.9	1275	278
0.098	0.22	0.053	0.12	20	3.1	2.9	2.2	1530	333
0.107	0.24	0.058	0.13	20	3.2	3.0	2.6	1785	388
0.116	0.26	0.063	0.14	20	3.3	3.1	3.0	2040	444
0.099	0.22	0.054	0.12	20	3.0	2.8	1.9	1275	278
0.117	0.26	0.064	0.14	20	3.1	2.9	2.2	1530	333
0.126	0.28	0.069	0.15	20	3.2	3.0	2.6	1785	388
0.135	0.30	0.074	0.16	20	3.3	3.1	3.0	2040	444
0.158	0.35	0.088	0.19	20	4.1	3.9	6.2	4250	278
0.177	0.39	0.099	0.22	20	4.4	4.1	7.5	5100	333
0.195	0.43	0.109	0.24	20	4.5	4.3	8.8	5950	388
0.205	0.45	0.114	0.25	20	4.8	4.5	10.0	6800	444
0.218	0.48	0.130	0.29	22	10.9	7.8	6.2	4250	278
0.239	0.53	0.143	0.32	22	11.5	8.2	7.5	5100	333
0.260	0.57	0.155	0.34	22	12.0	8.6	8.8	5950	388
0.270	0.60	0.161	0.35	22	12.6	9.0	10.0	6800	444
0.276	0.61	0.173	0.38	25	12.5	8.9	6.2	4250	278
0.299	0.66	0.187	0.41	25	13.0	9.4	7.5	5100	333
0.322	0.71	0.202	0.44	25	13.4	9.5	8.8	5950	388
0.345	0.76	0.216	0.48	25	14.4	10.2	10.0	6800	444
0.130	0.29	0.075	0.16	21	5.9	3.3	1.9	1275	166
0.133	0.29	0.074	0.16	20	6.1	3.4	2.5	1700	222
0.133	0.29	0.073	0.16	20	6.3	3.5	3.1	2125	278
0.133	0.29	0.072	0.16	20	6.5	3.6	3.8	2550	333
0.184	0.41	0.108	0.24	21	6.3	3.5	2.5	1700	222
0.219	0.48	0.122	0.27	20	7.2	4.0	3.8	2550	333
0.249	0.55	0.134	0.30	20	7.9	4.4	5.0	3400	444
0.249	0.55	0.134	0.30	20	8.3	4.6	6.2	4250	555
0.231	0.51	0.125	0.28	20	8.5	4.7	7.5	5100	666
0.143	0.32	0.087	0.19	24	4.8	3.0	1.1	765	166
0.154	0.34	0.094	0.21	23	4.8	3.0	1.3	892	194
0.176	0.39	0.107	0.24	22	5.0	3.1	1.5	1020	222
0.187	0.41	0.114	0.25	22	5.0	3.1	1.7	1148	250
0.198	0.44	0.121	0:27	22	5.0	3.1	1.9	1275	278
0.209	0.46	0.127	0.28	22	5.0	3.1	2.1	1402	305
0.096	0.21	0.069	0.15	36	2.4	1.6	1.2	850	67
0.160	0.35	0.115	0.25	36	3.6	2.4	2.5	1700	133
0.196	0.43	0.133	0.29	30	5.0	3.3	3.8	2550	200

[e] Values in parentheses are for ewes suckling singles last 8 weeks of lactation.

[f] Values in parentheses are for ewes suckling twins last 8 weeks of lactation.

[g] Requirements for replacement lambs (ewe and ram) start when the lambs are weaned.

[h] Maximum gains expected. If lambs are held for later market, they should be fed as replacement ewe lambs are fed. Lambs capable of gaining faster than indicated should be fed at a higher level. Lambs finish at the maximum rate if they are self-fed.

[i] A 40-kg (88-pound) early weaned lamb should be fed the same as a finishing lamb of the same weight.

Table 8 Nutrient Content of Diets for Sheep
(Nutrient Concentration in Diet Dry Matter)

Body Weight (kg)	(lb)	Daily Gain or Loss (kg)	(lb)	Daily Dry Matter Per Animal (kg)	(lb)	% Live-weight	TDN (%)	Energy DE[b] (Mcal/kg)	(Mcal/lb)
EWES[d]									
Maintenance									
50	110	0.01	0.02	1.0	2.2	2.0	55	2.4	1.09
60	132	0.01	0.02	1.1	2.4	1.8	55	2.4	1.09
70	154	0.01	0.02	1.2	2.6	1.7	55	2.4	1.09
80	176	0.01	0.02	1.3	2.9	1.6	55	2.4	1.09
Nonlactating and first 15 weeks of gestation									
50	110	0.03	0.07	1.1	2.4	2.2	55	2.4	1.09
60	132	0.03	0.07	1.3	2.9	2.1	55	2.4	1.09
70	154	0.03	0.07	1.4	3.1	2.0	55	2.4	1.09
80	176	0.03	0.07	1.5	3.3	1.9	55	2.4	1.09
Last 6 weeks of gestation or last 8 weeks of lactation suckling singles[e]									
50	110	0.175(+0.045)	0.39(+0.1)	1.7	3.7	3.3	58	2.6	1.18
60	132	0.180(+0.045)	0.40(+0.1)	1.9	4.2	3.2	58	2.6	1.18
70	154	0.185(+0.045)	0.41(+0.1)	2.1	4.6	3.0	58	2.6	1.18
80	176	0.190(+0.045)	0.42(+0.1)	2.2	4.8	2.8	58	2.6	1.18
First 8 weeks of lactation suckling singles or last 8 weeks of lactation suckling twins[f]									
50	110	−0.025(+0.08)	−0.06(+0.18)	2.1	4.6	4.2	65	2.9	1.32
60	132	−0.025(+0.08)	−0.06(+0.18)	2.3	5.1	3.9	65	2.9	1.32
70	154	−0.025(+0.08)	−0.06(+0.18)	2.5	5.5	3.6	65	2.9	1.32
80	176	−0.025(+0.08)	−0.06(+0.18)	2.6	5.7	3.2	65	2.9	1.32
First 8 weeks of lactation suckling twins									
50	110	0.06	0.13	2.4	5.3	4.8	65	2.9	1.32
60	132	0.06	0.13	2.6	5.7	4.3	65	2.9	1.32
70	154	0.06	0.13	2.8	6.2	4.0	65	2.9	1.32
80	176	0.06	0.13	3.0	6.6	3.7	65	2.9	1.32
Replacement lambs and yearlings[g]									
30	66	0.18	0.40	1.3	2.9	4.3	62	2.7	1.22
40	88	0.12	0.26	1.4	3.1	3.5	60	2.6	1.18
50	110	0.08	0.18	1.5	3.3	3.0	55	2.4	1.09
60	132	0.04	0.09	1.5	3.3	2.5	55	2.4	1.09
RAMS									
Replacement lambs and yearlings[g]									
40	88	0.25	0.55	1.8	4.0	4.5	65	2.9	1.32
60	132	0.20	0.44	2.3	5.1	3.8	60	2.6	1.18
80	176	0.15	0.33	2.8	6.2	3.5	55	2.4	1.09
100	220	0.10	0.22	2.8	6.2	2.8	55	2.4	1.09
120	265	0.05	0.11	2.6	5.7	2.2	55	2.4	1.09
LAMBS									
Finishing[h]									
30	66	0.20	0.44	1.3	2.9	4.3	64	2.8	1.27
35	77	0.22	0.48	1.4	3.1	4.0	67	3.0	1.36
40	88	0.25	0.55	1.6	3.5	4.0	70	3.1	1.41
45	99	0.25	0.55	1.7	3.7	3.8	70	3.1	1.41
50	110	0.22	0.48	1.8	4.0	3.6	70	3.1	1.41
55	121	0.20	0.44	1.9	4.2	3.5	70	3.1	1.41
Early weaned[i]									
10	22	0.25	0.55	0.6	1.3	6.0	73	3.2	1.45
20	44	0.275	0.60	1.0	2.2	5.0	73	3.2	1.45
30	66	0.30	0.66	1.4	3.1	4.7	73	3.2	1.45

Source: *Nutrient Requirements of Sheep*, fifth revised edition, National Research Council, National Academy of Sciences, Washington, D.C., 1975.

[a] To convert dry matter to an as fed basis, divide dry matter by percentage of dry matter.

[b] 1 kg TDN=4.4 Mcal DE (digestible energy). DE may be converted to ME (Metabolizable energy) by multiplying by 82%. Because of rounding errors, calculations between Table 7 and Table 8 may not give the same values.

[c] DP=Digestible Protein.

Energy ME (Mcal/kg)	(Mcal/lb)	Total Protein (%)	DP[c] (%)	Ca (%)	P (%)	Carotene (mg/kg)	(mg/lb)	Vitamin A (IU/kg)	(IU/lb)	Vitamin D (IU/kg)	(IU/lb)
2.0	0.91	8.9	4.8	0.30	0.28	1.9	0.86	1275	578	278	126
2.0	0.91	8.9	4.8	0.28	0.26	2.0	0.91	1391	631	303	137
2.0	0.91	8.9	4.8	0.27	0.25	2.2	1.00	1488	675	323	146
2.0	0.91	8.9	4.8	0.25	0.24	2.3	1.04	1569	712	342	155
2.0	0.91	9.0	4.9	0.27	0.25	1.7	0.77	1159	526	253	115
2.0	0.91	9.0	4.9	0.24	0.22	1.7	0.77	1177	534	256	116
2.0	0.91	9.0	4.9	0.23	0.21	1.9	0.86	1275	578	277	126
2.0	0.91	9.0	4.9	0.22	0.26	2.0	0.91	1360	617	296	134
2.1	0.95	9.3	5.2	0.24	0.23	3.6	1.63	2500	1134	164	74
2.1	0.95	9.3	5.2	0.23	0.22	3.9	1.77	2684	1217	175	79
2.1	0.95	9.3	5.2	0.21	0.20	4.2	1.90	2833	1285	185	84
2.1	0.95	9.3	5.2	0.21	0.20	4.5	2.04	3091	1402	202	92
2.4	1.09	10.4	6.2	0.52	0.37	3.0	1.36	2024	918	132	60
2.4	1.09	10.4	6.2	0.50	0.36	3.3	1.50	2217	1006	145	66
2.4	1.09	10.4	6.2	0.48	0.32	3.5	1.59	2380	1080	155	70
2.4	1.09	10.4	6.2	0.48	0.34	3.8	1.72	2615	1186	171	78
2.4	1.09	11.5	7.2	0.52	0.37	2.6	1.18	1771	803	116	53
2.4	1.09	11.5	7.2	0.50	0.36	2.9	1.32	1962	890	128	58
2.4	1.09	11.5	7.2	0.48	0.34	3.1	1.41	2125	964	139	63
2.4	1.09	11.5	7.2	0.48	0.34	3.3	1.50	2267	1028	148	67
2.2	1.00	10.0	5.8	0.45	0.25	1.5	0.54	981	445	128	58
2.1	0.95	9.5	5.3	0.44	0.24	1.8	0.82	1214	551	159	72
2.0	0.91	8.9	4.8	0.42	0.23	2.1	0.95	1417	643	185	84
2.0	0.91	8.9	4.8	0.43	0.24	2.5	1.1	1700	771	222	101
2.4	1.09	10.2	6.0	0.35	0.19	1.4	0.64	944	428	123	56
2.1	0.95	9.5	5.3	0.31	0.17	1.7	0.77	1109	503	145	66
2.0	0.91	8.9	4.8	0.28	0.16	1.8	0.82	1214	551	159	72
2.0	0.91	8.9	4.8	0.30	0.17	2.2	1.00	1518	689	198	90
2.0	0.91	8.9	4.8	0.33	0.18	2.9	1.32	1962	890	256	116
2.3	1.04	11.0	6.7	0.37	0.23	0.8	0.36	588	267	128	58
2.4	1.09	11.0	6.7	0.34	0.21	0.9	0.41	637	289	139	63
2.5	1.13	11.0	6.7	0.31	0.19	0.9	0.41	638	289	139	63
2.5	1.13	11.0	6.7	0.29	0.18	1.0	0.45	675	306	147	67
2.5	1.13	11.0	6.7	0.28	0.17	1.1	0.50	708	321	154	70
2.5	1.13	11.0	6.7	0.26	0.16	1.1	0.50	738	335	161	73
2.6	1.18	16	11.5	0.40	0.27	2.0	0.91	1417	643	112	51
2.6	1.18	16	11.5	0.36	0.24	2.5	1.13	1700	771	133	60
2.6	1.18	14	9.5	0.36	0.24	2.7	1.22	1821	826	143	65

d Values are for ewes in moderate condition, not excessively fat or thin. Fat ewes should be fed at the next lower weight, thin ewes at the next higher weight. Once maintenance weight is established, such weight would follow through all production phases.

e Values in parentheses are for ewes suckling singles last 8 weeks of lactation.

f Values in parentheses are for ewes suckling twins last 8 weeks of lactation.

g Requirements for replacement lambs (ewe and ram) start when the lambs are weaned.

h Maximum gains expected. If lambs are held for later market, they should be fed as replacement ewe lambs are fed. Lambs capable of gaining faster than indicated should be fed at a higher level. Lambs finish at the maximum rate if they are self-fed.

i A 40-kg (88-pound) early weaned lamb should be fed the same as a finishing lamb of the same weight.

Table 9 Daily Nutrient Requirements of Goats

Body Weight (kg)	Feed Energy TDN (g)	DE (Mcal)	ME (Mcal)	NE (Mcal)	Crude Protein TP (g)	DP (g)	Ca (g)	P (g)	Vita-min A (1000 IU)	Vita-min D IU	Dry Matter per Animal 1 kg = 2.0 Mcal ME Total (kg)	% of kg BW	1 kg = 2.4 Mcal ME Total (kg)	% of kg BW
Maintenance only (includes stable feeding conditions, minimal activity, and early pregnancy)														
10	159	0.70	0.57	0.32	22	15	1	0.7	0.4	84	0.28	2.8	0.24	2.4
20	267	1.18	0.96	0.54	38	26	1	0.7	0.7	144	0.48	2.4	0.40	2.0
30	362	1.59	1.30	0.73	51	35	2	1.4	0.9	195	0.65	2.2	0.54	1.8
40	448	1.98	1.61	0.91	63	43	2	1.4	1.2	243	0.81	2.0	0.67	1.7
50	530	2.34	1.91	1.08	75	51	3	2.1	1.4	285	0.95	1.9	0.79	1.6
60	608	2.68	2.19	1.23	86	59	3	2.1	1.6	327	1.09	1.8	0.91	1.5
70	682	3.01	2.45	1.38	96	66	4	2.8	1.8	369	1.23	1.8	1.02	1.5
80	754	3.32	2.71	1.53	106	73	4	2.8	2.0	408	1.36	1.7	1.13	1.4
90	824	3.63	2.96	1.67	116	80	4	2.8	2.2	444	1.48	1.6	1.23	1.4
100	891	3.93	3.21	1.81	126	86	5	3.5	2.4	480	1.60	1.6	1.34	1.3
Maintenance plus low activity (= 25% increment, intensive management, tropical range and early pregnancy)														
10	199	0.87	0.71	0.40	27	19	1	0.7	0.5	108	0.36	3.6	0.30	3.0
20	334	1.47	1.20	0.68	46	32	2	1.4	0.9	180	0.60	3.0	0.50	2.5
30	452	1.99	1.62	0.92	62	43	2	1.4	1.2	243	0.81	2.7	0.67	2.2
40	560	2.47	2.02	1.14	77	54	3	2.1	1.5	303	1.01	2.5	0.84	2.1
50	662	2.92	2.38	1.34	91	63	4	2.8	1.8	357	1.19	2.4	0.99	2.0
60	760	3.35	2.73	1.54	105	73	4	2.8	2.0	408	1.36	2.3	1.14	1.9
70	852	3.76	3.07	1.73	118	82	5	3.5	2.3	462	1.54	2.2	1.28	1.8
80	942	4.16	3.39	1.91	130	90	5	3.5	2.6	510	1.70	2.1	1.41	1.8
90	1030	4.54	3.70	2.09	142	99	6	4.2	2.8	555	1.85	2.1	1.54	1.7
100	1114	4.91	4.01	2.26	153	107	6	4.2	3.0	600	2.00	2.0	1.67	1.7
Maintenance plus medium activity (= 50% increment, semiarid rangeland, slightly hilly pastures, and early pregnancy)														
10	239	1.05	0.86	0.48	33	23	1	0.7	0.6	129	0.43	4.3	0.36	3.6
20	400	1.77	1.44	0.81	55	38	2	1.4	1.1	216	0.72	3.6	0.60	3.0
30	543	2.38	1.95	1.10	74	52	3	2.1	1.5	294	0.98	3.3	0.81	2.7
40	672	2.97	2.42	1.36	93	64	4	2.8	1.8	363	1.21	3.0	1.01	2.5
50	795	3.51	2.86	1.62	110	76	4	2.8	2.1	429	1.43	2.9	1.19	2.4
60	912	4.02	3.28	1.84	126	87	5	3.5	2.5	492	1.64	2.7	1.37	2.3
70	1023	4.52	3.68	2.07	141	98	6	4.2	2.8	552	1.84	2.6	1.53	2.2
80	1131	4.98	4.06	2.30	156	108	6	4.2	3.0	609	2.03	2.5	1.69	2.1
90	1236	5.44	4.44	2.50	170	118	7	4.9	3.3	666	2.22	2.5	1.85	2.0
100	1336	5.90	4.82	2.72	184	128	7	4.9	3.6	723	2.41	2.4	2.01	2.0

Body Weight (kg)	TDN (g)	Feed Energy DE (Mcal)	ME (Mcal)	NE (Mcal)	Crude Protein TP (g)	DP (g)	Ca (g)	P (g)	Vita-min A (1000 IU)	Vita-min D IU	Dry Matter per Animal 1 kg = 2.0 Mcal ME Total (kg)	% of kg BW	1 kg = 2.4 Mcal ME Total (kg)	% of kg BW
Maintenance plus high activity (= 75% increment, arid rangeland, sparse vegetation, mountainous pastures, and early pregnancy)														
10	278	1.22	1.00	0.56	38	26	2	1.4	0.8	150	0.50	5.0	0.42	4.2
20	467	2.06	1.68	0.94	64	45	2	1.4	1.3	252	0.84	4.2	0.70	3.5
30	634	2.78	2.28	1.28	87	60	3	2.1	1.7	342	1.14	3.8	0.95	3.2
40	784	3.46	2.82	1.59	108	75	4	2.8	2.1	423	1.41	3.5	1.18	3.0
50	928	4.10	3.34	1.89	128	89	5	3.5	2.5	501	1.67	3.3	1.39	2.7
60	1064	4.69	3.83	2.15	146	102	6	4.2	2.9	576	1.92	3.2	1.60	2.7
70	1194	5.27	4.29	2.42	165	114	6	4.2	3.2	642	2.14	3.0	1.79	2.6
80	1320	5.81	4.74	2.68	182	126	7	4.9	3.6	711	2.37	3.0	1.98	2.5
90	1442	6.35	5.18	2.92	198	138	8	5.6	3.9	777	2.59	2.9	2.16	2.4
100	1559	6.88	5.62	3.17	215	150	8	5.6	4.2	843	2.81	2.8	2.34	2.3
Additional requirements for late pregnancy (for all goat sizes)														
	397	1.74	1.42	0.80	82	57	2	1.4	1.1	213	0.71		0.59	
Additional requirements for growth - weight gain at 50 g per day (for all goat sizes)														
	100	0.44	0.36	0.20	14	10	1	0.7	0.3	54	0.18		0.15	
Additional requirements for growth - weight gain at 100 g per day (for all goat sizes)														
	200	0.88	0.72	0.40	28	20	1	0.7	0.5	108	0.36		0.30	
Additional requirements for growth - weight gain at 150 g per day (for all goat sizes)														
	300	1.32	1.08	0.60	42	30	2	1.4	0.8	162	0.54		0.45	
Additional requirements for milk production per kg at different fat percentages (including requirements for nursing single, twin or triplet kids at the respective milk production level)														
(% fat)														
2.5	333	1.47	1.20	0.68	59	42	2	1.4	3.8	760				
3.0	337	1.49	1.21	0.68	64	45	2	1.4	3.8	760				
3.5	342	1.51	1.23	0.69	68	48	2	1.4	3.8	760				
4.0	346	1.53	1.25	0.70	72	51	3	2.1	3.8	760				
4.5	351	1.55	1.26	0.71	77	54	3	2.1	3.8	760				
5.0	356	1.57	1.28	0.72	82	57	3	2.1	3.8	760				
5.5	360	1.59	1.29	0.73	86	60	3	2.1	3.8	760				
6.0	365	1.61	1.31	0.74	90	63	3	2.1	3.8	760				
Additional requirements for mohair production by Angora at different production levels														
Annual Fleece Yield (kg)														
2	16	0.07	0.06	0.03	9	6								
4	34	0.15	0.12	0.07	17	12								
6	50	0.22	0.18	0.10	26	18								
8	66	0.29	0.24	0.14	34	24								

Source: *Nutrient Requirements of Goats: Angora, Dairy, and Meat Goats in Temperate and Tropical Countries*, National Research Council, National Academy of Sciences, Washington, D.C. 1981.

Table 10 Nutrient Requirements of Growing-Finishing Swine Fed *Ad Libitum* (Percent or Amount per Kilogram of Diet)[a]

Liveweight (kg)		1-5[a]	5-10	10-20	20-35	35-60	60-100
Expected Daily Gain (g)		200	300	500	600	700	800
Expected Efficiency (g gain/kg feed)		800	600	500	400	350	270
Expected Efficiency (feed/gain)		1.25	1.67	2.00	2.50	2.86	3.75
Digestible energy[b]	kcal	3,700	3,500	3,370	3,380	3,390	3,395
Metabolizable energy[b]	kcal	3,600	3,400	3,160	3,175	3,190	3,195
Crude Protein[c]	%	27	20	18	16	14	13
Indispensable amino acids							
Lysine	%	1.28	0.95	0.79	0.70	0.61	0.57
Arginine	%	0.33	0.25	0.23	0.20	0.18	0.16
Histidine	%	0.31	0.23	0.20	0.18	0.16	0.15
Isoleucine	%	0.85	0.63	0.56	0.50	0.44	0.41
Leucine	%	1.01	0.75	0.68	0.60	0.52	0.48
Methionine + cystine[d]	%	0.76	0.56	0.51	0.45	0.40	0.30
Phenylalanine + tyrosine[e]	%	1.18	0.88	0.79	0.70	0.61	0.57
Threonine	%	0.76	0.56	0.51	0.45	0.39	0.37
Tryptophan[f]	%	0.20	0.15	0.13	0.12	0.11	0.10
Valine	%	0.85	0.63	0.56	0.50	0.44	0.41
Mineral elements							
Calcium	%	0.90	0.80	0.65	0.60	0.55	0.50
Phosphorus[g]	%	0.70	0.60	0.55	0.50	0.45	0.40
Sodium	%	0.10	0.10	0.10	0.10	0.10	0.10
Chlorine	%	0.13	0.13	0.13	0.13	0.13	0.13
Potassium	%	0.30	0.26	0.26	0.23	0.20	0.17
Magnesium	%	0.04	0.04	0.04	0.04	0.04	0.04
Iron	mg	150	140	80	60	50	40
Zinc	mg	100	100	80	60	50	50
Managanese	mg	4.0	4.0	3.0	2.0	2.0	2.0
Copper	mg	6.0	6.0	5.0	4.0	3.0	3.0
Iodine	mg	0.14	0.14	0.14	0.14	0.14	0.14
Selenium	mg	0.15	0.15	0.15	0.15	0.15	0.10
Vitamins							
Vitamin A	IU	2,200	2,200	1,750	1,300	1,300	1,300
or Beta-carotene	mg	8.8	8.8	7.0	5.2	5.2	5.2
Vitamin D	IU	220	220	200	200	150	125
Vitamin E	IU	11	11	11	11	11	11
Vitamin K (menadione)	mg	2.0	2.0	2.0	2.0	2.0	2.0
Riboflavin	mg	3.0	3.0	3.0	2.6	2.2	2.2
Niacin[h]	mg	22	22	18	14	12	10
Pantothenic acid	mg	13	13	11	11	11	11
Vitamin B_{12}	micro-g	22	22	15	11	11	11
Choline[i]	mg	1,100	1,100	900	700	550	400
Thiamin	mg	1.3	1.3	1.1	1.1	1.1	1.1
Vitamin B_6	mg	1.5	1.5	1.5	1.1	1.1	1.1
Biotin[j]	mg	0.10	0.10	0.10	0.10	0.10	0.10
Folacin[j]	mg	0.60	0.60	0.60	60	0.60	0.60

Source: *Nutrient Requirements of Swine*, eighth revised edition, National Research Council, National Academy of Sciences, Washington, D.C. 1979.

[a]Requirements reflect the estimated levels of each nutrient needed for optimal performance when a fortified grain-soybean meal diet is fed, except that a substantial level of milk products should be included in the diet of the 1-5 kg pig. Concentrations are based upon amounts per unit of air-dry diet (i.e., 90 percent dry matter).

[b]These are not absolute requirements but are suggested energy levels derived from diets containing corn and soybean meal (44 percent crude protein). When lower energy grains are fed, these energy levels will not be met, consequently, feed efficiency would be lowered.

[c]Approximate protein levels required to meet the need for indispensable amino acids when a fortified grain-soybean meal diet is fed to pigs weighing more than 5 kg.

[d]Methionine can fulfill the total requirement; cystine can meet at least 50 percent of the total requirement.

[e]Phenylalanine can fulfill the total requirement; tyrosine can meet at least 50 percent of the total requirement.

[f]It is assumed that usable tryptophan content of corn does not exceed 0.05 percent.

[g]At least 30 percent of the phosphorus requirement should be provided by inorganic and/or animal product sources.

[h]It is assumed that most of the niacin present in cereal grains and their by-products is in bound form and thus unavailable to swine. The niacin contributed by these sources is not included in the requirement listed. In excess of its requirement for protein synthesis, tryptophan can be converted to niacin (50 mg tryptophan yields 1 mg niacin).

[i]In excess of its requirement for protein synthesis, methionine can spare dietary choline (4.3 mg methionine is equal in methylating capacity to 1 mg choline).

[j]These levels are suggested. No requirements have been established.

Table 11 Daily Nutrient Requirements of Growing-Finishing Swine Fed *Ad Libitum*[a]

Liveweight (kg) Air Dry Feed Intake (g)		1-5[a] 250	5-10 500	10-20 1,000	20-35 1,500	35-60 2,000	60-100 3,000
Digestible energy[b]	kcal	925	1,750	3,370	5,055	6,740	10,110
Metabolizable energy[b]	kcal	900	1,700	3,160	4,740	6,320	9,480
Crude Protein[c]	g	67.5	100	180	240	280	390
Indispensable amino acids							
Lysine	g	3.2	4.8	7.9	10.5	12.2	17.1
Arginine	g	0.8	1.3	2.3	3.0	3.6	4.8
Histidine	g	0.8	1.2	2.0	2.7	3.2	4.5
Isoleucine	g	2.1	3.2	5.6	7.5	8.8	12.3
Leucine	g	2.5	3.8	6.8	9.0	10.4	14.4
Methionine + cystine[d]	g	1.9	2.8	5.1	6.8	8.0	9.0
Phenylalanine + tyrosine[e]	g	3.0	4.4	7.9	10.5	12.2	17.1
Threonine	g	1.9	2.8	5.1	6.8	7.8	11.1
Tryptophan[f]	g	0.5	0.8	1.3	1.8	2.2	3.0
Valine	g	2.1	3.2	5.6	7.5	8.8	12.3
Mineral elements							
Calcium	g	2.3	4.0	6.5	9.0	11.0	15.0
Phosphorus[g]	g	1.8	3.0	5.5	7.5	9.0	12.0
Sodium	g	0.25	0.5	1.0	1.5	2.0	3.0
Chlorine	g	0.33	0.7	1.3	2.0	2.6	3.9
Potassium	g	0.75	1.3	2.6	3.5	4.0	5.1
Magnesium	g	0.10	0.2	0.4	0.6	0.8	1.2
Iron	mg	38	70	80	90	100	120
Zinc	mg	25	50	80	90	100	150
Managanese	mg	1.0	2	3	3	4	6
Copper	mg	1.5	3	5	6	6	9
Iodine	mg	0.04	0.07	0.14	0.21	0.28	0.42
Selenium	mg	0.04	0.08	0.15	0.22	0.30	0.30
Vitamins							
Vitamin A	IU	550	1,100	1,750	1,950	2,600	3,900
or Beta-carotene	mg	2.2	4.4	7.0	7.8	10.4	15.6
Vitamin D	IU	55	110	200	300	300	375
Vitamin E	IU	2.8	5.5	11	17	22	33
Vitamin K (menadione)	mg	0.50	1.1	2.2	3.3	4.4	6
Riboflavin	mg	0.75	1.5	3.0	3.9	4.4	7
Niacin[h]	mg	5.5	11	18	21	24	30
Pantothenic acid	mg	3.3	6.5	11	17	22	33
Vitamin B$_{12}$	micro-g	5.5	11	15	17	22	33
Choline[i]	mg	275	550	900	1,050	1,100	1,200
Thiamin	mg	0.33	0.65	1.1	1.7	2.2	3.3
Vitamin B$_6$	mg	0.38	0.75	1.5	1.7	2.2	3.3
Biotin[j]	mg	0.03	0.05	0.10	0.15	0.20	0.30
Folacin[j]	mg	0.15	0.30	0.60	0.90	1.2	1.8

Source: *Nutrient Requirements of Swine*, eighth revised edition, National Research Council, National Academy of Sciences, Washington, D.C. 1979.

[a]Requirements reflect the estimated levels of each nutrient needed for optimal performance when a fortified grain-soybean meal diet is fed, except that a substantial level of milk products should be included in the diet of the 1-5 kg pig. Concentrations are based upon amounts per unit of air-dry diet (i.e., 90 percent dry matter).

[b]These are not absolute requirements but are suggested energy levels derived from diets containing corn and soybean meal (44 percent crude protein). When lower energy grains are fed, these energy levels will not be met, consequently, feed efficiency would be lowered.

[c]Approximate protein levels required to meet the need for indispensable amino acids when a fortified grain-soybean meal diet is fed to pigs weighing more than 5 kg.

[d]Methionine can fulfill the total requirement; cystine can meet at least 50 percent of the total requirement.

[e]Phenylalanine can fulfill the total requirement; tyrosine can meet at least 50 percent of the total requirement.

[f]It is assumed that usable tryptophan content of corn does not exceed 0.05 percent.

[g]At least 30 percent of the phosphorus requirement should be provided by inorganic and/or animal product sources.

[h]It is assumed that most of the niacin present in cereal grains and their by-products is in bound form and thus unavailable to swine. The niacin contributed by these sources is not included in the requirement listed. In excess of its requirement for protein synthesis, tryptophan can be converted to niacin (50 mg tryptophan yields 1 mg niacin).

[i]In excess of its requirement for protein synthesis, methionine can spare dietary choline (4.3 mg methionine is equal in methylating capacity to 1 mg choline).

[j]These levels are suggested. No requirements have been established.

Table 12 Nutrient Requirements of Breeding Swine: Percent or Amount per Kilogram of Diet[a]

		Bred Gilts and Sows; Young and Adult Boars[b]	Lactating Gilts and Sows
Digestible energy	kcal	3,400	3,395
Metabolizable energy	kcal	3,200	3,195
Crude Protein[c]	%	12	13
Indispensable amino acids			
Arginine	%	0	0.40
Histidine	%	0.15	0.25
Isoleucine	%	0.37	0.39
Leucine	%	0.42	0.70
Lysine	%	0.43	0.58
Methionine + cystine[d]	%	0.23	0.36
Phenylalanine + tyrosine[e]	%	0.52	0.85
Threonine	%	0.34	0.43
Tryptophan[f]	%	0.09	0.12
Valine	%	0.46	0.55
Mineral elements			
Calcium	%	0.75	0.75
Phosphorus[g]	%	0.60	0.50
Sodium	%	0.15	0.20
Chlorine	%	0.25	0.30
Potassium	%	0.20	0.20
Magnesium	%	0.04	0.04
Iron	mg	80	80
Zinc	mg	50	50
Managanese	mg	10	10
Copper	mg	5	5
Iodine	mg	0.14	0.14
Selenium	mg	0.15	0.15
Vitamins			
Vitamin A	IU	4,000	2,000
or Beta-carotene	mg	16	8
Vitamin D	IU	200	200
Vitamin E	IU	10	10
Vitamin K (menadione)	mg	2	2
Riboflavin	mg	3	3
Niacin[h]	mg	10	10
Pantothenic acid	mg	12	12
Vitamin B_{12}	micro-g	15	15
Choline	mg	1,250	1,250
Thiamin	mg	1	1
Vitamin B_6	mg	1	1
Biotin[i]	mg	0.1	0.1
Folacin[i]	mg	0.6	0.6

Source: *Nutrient Requirements of Swine*, eighth revised edition, National Research Council, National Academy of Sciences, Washington, D.C. 1979.

[a]Requirements reflect the estimated levels of each nutrient needed for optimal performance when a fortified grain-soybean meal diet is fed. Concentrations are based upon amounts per unit of air-dry diet (i.e., 90 percent dry matter).

[b]Requirements for boars of breeding age have not been established. It is suggested that the requirements will not differ significantly from that of bred gilts and sows.

[c]Approximate protein levels required to meet the need for indispensable amino acids when a fortified grain-soybean meal diet is fed. The true digestibilities of the amino acids were assumed to be 90 percent.

[d]Methionine can fulfill the total requirement; cystine can meet at least 50 percent of the total requirement.

[e]Phenylalanine can fulfill the total requirement; tyrosine can meet at least 50 percent of the total requirement.

[f]It is assumed that usable tryptophan content of corn does not exceed 0.05 percent.

[g]At least 30 percent of the phosphorus requirement should be provided by inorganic and/or animal product sources.

[h]It is assumed that most of the niacin present in cereal grains and their by-products is in bound form and thus unavailable to swine. The niacin contributed by these sources is not included in the requirement listed. In excess of its requirement for protein synthesis, tryptophan can be converted to niacin (50 mg tryptophan yields 1 mg niacin).

[i]These levels are suggested. No requirements have been established.

Table 13 Daily Nutrient Requirements of Breeding Swine

Air Dry Feed Intake (g)		Bred Gilts and Sows; Young and Adult Boars 1,800[b]	Lactating Gilts and Sows 4,000	4,750	5,500
Digestible energy	kcal	6,120[c]	13,580	16,130	18,670
Metabolizable energy	kcal	5,760[c]	12,780	15,180	17,570
Crude Protein	g	216	520	618	715
Indispensable amino acids					
Arginine	g	0	16.0	19.0	22.0
Histidine	g	2.7	10.0	11.9	13.8
Isoleucine	g	6.7	15.6	18.5	21.4
Leucine	g	7.6	28.0	33.2	38.5
Lysine	g	7.7	23.2	27.6	31.9
Methionine + cystine[d]	g	4.1	14.4	17.1	19.8
Phenylalanine + tyrosine[e]	g	9.4	34.0	40.4	46.8
Threonine	g	6.1	17.2	20.4	23.6
Tryptophan[f]	g	1.6	4.8	5.7	6.6
Valine	g	8.3	22.0	26.1	30.2
Mineral elements					
Calcium	g	13.5	30.0	35.6	41.2
Phosphorus[g]	g	10.8	20.0	23.8	27.5
Sodium	g	2.7	8.0	9.5	11.0
Chlorine	g	4.5	12.0	14.2	16.5
Potassium	g	3.6	8.0	9.5	11.0
Magnesium	g	0.7	1.6	1.9	2.2
Iron	mg	144	320	380	440
Zinc	mg	90	200	238	275
Managanese	mg	18	40	48	55
Copper	mg	9	20	24	28
Iodine	mg	0.25	0.56	0.66	0.77
Selenium	mg	0.27	0.40	0.48	0.55
Vitamins					
Vitamin A	IU	7,200	8,000	9,500	11,000
or Beta-carotene	mg	28.8	32.0	38.0	44.0
Vitamin D	IU	360	800	950	1,100
Vitamin E	IU	18.0	40.0	47.5	55.0
Vitamin K (menadione)	mg	3.6	8.0	9.5	11.0
Riboflavin	mg	5.4	12.0	14.2	16.5
Niacin[h]	mg	18	40	47.5	55.0
Pantothenic acid	mg	21.6	48.0	57.0	66.0
Vitamin B_{12}	micro-g	27.0	60.0	71.2	82.5
Choline	mg	2,250	5,000	5,940	6,875
Thiamin	mg	1.8	4.0	4.8	5.5
Vitamin B_6	mg	1.8	4.0	4.8	5.5
Biotin[i]	mg	0.18	0.4	0.48	0.55
Folacin[i]	mg	1.08	2.4	2.8	3.3

Source: *Nutrient Requirements of Swine*, eighth revised edition, National Research Council, National Academy of Sciences, Washington, D.C. 1979.

[a]Requirements reflect the estimated levels of each nutrient needed for optimal performance when a fortified grain-soybean meal diet is fed. Concentrations are based upon amounts per unit of air-dry diet (i.e., 90 percent dry matter).

[b]An additional 25 percent should be fed to working boars.

[c]Individual feeding and moderate climatic conditions are assumed. An energy reduction of about 10 percent is possible when gilts and sows are tethered or individually penned in a stall in onvironmentally controlled housing. An energy increase of about 25 percent is suggested for cold climatic (winter) conditions.

[d]Methionine can fulfill the total requirement; cystine can meet at least 50 percent of the total requirement.

[e]Phenylalanine can fulfill the total requirement; tyrosine can meet at least 50 percent of the total requirement.

[f]It is assumed that usable tryptophan content of corn does not exceed 0.05 percent.

[g]At least 30 percent of the phosphorus requirement should be provided by inorganic and/or animal product sources.

[h]It is assumed that most of the niacin present in cereal grains and their by-products is in bound form and thus unavailable to swine. The niacin contributed by these sources is not included in the requirement listed. In excess of its requirement for protein synthesis, tryptophan can be converted to niacin (50 mg tryptophan yields 1 mg niacin).

[i]These levels are suggested. No requirements have been established.

Table 14 Nutrient Requirements of Horses (Daily Nutrients per Horse), Ponies, 200 kg (440 lb) Mature Weight

	Weight (kg)	Weight (lb)	Daily Gain (kg)	Daily Gain (lb)	Digestible Energy (Mcal)	TDN (kg)	TDN (lb)	Crude Protein (kg)	Crude Protein (lb)	Digestible Protein (kg)	Digestible Protein (lb)	Calcium (g)	Phosphorus (g)	Vitamin A Activity (1,000 IU)	Daily Feed[a] (kg)	Daily Feed[a] (lb)
Mature ponies, maintenance	200	440	0.0		8.24	1.87	4.12	0.32	0.70	0.14	0.31	9	6	5.0	3.75	8.2
Mares, last 90 days gestation			0.27	0.594	9.23	2.10	4.62	0.39	0.86	0.20	0.44	14	9	10.0	3.70	8.1
Lactating mare, first 3 months (8 kg milk per day)			0.0		14.58	3.31	7.29	0.71	1.56	0.54	1.19	24	16	13.0	5.20	11.5
Lactating mare, 3 months to weanling (6 kg milk per day)			0.0		12.99	2.95	6.50	0.60	1.32	0.34	0.75	20	13	11.0	5.00	11.0
Nursing foal (3 months of age)	60	132	0.70	1.54	7.35	1.67	3.68	0.41	0.90	0.38	0.84	18	11	2.4	2.25	5.0
Requirements above milk					3.74	0.85	1.87	0.17	0.37	0.20	0.44	10	7	0.0	1.20	2.7
Weanling (6 months of age)	95	209	0.50	1.10	8.80	2.0	4.40	0.47	1.03	0.31	0.68	19	14	3.8	2.85	6.3
Yearling (12 months of age)	140	308	0.20	0.44	8.15	1.85	4.07	0.35	0.77	0.20	0.44	12	9	5.5	2.90	6.4
Long yearling (18 months of age)	170	374	0.10	0.22	8.10	1.84	4.05	0.32	0.70	0.17	0.37	11	7	6.0	3.10	6.8
Two year old (24 months of age)	185	407	0.05	0.11	8.10	1.84	4.05	0.30	0.66	0.15	0.33	10	7	5.5	3.10	6.8

Source: *Nutrient Requirements of Horses*, fourth revised edition, National Research Council, National Academy of Sciences, Washington, D.C., 1978.

a Dry matter basis

Table 15 Nutrient Requirements of Horses (Daily Nutrients per Horse), 400 kg (880 lb) Mature Weight

	Weight (kg)	Weight (lb)	Daily Gain (kg)	Daily Gain (lb)	Digestible Energy (Mcal)	TDN (kg)	TDN (lb)	Crude Protein (kg)	Crude Protein (lb)	Digestible Protein (kg)	Digestible Protein (lb)	Calcium (g)	Phosphorus (g)	Vitamin A Activity (1,000 IU)	Daily Feed[a] (kg)	Daily Feed[a] (lb)
Mature horses, maintenance	400	880	0.0		13.86	3.15	6.93	0.54	1.19	0.24	0.53	18	11	10.0	6.30	13.9
Mares, last 90 days gestation			0.53	1.17	15.52	3.53	7.76	0.64	1.41	0.34	0.75	27	19	20.0	6.20	13.7
Lactating mare, first 3 months (12 kg milk per day)			0.0		23.36	5.31	11.68	1.12	2.46	0.68	1.50	40	27	22.0	8.35	18.4
Lactating mare, 3 months to weanling (8 kg milk per day)			0.0		20.20	4.59	10.10	0.91	2.00	0.51	1.12	33	22	18.0	7.75	17.1
Nursing foal, (3 months of age)	125	275	1.00	2.2	11.51	2.62	5.76	0.65	1.43	0.50	1.10	27	17	5.0	3.55	7.8
Requirements above milk					6.10	1.39	3.05	0.40	0.88	0.30	0.66	15	12	0.0	1.95	4.3
Weanling (6 months of age)	185	407	0.65	1.43	13.03	2.96	6.51	0.66	1.45	0.43	0.95	27	20	7.4	4.20	9.2
Yearling (12 months of age)	265	583	0.40	0.88	13.80	3.14	6.91	0.60	1.32	0.35	0.77	24	17	10.0	4.95	10.9
Long yearling (18 months of age)	330	726	0.25	0.55	14.36	3.26	7.17	0.59	1.30	0.32	0.70	22	15	11.5	5.50	12.2
Two year old (24 months of age)	365	803	0.10	0.22	13.89	3.16	6.95	0.52	1.14	0.27	0.59	20	13	11.0	5.35	11.8

Source: *Nutrient Requirements of Horses*, fourth revised edition, National Research Council, National Academy of Sciences, Washington, D.C., 1978.

a Dry matter basis

Table 16 Nutrient Requirements of Horses (Daily Nutrients per Horse), 500 kg (1,100 lb) Mature Weight

	Weight (kg)	(lb)	Daily Gain (kg)	(lb)	Digestible Energy (Mcal)	TDN (kg)	(lb)	Crude Protein (kg)	(lb)	Digestible Protein (kg)	(lb)	Cal-cium (g)	Phos-phorus (g)	Vita-min A Activity (1,000 IU)	Daily Feed[a] (kg)	(lb)
Mature horses, maintenance	500	1,100	0.0	0.0	16.39	3.73	8.20	0.63	1.39	0.29	0.64	23	14	12.5	7.45	16.4
Mares, last 90 days gestation			0.55	1.21	18.36	4.17	9.18	0.75	1.65	0.39	0.86	34	23	25.0	7.35	16.2
Lactating mare, first 3 months (15 kg milk per day)					28.27	6.43	14.14	1.36	2.99	0.84	1.85	50	34	27.5	10.10	22.2
Lactating mare, 3 months to weaning (10 kg milk per day)				0.0	24.31	5.53	12.16	1.10	2.42	0.62	1.36	41	27	22.5	9.35	20.6
Nursing foal (3 months of age)	155	341	1.20	2.64	13.66	3.10	6.83	0.75	1.65	0.54	1.19	33	20	6.2	4.20	9.2
Requirements above milk					6.89	1.57	3.45	0.41	0.90	0.31	0.68	18	13	0.0	2.25	4.9
Weanling (6 months of age)	230	506	0.80	1.76	15.60	3.55	7.80	0.79	1.74	0.52	1.14	34	25	9.2	5.00	11.0
Yearling (12 months of age)	325	715	0.55	1.21	16.81	3.82	8.41	0.76	1.67	0.45	0.99	31	22	12.0	6.00	13.2
Long yearling (18 months of age)	400	880	0.35	0.77	17.00	3.90	8.58	0.71	1.56	0.39	0.86	28	19	14.0	6.50	14.3
Two year old (24 months of age)	450	990	0.15	0.33	16.45	3.74	8.23	0.63	1.39	0.33	0.72	25	17	13.0	6.60	14.5

Source: *Nutrient Requirements of Horses*, fourth revised edition, National Research Council, National Academy of Sciences, Washington, D.C., 1978.

a Dry matter basis

Table 17 Nutrient Requirements of Horses (Daily Nutrients per Horse), 600 kg (1,320 lb) Mature Weight

	Weight (kg)	Weight (lb)	Daily Gain (kg)	Daily Gain (lb)	Digestible Energy (Mcal)	TDN (kg)	TDN (lb)	Crude Protein (kg)	Crude Protein (lb)	Digestible Protein (kg)	Digestible Protein (lb)	Calcium (g)	Phosphorus (g)	Vitamin A Activity (1,000 IU)	Daily Feed[a] (kg)	Daily Feed[a] (lb)
Mature horses, maintenance	600	1,320	0.0		18.79	4.27	9.40	0.73	1.61	0.33	0.73	27	17	15.0	8.50	18.8
Mares, last 90 days gestation			0.67	1.47	21.04	4.78	10.52	0.87	1.91	0.46	1.01	40	27	30.0	8.40	18.5
Lactating mare, first 3 months (18 kg milk per day)			0.0		33.05	7.51	16.53	1.60	3.52	0.99	2.18	60	40	33.0	11.80	26.0
Lactating mare, 3 months to weanling (12 kg milk per day)			0.0		28.29	6.43	14.15	1.29	2.84	0.73	1.61	49	30	27.0	10.90	23.9
Nursing foal (3 months of age)	170	374	1.40	3.08	15.05	3.42	7.53	0.84	1.85	0.78	1.72	36	23	6.8	4.65	10.2
Requirements above milk					6.93	1.58	3.47	0.51	1.12	0.38	0.84	18	15	0.0	2.25	4.9
Weanling (6 months of age)	265	583	0.85	1.87	16.92	3.85	8.47	0.86	1.89	0.57	1.25	37	27	10.6	5.45	12.0
Yearling (12 months of age)	385	847	0.60	1.32	18.85	4.28	9.42	0.90	1.98	0.50	1.10	35	25	14.0	6.75	14.8
Long yearling (18 months of age)	475	1,045	0.35	0.77	19.06	4.33	9.53	0.75	1.65	0.43	0.95	32	22	13.5	7.35	16.2
Two year old (24 months of age)	540	1,188	0.20	0.44	19.26	4.38	9.64	0.74	1.63	0.39	0.86	31	20	13.0	7.40	16.3

Source: *Nutrient Requirements of Horses*, fourth revised edition, National Research Council, National Academy of Sciences, Washington, D.C., 1978.

a Dry matter basis

Table 18 Nutrient Concentrations in Diets for Horses and Ponies Expressed on 100 percent Dry Matter Basis[a]

	Digestible Energy		Example Diet Proportions				Crude Protein (%)	Calcium (%)	Phosphorus (%)	Vitamin A Activity	
			Hay containing 2.2 Mcal/kg (1.0 Mcal/lb)		Hay Containing 2.0 Mcal/kg (0.9 Mcal/lb)						
	(Mcal/kg)	(Mcal/lb)	Concentrate[b]	Roughage	Concentrate[b]	Roughage				(IU/kg)	(IU/lb)
Mature horses and ponies at maintenance	2.2	1.0	0	100	10	95	8.5	0.30	0.20	1,600	725
Mares, last 90 days of gestation	2.5	1.1	25	75	35	65	11.0	0.50	0.35	3,400	1,550
Lactating mare, first 3 months	2.8	1.3	45	55	55	45	14.0	0.50	0.35	2,800	1,275
Lactating mare, 3 months to weanling	2.6	1.2	30	70	40	60	12.0	0.45	0.30	2,450	1,150
Creep feed	3.5	1.6	100	0	100	0	18.0	0.85	0.60		
Foal (3 months of age)	3.25	1.5	75	25	80	20	18.0	0.85	0.60	2,000	900
Weanling (6 months of age)	3.1	1.4	65	35	70	30	16.0	0.70	0.50	2,000	900
Yearling (12 months of age)	2.8	1.3	45	55	55	45	13.5	0.55	0.40	2,000	900
Long yearling (18 months of age)	2.6	1.2	30	70	40	60	11.0	0.45	0.35	2,000	900
Two year old (light training)	2.6	1.3	30	70	40	60	10.0	0.45	0.35	2,000	900
Mature working horses (light work[c])	2.5	1.1	25	75	35	65	8.5	0.30	0.20	1,600	725
(moderate work[d])	2.9	1.3	50	50	60	40	8.5	0.30	0.20	1,600	725
(intense work[e])	3.1	1.4	65	35	70	30	8.5	0.30	0.20	1,600	725

Source: *Nutrient Requirements of Horses*, fourth revised edition, National Research Council, National Academy of Sciences, Washington, D.C., 1978.

[a] Values are rounded to account for differences among Tables 14-17 and for greater practical application.

[b] Concentrate containing 3.6 Mcal/kg (1.6 Mcal/lb).

[c] Examples are horses used in western pleasure, bridle path hack, equitation, etc.

[d] Examples are ranch work, roping, cutting, barrel racing, jumping, etc.

[e] Examples are race training, polo, etc.

Table 19 Nutrient Concentration in Diets for Horses and Ponies Expressed on 90 Percent Dry Matter Basis[a]

	Digestible Energy		Crude Protein (%)	Calcium (%)	Phosphorus (%)	Vitamin A Activity	
	(Mcal/kg)	(Mcal/lb)				(IU/kg)	(IU/lb)
Mature horses and ponies at maintenance	2.0	0.9	7.7	0.27	0.18	1,450	650
Mares, last 90 days of gestation	2.25	1.0	10.0	0.45	0.30	3,000	1,400
Lactating mare, first 3 months	2.6	1.2	12.5	0.45	0.30	2,500	1,150
Lactating mare, 3 months to weanling	2.3	1.1	11.0	0.40	0.25	2,200	1,000
Creep feed	3.15	1.4	16.0	0.80	0.55		
Foal (3 months of age)	2.9	1.35	16.0	0.80	0.55	1,800	800
Weanling (6 months of age)	2.8	1.25	14.5	0.60	0.45	1,800	800
Yearling (12 months of age)	2.6	1.2	12.0	0.50	0.35	1,800	800
Long yearling (18 months of age)	2.3	1.1	10.0	0.40	0.30	1,800	800
Two year old (light training)	2.6	1.2	9.0	0.40	0.30	1,800	800
Mature working horses (light work[b])	2.25	1.0	7.7	0.27	0.18	1,450	650
(moderate work[c])	2.6	1.2	7.7	0.27	0.18	1,450	650
(intense work[d])	2.8	1.25	7.7	0.27	0.18	1,450	650

Source: *Nutrient Requirements of Horses,* fourth revised edition, National Research Council, National Academy of Sciences, Washington, D.C., 1978.

a Values are rounded to account for differences among Tables 14-17 and for greater practical application.

b Examples are horses used in western pleasure, bridle path hack, equitation, etc.

c Examples are ranch work, roping, cutting, barrel racing, jumping, etc.

d Examples are race training, polo, etc.

Table 20 Daily Nutrient Requirements per Female
(Single-Comb White Leghorns and Similar Breeds)

| Body Weight-Growing Chicken | | Total Daily Feed | | Crude Protein | | AMINO ACIDS | | | | | |
| | | | | | | Methionine | | Methionine and Cystine | | Lysine | |
(kg)	(lb)	(kg)	(lb)	(kg)	(lb)	(g)	(lb)	(g)	(lb)	(g)	(lb)
0.25	0.55	0.021	0.046	0.0038	0.0084	0.07	0.00015	0.12	0.00026	0.18	0.00040
0.50	1.10	0.043	0.095	0.0064	0.0141	0.12	0.00026	0.22	0.00048	0.26	0.00057
0.75	1.65	0.052	0.115	0.0078	0.0172	0.14	0.00031	0.26	0.00057	0.31	0.00068
1.00	2.20	0.060	0.132	0.0090	0.0198	0.16	0.00035	0.30	0.00066	0.36	0.00079
1.25	2.76	0.071	0.156	0.0085	0.0187	0.15	0.00033	0.28	0.00062	0.32	0.00070
1.50	3.31	0.078	0.172	0.0094	0.0207	0.16	0.00035	0.31	0.00068	0.35	0.00077
Mature Hen											
Maintenance											
1.8	3.97	0.070	0.154	?	?	?	?	?	?	?	?
Laying[a]											
1.8	3.97	0.110	0.242	0.0165	0.0364	0.30	0.66	0.55	0.00121	0.66	0.00146
Breeding[a]											
1.8	3.97	0.110	0.242	0.0165	0.0364	0.30	0.66	0.55	0.00121	0.66	0.00146

| Body Weight-Growing Chicken | | MINERALS | | VITAMINS | | | |
| | | Manganese | Iodine | | | | |
(kg)	(lb)	(mg)	(mg)	Vitamin A (IU)	Vitamin D (ICU[b])	Thiamine (mg)	Riboflavin (mg)
0.25	0.55	1.2	0.01	32	4.2	0.04	0.08
0.50	1.10	2.4	0.02	65	8.6	0.08	0.16
0.75	1.65	1.3	0.02	78	10.4	0.07	0.10
1.00	2.20	1.5	0.02	90	12.0	0.08	0.11
1.25	2.76	1.8	0.03	106	14.2	0.09	0.13
1.50	3.31	2.0	0.03	117	15.6	0.10	0.14
Mature Hen							
Maintenance							
1.8	3.97	?	?	?	?	?	?
Laying[a]							
1.8	3.97	2.75	0.03	440	55	?	0.24
Breeding[a]							
1.8	3.97	3.6	0.03	440	55	0.09	0.42

Source: *Nutrient Requirements of Poultry*, seventh revised edition, National Research Council, National Academy of Sciences, Washington, D.C., 1973.

[a] Sixty-five percent production.

[b] ICU — International Chick Unit.

MINERALS

Calcium		Phosphorus		Sodium		Potassium		Magnesium
(g)	(lb)	(g)	(lb)	(g)	(lb)	(g)	(lb)	(mg)
0.19	0.00042	0.15	0.00033	0.03	0.00007	0.04	0.00009	12.6
0.36	0.00079	0.30	0.00066	0.07	0.00015	0.09	0.00020	25.8
0.31	0.00068	0.21	0.00046	0.08	0.00018	0.08	0.00018	20.8
0.36	0.00079	0.24	0.00053	0.09	0.00020	0.10	0.00022	24.0
0.43	0.00095	0.28	0.00062	0.11	0.00024	0.11	0.00024	28.4
0.47	0.00104	0.31	0.00068	0.12	0.00026	0.13	0.00029	31.2
?	?	?	?	?	?	?	?	?
3.6	0.0079	0.55	0.00121	0.16	0.00035	0.11	0.00024	55
3.0	0.0066	0.55	0.00121	0.16	0.00035	0.11	0.00024	55

VITAMINS

Pantothenic Acid (mg)	Niacin (mg)	Pyridoxine (mg)	Biotin (mg)	Choline (mg)	Folacin (mg)	Vitamin B$_{12}$ (mg)
0.21	0.57	0.06	0.003	27	0.012	0.0002
0.43	1.16	0.13	0.006	56	0.024	0.0004
0.52	0.57	0.16	0.005	26	0.013	0.0002
0.60	0.66	0.18	0.006	30	0.015	0.0002
0.71	0.78	0.21	0.007	36	0.017	0.0002
0.78	0.86	0.23	0.008	39	0.020	0.0002
?	?	?	?	?	?	?
0.24	1.1	0.33	?	?	0.03	?
1.10	1.1	0.49	0.024	?	0.04	0.0003

**Table 21 Daily Nutrient Requirements per Chicken
(Chickens of Broiler Strains)**

Body Weight-Growing Chicken		Total Daily Feed		Crude Protein		AMINO ACIDS					
						Methionine		Methionine and Cystine		Lysine	
(kg)	(lb)	(kg)	(lb)	(kg)	(lb)	(g)	(lb)	(g)	(lb)	(g)	(lb)
0.25	0.55	0.028	0.06	0.006	0.013	0.14	0.00031	0.26	0.00057	0.34	0.00075
0.50	1.10	0.051	0.11	0.012	0.026	0.25	0.00055	0.47	0.00104	0.61	0.00134
0.75	1.65	0.073	0.16	0.015	0.033	0.28	0.00062	0.53	0.00117	0.73	0.00161
1.00	2.20	0.098	0.22	0.020	0.044	0.37	0.00082	0.70	0.00154	0.98	0.00216
1.50	3.31	0.113	0.25	0.020	0.044	0.36	0.00079	0.68	0.00150	0.96	0.00212
2.00	4.41	0.120	0.26	0.022	0.048	0.38	0.00084	0.72	0.00159	1.02	0.00225
Mature Chicken											
Maintenance											
2.5	5.51	0.087	0.19	?	?	?	?	?	?	?	?
Laying[a]											
2.5	5.51	0.135	0.30	0.020	0.044	0.36	0.00079	0.68	0.00150	0.81	0.00179
Breeding[a]											
2.5	5.51	0.135	0.30	0.020	0.044	0.36	0.00079	0.68	0.00150	0.81	0.00179

Body Weight-Growing Chicken		MINERALS		VITAMINS			
		Manganese	Iodine	Vitamin A	Vitamin D	Thiamine	Riboflavin
(kg)	(lb)	(mg)	(mg)	(IU)	(ICU)	(mg)	(mg)
0.25	0.55	1.5	0.01	42	5.6	0.05	0.10
0.50	1.10	2.8	0.02	76	10	0.09	0.18
0.75	1.65	4.0	0.02	110	15	0.13	0.26
1.00	2.20	5.4	0.03	150	20	0.18	0.36
1.50	3.31	6.1	0.04	170	23	0.20	0.41
2.00	4.41	3.0	0.04	180	24	0.16	0.22
Mature Chicken							
Maintenance							
2.5	5.51	?	?	?	?	?	?
Laying[a]							
2.5	5.51	3.4	0.04	540	68	?	0.30
Breeding[a]							
2.5	5.51	4.4	0.04	540	68	0.11	0.51

Source: *Nutrient Requirements of Poultry,* seventh revised edition, National Research Council, National Academy of Sciences, Washington, D.C., 1977.

[a] Sixty percent production.

MINERALS

Calcium (g)	Calcium (lb)	Phosphorus (g)	Phosphorus (lb)	Sodium (g)	Sodium (lb)	Potassium (g)	Potassium (lb)	Magnesium (mg)
0.25	0.00055	0.19	0.00042	0.04	0.00009	0.06	0.00013	17
0.46	0.00101	0.36	0.00079	0.07	0.00015	0.10	0.00022	31
0.66	0.00146	0.51	0.00112	0.11	0.00024	0.14	0.00031	44
0.87	0.00192	0.68	0.00150	0.14	0.00031	0.19	0.00042	59
1.01	0.00223	0.79	0.00174	0.19	0.00042	0.23	0.00051	67
0.71	0.00156	0.48	0.00106	0.18	0.00040	0.20	0.00044	48
?	?	?	?	?	?	?	?	?
4.4	0.00970	0.68	0.00150	0.20	0.00044	0.13	0.00029	68
3.7	0.00816	0.68	0.00150	0.20	0.00044	0.13	0.00029	68

VITAMINS

Pantothenic Acid (mg)	Niacin (mg)	Pyridoxine (mg)	Biotin (mg)	Choline (mg)	Folacin (mg)	Vitamin B_{12} (mg)	Approximate Age (weeks) (Male & Female)
0.28	0.62	0.08	0.004	41	0.015	0.0003	2.1
0.51	1.37	0.15	0.008	66	0.028	0.0005	3.2
0.73	1.97	0.22	0.011	95	0.040	0.0007	4.3
0.98	2.64	0.29	0.014	129	0.054	0.0009	5.3
1.13	3.08	0.34	0.017	128	0.061	0.0010	6.6
1.20	1.30	0.36	0.012	60	0.063	0.0004	8.2
?	?	?	?	?	?	?	____
0.30	1.35	0.40	?	?	0.03	?	____
1.35	1.35	0.40	0.024	?	0.05	0.0004	____

Table 22 Protein and Amino Acid Requirements of Egg-type and Meat-type Chickens

Nutrient	Replacement Pullets[a] (Egg- or meat-type) 0-6 weeks %	g/Mcal	lb/Mcal	6-14 weeks %	g/Mcal	lb/Mcal	14-20 weeks %	g/Mcal	lb/Mcal	Laying and Breeding Hens (Egg- or meat-type) % of diet[b]	Daily intake per hen[c] mg	lb
Protein	18	—	—	15	—	—	12	—	—	15	16,500	0.0364
Arginine	1.00	3.45	0.0076	0.83	2.86	0.0063	0.67	2.31	0.0051	0.8	880	0.0019
Glycine + Serine	0.70	2.42	0.0053	0.58	2.00	0.0044	0.47	1.62	0.0036	0.5[d]	550[d]	0.0012[d]
Histidine	0.26	0.90	0.0020	0.22	0.76	0.0017	0.17	0.59	0.0013	0.22[d]	240[d]	0.0005[d]
Isoleucine	0.60	2.07	0.0060	0.50	1.73	0.0038	0.40	1.38	0.0030	0.5	550	0.0012
Leucine	1.00	3.45	0.0076	0.83	2.86	0.0063	0.67	2.31	0.0051	1.2	1,320	0.0029
Lysine	0.85	2.93	0.0065	0.60	2.07	0.0046	0.45	1.55	0.0034	0.60[e]	660[e]	0.0015[e]
Methionine + Cystine	0.60	2.07	0.0046	0.50	1.73	0.0038	0.40	1.38	0.0030	0.50	550	0.0012
Methionine	0.32	1.10	0.0024	0.27	0.93	0.0020	0.21	0.72	0.0016	0.27	300	0.0007
Phenylalanine + Tyrosine	1.00	3.45	0.0076	0.83	2.86	0.0063	0.67	2.31	0.0051	0.8[d]	880[d]	0.0019[d]
Phenylalanine	0.54	1.86	0.0041	0.45	1.55	0.0034	0.36	1.24	0.0027	0.4[d]	440[d]	0.0010[d]
Threonine	0.56	1.93	0.0042	0.47	1.62	0.0036	0.37	1.28	0.0028	0.4	440	0.0010
Tryptophan	0.17	0.59	0.0013	0.14	0.48	0.0011	0.11	0.38	0.0008	0.11	120	0.0003
Valine	0.62	2.14	0.0047	0.52	1.79	0.0040	0.41	1.41	0.0031	0.5[d]	550[d]	0.0012[d]

Source: *Nutrient Requirements of Poultry*, seventh revised edition, National Research Council, National Academy of Sciences, Washington, D.C., 1977.

a The requirements shown are for reasonable growth and development at 20 weeks of age when rations are fed on an ad libitum basis. Most of the values are not determined but are based on values for young birds and extrapolated for the older birds. Values under percent are based on 2,900 kcal of metabolizable energy per kg (1,315 kcal of ME/lb).

b Based on 2,850 kcal of metabolizable energy/kg diet (1,293 kcal ME/lb).

c Based on a feed consumption of 110 g (0.24 lb) per day.

d Estimated values.

e Determined with corn-soybean meal diets. The requirements may be higher on wheat-based diets.

Table 23 Protein and Amino Acid Requirements of Broilers

	Broilers								
	0-3 weeks			3-6 weeks			6-9 weeks		
Nutrient	%ᵃ	g/Mcal	lb/Mcal	%ᵃ	g/Mcal	lb/Mcal	%ᵃ	g/Mcal	lb/Mcal
Protein	23	—	—	20	—	—	18	—	—
Arginine	1.44	4.50	0.0099	1.20	3.75	0.0083	1.00	3.13	0.0069
Glycine + Serine	1.50	4.69	0.0103	1.00	3.13	0.0069	0.70	2.19	0.0048
Histidine	0.35	1.09	0.0024	0.30	0.94	0.0021	0.26	0.81	0.0018
Isoleucine	0.80	2.50	0.0055	0.70	2.19	0.0048	0.60	1.88	0.0041
Leucine	1.35	4.22	0.0093	1.18	3.69	0.0081	1.00	3.13	0.0069
Lysine	1.20	3.75	0.0083	1.00	3.13	0.0069	0.85	2.66	0.0059
Methionine + Cystine	0.93	2.90	0.0064	0.72	2.25	0.0050	0.60	1.88	0.0041
Methionine	0.50	1.56	0.0034	0.38	1.19	0.0026	0.32	1.00	0.0022
Phenylalanine + Tyrosine	1.34	4.19	0.0092	1.17	3.66	0.0081	1.00	3.13	0.0069
Phenylalanine	0.72	2.25	0.0050	0.63	1.97	0.0043	0.54	1.69	0.0037
Threonine	0.75	2.34	0.0052	0.65	2.03	0.0045	0.56	1.75	0.0039
Tryptophan	0.23	0.72	0.0016	0.20	0.63	0.0014	0.17	0.53	0.0012
Valine	0.82	2.56	0.0056	0.72	2.25	0.0050	0.62	1.94	0.0043

Source: *Nutrient Requirements of Poultry*, seventh revised edition, National Research Council, National Academy of Sciences, Washington, D.C., 1977.

a Requirement for diets containing 3,200 kcal metabolizable energy per kg (1,452 kcal/lb).

Table 24 Vitamin, Linoleic Acid, and Mineral Requirements of Chickens
(In percentage or amount per kilogram or pound of feed — as-fed moisture basis)

Nutrient	Starting chickens (0-8 weeks)			Growing chickens (8-18 weeks)			Laying hens			Breeding hens		
	per kg	per lb	%	per kg	per lb	%	per kg	per lb	%	per kg	per lb	%
Vitamin A activity (IU)	1,500	680	—	1,500	680	—	4,000	1,814	—	4,000	1,814	—
Vitamin D[a] (ICU)	200	91	—	200	91	—	500	227	—	500	227	—
Vitamin E (IU)	10	4.5	—	5[b]	2.3	—	5[b]	2.3	—	10[b]	4.5	—
Vitamin K$_1$ or equivalent activity (mg)	0.5	0.23	—	0.5[b]	0.23	—	0.5[b]	0.23	—	0.5[b]	0.23	—
Thiamine (mg)	1.8	0.82	—	1.3[b]	0.59	—	0.8[b]	0.36	—	0.8	0.36	—
Riboflavin (mg)	3.6	1.6	—	1.8	0.82	—	2.2	1.00	—	3.8	1.7	—
Pantothenic acid (mg)	10	4.5	—	10	4.5	—	2.2	1.00	—	10	4.5	—
Niacin (mg)	27	12	—	11	5.0	—	10[b]	4.5	—	10[b]	4.5	—
Pyridoxine (mg)	3	1.4	—	3[b]	1.4	—	3	1.4	—	4.5	2.04	—
Biotin (mg)	0.15	0.07	—	0.10[b]	0.04	—	0.10[b]	0.04	—	0.15	0.07	—
Choline (mg)	1,300	590	—	500[b]	227	—	500[b]	227	—	500[b]	227	—
Folacin[c] (mg)	0.55	0.25	—	0.25[b]	0.11	—	0.25	0.11	—	0.35	0.16	—
Vitamin B$_{12}$ (mg)	0.009	0.004	—	0.003[b]	0.001	—	0.003[b]	0.001	—	0.003	0.001	—
Linoleic Acid (%)	—	—	1.0	—	—	0.8[b]	—	—	1.0	—	—	1.0
Calcium (%)	—	—	0.9	—	—	0.6	—	—	3.25	—	—	2.75
Phosphorus (%)	—	—	0.7	—	—	0.4	—	—	0.5	—	—	0.5
Potassium (%)	—	—	0.2	—	—	0.16	—	—	0.1	—	—	0.1
Sodium (%)	—	—	0.15	—	—	0.15	—	—	0.15	—	—	0.15
Chlorine (mg)	800[b]	363	—	800[b]	363	—	800[b]	363	—	800[b]	363	—
Copper (mg)	4	1.8	—	3[b]	1.4	—	3[b]	1.4	—	4[b]	1.8	—
Iodine (mg)	0.35	0.16	—	0.35	0.16	—	0.3	0.14	—	0.3	0.14	—
Iron (mg)	80	36	—	40[b]	18	—	50[b]	23	—	80[b]	36	—
Magnesium (mg)	600	272	—	400[b]	181	—	500	227	—	500[b]	227	—
Manganese (mg)	55	25	—	25[b]	11	—	25[b]	11	—	33	15	—
Selenium (mg)	0.1	0.04	—	0.1[b]	0.04	—	0.1[b]	0.04	—	0.1[b]	0.04	—
Zinc (mg)	40	18	—	35[b]	15.9	—	50[b]	23	—	65[b]	29.5	—

Source: *Nutrient Requirements of Poultry*, seventh revised edition, National Research Council, National Academy of Sciences, Washington, D.C., 1977.

a These levels of vitamin D are satisfactory when levels of calcium and readily available phosphorus conform to this table.

b Estimated value.

c With sucrose, diet requirement is 1.2 mg/kg (0.5 mg/lb).

Table 25 Protein and Amino Acid Requirements of Turkeys

Nutrient			Age (weeks)				Holding	Breeding Hens
Male:	0-4	4-8	8-12	12-16	16-20	20-24		
Female:	0-4	4-8	8-11	11-14	14-17	17-20		
Metabolizable energy[a]								
kcal/kg	2,800	2,900	3,000	3,100	3,200	3,300	2,900	2,900
kcal/lb	1,270	1,315	1,361	1,406	1,452	1,497	1,315	1,315
Protein (%)	28	26	22	19	16.5	14	12	14
Arginine (%)	1.6	1.5[b]	1.25[b]	1.1[b]	0.95[b]	0.8[b]	0.6[b]	0.6[b]
Glycine + Serine (%)	1.0	0.9[b]	0.8[b]	0.7[b]	0.6[b]	0.5[b]	0.4[b]	0.5[b]
Histidine (%)	0.58	0.54[b]	0.46[b]	0.39[b]	0.35[b]	0.29[b]	0.25[b]	0.3[b]
Isoleucine (%)	1.1	1.0[b]	0.85[b]	0.75[b]	0.65[b]	0.55[b]	0.45[b]	0.5[b]
Leucine (%)	1.9	1.75[b]	1.5[b]	1.3[b]	1.1[b]	0.95[b]	0.5[b]	0.5[b]
Lysine (%)	1.7	1.6	1.35	1.0	0.80	0.65	0.5[b]	0.6[b]
Methionine + Cystine (%)	1.05	0.90	0.75	0.65	0.55[b]	0.45[b]	0.4[b]	0.4[b]
Methionine (%)	0.53	0.45	0.38	0.33	0.28	0.23	0.2[b]	0.2[b]
Phenylalanine + Tyrosine (%)	1.8	1.65[b]	1.4[b]	1.2[b]	1.05[b]	0.9[b]	0.8[b]	1.0[b]
Phenylalanine (%)	1.0	0.9[b]	0.8[b]	0.7[b]	0.6[b]	0.5[b]	0.4[b]	0.55[b]
Threonine (%)	1.0	0.93[b]	0.79[b]	0.68[b]	0.59[b]	0.5[b]	0.4[b]	0.45[b]
Tryptophan (%)	0.26	0.24[b]	0.20[b]	0.18[b]	0.15[b]	0.13[b]	0.10[b]	0.13[b]
Valine (%)	1.2	1.1[b]	0.94[b]	0.8[b]	0.7[b]	0.6[b]	0.5[b]	0.58[b]

Source: *Nutrient Requirements of Poultry*, seventh revised edition, National Research Council, National Academy of Sciences, Washington, D.C., 1977.

a Requirement for diets containing the stated levels of metabolizable energy.

b Estimated value.

Table 26 Vitamin, Linoleic Acid, and Mineral Requirements of Turkeys
(In percentage or amount per kilogram or pound of feed — as-fed moisture basis)

Nutrient	0-8 weeks			8 weeks			Breeding		
	per kg	per lb	%	per kg	per lb	%	per kg	per lb	%
Vitamin A (IU)	4,000	1,814	—	4,000[b]	1,814	—	4,000	1,814	—
Vitamin D[c] (ICU)	900	408	—	900[b]	408[b]	—	900	408[b]	—
Vitamin E (IU)	12[b]	5.4[b]	—	10[b]	4.5[b]	—	25	11	—
Vitamin K[1] or equivalent activity (mg)	1[b]	0.45[b]	—	0.8[b]	0.36[b]	—	1[b]	0.45[b]	—
Thiamine (mg)	2	0.9	—	2	0.9	—	2[b]	0.9[b]	—
Riboflavin (mg)	3.6	1.6	—	3.0[b]	1.4	—	4	1.8	—
Pantothenic acid (mg)	11	5.0	—	9[b]	4.1	—	16	7.3	—
Niacin (mg)	70	32	—	50[b]	23	—	30[b]	14[b]	—
Vitamin B[6] (mg)	4.5	2	—	3.5[b]	1.6[b]	—	4[b]	1.8[b]	—
Biotin[d] (mg)	0.2	0.09	—	0.1	0.04	—	0.15	0.07	—
Choline (mg)	1,900	862	—	1,100[b]	499[b]	—	1,000[b]	454[b]	—
Folacin (mg)	1.0	0.45	—	0.8[b]	0.36[b]	—	1.0	0.45	—
Vitamin B[12] (mg)	0.003	0.001	—	0.003[b]	0.001[b]	—	0.003[b]	0.001[b]	—
Linoleic acid (%)	—	—	1.0	—	—	0.8[b]	—	—	1.0[b]
Calcium (%)	—	—	1.2	—	—	0.8	—	—	2.25
Phosphorus (%)	—	—	0.8	—	—	0.7	—	—	0.7
Potassium (%)	—	—	0.4	—	—	0.4[b]	—	—	0.4[b]
Sodium (%)	—	—	0.15	—	—	0.15	—	—	0.15
Chlorine (mg)	800[b]	363[b]	—	800[b]	363[b]	—	800[b]	363[b]	—
Copper (mg)	6	2.7	—	4	1.8[b]	—	6[b]	2.7[b]	—
Iodine (mg)	0.4[b]	0.18[b]	—	0.4[b]	0.18[b]	—	0.4[b]	0.18[b]	—
Iron (mg)	60	27	—	40[b]	18[b]	—	60[b]	27[b]	—
Magnesium (mg)	500[b]	227[b]	—	500[b]	227[b]	—	500[b]	227[b]	—
Manganese (mg)	55	25	—	25	11	—	35[b]	16[b]	—
Selenium (mg)	0.2	0.09	—	0.2	0.09[b]	—	0.2[b]	0.09[b]	—
Zinc (mg)	75	34	—	40[b]	18[b]	—	65[b]	29.5[b]	—

Source: *Nutrient Requirements of Poultry*, seventh revised edition, National Research Council, National Academy of Sciences, Washington, D.C., 1977.

a From 8 weeks to market or prebreeding.

b Estimated value.

c These levels of vitamin D are satisfactory when levels of calcium and readily available phosphorus conform to this table.

d Requirements may be 50 percent greater for wheat- or barley-based diets.

Table 27 Nutrient Requirements of Ducks
(In percentage or amount per kilogram or pound of feed — as-fed moisture basis)

Nutrient[a]	Starting and Growing Ducks			Breeding Ducks		
	per kg	per lb	%	per kg	per lb	%
Metabolizable energy (kcal)	2,900	1,315	—	2,900	1,315	—
Protein (%)	—	—	16[b]	—	—	15[c]
Lysine (%)	—	—	0.9[c]	—	—	0.7[c]
Methionine + Cystine (%)	—	—	0.8[c]	—	—	0.55[c]
Vitamin A (IU)	4,000	1,814	—	4,000	1,814	—
Vitamin D (ICU)	220	100	—	500[c]	227[c]	—
Riboflavin (mg)	4	1.8	—	4[c]	1.8[c]	—
Pantothenic acid (mg)	11	5	—	10[c]	4.5[c]	—
Niacin (mg)	55	25	—	40[c]	18[c]	—
Pyridoxine (mg)	2.6	1.2	—	3[c]	1.4[c]	—
Calcium (%)	—	—	0.6	—	—	2.75[c]
Phosphorus (%)	—	—	0.6	—	—	0.6[c]
Sodium (%)	—	—	0.15	—	—	0.15[c]
Manganese (mg)	40	18	—	25[c]	11[c]	—
Magnesium (mg)	500[c]	227[c]	—	500[c]	227[c]	—

Source: *Nutrient Requirements of Poultry*, seventh revised edition, National Research Council, National Academy of Sciences, Washington, D.C., 1977.

[a] For nutrients not listed, see requirements for chickens (Table 24) as a guide.

[b] Increasing protein level to 22 percent for the first 2 weeks will increase early growth.

[c] Estimated value.

Table 28 Nutrient Requirements of Pheasants and Quail
(In percentage or amount per kilogram or pound of feed — as-fed moisture basis)

Nutrient[a]	Pheasant					
	Starting			Growing (6-20 weeks)		
	per kg	per lb	%	per kg	per lb	%
Metabolizable energy (kcal)	2,800	1,270	—	2,700	1,225	—
Protein (%)	—	—	30	—	—	16
Lysine (%)	—	—	1.5[c]	—	—	0.8[c]
Methionine + Cystine (%)	—	—	1.0[c]	—	—	0.6[c]
Glycine + Serine (%)	—	—	1.8[c]	—	—	1.0[c]
Vitamin A (IU)	3,000[c]	1,361[c]	—	3,000[c]	1,361[c]	—
Vitamin D (ICU)	1,200[c]	544[c]	—	900[c]	408[c]	—
Riboflavin (mg)	3.5[c]	1.6[c]	—	2.6[c]	1.2[c]	—
Pantothenic acid (mg)	10	4.5	—	10[c]	4.5[c]	—
Niacin (mg)	60	27.2	—	40[c]	18[c]	—
Choline (mg)	1,500[c]	680[c]	—	1,000[c]	454[c]	—
Linoleic acid (%)	—	—	1.0[c]	—	—	1.0[c]
Calcium (%)	—	—	1.0	—	—	0.7[c]
Chlorine (%)	—	—	0.11	—	—	0.11[c]
Phosphorus (%)	—	—	0.8	—	—	0.6[c]
Sodium (%)	—	—	0.1[c]	—	—	0.1[c]
Iodine (mg)	0.3[c]	0.14[c]	—	0.3[c]	0.14[c]	—
Magnesium (mg)	600[c]	272[c]	—	400[c]	181[c]	—
Manganese (mg)	90[c]	41[c]	—	70[c]	32[c]	—
Zinc (mg)	60[c]	27[c]	—	50[c]	23[c]	—

Nutrient[a]	Bobwhite Quail					
	Starting and Growing			Breeding		
	per kg	per lb	%	per kg	per lb	%
Metabolizable energy (kcal)	2,800	1,270	—	2,800	1,270	—
Protein (%)	—	—	28[b]	—	—	24[c]
Lysine (%)	—	—	1.4[c]	—	—	0.7[c]
Methionine + Cystine (%)	—	—	0.9[c]	—	—	0.6[c]
Glycine + Serine (%)	—	—	1.6[c]	—	—	0.9[c]
Vitamin A (IU)	3,000[c]	1,361[c]	—	3,000[c]	1,361[c]	—
Vitamin D (ICU)	900[c]	408[c]	—	900[c]	408[c]	—
Riboflavin (mg)	3.8	1.7	—	4.0[c]	1.8[c]	—
Pantothenic acid (mg)	12.6	5.7	—	15[c]	6.8[c]	—
Niacin (mg)	31	14	—	20[c]	9[c]	—
Choline (mg)	1,500	680	—	1,000[c]	454[c]	—
Linoleic acid (%)	—	—	1.0[c]	—	—	1.0[c]
Calcium (%)	—	—	0.65	—	—	2.3
Chlorine (%)	—	—	0.11	—	—	0.15[c]
Phosphorus (%)	—	—	0.65	—	—	1.0
Sodium (%)	—	—	0.085	—	—	0.15[c]
Iodine (mg)	0.30	0.14	—	0.30[c]	0.14[c]	—
Magnesium (mg)	600[c]	272[c]	—	400[c]	181[c]	—
Manganese (mg)	90[c]	41[c]	—	70[c]	32[c]	—
Zinc (mg)	50[c]	23[c]	—	50[c]	23[c]	—

Nutrient[a]	Japanese Quail					
	Starting and Growing			Breeding		
	per kg	per lb	%	per kg	per lb	%
Metabolizable energy (kcal)	3,000	1,361	—	2,800	1,270	—
Protein (%)	—	—	24[d]	—	—	24[c]
Lysine (%)	—	—	1.4	—	—	1.1[c]
Methionine + Cystine (%)	—	—	0.75	—	—	0.8[c]
Glysine + Serine (%)	—	—	1.7	—	—	0.9[c]
Vitamin A (IU)	5,000[c]	2,268[c]	—	5,000[c]	2,268[c]	—
Vitamin D (ICU)	480[c]	218[c]	—	1,200[c]	544[c]	—
Riboflavin (mg)	4.0[c]	1.8[c]	—	4.0[c]	1.8[c]	—
Pantothenic acid (mg)	10[c]	4.5[c]	—	15[c]	6.8[c]	—
Niacin (mg)	40	18	—	20[c]	9[c]	—
Choline (mg)	2,000	907	—	1,500[c]	680[c]	—
Linoleic acid (%)	—	—	1.0[c]	—	—	1.0[c]
Calcium (%)	—	—	0.8	—	—	2.5
Chlorine (%)	—	—	0.15[c]	—	—	0.15[c]
Phosphorus (%)	—	—	0.65	—	—	0.8
Sodium (%)	—	—	0.15[c]	—	—	0.15[c]
Iodine (mg)	0.30[c]	0.14[c]	—	0.30[c]	0.14[c]	—
Magnesium (mg)	150	68	—	500[c]	227[c]	—
Manganese (mg)	90[c]	41[c]	—	70[c]	32[c]	—
Zinc (mg)	25	11	—	50[c]	23[c]	—

Source: *Nutrient Requirements of Poultry*, seventh revised edition, National Research Council, National Academy of Sciences, Washington, D.C., 1977.

[a] For nutrients not listed see requirements for chickens (Table 24) as a guide.

[b] May be reduced to 20 percent at 6 weeks of age.

[c] Estimated value.

[d] May be reduced to 20 percent at 3 weeks of age.

Table 29 Nutrient Requirements of Geese
(In percentage or amount per kilogram or pound of feed — as-fed moisture basis)

Nutrient[a]	Starting (0-6 weeks)			Growing (after 6 weeks)			Breeding		
	per kg	per lb	%	per kg	per lb	%	per kg	per lb	%
Metabolizable energy (kcal)	2,900	1,315	—	2,900	1,315	—	2,900	1,315	—
Protein (%)	—	—	22	—	—	15[b]	—	—	15[b]
Lysine (%)	—	—	0.9	—	—	0.6[b]	—	—	0.6[b]
Vitamin A (IU)	1,500[b]	680[b]	—	1,500[b]	680[b]	—	4,000[b]	1,814[b]	—
Vitamin D (ICU)	200[b]	91[b]	—	200[b]	91[b]	—	200[b]	91[b]	—
Riboflavin (mg)	4	1.8	—	2.5[b]	1.1[b]	—	4[b]	1.8[b]	—
Niacin, available (mg)	55	25	—	35[b]	16[b]	—	20[b]	9[b]	—
Calcium (%)	—	—	0.8[b]	—	—	0.6[b]	—	—	2.25[b]
Phosphorus (%)	—	—	0.6[b]	—	—	0.4[b]	—	—	0.6[b]

Source: *Nutrient Requirements of Poultry*, seventh revised edition, National Research Council, National Academy of Sciences, Washington, D.C., 1977.

a For nutrients not listed, see requirements for chickens (Table 24) as a guide.

b Estimated value.

Table 30 Feed Composition

Feed Name	International Feed Number	Moisture Basis	Dry Matter (%)	Ruminants TDN (%)	DE (Mcal/kg)	ME (Mcal/kg)	NE$_m$ (Mcal/kg)	NE$_g$ (Mcal/kg)	Dairy Cattle NE$_l$ (Mcal/kg)	Chickens ME$_n$ (kcal/kg)	TME (kcal/kg)	NE$_p$ (kcal/kg)	Horses TDN (%)
ALFALFA *Medicago sativa*													
1 fresh, full bloom	2-00-188	As fed	25.0	14.0	0.61	0.50	0.30	0.12	0.31	—	—	—	—
		Dry	100.0	55.0	2.43	2.00	1.19	0.47	1.23	—	—	—	—
2 hay, sun-cured, late bloom	1-20-681	As fed	90.0	47.0	2.06	1.68	1.01	0.32	1.04	—	—	—	—
		Dry	100.0	52.0	2.29	1.87	1.12	0.36	1.15	—	—	—	—
3 hay, sun-cured,	1-00-071	As fed	91.0	46.0	2.01	1.62	0.98	0.25	1.01	—	—	—	42.0
		Dry	100.0	50.0	2.21	1.78	1.07	0.28	1.11	—	—	—	46.0
4 meal, dehydrated, 17% protein	1-00-023	As fed	92.0	55.0	2.47	2.08	1.22	0.63	1.26	1,504	1,393	770	45.0
		Dry	100.0	61.0	2.69	2.27	1.33	0.69	1.38	1,640	1,519	840	49.0
5 wilted silage, midbloom	3-00-217	As fed	38.0	22.0	0.97	0.81	0.48	0.22	0.50	—	—	—	—
		Dry	100.0	58.0	2.56	2.13	1.26	0.58	1.30	—	—	—	—
6 hay, sun-cured, early bloom (North)	1-00-059	As fed	90.0	54.0	2.38	2.00	1.18	0.59	1.22	—	—	—	46.0
		Dry	100.0	60.0	2.65	2.22	1.31	0.65	1.35	—	—	—	51.0
7 hay, sun-cured, midbloom (North)	1-00-063	As fed	90.0	52.0	2.30	1.92	1.13	0.52	1.17	—	—	—	42.0
		Dry	100.0	58.0	2.56	2.13	1.26	0.58	1.30	—	—	—	47.0
8 hay, sun-cured, early bloom (South)	1-00-059	As fed	90.0	53.0	2.34	1.96	1.16	0.56	1.19	—	—	—	46.0
		Dry	100.0	59.0	2.60	2.18	1.28	0.62	1.33	—	—	—	51.0
9 hay, sun-cured, midbloom (South)	1-00-063	As fed	90.0	51.0	2.26	1.88	1.11	0.49	1.15	—	—	—	42.0
		Dry	100.0	57.0	2.51	2.09	1.23	0.55	1.28	—	—	—	47.0
ANIMAL													
10 by-product, meal rendered	5-08-786	As fed	93.0	68.0	2.99	2.61	1.55	0.98	1.55	2,714	—	—	—
		Dry	100.0	73.0	3.22	2.80	1.67	1.06	1.67	2,918	—	—	—
BAHIAGRASS *Paspalum notatum*													
11 fresh	2-00-464	As fed	30.0	16.0	0.70	0.58	0.34	0.13	0.36	—	—	—	—
		Dry	100.0	54.0	2.38	1.96	1.16	0.43	1.20	—	—	—	—
12 hay, sun-cured	1-00-462	As fed	91.0	46.0	2.05	1.66	1.00	0.29	1.03	—	—	—	40.0
		Dry	100.0	51.0	2.25	1.82	1.10	0.32	1.13	—	—	—	44.0
13 hay, sun-cured, early bloom	1-06-138	As fed	91.0	36.0	1.61	1.21	0.81	—	0.78	—	—	—	—
		Dry	100.0	40.0	1.76	1.33	0.89	—	0.86	—	—	—	—
BAKERY													
14 waste, dehydrated (Dried bakery product)	4-00-466	As fed	92.0	82.0	3.61	3.23	1.98	1.35	1.89	3,862	—	2,879	—
		Dry	100.0	89.0	3.92	3.51	2.15	1.47	2.06	4,203	—	3,133	—
BARLEY *Hordeum vulgare*													
15 grain	4-00-549	As fed	88.0	74.0	3.27	2.90	1.76	1.19	1.71	2,508	3,011	1,803	72.0
		Dry	100.0	84.0	3.70	3.29	2.00	1.35	1.94	2,843	3,413	2,044	82.0
16 hay, sun-cured	1-00-495	As fed	87.0	49.0	2.16	1.79	1.06	0.45	1.09	—	—	—	39.0
		Dry	100.0	56.0	2.47	2.04	1.21	0.51	1.25	—	—	—	44.0
BEET, SUGAR *Beta vulgaris altissima*													
17 pulp, dehydrated	4-00-669	As fed	91.0	67.0	2.96	2.58	1.54	0.98	1.54	646	—	438	—
		Dry	100.0	74.0	3.26	2.85	1.70	1.08	1.69	713	—	483	—
BENTGRASS, CREEPING *Agrostis palustris*													
18 hay, sun-cured, postripe	1-00-688	As fed	92.0	48.0	2.11	1.72	1.03	0.33	1.06	—	—	—	31.0
		Dry	100.0	52.0	2.29	1.87	1.12	0.36	1.15	—	—	—	34.0
BERMUDAGRASS *Cynodon dactylon*													
19 fresh	2-00-712	As fed	34.0	20.0	0.89	0.75	0.44	0.22	0.45	—	—	—	—
		Dry	100.0	60.0	2.65	2.22	1.31	0.65	1.35	—	—	—	—
BERMUDAGRASS, COASTAL *Cynodon dactylon*													
20 fresh	2-00-719	As fed	29.0	19.0	0.82	0.70	0.41	0.23	0.42	—	—	—	—
		Dry	100.0	64.0	2.82	2.40	1.41	0.78	1.45	—	—	—	—
21 hay, sun-cured	1-00-716	As Fed	90.0	49.0	2.15	1.77	1.05	0.39	1.09	—	—	—	43.0
		Dry	100.0	54.0	2.38	1.96	1.16	0.43	1.20	—	—	—	48.0
BLOOD													
22 meal	5-00-380	As fed	92.0	61.0	2.66	2.28	1.34	0.78	1.37	2,833	2,361	2,179	—
		Dry	100.0	66.0	2.91	2.49	1.47	0.85	1.50	3,096	2,580	2,381	—
BLUEGRASS, CANADA *Poa compressa*													
23 hay, sun-cured, late vegetative	1-20-889	As fed	97.0	69.0	3.04	2.63	1.56	0.97	1.57	—	—	—	47.0
		Dry	100.0	71.0	3.13	2.71	1.61	1.00	1.62	—	—	—	48.0
BLUEGRASS, KENTUCKY *Poa pratensis*													
24 fresh	2-00-786	As fed	35.0	23.0	1.00	0.85	0.50	0.28	0.51	—	—	—	—
		Dry	100.0	64.0	2.82	2.40	1.41	0.78	1.45	—	—	—	—
25 hay, sun-cured	1-00-776	As fed	89.0	54.0	2.40	2.02	1.19	0.61	1.22	—	—	—	44.0
		Dry	100.0	61.0	2.69	2.27	1.33	0.69	1.38	—	—	—	50.0
BREWERS													
26 grains, dehydrated	5-02-141	As fed	92.0	81.0	3.57	2.46	2.10	1.25	1.47	2,293	3,056	1,969	48.0
		Dry	100.0	88.0	3.88	2.67	2.28	1.36	1.60	2,491	3,319	2,139	52.0
BROME, SMOOTH *Bromus inermis*													
27 fresh, mature	2-08-364	As fed	55.0	29.0	1.28	1.05	0.63	0.22	0.65	—	—	—	—
		Dry	100.0	53.0	2.34	1.91	1.14	0.40	1.18	—	—	—	—
28 hay, sun-cured, midbloom	1-05-633	As fed	90.0	51.0	2.23	1.85	1.09	0.46	1.13	—	—	—	45.0
		Dry	100.0	56.0	2.47	2.04	1.21	0.51	1.25	—	—	—	49.0
29 hay, sun-cured, mature	1-00-944	As fed	93.0	48.0	2.12	1.73	1.03	0.33	1.07	—	—	—	45.0
		Dry	100.0	52.0	2.29	1.87	1.12	0.36	1.15	—	—	—	49.0

Horses DE (Mcal/kg)	Horses ME (Mcal/kg)	Swine TDN (%)	Swine DE (kcal/kg)	Swine ME (kcal/kg)	Crude Protein (%)	Cell Walls (%)	Cellulose (%)	Hemicellulose (%)	Lignin (%)	Acid Detergent Fiber (%)	Crude Fiber (%)	Ether Extract (%)	Ash (%)	Calcium (%)	Phosphorus (%)	Potassium (%)	Carotene (Provitamin A) (mg/kg)	Vitamin D12 (IU/g)	Vitamin E (mg/kg)
—	—	—	—	—	3.5	13.0	7.0	3.0	2.0	9.0	7.7	0.7	2.1	—	—	—	—	—	—
—	—	—	—	—	14.0	52.0	27.0	13.0	10.0	37.0	31.0	2.8	8.5	—	—	—	—	—	—
—	—	—	—	—	12.6	47.0	23.0	11.0	11.0	35.0	28.8	1.6	7.0	—	—	—	—	—	—
—	—	—	—	—	14.0	52.0	26.0	12.0	12.0	39.0	32.0	1.8	7.8	—	—	—	—	—	—
1.68	1.38	—	—	—	11.7	53.0	26.0	12.0	13.0	40.0	34.4	1.2	6.9	1.03	0.17	1.62	11	1,287	—
1.84	1.51	—	—	—	12.9	58.0	29.0	13.0	14.0	44.0	37.7	1.3	7.5	1.13	0.18	1.78	12	1,411	—
1.79	1.47	44.0	1,418	1,196	17.3	41.0	22.0	—	10.0	32.0	24.0	2.7	9.7	1.40	0.23	2.39	120	—	111
1.95	1.60	48.0	1,546	1,304	18.9	45.0	24.0	—	11.0	35.0	26.2	3.0	10.6	1.52	0.25	2.60	131	—	121
—	—	—	—	—	5.9	18.0	9.0	4.0	4.0	13.0	11.4	1.2	3.0	—	—	—	—	—	—
—	—	—	—	—	15.5	47.0	24.0	10.0	11.0	35.0	30.0	3.1	7.9	—	—	—	—	—	—
1.83	1.50	—	—	—	16.2	38.0	22.0	8.0	7.0	28.0	20.7	2.7	8.6	1.27	0.20	2.27	126	1,796	23
2.04	1.67	—	—	—	18.0	42.0	24.0	9.0	8.0	31.0	23.0	3.0	9.6	1.41	0.22	2.52	140	1,996	26
1.70	1.39	—	—	—	15.3	41.0	23.0	9.0	8.0	32.0	23.4	2.3	8.2	1.27	0.22	1.54	—	—	—
1.89	1.55	—	—	—	17.0	46.0	26.0	10.0	9.0	35.0	26.0	2.6	9.1	1.41	0.24	1.71	—	—	—
1.83	1.50	—	—	—	20.7	36.0	18.0	7.0	8.0	27.0	20.7	3.2	8.0	—	—	—	—	—	—
2.04	1.67	—	—	—	23.0	40.0	20.0	8.0	9.0	30.0	23.0	3.6	8.9	—	—	—	—	—	—
1.70	1.39	—	—	—	17.1	40.0	19.0	9.0	9.0	29.0	22.5	2.9	7.6	—	—	—	—	—	—
1.89	1.55	—	—	—	19.0	44.0	21.0	10.0	10.0	32.0	25.0	3.2	8.5	—	—	—	—	—	—
—	—	—	—	—	60.2	—	—	—	—	—	2.2	9.0	21.9	—	—	—	—	—	—
—	—	—	—	—	64.7	—	—	—	—	—	2.4	9.7	23.6	—	—	—	—	—	—
—	—	—	—	—	2.6	20.0	—	—	2.0	11.0	9.0	0.5	3.3	0.14	0.06	0.43	54	—	—
—	—	—	—	—	8.9	68.0	—	—	7.0	38.0	30.4	1.6	11.1	0.46	0.22	1.45	183	—	—
1.61	1.32	—	—	—	7.4	66.0	29.0	27.0	7.0	37.0	29.2	1.9	5.9	0.46	0.20	—	—	—	—
1.77	1.45	—	—	—	8.2	72.0	32.0	30.0	8.0	41.0	32.0	2.1	6.4	0.50	0.22	—	—	—	—
—	—	—	—	—	6.4	69.0	31.0	—	6.0	38.0	30.9	1.4	8.5	0.24	0.18	1.46	—	—	—
—	—	—	—	—	7.0	76.0	34.0	—	7.0	42.0	34.0	1.5	9.3	0.26	0.20	1.60	—	—	—
—	—	90.0	3,983	3,738	9.8	—	—	—	—	—	1.2	11.7	4.0	0.13	0.24	0.49	4	—	41
—	—	98.0	4,335	4,068	10.7	—	—	—	—	—	1.3	12.7	4.4	0.14	0.26	0.53	5	—	45
—	—	70.0	3,108	2,910	11.9	17.0	4.0	—	2.0	6.0	5.0	1.9	2.3	0.04	0.34	0.41	2	—	22
—	—	79.0	3,523	3,299	13.5	19.0	5.0	—	2.0	7.0	5.7	2.1	2.6	0.05	0.38	0.47	2	—	25
1.56	1.28	—	—	—	7.6	—	—	—	—	—	24.1	1.9	6.6	0.20	0.23	1.03	46	963	—
1.79	1.47	—	—	—	8.7	—	—	—	—	—	27.5	2.1	7.6	0.23	0.26	1.18	53	1,103	—
—	—	67.0	2,932	2,693	8.8	49.0	—	—	2.0	30.0	18.0	0.5	4.9	0.63	0.09	0.18	0	577	—
—	—	73.0	3,235	2,971	9.7	54.0	—	—	2.0	33.0	19.8	0.6	5.4	0.69	0.10	0.20	0	637	—
1.31	1.07	—	—	—	4.0	—	—	—	—	—	28.3	1.3	5.3	—	—	—	—	—	—
1.42	1.16	—	—	—	4.3	—	—	—	—	—	30.8	1.4	5.8	—	—	—	—	—	—
—	—	23.0	1,019	954	4.1	—	—	—	—	—	8.9	0.8	3.4	0.18	0.07	0.57	104	—	—
—	—	69.0	3,030	2,835	12.0	—	—	—	—	—	26.4	2.2	10.2	0.53	0.21	1.70	310	—	—
—	—	—	—	—	4.4	—	—	—	—	—	8.3	1.1	1.8	0.14	0.08	—	96	—	—
—	—	—	—	—	15.0	—	—	—	—	—	28.4	3.8	6.3	0.49	0.27	—	331	—	—
1.73	1.42	—	—	—	5.4	70.0	—	—	5.0	34.0	27.7	2.1	5.9	—	—	—	—	—	—
1.92	1.57	—	—	—	6.0	78.0	—	—	6.0	38.0	30.7	2.3	6.6	—	—	—	—	—	—
—	—	61.0	2,739	2,313	79.8	—	—	—	—	—	1.0	1.3	5.3	0.29	0.24	0.09	—	—	—
—	—	67.0	2,993	2,527	87.2	—	—	—	—	—	1.1	1.4	5.8	0.32	0.26	0.10	—	—	—
1.86	1.53	—	—	—	—	—	—	—	—	—	—	—	—	—	—	—	—	—	—
1.92	1.58	—	—	—	—	—	—	—	—	—	—	—	—	—	—	—	—	—	—
—	—	—	—	—	5.2	—	7.0	6.0	—	8.0	8.1	1.6	2.8	0.12	0.12	0.70	87	—	—
—	—	—	—	—	14.9	—	20.0	17.0	—	24.0	23.0	4.5	8.0	0.33	0.34	1.98	248	—	—
1.77	1.45	—	—	—	11.6	—	—	—	—	—	27.6	3.1	5.9	0.29	0.22	1.51	—	—	—
1.99	1.63	—	—	—	13.0	—	—	—	—	—	31.0	3.5	6.6	0.33	0.25	1.69	—	—	—
—	—	58.0	2,487	2,285	27.1	42.0	—	—	6.0	22.0	13.2	6.6	3.6	0.30	0.51	0.08	0	—	26
—	—	63.0	2,701	2,482	29.4	46.0	—	—	6.0	24.0	14.4	7.2	3.9	0.33	0.55	0.09	0	—	29
—	—	—	—	—	3.3	—	—	—	—	—	19.1	1.3	3.8	0.14	0.09	—	—	—	—
—	—	—	—	—	6.0	—	—	—	—	—	34.8	2.4	6.9	0.26	0.16	—	—	—	—
1.78	1.46	—	—	—	13.2	55.0	28.0	20.0	4.0	33.0	28.8	2.3	9.0	—	—	—	—	—	—
1.97	1.61	—	—	—	14.6	61.0	31.0	22.0	4.0	37.0	31.8	2.6	10.0	—	—	—	—	—	—
1.80	1.48	—	—	—	5.4	65.0	33.0	23.0	7.0	42.0	29.8	2.8	6.7	—	—	—	—	—	—
1.94	1.59	—	—	—	5.8	71.0	36.0	25.0	8.0	45.0	32.2	3.0	7.2	—	—	—	—	—	—

Table 30 Feed Composition *continued*

Feed Name	International Feed Number	Moisture Basis	Dry Matter (%)	Ruminants TDN (%)	DE (Mcal/kg)	ME (Mcal/kg)	NEm (Mcal/kg)	NEg (Mcal/kg)	Dairy Cattle NEl (Mcal/kg)	Chickens MEn (kcal/kg)	TME (kcal/kg)	NEp (kcal/kg)	Horses TDN (%)
BUTTERMILK													
30 dehydrated (Cattle)	5-01-160	As fed	92.0	82.0	3.62	3.24	1.99	1.36	1.90	2,752	—	1,731	—
		Dry	100.0	89.0	3.92	3.51	2.15	1.47	2.06	2,982	—	1,876	—
CANARYGRASS, REED *Phalaris arundinacea*													
31 fresh	2-01-113	As fed	27.0	17.0	0.76	0.65	0.38	0.22	0.39	—	—	—	—
		Dry	100.0	65.0	2.87	2.45	1.44	0.82	1.47	—	—	—	—
32 hay, sun-cured	1-01-104	As fed	91.0	50.0	2.21	1.82	1.08	0.43	1.12	—	—	—	44.0
		Dry	100.0	55.0	2.43	2.00	1.19	0.47	1.23	—	—	—	48.0
CASEIN													
33 dehydrated (Cattle)	5-01-162	As fed	91.0	81.0	3.56	3.18	1.95	1.33	1.87	4,117	—	—	—
		Dry	100.0	89.0	3.92	3.51	2.15	1.47	2.06	4,544	—	—	—
CATTLE *Bos taurus*													
34 manure, dehydrated, all forage	1-28-274	As fed	92.0	23.0	1.01	0.61	0.70	—	0.45	—	—	—	—
		Dry	100.0	25.0	1.10	0.66	0.76	—	0.49	—	—	—	—
35 manure, dehydrated, feedlot (High concentrate)	1-28-213	As fed	92.0	40.0	1.74	1.35	0.87	—	0.86	—	—	—	—
		Dry	100.0	43.0	1.90	1.47	0.94	—	0.93	—	—	—	—
36 manure, dehydrated, forage and concentrate	1-28-214	As fed	92.0	28.0	1.22	0.82	0.72	—	0.57	—	—	—	—
		Dry	100.0	30.0	1.32	0.89	0.78	—	0.62	—	—	—	—
CITRUS *Citrus* spp													
37 pulp without fines, dehydrated (Dried citrus pulp)	4-01-237	As fed	91.0	75.0	3.30	2.92	1.77	1.18	1.72	1,337	—	943	—
		Dry	100.0	82.0	3.62	3.20	1.94	1.30	1.89	1,467'	—	1,035	—
CLOVER, ALSIKE *Trifolium hybridum*													
38 fresh, early vegetative	2-01-314	As fed	19.0	12.0	0.55	0.47	0.28	0.16	0.28	—	—	—	—
		Dry	100.0	66.0	2.91	2.49	1.47	0.85	1.50	—	—	—	—
39 hay, sun-cured	1-01-313	As fed	88.0	51.0	2.25	1.87	1.11	0.51	1.14	—	—	—	42.0
		Dry	100.0	58.0	2.56	2.13	1.26	0.58	1.30	—	—	—	48.0
CLOVER, CRIMSON *Trifolium incarnatum*													
40 fresh, early vegetative	2-20-890	As fed	18.0	11.0	0.50	0.42	0.25	0.14	0.26	—	—	—	—
		Dry	100.0	63.0	2.78	2.36	1.39	0.75	1.42	—	—	—	45.0
41 hay, sun-cured	1-01-328	As fed	87.0	50.0	2.19	1.82	1.08	0.48	1.11	—	—	—	52.0
		Dry	100.0	57.0	2.51	2.09	1.23	0.55	1.28	—	—	—	
CLOVER, LADINO *Trifolium repens*													
42 fresh, early vegetative	2-01-380	As fed	19.0	13.0	0.58	0.50	0.30	0.18	0.30	—	—	—	—
		Dry	100.0	68.0	3.00	2.58	1.52	0.91	1.55	—	—	—	45.0
43 hay, sun-cured	1-01-378	As fed	90.0	58.0	2.57	2.19	1.29	0.73	1.32	—	—	—	51.0
		Dry	100.0	65.0	2.87	2.45	1.44	0.82	1.47	—	—	—	
CLOVER, RED *Trifolium pratense*													
44 fresh, full bloom	2-01-429	As fed	26.0	17.0	0.74	0.63	0.37	0.21	0.38	—	—	—	—
		Dry	100.0	64.0	2.82	2.40	1.41	0.78	1.45	—	—	—	44.0
45 hay, sun-cured	1-01-415	As fed	89.0	49.0	2.15	1.77	1.05	0.42	1.09	—	—	—	50.0
		Dry	100.0	55.0	2.43	2.00	1.19	0.47	1.23	—	—	—	
CORN, DENT YELLOW *Zea mays indentata*													
46 aerial part without ears/husks, sun-cured (Stover) (Straw)	1-28-233	As fed	85.0	51.0	2.21	1.85	1.09	0.53	1.13	—	—	—	32.0
		Dry	100.0	59.0	2.60	2.18	1.28	0.62	1.33	—	—	—	37.0
47 cobs, ground	1-28-234	As fed	90.0	45.0	1.99	1.60	0.97	0.25	1.00	1,651	—	—	28.0
		Dry	100.0	50.0	2.21	1.78	1.07	0.28	1.11	1,830	—	—	31.0
48 distillers grains with solubles, dehydrated	5-28-236	As fed	92.0	80.0	3.56	3.19	1.95	1.33	1.87	2,535	2,970	1,951	—
		Dry	100.0	88.0	3.88	3.47	2.12	1.45	2.04	2,760	3,234	2,124	—
49 ears, ground (Corn and cob meal)	4-28-238	As fed	87.0	72.0	3.17	2.81	1.70	1.14	1.66	2,730	—	1,968	—
		Dry	100.0	83.0	3.66	3.25	1.97	1.32	1.91	3,155	—	2,274	—
50 gluten, meal, 60% protein	5-28-242	As fed	90.0	81.0	3.54	3.17	1.95	1.33	1.86	3,689	4,003	2,724	—
		Dry	100.0	89.0	3.92	3.51	2.15	1.47	2.06	4,086	4,434	3,017	—
51 grain	4-02-935	As fed	89.0	77.0	3.40	3.03	1.85	1.26	1.78	3,383	3,671	2,491	—
		Dry	100.0	87.0	3.84	3.42	2.09	1.42	2.01	3,818	4,143	2,812	—
53 grain, high moisture	4-20-770	As fed	77.0	71.0	3.12	2.80	1.73	1.18	1.64	—	—	—	—
		Dry	100.0	92.0	4.06	3.65	2.25	1.54	2.13	—	—	—	—
53 grain, opaque 2 (High lysine)	4-28-253	As fed	90.0	80.0	3.54	3.17	1.94	1.32	1.86	3,369	—	2,484	—
		Dry	100.0	89.0	3.92	3.51	2.15	1.47	2.06	3,738	—	2,756	—
54 silage	3-02-912	As fed ~	30.0 [2]	21.0	0.91	0.79	0.47	0.28	0.47	—	—	—	—
		Dry	100.0	69.0	3.04	2.62	1.55	0.94	1.57	—	—	—	—
55 silage, aerial part without ears/husks (Stalkage) (Stover)	3-28-251	As fed	31.0	17.0	0.74	0.61	0.36	0.15	0.38	—	—	—	—
		Dry	100.0	55.0	2.43	2.00	1.19	0.47	1.23	—	—	—	—
COTTON *Gossypium*													
56 hulls	1-01-599	As fed	91.0	41.0	1.80	1.41	0.88	0.05	0.89	—	—	—	29.0
		Dry	100.0	45.0	1.98	1.55	0.98	0.06	0.98	—	—	—	32.0
57 seeds, meal prepressed, solvent extracted, 41% protein	5-07-872	As fed	91.0	72.0	3.19	2.82	1.70	1.13	1.67	2,144	—	1,286	—
		Dry	100.0	80.0	3.53	3.11	1.88	1.24	1.84	2,368	—	1,420	—
58 seeds, meal, solvent extracted, low gossypol	5-01-633	As fed	93.0	66.0	2.90	2.51	1.49	0.98	1.50	—	—	—	—
		Dry	100.0	71.0	3.13	2.71	1.61	1.00	1.62	—	—	—	—

Horses DE (Mcal/kg)	Horses ME (Mcal/kg)	Swine TDN (%)	Swine DE (kcal/kg)	Swine ME (kcal/kg)	Crude Protein (%)	Cell Walls (%)	Cellulose (%)	Hemicellulose (%)	Lignin (%)	Acid Detergent Fiber (%)	Crude Fiber (%)	Ether Extract (%)	Ash (%)	Calcium (%)	Phosphorus (%)	Potassium (%)	Carotene (Provitamin A) (mg/kg)	Vitamin D12 (IU/g)	Vitamin E (mg/kg)
—	—	77.0	3,411	3,046	31.7	—	—	—	—	—	0.3	4.7	9.1	1.33	0.94	0.83	—	—	6
—	—	84.0	3,696	3,300	34.4	—	—	—	—	—	0.4	5.0	9.9	1.44	1.01	0.90	—	—	7
—	—	—	—	—	3.1	—	—	5.0	—	—	7.8	0.9	2.2	0.11	0.09	0.97	—	—	—
—	—	—	—	—	11.6	—	—	20.0	—	—	29.5	3.5	8.3	0.41	0.35	3.64	—	—	—
1.83	1.50	—	—	—	9.4	58.0	24.0	21.0	3.0	33.0	30.1	2.8	7.2	0.35	0.23	2.51	23	—	—
2.01	1.65	—	—	—	10.3	64.0	26.0	23.0	4.0	36.0	33.0	3.1	7.9	0.38	0.25	2.76	26	—	—
—	—	80.0	3,507	2,710	84.0	—	—	—	—	—	0.2	0.6	2.2	0.61	0.82	0.01	—	—	—
—	—	88.0	3,870	2,991	92.7	—	—	—	—	—	0.2	0.7	2.4	0.67	0.90	0.01	—	—	—
—	—	—	—	—	15.6	63.0	—	—	25.0	42.0	—	—	—	—	—	—	—	—	—
—	—	—	—	—	17.0	69.0	—	—	27.0	46.0	—	—	—	—	—	—	—	—	—
—	—	—	—	—	23.0	29.0	—	—	5.0	24.0	—	—	18.2	—	—	—	—	—	—
—	—	—	—	—	25.0	32.0	—	—	5.0	26.0	—	—	19.8	—	—	—	—	—	—
—	—	—	—	—	15.6	53.0	—	—	7.0	31.0	—	—	—	—	—	—	—	—	—
—	—	—	—	—	17.0	58.0	—	—	7.0	34.0	—	—	—	—	—	—	—	—	—
—	—	46.0	2,286	2,163	6.1	21.0	—	—	3.0	20.0	11.6	3.4	6.0	1.67	0.11	0.72	0	—	—
—	—	50.0	2,508	2,374	6.7	23.0	—	—	3.0	22.0	12.7	3.7	6.6	1.84	0.12	0.79	0	—	—
—	—	—	—	—	4.5	—	—	—	—	—	3.3	0.6	2.4	±	—	—	73	—	—
—	—	—	—	—	24.1	—	—	—	—	—	17.5	3.2	12.8	—	—	—	385	—	—
1.70	1.39	—	—	—	13.1	—	—	11.0	—	—	26.5	2.7	7.7	1.13	0.23	2.17	164	—	—
1.93	1.59	—	—	—	14.9	—	—	13.0	—	—	30.1	3.0	8.7	1.29	0.26	2.46	187	—	—
—	—	—	—	—	3.1	—	—	—	—	—	5.0	—	—	—	—	—	43	—	—
—	—	—	—	—	17.0	—	—	—	—	—	28.0	—	—	—	—	—	238	—	—
1.80	1.47	—	—	—	16.1	—	—	—	—	—	26.3	2.1	9.6	1.22	0.19	2.09	20	—	—
2.06	1.69	—	—	—	18.4	—	—	—	—	—	30.1	2.4	11.0	1.40	0.22	2.40	23	—	—
—	—	—	—	—	5.3	—	—	—	—	—	2.7	0.5	2.6	—	—	—	68	—	—
—	—	—	—	—	27.2	—	—	—	—	—	14.0	2.5	13.5	—	—	—	353	—	—
1.81	1.49	—	—	—	19.7	32.0	—	—	6.0	29.0	19.1	2.4	9.0	1.21	0.28	2.35	75	—	—
2.02	1.66	—	—	—	22.0	36.0	—	—	7.0	32.0	21.2	2.7	10.1	1.35	0.31	2.62	83	—	—
—	—	—	—	—	3.8	—	—	—	—	—	6.8	0.8	2.0	—	—	—	—	—	—
—	—	—	—	—	14.6	—	—	—	—	—	26.1	2.9	7.8	—	—	—	—	—	—
1.76	1.45	—	—	—	14.2	50.0	23.0	8.0	9.0	—	25.5	2.4	7.5	1.35	0.22	1.44	18	1,694	—
1.99	1.63	—	—	—	16.0	56.0	26.0	9.0	10.0	—	28.8	2.8	8.5	1.53	0.25	1.62	20	1,914	—
1.31	1.07	—	—	—	5.6	57.0	21.0	—	9.0	33.0	29.3	1.1	6.1	0.49	0.08	1.24	—	—	—
1.53	1.26	—	—	—	6.6	67.0	25.0	—	11.0	39.0	34.4	1.3	7.2	0.57	0.10	1.45	—	—	—
1.17	0.96	—	—	—	2.8	80.0	25.0	—	6.0	32.0	32.7	0.7	1.5	0.11	0.04	0.79	1	—	—
1.30	1.07	—	—	—	3.2	89.0	28.0	—	7.0	35.0	36.2	0.7	1.7	0.12	0.04	0.87	1	—	—
—	—	—	3,131	2,819	23.0	40.0	13.0	—	4.0	17.0	9.1	9.4	4.4	0.14	0.65	0.40	3	551	40
—	—	—	3,410	3,070	25.0	44.0	14.0	—	4.0	18.0	9.9	10.3	4.8	0.15	0.71	0.44	3	600	43
—	—	69.0	3,109	2,779	7.8	—	—	—	—	—	8.2	3.2	1.7	0.06	0.24	0.46	3	—	18
—	—	79.0	3,593	3,212	9.0	—	—	—	—	—	9.4	3.7	1.9	0.07	0.27	0.53	4	—	20
—	—	80.0	3,981	3,528	60.7	13.0	4.0	—	1.0	5.0	2.0	2.2	1.6	0.07	0.48	0.19	30	—	24
—	—	89.0	4,409	3,907	67.2	14.0	4.0	—	1.0	5.0	2.2	2.4	1.8	0.08	0.54	0.21	34	—	26
—	—	80.0	3,399	3,300	9.6	8.0	2.0	5.0	1.0	3.0	2.6	3.8	1.3	0.03	0.26	0.33	2	—	22
—	—	90.0	3,837	3,724	10.9	9.0	2.0	6.0	1.0	3.0	2.9	4.3	1.5	0.03	0.29	0.37	3	—	25
—	—	66.0	2,891	2,713	8.2	—	—	—	—	4.0	2.0	3.3	1.2	0.01	0.25	0.27	—	—	—
—	—	85.0	3,765	3,534	10.1	—	—	—	—	5.0	2.6	4.3	1.6	0.02	0.32	0.35	—	—	—
—	—	77.0	3,664	3,434	10.1	—	—	—	—	—	3.3	4.3	1.6	0.03	0.20	0.35	5	—	—
—	—	85.0	4,065	3,810	11.3	—	—	—	—	—	3.7	4.8	1.8	0.03	0.22	0.39	5	—	—
—	—	22.0	950	896	2.5	—	9.0	—	1.0	9.0	7.5	1.0	1.7	0.09	0.08	0.36	13	132	—
—	—	72.0	3,161	2,981	8.3	—	29.0	—	5.0	30.0	25.1	3.3	5.5	0.29	0.26	1.18	43	439	—
—	—	—	—	—	1.9	21.0	8.0	—	2.0	17.0	9.6	0.7	3.5	0.12	0.09	0.47	5	—	—
—	—	—	—	—	6.3	68.0	25.0	—	7.0	55.0	31.3	2.1	11.6	0.38	0.31	1.54	15	—	—
1.22	1.00	—	—	—	3.7	82.0	53.0	—	22.0	66.0	43.3	1.5	2.6	0.13	0.09	0.79	—	—	—
1.35	1.11	—	—	—	4.1	90.0	59.0	—	24.0	73.0	47.8	1.7	2.8	0.15	0.09	0.87	—	—	—
—	—	61.0	2,615	2,485	41.3	24.0	11.0	—	5.0	17.0	12.8	1.1	6.4	0.20	1.09	1.26	—	—	—
—	—	68.0	2,888	2,745	45.6	26.0	12.0	—	6.0	19.0	14.1	1.3	7.0	0.22	1.21	1.39	—	—	—
—	—	—	—	—	41.5	—	—	—	—	—	12.7	1.2	5.8	—	—	—	—	—	—
—	—	—	—	—	44.8	—	—	—	—	—	13.7	1.3	6.3	—	—	—	—	—	—

Table 30 Feed Composition *continued*

Feed Name	International Feed Number	Moisture Basis	Dry Matter (%)	Ruminants					Dairy Cattle NE$_l$ (Mcal/kg)	Chickens			Horses TDN (%)
				TDN (%)	DE (Mcal/kg)	ME (Mcal/kg)	NE$_m$ (Mcal/kg)	NE$_g$ (Mcal/kg)		ME$_n$ (kcal/kg)	TME (kcal/kg)	NE$_p$ (kcal/kg)	
COTTON *Gossypium* spp													
59 seeds, meal, solvent extracted, 41% protein	5-01-621	As fed	91.0	70.0	3.06	2.68	1.60	1.04	1.59	1,943	—	1,410	—
		Dry	100.0	76.0	3.35	2.93	1.76	1.14	1.74	2,131	—	1,546	—
COWPEA, COMMON *Vigna sinensis*													
60 hay, sun-cured	1-01-645	As fed	90.0	53.0	2.34	1.96	1.15	0.56	1.19	—	—	—	43.0
		Dry	100.0	59.0	2.60	2.18	1.28	0.62	1.33	—	—	—	48.0
FATS AND OILS													
61 fat, animal, hydrolyzed	4-00-376	As fed	99.0	223.0	9.84	9.49	6.55	4.19	5.35	8,164	—	5,317	—
		Dry	100.0	225.0	9.92	9.57	6.61	4.23	5.39	8,232	—	5,362	—
FESCUE, ALTA *Festuca arundinacea*													
62 hay, sun-cured	1-05-684	As fed	91.0	57.0	2.49	2.10	1.24	0.65	1.27	—	—	—	41.0
		Dry	100.0	62.0	2.73	2.31	1.36	0.72	1.40	—	—	—	45.0
FESCUE, KENTUCKY 31 *Festuca arundinacea*													
63 fresh, vegetative	2-01-902	As fed	29.0	19.0	0.85	0.72	0.43	0.25	0.44	—	—	—	—
		Dry	100.0	67.0	2.91	2.49	1.47	0.85	1.50	—	—	—	—
FESCUE, MEADOW *Festuca elatior*													
64 hay, sun-cured	1-01-912	As fed	88.0	52.0	2.29	1.91	1.13	0.54	1.17	—	—	—	41.0
		Dry	100.0	59.0	2.60	2.18	1.28	0.62	1.33	—	—	—	46.0
FISH													
65 solubles, condensed	5-01-969	As fed	50.0	42.0	1.86	1.65	1.00	0.68	0.97	1,786	—	1,048	—
		Dry	100.0	84.0	3.70	3.29	2.00	1.35	1.94	3,562	—	2,091	—
FISH, MENHADEN *Brevoortia tyrannus*													
66 meal mechanical extracted	5-02-009	As fed	92.0	67.0	2.95	2.57	1.53	0.97	1.53	2,849	2,744	2,037	—
		Dry	100.0	73.0	3.22	2.80	1.67	1.06	1.67	3,110	2,995	2,225	—
FLAX *Linum usitatissimum*													
67 seeds, meal mechanical extracted (Linseed meal)	5-02-045	As fed	91.0	74.0	3.28	2.90	1.76	1.18	1.71	1,518	—	1,123	47.0
		Dry	100.0	82.0	3.62	3.20	1.94	1.30	1.89	1,673	—	1,237	51.0
68 seeds, meal solvent extracted (Linseed meal)	5-02-048	As fed	90.0	70.0	3.10	2.73	1.64	1.08	1.62	1,411	2,644	991	—
		Dry	100.0	78.0	3.44	3.02	1.82	1.19	1.79	1,565	2,931	1,099	—
LESPEDEZA, COMMON *Lespedeza striata*													
69 hay, sun-cured, midbloom	1-02-554	As fed	92.0	46.0	2.02	1.63	0.98	0.25	1.01	—	—	—	43.0
		Dry	100.0	50.0	2.21	1.78	1.07	0.28	1.11	—	—	—	47.0
70 hay, sun-cured, full bloom	1-20-887	As fed	89.0	42.0	1.85	1.47	0.90	0.13	0.92	—	—	—	43.0
		Dry	100.0	47.0	2.07	1.64	1.01	0.15	1.03	—	—	—	48.0
MEADOW PLANTS, INTERMOUNTAIN													
71 hay, sun-cured	1-03-181	As fed	95.0	55.0	2.43	2.03	1.20	0.55	1.24	—	—	—	43.0
		Dry	100.0	58.0	2.56	2.13	1.26	0.58	1.30	—	—	—	46.0
MEAT													
72 with blood, meal rendered (Tankage)	5-00-386	As fed	92.0	67.0	2.92	2.54	1.51	0.95	1.51	2,672	2,981	1,781	—
		Dry	100.0	72.0	3.17	2.76	1.64	1.03	1.64	2,901	3,238	1,934	—
73 with blood/bone, meal, rendered (Tankage)	5-00-387	As fed	93.0	63.0	2.79	2.40	1.42	0.84	1.44	1,791	—	1,514	—
		Dry	100.0	68.0	3.00	2.58			1.52	0.91 1,928	—	1,629	—
MILK													
74 dehydrated (Cattle)	5-01-167	As fed	96.0	114.0	5.03	4.64	2.95	2.08	2.68	—	—	—	—
		Dry	100.0	119.0	5.25	4.85	3.08	2.17	2.80	—	—	—	—
75 fresh (Cattle)	5-01-168	As fed	12.0	16.0	0.70	0.65	0.42	0.30	0.38	—	—	—	—
		Dry	100.0	129.0	5.69	5.29	3.37	2.41	3.04	—	—	—	—
MILLET, FOXTAIL *Setaria italica*													
76 hay, sun-cured	1-03-099	As fed	87.0	51.0	2.27	1.90	1.12	0.54	1.16	—	—	—	42.0
							1.28	0.62	1.33	—	—	—	48.0
MILLET, PROSO *Panicum miliaceum*													
77 grain	4-03-120	As fed	90.0	75.0	3.33	2.96				2,898	—	—	—
		Dry	100.0	84.0	3.70	3.29	1.80	1.21	1.74	3,222	—	—	—
							2.00	1.35	1.94				
MOLASSES AND SYRUP													
78 beet, sugar, molasses, more than 48% invert sugar, more than 79.5 degrees brix	4-00-668	As fed	78.0	61.0	2.71	2.38				1,925	—	1,568	—
		Dry	100.0	79.0	3.48	3.07	1.44	0.95	1.41	2,477	—	2,018	—
							1.85	1.22	1.82				
79 citrus, syrup (Citrus molasses)	4-01-241	As fed	68.0	51.0	2.24	1.96	1.17	0.75	1.16	—	—	—	—
		Dry	100.0	75.0	3.31	2.89	1.73	1.11	1.72	—	—	—	—
80 sugarcane, molasses dehydrated	4-04-695	As fed	94.0	66.0	2.91	2.52	1.49	0.91	1.51	2,706	—	—	—
		Dry	100.0	70.0	3.09	2.67	1.58	0.97	1.60	2,866	—	—	—
81 sugarcane, molasses more than 46% invert sugar, more than 79.5 degrees brix (Black strap)	4-04-696	As fed	75.0	54.0	2.37	2.05	1.22	0.77	1.23	1,927	—	1,563	—
		Dry	100.0	72.0	3.17	2.76	1.64	1.03	1.64	2,585	—	2,098	—

Horses DE (Mcal/kg)	ME (Mcal/kg)	Swine TDN (%)	DE (kcal/kg)	ME (kcal/kg)	Crude Protein (%)	Cell Walls (%)	Cellulose (%)	Hemicellulose (%)	Lignin (%)	Acid Detergent Fiber (%)	Crude Fiber (%)	Ether Extract (%)	Ash (%)	Calcium (%)	Phosphorus (%)	Potassium (%)	Carotene (Provitamin A) (mg/kg)	Vitamin D12 (IU/g)	Vitamin E (mg/kg)
—	—	61.0	2,675	2,364	41.2	—	—	—	—	—	12.1	1.4	6.5	0.17	1.10	1.39	—	—	16
—	—	67.0	2,933	2,592	45.2	—	—	—	—	—	13.3	1.6	7.1	0.18	1.21	1.52	—	—	17
1.73	1.42	—	—	—	17.5	—	—	—	—	—	24.0	2.8	10.2	1.26	0.31	2.03	31	—	—
1.92	1.58	—	—	—	19.4	—	—	—	—	—	26.7	3.1	11.3	1.40	0.35	2.26	35	—	—
—	—	209.0	9,272	8,274	—	—	—	—	—	—	—	98.7	—	—	—	—	—	—	—
—	—	211.0	9,350	8,343	—	—	—	—	—	—	—	99.5	—	—	—	—	—	—	—
1.75	1.44	—	—	—	9.3	64.0	—	24.0	—	37.0	32.5	2.0	7.0	0.36	0.20	2.23	—	—	—
1.92	1.58	—	—	—	10.2	70.0	—	26.0	—	41.0	35.7	2.2	7.7	0.40	0.22	2.45	*—	—	—
—	—	—	—	—	4.2	—	—	—	—	—	7.1	1.6	2.9	0.15	0.11	—	—	—	—
—	—	—	—	—	14.5	—	—	—	—	—	24.6	5.5	9.9	0.51	0.37	—	—	—	—
1.63	1.34	—	—	—	8.0	57.0	33.0	—	6.0	38.0	29.1	2.1	7.2	0.35	0.27	1.61	64	—	119
1.86	1.53	—	—	—	9.1	65.0	38.0	—	7.0	43.0	33.1	2.4	8.2	0.40	0.31	1.84	73	—	136
—	—	44.0	1,898	1,613	32.7	—	—	—	—	—	0.5	5.6	9.6	0.22	0.59	1.61	1	—	—
—	—	88.0	3,784	3,217	65.3	—	—	—	—	—	0.9	11.2	19.2	0.43	1.18	3.22	3	—	—
—	—	61.0	3,480	2,633	61.1	—	—	—	—	—	0.9	9.6	19.0	5.18	2.89	0.70	—	—	12
—	—	67.0	3,799	2,875	66.7	—	—	—	—	—	1.0	10.5	20.8	5.65	3.16	0.76	—	—	13
—	—	73.0	3,381	2,761	34.3	23.0	—	—	6.0	15.0	8.8	5.4	5.7	0.41	0.87	1.22	0	—	8
—	—	81.0	3,727	3,044	37.9	25.0	—	—	7.0	17.0	9.6	6.0	6.3	0.45	0.96	1.34	0	—	9
—	—	65.0	2,883	2,523	34.6	23.0	—	—	5.0	17.0	9.1	1.4	5.8	0.39	0.80	1.38	—	—	14
—	—	72.0	3,196	2,797	38.3	25.0	—	—	6.0	19.0	10.1	1.5	6.5	0.43	0.89	1.53	—	—	15
1.74	1.43	—	—	—	11.5	—	—	—	—	—	26.4	2.3	4.5	1.08	0.22	0.92	—	—	—
1.90	1.56	—	—	—	12.6	—	—	—	—	—	28.8	2.5	4.9	1.18	0.24	1.01	—	—	—
1.73	1.42	—	—	—	12.8	—	—	—	—	—	27.4	1.9	5.0	1.02	0.19	0.93	—	—	—
1.93	1.58	—	—	—	14.3	—	—	—	—	—	30.7	2.1	5.6	1.14	0.21	1.04	—	—	—
1.75	1.43	—	—	—	8.3	—	—	—	—	—	30.7	2.4	8.0	0.58	0.17	1.50	—	—	—
1.84	1.51	—	—	—	8.7	—	—	—	—	—	32.3	2.5	8.5	0.61	0.18	1.58	—	—	—
—	—	67.0	2,450	2,095	59.4	—	—	—	—	—	2.0	8.9	21.5	5.86	3.07	0.55	—	—	—
—	—	73.0	2,660	2,275	64.5	—	—	—	—	—	2.2	9.7	23.4	6.37	3.33	0.60	—	—	—
—	—	68.0	2,992	2,644	46.6	—	—	—	—	—	2.2	12.8	28.2	11.16	5.41	—	—	—	—
—	—	73.0	3,220	2,846	50.2	—	—	—	—	—	2.4	13.7	30.4	12.01	5.82	—	—	—	—
—	—	—	—	—	25.4	—	—	—	—	—	0.2	26.6	5.4	0.91	0.71	1.04	—	338	—
—	—	—	—	—	26.5	—	—	—	—	—	0.2	27.8	5.7	0.95	0.74	1.08	—	353	—
—	—	15.0	680	616	3.3	—	—	—	—	—	—	3.6	0.8	0.12	0.09	0.14	—	—	—
—	—	125.0	5,512	4,994	26.7	—	—	—	—	—	—	29.5	6.3	0.95	0.76	1.12	—	—	—
1.70	1.39	—	—	—	7.5	—	—	—	—	—	25.8	2.6	7.5	0.29	0.17	1.69	—	—	—
1.94	1.59	—	—	—	8.6	—	—	—	—	—	29.6	2.9	8.6	0.33	0.19	1.94	—	—	—
—	—	74.0	3,273	3,057	11.6	—	—	—	3.0	15.0	6.1	3.5	2.6	0.03	0.30	0.43	—	—	—
—	—	83.0	3,639	3,399	12.9	—	—	—	4.0	17.0	6.8	3.9	2.9	0.03	0.34	0.48	—	—	—
—	—	57.0	2,513	2,333	6.6	—	—	—	—	—	—	0.2	8.8	0.13	0.03	4.72	—	—	4
—	—	73.0	3,233	3,002	8.5	—	—	—	—	—	—	0.2	11.3	0.17	0.03	6.07	—	—	5
—	—	54.0	2,379	2,262	5.5	—	—	—	—	—	—	0.2	5.3	1.16	0.09	0.09	—	—	—
—	—	80.0	3,517	3,344	8.2	—	—	—	—	—	—	0.9	7.9	1.72	0.13	0.14	—	—	—
—	—	70.0	3,079	2,485	9.7	—	—	—	—	—	6.3	0.9	12.5	1.04	0.14	3.40	—	—	—
—	—	74.0	3,261	2,632	10.3	—	—	—	—	—	6.7	0.9	13.3	1.10	0.15	3.60	—	—	—
—	—	56.0	2,507	2,199	4.4	—	—	—	—	—	—	0.1	9.8	0.75	0.08	2.86	—	—	5
—	—	76.0	3,364	2,951	5.8	—	—	—	—	—	—	0.1	13.1	1.00	0.11	3.84	—	—	7

Table 30 Feed Composition, *continued*

Feed Name	International Feed Number	Moisture Basis	Dry Matter (%)	Ruminants						Dairy Cattle	Chickens			Horses
				TDN (%)	DE (Mcal/kg)	ME (Mcal/kg)	NE$_m$ (Mcal/kg)	NE$_g$ (Mcal/kg)		NE$_l$ (Mcal/kg)	ME$_n$ (kcal/kg)	TME (kcal/kg)	NE$_p$ (kcal/kg)	TDN (%)
NAPIERGRASS *Pennisetum purpureum*														
82 fresh, late bloom	2-03-162	As fed	23.0	12.0	0.54	0.44	0.26	0.09		0.27	—	—	—	—
		Dry	100.0	53.0	2.34	1.91	1.14	0.40		1.18	—	—	—	—
OATS *Avena sativa*														
83 breakfast cereal by-product, less than 4% fiber (Feeding oat meal) (Oat middlings)	4-03-303	As fed	91.0	86.0	3.80	3.43	2.12	1.46		2.00	3,158	3,800	2,536	—
		Dry	100.0	95.0	4.19	3.78	2.34	1.61		2.21	3,483	4,191	2,796	—
84 grain	4-03-309	As fed	89.0	68.0	3.02	2.65	1.59	1.04		1.57	2,543	3,264	1,735	66.0
		Dry	100.0	77.0	3.40	2.98	1.79	1.17		1.77	2,862	3,674	1,953	74.0
85 grain, Pacific Coast	4-07-999	As fed	91.0	71.0	3.13	2.75	1.65	1.08		1.63	2,645	3,469	1,767	—
		Dry	100.0	78.0	3.44	3.02	1.82	1.19		1.79	2,909	3,816	1,944	—
86 hay, sun-cured	1-03-280	As fed	91.0	56.0	2.46	2.07	1.22	0.63		1.26	—	—	—	43.0
		Dry	100.0	61.0	2.69	2.27	1.33	0.69		1.38	—	—	—	47.0
87 silage, dough stage	3-03-296	As fed	35.0	20.0	0.88	0.73	0.43	0.19		0.45	—	—	—	—
		Dry	100.0	57.0	2.51	2.09	1.23	0.55		1.28	—	—	—	—
88 straw	1-03-283	As fed	92.0	46.0	2.03	1.64	0.99	0.25		1.02	—	—	—	44.0
		Dry	100.0	50.0	2.21	1.78	1.07	0.28		1.11	—	—	—	48.0
ORCHARDGRASS *Dactylis glomerata*														
89 fresh, midbloom	2-03-443	As fed	31.0	17.0	0.77	0.64	0.38	0.17		0.39	—	—	—	—
		Dry	100.0	57.0	2.51	2.09	1.23	0.55		1.28	—	—	—	—
90 hay, sun-cured, early bloom	1-03-425	As fed	89.0	58.0	2.55	2.18	1.28	0.73		1.31	—	—	—	42.0
		Dry	100.0	65.0	2.87	2.45	1.44	0.82		1.47	—	—	—	48.0
91 hay, sun-cured, late bloom	1-03-428	As fed	91.0	49.0	2.16	1.77	1.05	0.39		1.09	—	—	—	44.0
		Dry	100.0	54.0	2.38	1.96	1.16	0.43		1.20	—	—	—	49.0
PANGOLAGRASS *Digitaria decumbens*														
92 fresh	2-03-493	As fed	21.0	12.0	0.51	0.42	0.25	0.10		0.26	—	—	—	—
		Dry	100.0	55.0	2.43	2.00	1.19	0.47		1.23	—	—	—	—
93 hay, sun-cured	1-09-459	As fed	88.0	43.0	1.90	1.53	0.93	0.21		0.95	—	—	—	38.0
		Dry	100.0	49.0	2.16	1.73	1.05	0.23		1.08	—	—	—	43.0
PEANUT *Arachis hypogaea*														
94 kernels, meal solvent extracted (Peanut meal)	5-03-650	As fed	92.0	71.0	3.12	2.74	1.64	1.07		1.63	2,693	—	1,967	—
		Dry	100.0	77.0	3.40	2.98	1.79	1.17		1.77	2,928	—	2,138	—
PEARLMILLET *Pennisetum glaucum*														
95 fresh	2-03-115	As fed	21.0	13.0	0.57	0.48	0.28	0.14		0.29	—	—	—	—
		Dry	100.0	61.0	2.69	2.27	1.33	0.69		1.38	—	—	—	—
POTATO *Solanum tuberosum*														
96 process residue, dehydrated	4-03-775	As fed	89.0	79.0	3.52	3.16	1.94	1.33		1.85	—	—	—	—
		Dry	100.0	90.0	3.97	3.56	2.19	1.49		2.09	—	—	—	—
POULTRY														
97 feathers, hydrolyzed	5-03-795	As fed	93.0	65.0	2.87	2.48	1.47	0.90		1.48	2,427	3,941	1,521	—
		Dry	100.0	70.0	3.09	2.67	1.58	0.97		1.60	2,609	4,238	1,636	—
98 manure, dehydrated	5-14-015	As fed	90.0	52.0	2.31	1.93	1.14	0.53		1.17	1,031	—	—	—
		Dry	100.0	58.0	2.56	2.13	1.26	0.58		1.30	1,142	—	—	—
PRAIRIE PLANTS, MIDWEST														
99 hay, sun-cured	1-03-191	As fed	92.0	47.0	2.06	1.67	1.00	0.29		1.04	—	—	—	40.0
		Dry	100.0	51.0	2.25	1.82	1.10	0.32		1.13	—	—	—	44.0
PRICKLYPEAR *Opuntia spp*														
100 fresh	2-01-061	As fed	17.0	9.0	0.42	0.35	0.21	0.09		0.21	—	—	—	—
		Dry	100.0	57.0	2.51	2.09	1.23	0.55		1.28	—	—	—	—
PROPYLENE GLYCOL														
101	8-03-809	As fed	100.0	158.0	6.95	6.57	4.15	3.15		3.74	—	—	—	—
		Dry	100.0	158.0	6.97	6.59	4.16	3.16		3.75	—	—	—	—
RAPE *Brassica spp*														
102 fresh, early bloom	2-03-866	As fed	11.0	8.0	0.37	0.33	0.20	0.13		0.19	—	—	—	—
		Dry	100.0	75.0	3.31	2.89	1.73	1.11		1.72	—	—	—	—
103 seeds, meal solvent extracted	5-03-871	As fed	91.0	63.0	2.77	2.39	1.41	0.86		1.43	1,751	2,103	1,088	—
		Dry	100.0	69.0	3.04	2.62	1.55	0.94		1.57	1,924	2,310	1,196	—
REDTOP *Agrostis alba*														
104 fresh	2-03-897	As fed	29.0	18.0	0.81	0.69	0.41	0.22		0.42	—	—	—	—
		Dry	100.0	63.0	2.78	2.36	1.39	0.75		1.42	—	—	—	—
105 hay, sun-cured, midbloom	1-03-886	As fed	94.0	54.0	2.37	1.97	1.17	0.52		1.21	—	—	—	45.0
		Dry	100.0	57.0	2.51	2.09	1.23	0.55		1.28	—	—	—	47.0
RUSSIANTHISTLE, TUMBLING *Salsola kali tenuifolia*														
106 hay, sun-cured	1-03-988	As fed	86.0	39.0	1.74	1.37	0.85	0.09		0.87	—	—	—	43.0
		Dry	100.0	46.0	2.03	1.60	0.99	0.10		1.01	—	—	—	50.0

Horses DE (Mcal/kg)	ME (Mcal/kg)	Swine TDN (%)	DE (kcal/kg)	ME (kcal/kg)	Crude Protein (%)	Cell Walls (%)	Cellulose (%)	Hemicellulose (%)	Lignin (%)	Acid Detergent Fiber (%)	Crude Fiber (%)	Ether Extract (%)	Ash (%)	Calcium (%)	Phosphorus (%)	Potassium (%)	Carotene (Provitamin A) (mg/kg)	Vitamin D12 (IU/g)	Vitamin E (mg/kg)
—	—	—	—	—	1.8	17.0	8.0	—	3.0	11.0	9.0	0.3	1.2	—	—	—	—	—	—
—	—	—	—	—	7.8	75.0	35.0	—	14.0	47.0	39.0	1.1	5.3	—	—	—	—	—	—
—	—	79.0	3,480	3,427	14.8	—	—	—	—	—	3.5	6.4	2.3	0.07	0.44	0.50	—	—	24
—	—	87.0	3,838	3,779	16.4	—	—	—	—	—	3.9	7.0	2.5	0.08	0.49	0.55	—	—	26
—	—	64.0	2,825	2,676	11.8	28.0	10.0	13.0	2.0	14.0	10.8	4.8	3.1	0.07	0.33	0.39	—	—	14
—	—	72.0	3,180	3,012	13.3	32.0	11.0	15.0	3.0	16.0	12.1	5.4	3.4	0.07	0.38	0.44	—	—	15
—	—	69.0	3,030	2,623	9.1	—	—	—	—	—	11.2	5.0	3.8	0.10	0.31	0.38	—	—	20
—	—	76.0	3,333	2,886	10.0	—	—	—	—	—	12.3	5.5	4.2	0.11	0.34	0.42	—	—	22
1.73	1.42	—	—	—	8.5	60.0	—	24.0	5.0	33.0	27.8	2.4	7.0	0.22	0.20	1.38	25	1,410	—
1.89	1.55	—	—	—	9.3	66.0	—	26.0	6.0	36.0	30.4	2.6	7.6	0.24	0.22	1.51	28	1,544	—
—	—	—	—	—	3.5	—	—	—	—	—	11.6	1.4	2.4	—	—	—	—	—	—
—	—	—	—	—	10.0	—	—	—	—	—	33.0	4.1	6.9	—	—	—	—	—	—
1.77	1.45	—	—	—	4.1	64.0	37.0	—	13.0	43.0	37.3	2.1	7.2	0.22	0.06	2.37	4	609	—
1.92	1.58	—	—	—	4.4	70.0	40.0	—	14.0	47.0	40.5	2.2	7.8	0.24	0.06	2.57	4	662	—
—	—	—	—	—	3.4	21.0	10.0	8.0	2.0	13.0	10.2	1.1	2.3	—	—	—	—	—	—
—	—	—	—	—	11.0	68.0	33.0	27.0	6.0	41.0	33.5	3.5	7.5	—	—	—	—	—	—
1.70	1.40	—	—	—	13.4	54.0	26.0	24.0	4.0	30.0	27.6	2.5	7.8	—	—	—	—	—	—
1.91	1.57	—	—	—	15.0	61.0	29.0	27.0	5.0	34.0	31.0	2.8	8.7	—	—	—	—	—	—
1.77	1.45	—	—	—	7.6	65.0	35.0	24.0	8.0	41.0	33.6	3.1	9.2	—	—	—	—	—	—
1.96	1.61	—	—	—	8.4	72.0	39.0	27.0	9.0	45.0	37.1	3.4	10.1	—	—	—	—	—	—
—	—	11.0	499	469	2.1	—	—	—	1.0	8.0	6.4	0.5	2.0	0.09	0.04	—	13	—	—
—	—	54.0	2,389	2,244	10.3	—	—	—	5.0	38.0	30.5	2.3	9.6	0.43	0.18	—	62	—	—
1.54	1.26	—	—	—	6.7	—	—	—	—	—	27.4	1.5	11.7	—	—	—	—	—	—
1.75	1.43	—	—	—	7.6	—	—	—	—	—	31.3	1.7	13.3	—	—	—	—	—	—
—	—	79.0	3,496	3,031	48.1	—	—	—	—	—	9.9	1.3	5.8	0.27	0.62	1.13	—	—	—
—	—	86.0	3,800	3,295	52.3	—	—	—	—	—	10.8	1.4	6.3	0.29	0.68	1.23	—	—	—
—	—	—	—	—	1.8	—	—	—	—	—	6.6	0.5	2.1	—	—	—	38	—	—
—	—	—	—	—	8.5	—	—	—	—	—	31.5	2.2	10.0	—	—	—	183	—	—
—	—	76.0	3,367	3,175	7.4	—	—	—	—	—	6.5	0.3	3.0	0.14	0.23	—	—	—	—
—	—	86.0	3,791	3,575	8.4	—	—	—	—	—	7.3	0.4	3.4	0.16	0.25	—	—	—	—
—	—	62.0	2,731	2,215	84.9	—	—	—	—	—	1.4	2.9	3.5	0.26	0.67	0.29	—	—	—
—	—	67.0	2,936	2,382	91.3	—	—	—	—	—	1.5	3.2	3.8	0.28	0.72	0.31	—	—	—
—	—	—	—	—	25.5	34.0	—	—	2.0	14.0	11.9	2.2	27.2	8.40	2.28	2.03	—	—	—
—	—	—	—	—	28.2	38.0	—	—	2.0	15.0	13.2	2.4	30.1	9.31	2.52	2.25	—	—	—
1.63	1.33	—	—	—	5.3	—	—	—	—	—	31.1	2.2	6.5	0.39	0.14	0.99	22	1,158	—
1.78	1.46	—	—	—	5.8	—	—	—	—	—	34.0	2.4	7.1	0.43	0.15	1.08	24	1,264	—
—	—	10.0	432	411	0.8	5.0	—	—	1.0	4.0	2.3	0.3	3.4	—	—	—	—	—	—
—	—	58.0	2,578	2,450	4.8	30.0	—	—	8.0	23.0	13.5	1.9	20.1	—	—	—	—	—	—
—	—	—	—	—	—	—	—	—	—	—	—	—	—	—	—	—	—	—	—
—	—	—	—	—	—	—	—	—	—	—	—	—	—	—	—	—	—	—	—
—	—	—	—	—	2.7	—	—	—	—	—	1.8	0.4	1.6	—	—	—	—	—	—
—	—	—	—	—	23.5	—	—	—	—	—	15.8	3.8	14.0	—	—	—	—	—	—
—	—	65.0	2,878	2,672	37.0	—	—	—	—	—	12.0	1.7	6.8	0.61	0.95	1.24	—	—	—
—	—	72.0	3,161	2,935	40.6	—	—	—	—	—	13.2	1.8	7.5	0.67	1.04	1.36	—	—	—
—	—	—	—	—	3.4	19.0	—	6.0	2.0	—	7.8	1.2	2.4	0.14	0.09	0.69	64	—	—
—	—	—	—	—	11.6	64.0	—	19.0	8.0	—	26.7	3.9	8.1	0.46	0.29	2.35	217	—	—
1.80	1.48	—	—	—	11.0	—	—	—	—	—	29.0	2.5	6.1	0.60	0.33	1.60	—	—	—
1.91	1.56	—	—	—	11.7	—	—	—	—	—	30.7	2.6	6.5	0.63	0.35	1.69	—	—	—
1.70	1.4	—	—	—	10.7	—	—	—	—	—	24.4	1.8	13.2	1.41	0.19	5.88	—	—	—
1.98	1.63	—	—	—	12.4	—	—	—	—	—	28.4	2.1	15.4	1.64	0.22	6.85	—	—	—

Table 30 Feed Composition *continued*

Feed Name	International Feed Number	Moisture Basis	Dry Matter (%)	TDN (%)	DE (Mcal/kg)	ME (Mcal/kg)	NE$_m$ (Mcal/kg)	NE$_g$ (Mcal/kg)	NE$_l$ (Mcal/kg)	ME$_n$ (kcal/kg)	TME (kcal/kg)	NE$_p$ (kcal/kg)	TDN (%)
				Ruminants					**Dairy Cattle**	**Chickens**			**Horses**
RYE *Secale cereale*													
107 distillers grains,	5-04-023	As fed	92.0	56.0	2.47	2.08	1.23	0.63	1.26	—			—
dehydrated		Dry	100.0	61.0	2.69	2.27	1.33	0.69	1.38	—	—	—	—
108 grain	4-04-047	As fed	88.0	73.0	3.24	2.88	1.75	1.18	1.70	2,626	3,185	2,074	—
		Dry	100.0	84.0	3.70	3.29	2.00	1.35	1.94	3,001	3,640	2,370	—
109 flour by-product,	4-04-031	As fed	89.0	73.0	3.23	2.86	1.73	1.16	1.69	—	—	—	—
less than 8.5% fiber (Rye middlings)		Dry	100.0	82.0	3.62	3.20	1.94	1.30	1.89	—	—	—	—
RYEGRASS, ITALIAN *Lolium multiflorum*													
110 hay, sun-cured,	1-04-066	As fed	83.0	45.0	1.99	1.63	0.97	0.36	1.00	—	—	—	28.0
early bloom		Dry	100.0	54.0	2.38	1.96	1.16	0.43	1.20	—	—	—	33.0
RYEGRASS, PERENNIAL *Lolium perenne*													
111 fresh	2-04-086	As fed	27.0	18.0	0.80	0.69	0.41	0.24	0.41	—	—	—	—
		Dry	100.0	68.0	3.00	2.58	1.52	0.91	1.55	—	—	—	—
112 hay, sun-cured	1-04-077	As fed	86.0	55.0	2.43	2.07	1.22	0.68	1.25	—	—	—	40.0
		Dry	100.0	64.0	2.82	2.40	1.41	0.78	1.45	—	—	—	46.0
SAFFLOWER *Carthamus tinctorius*													
113 seeds, meal	5-04-110	As fed	92.0	52.0	2.32	1.93	1.14	0.50	1.18	1,193	—	1,035	—
solvent extracted		Dry	100.0	57.0	2.51	2.09	1.23	0.55	1.28	1,294	—	1,122	—
114 seeds, without	5-07-959	As fed	92.0	67.0	2.95	2.57	1.53	0.97	1.53	1,921	—	1,185	—
hulls, meal		Dry	100.0	73.0	3.22	2.80	1.67	1.06	1.67	2,096	—	1,293	—
SAGE, BLACK *Salvia mellifera*													
115 browse, fresh,	2-05-564	As fed	65.0	32.0	1.40	1.13							
stem-cured		Dry	100.0	49.0	2.16	1.73	0.68	0.15	0.70	—	—	—	—
							1.05	0.23	1.08				
SAGEBRUSH, BIG *Artemisia tridentata*													
116 browse, fresh,	2-07-992	As fed	65.0	33.0	1.43	1.16							
stem-cured		Dry	100.0	50.0	2.21	1.78	0.70	0.18	0.72	—	—	—	—
							1.07	0.28	1.11				
SORGHUM *Sorghum bicolor*													
117 aerial part with	1-07-960	As fed	89.0	52.0	2.28	1.90							
heads, sun-cured		Dry	100.0	58.0	2.56	2.13	1.12	0.52	1.16	—	—	—	41.0
(Fodder)							1.26	0.58	1.30	—	—	—	46.0
118 grain	4-04-383	As fed	90.0	78.0	3.40	3.03	1.85	1.25	1.78	3,311	3,359	2,564	—
		Dry	100.0	86.0	3.79	3.38	2.06	1.40	1.99	3,691	3,745	2,858	—
119 hay, sun-cured, late	1-06-141	As fed	92.0	49.0	2.15	1.76	1.05	0.36	1.08	—	—	—	—
vegetative (South)		Dry	100.0	53.0	2.34	1.91	1.14	0.40	1.18	—	—	—	—
120 silage	3-04-323	As fed	30.0	18.0	0.78	0.66	0.39	0.19	0.40	—	—	—	—
		Dry	100.0	60.0	2.65	2.22	1.31	0.65	1.35	—	—	—	—
SORGHUM, JOHNSONGRASS *Sorghum halepense*													
121 hay, sun-cured	1-04-407	As fed	89.0	48.0	2.09	1.71	1.02	0.35	1.05	—	—	—	41.0
		Dry	100.0	53.0	2.34	1.91	1.14	0.40	1.18	—	—	—	46.0
SORGHUM, SUDANGRASS *Sorghum bicolor sudanense*													
122 fresh, midbloom	2-04-485	As fed	23.0	14.0	0.63	0.54	0.32	0.17	0.32	—	—	—	—
		Dry	100.0	63.0	2.78	2.36	1.39	0.75	1.42	—	—	—	—
123 hay, sun-cured,	1-04-480	As fed	91.0	51.0	2.25	1.86	1.10	0.46	1.14	—	—	—	42.0
full bloom		Dry	100.0	56.0	2.47	2.04	1.21	0.51	1.25	—	—	—	46.0
SOYBEAN *Glycine max*													
124 hay, sun-cured	1-04-542	As fed	88.0	54.0	2.36	1.99	1.17	0.60	1.21	—	—	—	37.0
dough stage		Dry	100.0	61.0	2.69	2.27	1.33	0.69	1.38	—	—	—	42.0
125 seeds, meal	5-04-600	As fed	90.0	77.0	3.37	3.00	1.83	1.23	1.77	2,429	—	1,722	—
mechanical extracted		Dry	100.0	85.0	3.75	3.34	2.03	1.37	1.96	2,699	—	1,914	—
126 seeds, meal	5-20-637	As fed	89.0	75.0	3.31	2.94	1.79	1.20	1.73	2,219	2,639	1,589	—
solvent extracted 44% protein		Dry	100.0	84.0	3.70	3.29	2.00	1.35	1.94	2,485	2,956	-1,779	—
SUGARCANE *Saccharum officinarum*													
127 bagasse, dehydrated	1-04-686	As fed	91.0	44.0	1.93	1.54	0.94	0.17	0.96	—	—	—	—
		Dry	100.0	48.0	2.12	1.69	1.03	0.19	1.06	—	—	—	—
SUNFLOWER, COMMON *Helianthus annuus*													
128 seeds, meal	5-09-340	As fed	90.0	40.0	1.75	1.36	0.86	0.01	0.86	1,543	—	—	—
solvent extracted		Dry	100.0	44.0	1.94	1.51	0.96	0.01	0.96	1,715	—	—	—
TIMOTHY *Phleum pratense*													
129 fresh, midbloom	2-04-905	As fed	29.0	18.0	0.81	0.69	0.41	0.22	0.42	—	—	—	—
		Dry	100.0	63.0	2.78	2.36	1.39	0.75	1.42	—	—	—	—
130 hay, sun-cured,	1-04-883	As fed	89.0	54.0	2.39	2.01	1.19	0.61	1.22	—	—	—	51.0
midbloom		Dry	100.0	61.0	2.69	2.27	1.33	0.69	1.38	—	—	—	57.0
131 hay, sun-cured,	1-04-885	As fed	88.0	50.0	2.18	1.81	1.07	0.45	1.11	—	—	—	42.0
late bloom		Dry	100.0	56.0	2.47	2.04	1.21	0.51	1.25	—	—	—	47.0
TREFOIL, BIRDSFOOT *Lotus corniculatus*													
132 fresh	2-20-786	As fed	24.0	16.0	0.71	0.60	0.36	0.21	0.36	—	—	—	—
		Dry	100.0	66.0	2.91	2.49	1.47	0.85	1.50	—	—	—	—
133 hay, sun-cured	1-05-044	As fed	92.0	54.0	2.40	2.01	1.18	0.57	1.22	—	—	—	46.0
		Dry	100.0	59.0	2.60	2.18	1.28	0.62	1.33	—	—	—	49.0

Horses DE (Mcal/kg)	Horses ME (Mcal/kg)	Swine TDN (%)	Swine DE (kcal/kg)	Swine ME (kcal/kg)	Crude Protein (%)	Plant Cell Wall Constituents Cell Walls (%)	Cellulose (%)	Hemicellulose (%)	Lignin (%)	Acid Detergent Fiber (%)	Crude Fiber (%)	Ether Extract (%)	Ash (%)	Minerals Calcium (%)	Phosphorus (%)	Potassium (%)	Fat-Soluble Vitamins Carotene (Provitamin A) (mg/kg)	Vitamin D12 (IU/g)	Vitamin E (mg/kg)
—	—	—	—	—	21.6	—	—	—	—	—	12.3	7.2	2.3	0.15	0.48	0.07	—	—	—
—	—	—	—	—	23.5	—	—	—	—	—	13.4	7.8	2.5	0.16	0.52	0.08	—	—	—
—	—	75.0	3,251	2,911	12.1	—	—	—	—	—	2.2	1.5	1.6	0.06	0.32	0.46	0	—	15
—	—	86.0	3,716	3,327	13.8	—	—	—	—	—	2.5	1.7	1.9	0.07	0.37	0.52	0	—	17
—	—	73.0	3,220	2,972	16.2	—	—	—	—	—	4.8	3.2	3.7	0.06	0.62	0.62	—	—	—
—	—	82.0	3,610	3,332	18.2	—	—	—	—	—	5.4	3.6	4.2	0.07	0.70	0.70	—	—	—
1.15	0.95	—	—	—	4.6	—	—	—	—	—	30.3	0.8	7.0	—	—	—	—	—	—
1.38	1.14	—	—	—	5.5	—	—	—	—	—	36.3	0.9	8.4	—	—	—	—	—	—
—	—	—	—	—	2.8	—	—	—	—	—	6.2	0.7	2.3	0.15	0.07	0.51	59	—	47
—	—	—	—	—	10.4	—	—	—	—	—	23.2	2.7	8.6	0.55	0.27	1.91	222	—	178
1.59	1.30	—	—	—	7.4	35.0	—	—	2.0	26.0	26.1	1.9	9.9	0.56	0.28	1.44	103	—	182
1.85	1.51	—	—	—	8.6	41.0	—	—	2.0	30.0	30.3	2.2	11.5	0.65	0.32	1.67	120	—	211
—	—	—	—	—	23.4	53.0	—	—	13.0	38.0	30.0	1.4	5.4	0.34	0.75	0.76	—	—	1
—	—	—	—	—	25.4	58.0	—	—	14.0	41.0	32.5	1.5	5.9	0.37	0.81	0.82	—	—	1
—	—	—	3,361	2,908	43.0	—	—	—	—	—	13.5	1.3	7.5	0.35	1.29	1.10	—	—	1
—	—	—	3,666	3,172	46.9	—	—	—	—	—	14.7	1.4	8.2	0.38	1.40	1.19	—	—	1
—	—	—	—	—	5.5	—	—	—	—	—	—	7.0	3.6	0.53	0.11	—	—	—	—
—	—	—	—	—	8.5	—	—	—	—	—	—	10.8	5.5	0.81	0.17	—	—	—	—
—	—	—	—	—	6.1	27.0	—	—	8.0	20.0	—	7.2	4.3	0.46	0.12	—	10	—	—
—	—	—	—	—	9.3	42.0	—	—	12.0	30.0	—	11.0	6.6	0.71	0.18	—	16	—	—
1.64	1.34	—	—	—	6.7	—	—	—	—	—	23.9	2.2	8.4	0.35	0.18	1.31	46	—	—
1.84	1.51	—	—	—	7.5	—	—	—	—	—	26.9	2.4	9.4	0.40	0.21	1.47	52	—	—
—	—	78.0	3,431	3,216	11.1	21.0	3.0	16.0	—	5.0	2.4	2.8	1.8	0.03	0.29	0.35	1	26	10
—	—	87.0	3,824	3,585	12.4	23.0	3.0	18.0	—	5.0	2.6	3.1	2.0	0.04	0.33	0.39	1	29	12
—	—	—	—	—	11.0	64.0	29.0	—	5.0	36.0	30.4	2.4	10.1	0.37	0.18	1.75	—	—	—
—	—	—	—	—	12.0	70.0	31.0	—	5.0	39.0	33.0	2.6	11.0	0.40	0.19	1.90	—	—	—
—	—	16.0	722	682	2.2	—	—	—	2.0	11.0	8.2	0.9	2.6	0.10	0.06	0.40	5	196	—
—	—	55.0	2,441	2,306	7.5	—	—	—	6.0	38.0	27.9	3.0	8.7	0.35	0.21	1.37	15	662	—
1.66	1.36	—	—	—	8.5	—	—	—	—	—	29.9	2.1	7.3	0.75	0.25	1.21	35	—	—
1.86	1.52	—	—	—	9.5	—	—	—	—	—	33.5	2.4	8.2	0.84	0.28	1.35	39	—	—
—	—	14.0	619	584	2.0	15.0	8.0	6.0	1.0	9.0	6.8	0.4	2.4	0.10	0.08	0.49	42	—	—
—	—	62.0	2,719	2,562	8.8	65.0	34.0	25.0	5.0	40.0	30.0	1.8	10.5	0.43	0.36	2.14	183	—	—
1.69	1.38	—	—	—	7.3	62.0	32.0	24.0	5.0	38.0	32.8	1.6	8.7	0.50	0.28	1.70	54	—	—
1.85	1.52	—	—	—	8.0	68.0	35.0	26.0	6.0	42.0	36.0	1.8	9.6	0.55	0.30	1.87	59	—	—
1.50	1.23	—	—	—	14.7	—	—	—	—	—	25.0	3.6	6.0	—	—	—	—	—	—
1.71	1.40	—	—	—	16.8	—	—	—	—	—	28.5	4.1	6.8	—	—	—	—	—	—
—	—	79.0	3,610	2,972	42.9	—	—	—	—	—	5.9	4.8	6.0	0.26	0.61	1.79	0	—	7
—	—	88.0	4,013	3,304	47.7	—	—	—	—	—	6.6	5.3	6.7	0.29	0.68	1.98	0	—	7
—	—	75.0	3,318	2,817	44.6	—	—	—	—	—	6.2	1.4	6.5	—	—	—	—	—	—
—	—	84.0	3,716	3,155	49.9	—	—	—	—	—	7.0	1.5	7.3	—	—	—	—	—	—
—	—	—	—	—	1.5	—	—	—	—	—	43.9	0.7	2.9	0.82	0.27	0.46	—	—	—
—	—	—	—	—	1.6	—	—	—	—	—	48.1	0.7	3.2	0.90	0.29	0.50	—	—	—
—	—	—	1,991	1,807	23.3	36.0	—	—	11.0	30.0	31.6	1.1	5.6	0.21	0.93	0.96	—	—	—
—	—	—	2,213	2,009	25.9	40.0	—	—	12.0	33.0	35.1	1.2	6.3	0.23	1.03	1.06	—	—	—
—	—	—	—	—	2.7	19.0	9.0	—	1.0	11.0	9.8	0.9	1.9	—	—	—	—	—	—
—	—	—	—	—	9.1	64.0	31.0	—	4.0	37.0	33.5	3.0	6.6	—	—	—	—	—	—
2.00	1.64	—	—	—	8.1	60.0	29.0	28.0	4.0	32.0	27.6	2.3	5.6	0.43	0.20	1.41	—	—	—
2.25	1.84	—	—	—	9.1	67.0	33.0	31.0	5.0	36.0	31.0	2.6	6.3	0.48	0.22	1.59	—	—	—
1.67	1.37	—	—	—	6.9	62.0	30.0	26.0	6.0	35.0	28.7	2.5	4.8	0.34	0.16	1.42	—	—	—
1.89	1.55	—	—	—	7.8	70.0	34.0	29.0	7.0	40.0	32.5	2.8	5.4	0.38	0.18	1.61	—	—	—
—	—	—	—	—	5.1	—	—	—	—	—	6.0	0.7	2.2	0.46	0.05	0.48	—	—	—
—	—	—	—	—	21.0	—	—	—	—	—	24.7	2.7	9.0	1.91	0.22	1.99	—	—	—
1.82	1.49	—	—	—	15.0	43.0	22.0	—	8.0	33.0	28.3	2.3	6.5	1.57	0.25	1.77	173	1,421	—
1.98	1.62	—	—	—	16.3	47.0	24.0	—	9.0	36.0	30.7	2.5	7.0	1.70	0.27	1.92	188	1,544	—

Table 30 Feed Composition *continued*

Feed Name	International Feed Number	Moisture Basis	Dry Matter (%)	Ruminants					Dairy Cattle	Chickens			Horses
				TDN (%)	DE (Mcal/kg)	ME (Mcal/kg)	NE$_m$ (Mcal/kg)	NE$_g$ (Mcal/kg)	NE$_l$ (Mcal/kg)	ME$_n$ (kcal/kg)	TME (kcal/kg)	NE$_p$ (kcal/kg)	TDN (%)
TRITICALE *Triticale hexaploide*													
134 grain	4-20-362	As fed	90.0	76.0	3.33	2.96	1.80	1.21	1.74	3,163	3,256	2,200	—
		Dry	100.0	84.0	3.70	3.29	2.00	1.35	1.94	3,521	3,625	2,450	—
UREA													
135 45% nitrogen, 281% protein equivalent	5-05-070	As fed	99.0	0.0	0.0	0.0	0.0	0.0	0.0	—	—	—	—
		Dry	100.0	0.0	0.0	0.0	0.0	0.0	0.0	—	—	—	—
VETCH *Vicia* spp													
136 hay, sun-cured	1-05-106	As fed	89.0	51.0	2.24	1.86	1.10	0.49	1.14	—	—	—	—
		Dry	100.0	57.0	2.51	2.09	1.23	0.55	1.28	—	—	—	—
WHEAT *Triticum aestivum*													
137 bran	4-05-190	As fed	89.0	63.0	2.74	2.37	1.40	0.86	1.42	1,237	1,706	981	44.0
		Dry	100.0	70.0	3.09	2.67	1.58	0.97	1.60	1,393	1,921	1,105	50.0
138 flour by-product, less than 4.5% fiber (Middlings)	4-28-220	As fed	88.0	73.0	3.22	2.86	1.73	1.16	1.68	2,543	—	1,751	—
		Dry	100.0	83.0	3.66	3.25	1.97	1.32	1.91	2,890	—	1,990	—
139 grain	4-05-211	As fed	89.0	78.0	3.45	3.08	1.89	1.28	1.81	3,023	3,455	2,197	—
		Dry	100.0	88.0	3.88	3.47	2.12	1.45	2.04	3,401	3,887	2,471	—
140 hay, sun-cured	1-05-172	As fed	88.0	51.0	2.24	1.87	1.10	0.51	1.14	—	—	—	39.0
		Dry	100.0	58.0	2.56	2.13	1.26	0.58	1.30	—	—	—	44.0
141 straw	1-05-175	As fed	89.0	39.0	1.72	1.34	0.85	0.01	0.85	—	—	—	30.0
		Dry	100.0	44.0	1.94	1.51	0.96	0.01	0.96	—	—	—	34.0
WHEATGRASS, CRESTED *Agropyron desertorum*													
142 fresh, full bloom	2-05-424	As fed	45.0	27.0	1.21	1.02	0.60	0.31	0.62	—	—	—	—
		Dry	100.0	61.0	2.69	2.27	1.33	0.69	1.38	—	—	—	—
143 hay, sun-cured	1-05-418	As fed	93.0	49.0	2.17	1.77	1.06	0.37	1.09	—	—	—	44.0
		Dry	100.0	53.0	2.34	1.91	1.14	0.40	1.18	—	—	—	47.0
WHEY													
144 dehydrated (Cattle)	4-01-182	As fed	93.0	75.0	3.33	2.95	1.78	1.19	1.74	1,949	1,662	1,548	—
		Dry	100.0	81.0	3.57	3.16	1.91	1.27	1.87	2,087	1,780	1,659	—
YEAST *Saccharomyces cerevisiae*													
145 brewers, dehydrated	7-05-527	As fed	93.0	74.0	3.25	2.87	1.73	1.14	1.70	2,055	2,943	1,263	—
		Dry	100.0	79.0	3.48	3.07	1.85	1.22	1.82	2,199	3,150	1,351	—
146 irradiated, dehydrated	7-05-529	As fed	94.0	72.0	3.15	2.76	1.65	1.07	1.64	—	—	—	—
		Dry	100.0	76.0	3.35	2.93	1.76	1.14	1.74	—	—	—	—
YEAST, TORULA *Torulopsis utilis*													
147 torula, dehydrated	7-05-534	As fed	93.0	73.0	3.21	2.82	1.69	1.11	1.67	1,855	2,872	—	—
		Dry	100.0	78.0	3.44	3.02	1.82	1.19	1.79	1,989	3,080	—	—

Source: *United States-Canadian Tables of Feed Composition*, third revision, National Research Council, National Academy of Sciences, Washington, D.C., 1982

Horses DE (Mcal/kg)	ME (Mcal/kg)	Swine TDN (%)	DE (kcal/kg)	ME (kcal/kg)	Crude Protein (%)	Plant Cell Wall Constituents Cell Walls (%)	Cellulose (%)	Hemicellulose (%)	Lignin (%)	Acid Detergent Fiber (%)	Crude Fiber (%)	Ether Extract (%)	Ash (%)	Minerals Calcium (%)	Phosphorus (%)	Potassium (%)	Fat-Soluble Vitamins Carotene (Provitamin A) (mg/kg)	Vitamin D₁₂ (IU/g)	Vitamin E (mg/kg)
—	—	75.0	3,299	3,050	15.8	—	—	—	—	—	4.0	1.5	1.8	0.05	0.30	0.36	—	—	—
—	—	83.0	3,673	3,396	17.6	—	—	—	—	—	4.4	1.7	2.0	0.06	0.33	0.40	—	—	—
—	—	—	—	—	275.8	—	—	—	—	—	—	—	—	—	—	—	—	—	—
—	—	—	—	—	279.6	—	—	—	—	—	—	—	—	—	—	—	—	—	—
—	—	—	—	—	18.5	43.0	—	—	7.0	30.0	27.3	2.7	8.1	1.05	0.29	2.07	411	—	—
—	—	—	—	—	20.8	48.0	—	—	8.0	33.0	30.6	3.0	9.1	1.18	0.32	2.32	461	—	—
—	—	57.0	2,414	2,212	15.2	46.0	9.0	30.0	3.0	14.0	10.0	3.9	6.1	0.11	1.22	1.38	3	—	18
—	—	64.0	2,718	2,491	17.1	51.0	11.0	34.0	3.0	15.0	11.3	4.4	6.9	0.13	1.38	1.56	3	—	21
—	—	71.0	3,080	2,860	15.1	—	—	—	—	—	2.6	3.2	2.4	—	—	—	—	—	—
—	—	81.0	3,500	3,250	17.2	—	—	—	—	—	3.0	3.6	2.7	—	—	—	—	—	—
—	—	79.0	3,268	3,253	14.2	—	7.0	—	—	7.0	2.6	1.8	1.7	0.04	0.37	0.38	0	—	15
—	—	89.0	3,676	3,660	16.0	—	8.0	—	—	8.0	2.9	2.0	1.9	0.04	0.42	0.42	0	—	17
1.57	1.28	—	—	—	7.4	60.0	—	—	6.0	36.0	24.6	1.9	6.2	0.13	0.17	0.87	75	1,352	—
1.79	1.47	—	—	—	8.5	68.0	—	—	7.0	41.0	28.1	2.2	7.1	0.15	0.20	1.00	85	1,544	—
1.27	1.04	—	—	—	3.2	75.0	35.0	—	12.0	48.0	36.9	1.6	6.9	0.16	0.04	1.26	2	587	—
1.43	1.17	—	—	—	3.6	85.0	39.0	—	14.0	54.0	41.6	1.8	7.8	0.18	0.05	1.42	2	662	—
—	—	—	—	—	4.4	—	—	—	—	—	13.6	1.6	4.2	0.18	0.13	—	—	—	—
—	—	—	—	—	9.8	—	—	—	—	—	30.3	3.6	9.3	0.39	0.28	—	—	—	—
1.76	1.45	—	—	—	11.5	—	—	—	5.0	34.0	30.5	2.1	6.7	0.31	0.20	1.85	—	—	—
1.90	1.56	—	—	—	12.4	—	—	—	6.0	36.0	32.9	2.3	7.2	0.33	0.21	2.00	—	—	—
—	—	77.0	3,188	3,115	13.3	0.0	—	0.0	—	0.0	0.2	0.7	9.2	0.86	0.76	1.15	—	—	0
—	—	83.0	3,415	3,337	14.2	0.0	—	0.0	—	0.0	0.2	0.7	9.8	0.92	0.82	1.23	—	—	0
—	—	71.0	3,111	2,876	43.8	—	—	—	—	—	2.9	0.8	6.6	0.12	1.40	1.67	—	—	2
—	—	76.0	3,330	3,078	46.9	—	—	—	—	—	3.1	0.9	7.1	0.13	1.49	1.79	—	—	2
—	—	—	—	—	48.1	—	—	—	—	—	6.2	1.1	6.5	0.78	1.42	2.14	—	—	—
—	—	—	—	—	51.2	—	—	—	—	—	6.6	1.2	6.6	0.83	1.51	2.28	—	—	—
—	—	64.0	2,842	2,421	49.1	—	—	—	—	—	2.3	1.6	7.7	0.50	1.59	1.90	—	—	—
—	—	69.0	3,049	2,597	52.7	—	—	—	—	—	2.4	1.7	8.3	0.54	1.71	2.04	—	—	—

Table 31 Amino Acid Composition of Common Feed

Feed Name	International Feed Number	Moisture Basis	Arginine (%)	Glycine (%)	Histidine (%)	Isoleucine (%)
1 Alfalfa meal, dehydrated, 17% protein	1-00-023	As fed Dry	0.77 0.84	0.84 0.91	0.33 0.36	0.81 0.88
2 Barley, grain	4-00-549	As fed Dry	0.51 0.58	0.38 0.43	0.24 0.28	0.45 0.51
3 Blood meal	5-00-380	As fed Dry	3.25 3.55	3.42 3.74	3.97 4.34	0.87 0.95
4 Brewers grains, dehydrated	5-02-141	As fed Dry	1.27 1.38	1.08 1.18	0.52 0.56	1.54 1.68
5 Buttermilk, dehydrated	5-01-160	As fed Dry	1.08 1.17	0.47 0.51	0.85 0.92	2.42 2.62
6 Casein, dehydrated	5-01-162	As fed Dry	3.49 3.85	1.61 1.77	2.59 2.86	5.72 6.32
7 Corn distillers grains with solubles, dehydrated	5-28-236	As fed Dry	0.96 1.05	0.51 0.55	0.64 0.70	1.39 1.52
8 Corn and cob meal	4-28-238	As fed Dry	0.36 0.42	0.31 0.36	0.16 0.19	0.35 0.40
9 Corn gluten meal, 60% protein	5-28-242	As fed Dry	2.08 2.31	2.10 2.33	1.40 1.55	2.54 2.82
10 Corn grain	4-02-935	As fed Dry	0.43 0.48	0.37 0.42	0.26 0.29	0.35 0.39
11 Corn grain, opaque 2 (High lysine)	4-28-253	As fed Dry	0.66 0.73	0.48 0.53	0.35 0.39	0.35 0.38
12 Cotton seed meal, solv extd, 41% protein	5-07-872	As fed Dry	4.27 4.71	1.94 2.14	1.15 1.27	1.44 1.59
13 Fish solubles, condensed	5-01-969	As fed Dry	1.63 3.25	3.85 7.68	1.43 2.85	1.03 2.06
14 Fish meal, Menhaden	5-02-009	As fed Dry	3.75 4.09	4.19 4.57	1.45 1.58	2.88 3.15
15 Meat meal, rendered (Tankage)	5-00-386	As fed Dry	3.59 3.90	6.61 7.17	1.83 1.99	1.93 2.09
16 Meat meal, with bone, rendered (Tankage)	5-00-387	As fed Dry	2.82 3.03	6.58 7.08	1.76 1.90	1.87 2.01

Leucine (%)	Lysine (%)	Methionine (%)	Cystine (%)	Phenylalanine (%)	Tyrosine (%)	Serine (%)	Threonine (%)	Tryptophan (%)	Valine (%)
1.28	0.85	0.27	0.29	0.80	0.54	0.71	0.71	0.34	0.88
1.39	0.93	0.29	0.31	0.87	0.59	0.77	0.77	0.37	0.96
0.75	0.39	0.15	0.21	0.58	0.34	0.43	0.37	0.15	0.57
0.85	0.44	0.17	0.24	0.66	0.38	0.49	0.42	0.17	0.64
9.94	6.33	0.88	1.24	5.49	1.92	4.35	3.56	0.98	6.52
10.86	6.92	0.97	1.35	6.00	2.09	4.75	3.89	1.07	7.12
2.49	0.88	0.46	0.35	1.44	1.20	1.30	0.93	0.37	1.61
2.70	0.95	0.50	0.38	1.56	1.30	1.42	1.01	0.40	1.75
3.21	2.28	0.71	0.39	1.46	1.00	1.50	1.52	0.49	2.58
3.48	2.47	0.76	0.42	1.58	1.08	1.62	1.64	0.53	2.80
8.80	7.14	2.81	0.31	4.81	4.90	5.46	3.91	1.08	6.71
9.71	7.88	3.10	0.34	5.31	5.41	6.03	4.32	1.19	7.40
2.23	0.70	0.50	0.29	1.51	0.70	1.30	0.93	0.17	1.50
2.43	0.77	0.54	0.32	1.64	0.76	1.42	1.01	0.19	1.63
0.86	0.17	0.14	0.12	0.39	0.32	—	0.28	0.07	0.31
1.00	0.20	0.16	0.14	0.45	0.38	—	0.33	0.08	0.36
10.23	1.01	1.78	0.99	4.02	3.19	3.35	2.22	0.30	3.09
11.23	1.12	1.98	1.10	4.45	3.54	3.71	2.46	0.33	3.43
1.21	0.25	0.17	0.22	0.48	0.38	0.50	0.35	0.08	0.44
1.37	0.28	0.19	0.25	0.54	0.43	0.57	0.40	0.09	0.50
0.99	0.42	0.17	0.20	0.43	0.40	0.47	0.37	0.11	0.50
1.10	0.46	0.19	0.22	0.48	0.44	0.52	0.41	0.12	0.56
2.42	1.82	0.56	0.81	2.01	1.15	1.82	1.34	0.51	1.99
2.67	2.01	0.62	0.90	2.21	1.27	2.01	1.48	0.56	2.20
1.86	1.86	0.71	0.27	1.02	0.44	1.03	0.87	0.34	1.22
3.72	3.71	1.42	0.54	2.04	0.87	2.05	1.73	0.68	2.43
4.48	4.72	1.75	0.56	2.46	1.94	2.23	2.50	0.65	3.22
4.89	5.15	1.91	0.61	2.69	2.12	2.43	2.73	0.71	3.52
5.12	3.74	0.73	0.45	2.54	1.29	—	2.32	0.65	3.77
5.56	4.06	0.79	0.49	2.76	1.40	—	2.52	0.70	4.10
5.27	3.32	0.69	0.27	2.28	—	—	2.18	0.62	3.42
5.67	3.57	0.74	0.29	2.46	—	—	2.35	0.67	3.68

Table 31 Amino Acid Composition of Common Feed *continued*

Feed Name	International Feed Number	Moisture Basis	Arginine (%)	Glycine (%)	Histidine (%)	Isoleucine (%)
17 Milk, skimmed, dehydrated	5-01-175	As fed	1.15	0.29	0.86	2.18
		Dry	1.23	0.31	0.92	2.32
18 Oats, grain	4-03-309	As fed	0.70	0.46	0.18	0.43
		Dry	0.79	0.52	0.21	0.49
19 Oats, grain, Pacific Coast	4-07-999	As fed	0.58	0.40	0.17	0.38
		Dry	0.63	0.44	0.18	0.42
20 Peanut meal, solv extd (Peanut meal)	5-03-650	As fed	4.55	2.35	0.95	1.76
		Dry	4.95	2.56	1.03	1.91
21 Poultry feathers, hydrolyzed	5-03-795	As fed	7.05	6.44	0.99	4.06
		Dry	7.58	6.92	1.06	4.37
22 Rape meal, solv extd	5-03-871	As fed	2.06	1.79	0.99	1.35
		Dry	2.26	1.97	1.09	1.48
23 Rye, grain	4-04-047	As fed	0.53	0.49	0.26	0.47
		Dry	0.61	0.56	0.29	0.53
24 Safflower meal, solv extd	5-04-110	As fed	1.95	1.13	—	0.28
		Dry	2.11	1.22	—	0.30
25 Sorghum, grain	4-04-383	As fed	0.39	0.34	0.23	0.45
		Dry	0.43	0.38	0.26	0.50
26 Soybean meal, mech extd	5-04-600	As fed	3.07	2.38	1.14	2.63
		Dry	3.41	2.64	1.26	2.92
27 Soybean meal, solv extd	5-04-604	As fed	3.03	1.82	1.07	2.03
		Dry	3.38	2.03	1.19	2.27
28 Sunflower meal, solv extd	5-09-340	As fed	2.30	—	0.55	1.00
		Dry	2.56	—	0.61	1.11
29 Wheat bran	4-05-190	As fed	0.96	0.86	0.39	0.51
		Dry	1.09	0.97	0.44	0.57
30 Whey, dehydrated	4-01-182	As fed	0.34	0.49	0.17	0.79
		Dry	0.36	0.53	0.18	0.84
31 Yeast, brewers, dehydrated	7-05-527	As fed	2.20	1.75	1.09	2.21
		Dry	2.35	1.87	1.17	2.37
32 Yeast, torula, dehydrated	7-05-534	As fed	2.64	2.66	1.32	2.85
		Dry	2.83	2.85	1.42	3.06

Source: *United States-Canadian Tables of Feed Composition*, third revision, National Research Council, National Academy of Sciences, Washington, D.C., 1982.

Leu-cine (%)	Ly-sine (%)	Me-thio-nine (%)	Cys-tine (%)	Phenyl-ala-nine (%)	Tyro-sine (%)	Ser-ine (%)	Threo-nine (%)	Tryp-to-phan (%)	Va-line (%)
3.32	2.53	0.90	0.45	1.56	1.14	1.67	1.56	0.43	2.28
3.53	2.70	0.96	0.48	1.66	1.22	1.78	1.67	0.46	2.43
0.81	0.39	0.17	0.19	0.52	0.46	0.44	0.36	0.15	0.56
0.91	0.44	0.19	0.22	0.58	0.52	0.50	0.40	0.17	0.63
0.70	0.33	0.13	0.17	0.43	0.70	0.40	0.30	0.12	0.49
0.77	0.37	0.14	0.18	0.47	0.77	0.44	0.33	0.13	0.54
2.70	1.77	0.42	0.73	2.04	1.51	3.10	1.16	0.48	1.88
2.94	1.93	0.46	0.79	2.22	1.65	3.37	1.26	0.52	2.04
6.94	2.32	0.55	3.24	3.05	2.32	9.26	3.97	0.52	6.48
7.46	2.49	0.59	3.48	3.28	2.49	9.96	4.27	0.56	6.97
2.50	1.98	0.71	0.30	1.41	0.79	1.57	1.56	0.43	1.79
2.74	2.18	0.78	0.33	1.55	0.87	1.72	1.72	0.47	1.96
0.70	0.42	0.17	0.19	0.56	0.26	0.52	0.36	0.11	0.56
0.80	0.48	0.19	0.21	0.64	0.30	0.60	0.41	0.13	0.64
—	0.72	0.34	0.36	—	—	—	0.51	0.27	—
—	0.78	0.37	0.39	—	—	—	0.56	0.29	—
1.44	0.25	0.13	0.20	0.56	0.41	0.50	0.36	0.11	0.52
1.60	0.28	0.15	0.22	0.62	0.46	0.55	0.40	0.12	0.58
3.62	2.79	0.65	0.56	2.20	1.55	2.01	1.72	0.61	2.28
4.02	3.10	0.72	0.63	2.45	1.72	2.23	1.92	0.68	2.53
3.27	2.68	0.52	0.75	2.11	1.33	2.11	1.66	0.64	2.02
3.65	2.99	0.58	0.83	2.36	1.48	2.36	1.85	0.71	2.25
1.60	1.00	0.50	0.50	1.15	—	1.00	1.05	0.45	1.60
1.78	1.11	0.56	0.56	1.28	—	1.11	1.17	0.50	1.78
0.92	0.58	0.19	0.32	0.55	0.42	0.68	0.46	0.25	0.69
1.03	0.65	0.22	0.36	0.62	0.48	0.77	0.51	0.28	0.78
1.18	0.94	0.19	0.30	0.35	0.25	0.47	0.90	0.18	0.68
1.26	1.00	0.20	0.32	0.37	0.26	0.50	0.96	0.19	0.73
3.23	3.11	0.74	0.49	1.83	1.50	—	2.12	0.52	2.36
3.45	3.33	0.79	0.53	1.96	1.60	—	2.27	0.55	2.52
3.52	3.74	0.77	0.61	2.85	2.00	2.76	2.64	0.52	2.96
3.78	4.01	0.83	0.65	3.06	2.14	2.96	2.83	0.56	3.17

Table 32 Conversion Factors

If the measure is given in this unit	X	Multiply by this conversion factor	=	To obtain this unit
		MASS		
lb		0.453592		kg
kg		2.204624		lb
lb		453.592		g
g		0.0022046		lb
lb		453,592.		mg
mg		0.0000022046		lb
oz		0.02835		kg
kg		35.2734		oz
oz		28.35		g
g		0.03527		oz
ton (short, 2,000 lb)		907.185		kg
kg		0.0011		ton (short, 2,000 lb)
tonne		1,000		kg
kg		0.001		tonne
kg		1 000 000		mg
mg		0.000 001		kg
kg		1 000		g
g		0.001		kg
g		1 000		mg
mg		0.001		g
g		1 000 000		μg
μg		0.000 001		g
mg		1 000		g
μg		0.001		mg
mg/kg		0.453592		mg/lb
mg/lb		2.2046		mg/kg
mg/g		453.592		mg/lb
mg/lb		0.0022046		mg/g
g/kg		0.453592		g/lb
g/lb		2.2046		g/kg
kg/kg		0.453592		kg/lb
kg/lb		2.2046		kg/kg
kcal/kg		0.453592		kcal/lb
kcal/lb		2.2046		kcal/kg
Mcal/kg		0.453592		Mcal/lb
Mcal/lb		2.2046		Mcal/kg
IU/kg		0.453592		IU/lb
IU/lb		2.2046		IU/kg
ICU/kg		0.453592		ICU/lb
ICU/lb		2.2046		ICU/kg
μg/kg		0.453592		μg/lb
μg/lb		2.2046		μg/kg
g/Mcal		0.0022046		lb/Mcal
lb/Mcal		453.592		g/Mcal
Mcal		1 000		kcal
kcal		0.001		Mcal
ppm		1.0		μg/g
ppm		1.0		mg/g
ppm		0.453592		mg/lb
ppm		0.0001		%
mg/kg		0.0001		%
mg/g		0.1		%
g/kg		0.1		%

If the measure is given in this unit	X	Multiply by this conversion factor	=	To obtain this unit
LENGTH				
ft		12		in
yd		3		ft
rd		16.5		ft
furlong		220		yd
mi		5,280		ft
mi		1,760		yd
mi		320		rd
mi		8		furlong
cm		0.01		m
m		100		cm
m		0.001		km
km		1 000		m
in		2.54		cm
cm		0.3937		in
in		0.0254		m
m		39.37		in
ft		0.3048		m
m		3.2808		ft
yd		0.9144		m
m		1.0936		yd
rd		5.0292		m
m		0.1988		rd
furlong		201.1675		m
m		0.00497		furlong
mi		1,609.34		m
m		0.0006214		mi
mi		1.60934		km
km		0.6214		mi
AREA				
in^2		6.4516		cm^2
cm^2		0.1550		in^2
in^2		0.000645		m^2
m^2		1,550.3875		in^2
ft^2		929.03		cm^2
cm^2		0.001076		ft^2
ft^2		0.092903		m^2
m^2		10.764		ft^2
yd^2		0.836127		m^2
m^2		1.19599		yd^2
rd^2		25.2928		m^2
m^2		0.03954		rd^2
mi^2		2.58998		km^2
km^2		0.3861		mi^2
acre		4,046.86		m^2
m^2		0.0002471		acre
acre		0.404686		hectare
hectare		2.471		acre
ft^2		144		in^2
yd^2		1,296		in^2
yd^2		9		ft^2
rd^2		272.25		ft^2
rd^2		30.25		yd^2

Table 32 Conversion Factors *continued*

If the measure is given in this unit	X	Multiply by this conversion factor	=	To obtain this unit
		AREA (Continued)		
acre		43,560		ft^2
acre		4,840		yd^2
acre		160		rd^2
mi^2		640		acre
cm^2		0.000 1		m^2
m^2		10 000		cm^2
m^2		0.000 001		km^2
km^2		1 000 000		m^2
m^2		0.000 1		hectare
hectare		10,000		m^2
km^2		100		hectare
hectare		0.01		km^2
		VOLUME OR CAPACITY		
oz		1.8047		in^3
oz		6		teaspoon
tablespoon		3		teaspoon
tablespoon		0.5		oz
cup		8		oz
pint		16		oz
pint		2		cup
quart		2		pint
gallon		4		quart
litre		1 000		ml
litre		1 000		cm^3
teaspoon		5		cm^3
tablespoon		15		cm^3
oz		29.5735		cm^3
pint		0.473176		litre
quart		0.946353		litre
litre		1.05669		quart
gallon		3.78541		litre
litre		0.26417		gallon
peck		8		quart
bushel		4		peck
peck		0.0088		m^3
m^3		113.636		peck
bushel		0.035239		m^3
m^3		28.3776		bushel
quart		0.0011		m^3
m^3		909		quart
gallon		0.003785		m^3
m^3		264.2		gallon
ft^3		1,728		in^3
yd^3		27		ft^3
cm^3		1 000		mm^3
dm^3		1 000		cm^3
m^3		1 000		dm^3
in^3		16.387		cm^3
cm^3		0.061		in^3
ft^3		0.028317		m^3
m^3		35.31		ft^3
yd^3		0.764555		m^3
m^3		1.3079		yd^3

If the measure is given in this unit	Use this formula for conversion	To obtain this unit
TEMPERATURE		
Degrees Fahrenheit (F)	$\dfrac{(F-32)5}{9}$	Degrees Celsius (C)
Degrees Celsius (C)	$\left(\dfrac{C \times 9}{5}\right) + 32$	Degrees Fahrenheit (F)

Table 33 Abbreviations Used In Table 32

cm	centimeter	lb	pound
cm^2	square centimeter	m	meter
cm^3	cubic centimeter	m^2	square meter
dm^3	cubic decimeter	m^3	cubic meter
ft^2	square foot	Mcal	megacalorie
ft^3	cubic foot	mg	milligram
g	gram	mi^2	square mile
ICU	International Chick Unit	μg	microgram
in^2	square inch	oz	ounce
in^3	cubic inch	ppm	parts per million
IU	International Unit	rd	rod
kcal	kilocalorie	rd^2	square rod
kg	kilogram	yd	yard
km	kilometer	yd^2	square yard
km^2	square kilometer	yd^3	cubic yard

Table 34 Daily Nutrient Requirements of Dairy Cattle (Metric system)

Body Weight (kg)	Breed Size, Age (wk)	Daily Gain (g)	Feed DM (kg)	NE_m (Mcal)	NE_g (Mcal)	ME (Mcal)	DE (Mcal)	TDN (kg)	Total Crude Protein (g)	Ca (g)	P (g)	A (1,000 IU)	D (IU)
						Feed Energy				**Minerals**		**Vitamins**	
Growing Dairy Heifer and Bull Calves Fed Only Milk													
25	S-1[a,b]	300	0.45	0.85	0.53	2.14	2.38	0.54	111	6	4	1.1	165
30	S-3	350	0.52	0.95	0.63	2.49	2.77	0.63	128	7	4	1.3	200
42	L-1	400	0.63	1.25	0.70	2.98	3.31	0.75	148	8	5	1.8	280
50	L-3	500	0.76	1.40	0.90	3.61	4.01	0.91	180	9	6	2.1	330
Growing Dairy Heifer and Bull Calves Fed Mixed Diets													
50		300	1.31	1.45	0.57	3.91	4.45	1.01	150	9	6	2.1	330
50	S-10	400	1.40	1.45	0.76	4.36	4.94	1.12	176	9	6	2.1	330
50	L-3	500	1.45	1.45	0.96	4.82	5.42	1.23	198	10	6	2.1	330
50		600	1.45	1.45	1.16	5.01	5.69	1.29	221	11	7	2.1	330
50		700	1.45	1.45	1.35	5.36	5.95	1.35	243	12	7	2.1	330
75		300	2.10	1.96	0.58	5.17	6.05	1.37	232	11	7	3.2	495
75		400	2.10	1.96	0.77	5.56	6.53	1.46	254	12	7	3.2	495
75	S-19	500	2.10	1.96	0.98	5.96	6.94	1.55	275	13	7	3.2	495
75		600	2.10	1.96	1.17	6.36	7.31	1.64	296	14	8	3.2	495
75	L-10	700	2.10	1.96	1.37	6.71	7.67	1.72	318	15	8	3.2	495
75		800	2.10	1.96	1.56	7.08	7.94	1.80	341	16	8	3.2	495
Growing Dairy Heifers													
100		300	2.80	2.43	0.60	6.27	7.45	1.69	317	14	7	4.2	660
100		400	2.80	2.43	0.84	6.78	7.96	1.81	336	15	8	4.2	660
100	S-26	500	2.80	2.43	1.05	7.17	8.35	1.89	360	16	8	4.2	660
100		600	2.80	2.43	1.26	7.64	8.81	2.00	380	17	9	4.2	660
100	L-16	700	2.80	2.43	1.47	8.09	9.26	2.10	402	18	9	4.2	660
100		800	2.80	2.43	1.68	8.47	9.63	2.18	426	19	10	4.2	660
150		300	4.00	3.30	0.72	8.44	10.14	2.30	433	16	10	6.4	990
150		400	4.00	3.30	0.96	8.90	10.59	2.40	455	17	11	6.4	990
150	S-40	500	4.00	3.30	1.20	9.42	11.11	2.52	474	17	11	6.4	990
150		600	4.00	3.30	1.44	9.97	11.65	2.64	491	18	11	6.4	990
150	L-26	700	4.00	3.30	1.68	10.49	12.17	2.76	510	19	12	6.4	990
150		800	4.00	3.30	1.92	11.03	12.70	2.88	528	20	12	6.4	990
200		300	5.00	4.10	0.84	10.44	12.57	2.85	533	18	12	8.5	1320
200		400	5.20	4.10	1.12	11.20	13.41	3.04	571	19	13	8.5	1320
200	S-54	500	5.20	4.10	1.40	11.86	14.06	3.19	586	20	13	8.5	1320
200		600	5.20	4.10	1.68	12.39	14.59	3.31	604	21	14	8.5	1320
200	L-36	700	5.20	4.10	1.96	13.01	15.20	3.45	620	21	14	8.5	1320
200		800	5.20	4.10	2.24	13.52	15.70	3.56	640	22	15	8.5	1320
250		300	5.89	4.84	0.93	12.05	14.55	3.30	610	20	15	10.6	1650
250		400	6.30	4.84	1.24	13.15	15.83	3.59	665	21	15	10.6	1650
250	S-69	500	6.30	4.84	1.55	13.81	16.49	3.74	678	22	16	10.6	1650
250		600	6.30	4.84	1.86	14.57	17.24	3.91	689	22	16	10.6	1650
250	L-47	700	6.30	4.84	2.17	15.20	17.86	4.05	704	23	17	10.6	1650
250		800	6.30	4.84	2.48	15.82	18.47	4.19	719	23	17	10.6	1650
300		300	6.67	5.55	1.02	13.64	16.47	3.74	671	20	15	12.7	1980
300		400	7.00	5.55	1.36	14.80	17.77	4.03	713	22	17	12.7	1980
300	S-83	500	7.20	5.55	1.70	15.69	18.74	4.25	746	23	17	12.7	1980
300		600	7.20	5.55	2.04	16.49	19.53	4.43	755	23	17	12.7	1980
300	L-57	700	7.20	5.55	2.38	17.07	20.11	4.56	771	24	18	12.7	1980
300		800	7.20	5.55	2.72	17.83	20.86	4.73	782	24	18	12.7	1980

Body Weight (kg)	Breed Size, Age (wk)	Daily Gain (g)	Feed DM (kg)	Feed Energy					Total Crude Protein (g)	Minerals		Vitamins	
				NE$_m$ (Mcal)	NE$_g$ (Mcal)	ME (Mcal)	DE (Mcal)	TDN (kg)		Ca (g)	P (g)	A (1,000 IU)	D (IU)
350		300	7.23	6.24	1.08	15.27	18.34	4.16	701	22	16	14.8	2310
350	S-97	400	7.42	6.24	1.44	15.99	19.14	4.34	738	23	17	14.8	2310
350		500	8.00	6.24	1.80	17.42	20.81	4.72	804	25	18	14.8	2310
350		600	8.00	6.24	2.16	18.21	21.60	4.90	812	25	19	14.8	2310
350	L-67	700	8.00	6.24	2.52	18.88	22.26	5.05	826	25	19	14.8	2310
350		800	8.00	6.24	2.88	19.56	22.93	5.20	841	26	19	14.8	2310
400	S-115	200	7.26	6.89	0.76	14.85	17.94	4.07	692	21	16	17.0	2640
400		400	8.50	6.89	1.52	17.76	21.38	4.85	833	24	19	17.0	2640
400		600	8.60	6.89	2.28	19.61	23.24	5.27	856	25	20	17.0	2640
400	L-77	700	8.60	6.89	2.66	20.40	24.03	5.45	864	25	20	17.0	2640
400		800	8.60	6.89	3.04	21.11	24.73	5.61	876	26	21	17.0	2640
450		200	7.87	7.52	0.80	16.09	19.44	4.41	749	23	18	19.1	2970
450		400	9.00	7.52	1.60	19.02	22.84	5.18	867	26	20	19.1	2970
450		600	9.10	7.52	2.40	21.03	24.87	5.64	883	27	21	19.1	2970
450	L-87	700	9.10	7.52	2.80	21.82	25.66	5.82	892	27	21	19.1	2970
450		800	9.10	7.52	3.20	22.67	26.50	6.01	898	28	21	19.1	2970
500		200	8.46	8.14	0.84	17.30	20.90	4.74	788	24	19	21.2	3300
500		400	9.50	8.14	1.68	20.26	24.29	5.51	900	27	21	21.2	3300
500	L-98	600	9.50	8.14	2.52	22.26	26.28	5.96	903	27	21	21.2	3300
500		800	9.50	8.14	3.36	24.00	28.00	6.35	916	28	21	21.2	3300
550		200	9.05	8.75	0.88	18.50	22.34	5.07	835	25	19	23.3	3630
550	L-109	400	9.80	8.75	1.76	21.33	25.48	5.78	913	27	20	23.3	3630
550		600	9.80	8.75	2.64	23.38	27.51	6.24	914	27	20	23.3	3630
550		800	9.80	8.75	3.52	25.08	29.19	6.62	928	28	21	23.3	3630
600	L-127	200	9.58	9.33	0.90	19.60	23.68	5.37	879	25	18	25.4	3960
600		300	9.72	9.33	1.35	20.78	24.87	5.64	895	25	18	25.4	3960
600		400	10.00	9.33	1.80	22.22	26.45	6.00	918	26	19	25.4	3960
600		500	10.00	9.33	2.25	23.34	27.56	6.25	916	26	19	25.4	3960
Growing Dairy Bulls													
100		500	2.80	2.43	1.05	7.17	8.35	1.89	361	16	8	4.2	660
100	S-26	600	2.80	2.43	1.26	7.64	8.81	2.00	381	17	9	4.2	660
100		700	2.80	2.43	1.47	8.09	9.26	2.10	403	18	9	4.2	660
100	L-15	800	2.80	2.43	1.68	8.47	9.63	2.18	427	19	10	4.2	660
100		900	2.80	2.43	1.89	8.84	10.00	2.27	450	20	10	4.2	660
150		500	4.00	3.30	1.15	9.42	11.11	2.52	476	18	11	6.4	990
150		600	4.00	3.30	1.38	9.91	11.59	2.63	497	19	11	6.4	990
150	S-38	700	4.00	3.30	1.61	10.30	11.98	2.72	520	20	12	6.4	990
150		800	4.00	3.30	1.84	10.84	12.52	2.84	539	21	12	6.4	990
150		900	4.00	3.30	2.07	11.47	13.14	2.98	555	21	13	6.4	990
150	L-24	1000	4.00	3.30	2.30	11.73	13.40	3.04	583	22	13	6.4	990
200		500	5.20	4.10	1.25	11.46	13.66	3.10	602	20	13	8.5	1320
200		600	5.20	4.10	1.50	12.01	14.21	3.22	622	21	14	8.5	1320
200	S-48	700	5.20	4.10	1.75	12.59	14.78	3.35	640	21	14	8.5	1320
200		800	5.20	4.10	2.00	13.07	15.26	3.46	660	22	15	8.5	1320
200		900	5.20	4.10	2.25	13.52	15.70	3.56	688	23	16	8.5	1320
200	L-31	1000	5.20	4.10	2.50	14.05	16.23	3.68	702	23	16	8.5	1320
250		500	6.30	4.84	1.35	13.44	16.11	3.65	684	22	16	10.6	1650
250		600	6.30	4.84	1.62	14.00	16.67	3.78	702	23	16	10.6	1650
250	S-58	700	6.30	4.84	1.89	14.62	17.28	3.92	718	23	17	10.6	1650

Table 34 Daily Nutrient Requirements of Dairy Cattle (Metric System) *continued*

Body Weight (kg)	Breed Size, Age (wk)	Daily Gain (g)	Feed DM (kg)	NEm (Mcal)	NEg (Mcal)	ME (Mcal)	DE (Mcal)	TDN (kg)	Total Crude Protein (g)	Ca (g)	P (g)	A (1,000 IU)	D (IU)
250		800	6.30	4.84	2.16	15.20	17.86	4.05	736	24	17	10.6	1650
250		900	6.30	4.84	2.43	15.78	18.43	4.18	753	25	17	10.6	1650
250	L-38	1000	6.30	4.84	2.70	16.13	18.78	4.26	778	25	18	10.6	1650
300		500	7.33	5.69	1.48	15.45	18.56	4.21	777	24	18	12.7	1980
300		600	7.40	5.69	1.77	16.13	19.27	4.37	800	25	19	12.7	1980
300	S-68	700	7.40	5.69	2.07	16.89	20.02	4.54	811	26	19	12.7	1980
300		800	7.40	5.69	2.36	17.51	20.63	4.68	827	26	19	12.7	1980
300		900	7.40	5.69	2.66	18.09	21.21	4.81	845	27	19	12.7	1980
300	L-45	1000	7.40	5.69	2.95	18.67	21.78	4.94	862	27	20	12.7	1980
350		500	8.10	6.54	1.60	17.27	20.71	4.70	828	25	19	14.8	2310
350		600	8.30	6.54	1.92	18.13	21.65	4.91	863	26	20	14.8	2310
350	S-79	700	8.30	6.54	2.24	18.93	22.44	5.09	873	27	20	14.8	2310
350		800	8.30	6.54	2.56	19.60	23.10	5.24	887	27	20	14.8	2310
350		900	8.30	6.54	2.88	20.22	23.72	5.38	903	28	20	14.8	2310
350	L-52	1000	8.30	6.54	3.20	20.89	24.38	5.53	917	28	21	14.8	2310
400		500	9.00	7.41	1.75	19.24	23.06	5.23	891	27	21	17.0	2640
400		600	9.00	7.41	2.10	20.00	23.81	5.40	902	27	21	17.0	2640
400	S-89	700	9.00	7.41	2.45	20.84	24.64	5.59	910	28	22	17.0	2640
400		800	9.00	7.41	2.80	21.60	25.40	5.76	921	28	22	17.0	2640
400		900	9.00	7.41	3.15	22.36	26.15	5.93	932	28	22	17.0	2640
400	L-60	1000	9.00	7.41	3.50	22.93	26.72	6.06	947	29	23	17.0	2640
450		200	8.41	8.27	0.76	17.20	20.77	4.71	762	23	19	19.1	2970
450		400	9.33	8.27	1.52	19.90	23.86	5.41	868	27	21	19.1	2970
450	S-90	600	9.50	8.27	2.28	21.83	25.84	5.86	898	28	22	19.1	2970
450		800	9.50	8.27	3.04	23.52	27.52	6.24	914	28	22	19.1	2970
450	L-67	1000	9.50	8.27	3.80	25.08	29.07	6.59	934	29	23	19.1	2970
500		100	8.26	8.95	0.40	16.90	20.41	4.63	740	22	18	21.2	3300
500		300	9.30	8.95	1.20	19.83	23.77	5.39	855	25	21	21.2	3300
500	S-111	500	10.00	8.95	2.00	22.22	26.45	6.00	941	28	23	21.2	3300
500		700	10.00	8.95	2.80	23.60	27.82	6.31	967	29	23	21.2	3300
500	L-74	900	10.00	8.95	3.60	25.56	29.76	6.75	973	29	23	21.2	3300
550		100	8.86	9.62	0.42	18.11	21.87	4.96	789	24	18	23.3	3630
550	S-125	300	10.20	9.62	1.25	21.29	25.62	5.81	935	28	22	23.3	3630
550		500	10.50	9.62	2.08	23.56	28.00	6.35	967	29	22	23.3	3630
550	L-82	700	10.50	9.62	2.91	25.51	29.94	6.79	976	29	22	23.3	3630
550		900	10.50	9.62	3.74	27.16	31.57	7.16	994	30	23	23.3	3630
600	S-149	100	9.42	10.27	0.43	19.27	23.28	5.28	833	25	19	25.4	3960
600		300	10.52	10.27	1.29	22.44	26.90	6.10	947	28	22	25.4	3960
600		500	10.80	10.27	2.15	24.72	29.28	6.64	980	29	23	25.4	3960
600	L-92	700	10.80	10.27	3.01	26.58	31.13	7.06	988	29	23	25.4	3960
650		100	9.96	10.90	0.44	20.37	24.60	5.58	875	26	20	27.6	4290
650		300	10.69	10.90	1.32	23.29	27.82	6.31	947	28	22	27.6	4290
650	L-102	500	11.10	10.90	2.20	25.75	30.44	6.90	992	29	23	27.6	4290
650		700	11.10	10.90	3.08	27.78	32.45	7.36	995	29	23	27.6	4290
700		100	10.51	11.53	0.45	21.50	25.97	5.89	918	27	21	29.7	4620
700		300	11.40	11.53	1.35	24.61	29.45	6.68	1005	29	23	29.7	4620
700	L-117	500	11.40	11.53	2.25	26.94	31.75	7.20	998	30	23	29.7	4620
700		700	11.40	11.53	3.15	28.99	33.78	7.66	1001	30	23	29.7	4620
750		100	11.02	12.14	0.45	22.53	27.21	6.17	960	28	22	31.8	4950
750	L-131	300	11.70	12.14	1.35	25.48	30.44	6.90	1024	30	23	31.8	4950

Body Weight (kg)	Breed Size, Age (wk)	Daily Gain (g)	Feed DM (kg)	NE$_m$ (Mcal)	Feed Energy NE$_g$ (Mcal)	ME (Mcal)	DE (Mcal)	TDN (kg)	Total Crude Protein (g)	Minerals Ca (g)	P (g)	Vitamins A (1,000 IU)	D (IU)
750		500	11.70	12.14	2.25	27.86	32.80	7.44	1014	30	23	31.8	4950
800		100	11.52	12.74	0.45	23.55	28.44	6.45	999	29	23	33.9	5280
800		300	12.00	12.74	1.35	26.35	31.44	7.13	1040	30	23	33.9	5280
800		500	12.00	12.74	2.25	28.62	33.68	7.64	1035	30	23	33.9	5280

Growing Veal Calves Fed Only Milk

Body Weight (kg)	Breed Size, Age (wk)	Daily Gain (g)	Feed DM (kg)	NE$_m$ (Mcal)	NE$_g$ (Mcal)	ME (Mcal)	DE (Mcal)	TDN (kg)	Total Crude Protein (g)	Ca (g)	P (g)	A (1,000 IU)	D (IU)
35	—	500	0.67	0.98	0.90	3.17	3.52	0.80	173	7	4	1.5	231
45	L-1.0	800	1.06	1.36	1.52	5.04	5.60	1.27	259	8	5	1.9	297
55	L-2.8	900	1.20	1.55	1.73	5.74	6.38	1.45	292	11	7	2.3	363
65	L-4.4	1000	1.36	1.76	1.95	6.48	7.20	1.63	324	13	8	2.8	429
75	L-5.8	1050	1.48	1.96	2.10	7.05	7.83	1.78	334	15	9	3.2	495
100	L-9.2	1100	1.69	2.43	2.31	8.05	8.94	2.03	357	17	10	4.2	660
125	L-12.4	1200	1.95	2.88	2.64	9.30	10.33	2.34	392	19	11	5.3	825
150	L-15.4	1300	2.22	3.30	2.99	10.58	11.75	2.66	428	20	12	6.4	990

Maintenance of Mature Breeding Bulls

Body Weight (kg)	Breed Size, Age (wk)	Daily Gain (g)	Feed DM (kg)	NE$_m$ (Mcal)	NE$_g$ (Mcal)	ME (Mcal)	DE (Mcal)	TDN (kg)	Total Crude Protein (g)	Ca (g)	P (g)	A (1,000 IU)	D (IU)
500	—	—	7.80	9.36	—	15.95	19.27	4.37	673	20	15	21	—
600	—	—	8.95	10.74	—	18.29	22.09	5.01	766	23	17	25	—
700	—	—	10.04	12.05	—	20.52	24.78	5.62	852	26	19	30	—
800	—	—	11.10	13.32	—	22.52	27.20	6.17	942	29	21	34	—
900	—	—	12.13	14.55	—	24.79	29.94	6.79	1017	31	23	38	—
1000	—	—	13.12	15.75	—	26.83	32.41	7.35	1093	34	25	42	—
1100	—	—	14.10	16.91	—	28.84	34.83	7.90	1169	36	27	47	—
1200	—	—	15.05	18.05	—	30.77	37.17	8.43	1244	39	29	51	—
1300	—	—	15.98	19.17	—	32.67	39.46	8.95	1316	41	31	55	—
1400	—	—	16.88	20.27	—	34.49	41.66	9.45	1386	43	33	59	—

[a] Breed size: S for small breeds (e.g., Jersey); L is for large breeds (e.g., Holstein).

[b] Age in weeks indicates probable age of S or L animals when they reach the weight indicated.

Source: *Nutrient Requirements of Dairy Cattle,* Fifth revised edition, National Research Council, National Academy of Sciences, Washington, D.C., 1978.

Table 35 Daily Nutrient Requirements of Lactating and Pregnant Cows (Metric System)

Body Weight (kg)	Feed Energy				Total Crude Protein (g)	Calcium (g)	Phosphorus (g)	Vitamin A (1,000 IU)
	NE_1 (Mcal)	ME (Mcal)	DE (Mcal)	TDN (kg)				
Maintenance of Mature Lactating Cows[a]								
350	6.47	10.76	12.54	2.85	341	14	11	27
400	7.16	11.90	13.86	3.15	373	15	13	30
450	7.82	12.99	15.14	3.44	403	17	14	34
500	8.46	14.06	16.39	3.72	432	18	15	38
550	9.09	15.11	17.60	4.00	461	20	16	42
600	9.70	16.12	18.79	4.27	489	21	17	46
650	10.30	17.12	19.95	4.53	515	22	18	50
700	10.89	18.10	21.09	4.79	542	24	19	53
750	11.47	19.06	22.21	5.04	567	25	20	57
800	12.03	20.01	23.32	5.29	592	27	21	61
Maintenance Plus Last 2 Months of Gestation of Mature Dry Cows								
350	8.42	14.00	16.26	3.71	642	23	16	27
400	9.30	15.47	17.98	4.10	702	26	18	30
450	10.16	16.90	19.64	4.47	763	29	20	34
500	11.00	18.29	21.25	4.84	821	31	22	38
550	11.81	19.65	22.83	5.20	877	34	24	42
600	12.61	20.97	24.37	5.55	931	37	26	46
650	13.39	22.27	25.87	5.90	984	39	28	50
700	14.15	23.54	27.35	6.23	1035	42	30	53
750	14.90	24.79	28.81	6.56	1086	45	32	57
800	15.64	26.02	30.24	6.89	1136	47	34	61
Milk Production — Nutrients Per Kg Milk of Different Fat Percentages								
(% Fat)								
2.5	0.59	0.99	1.15	0.260	72	2.40	1.65	
3.0	0.64	1.07	1.24	0.282	77	2.50	1.70	
3.5	0.69	1.16	1.34	0.304	82	2.60	1.75	
4.0	0.74	1.24	1.44	0.326	87	2.70	1.80	
4.5	0.78	1.31	1.52	0.344	92	2.80	1.85	
5.0	0.83	1.39	1.61	0.365	98	2.90	1.90	
5.5	0.88	1.48	1.71	0.387	103	3.00	2.00	
6.0	0.93	1.56	1.81	0.410	108	3.10	2.05	
Body Weight Change During Lactation — Nutrients Per Kg Weight Change								
Weight loss	−4.92	−8.25	−9.55	−2.17	−320			
Weight gain	5.12	8.55	9.96	2.26	500			

[a]To allow for growth of young lactating cows, increase the maintenance allowances for all nutrients except vitamin A by 20 percent during the first lactation and 10 percent during the second lactation.

Source: *Nutrient Requirements of Dairy Cattle,* Fifth revised edition, National Research Council, National Academy of Sciences, Washington, D.C., 1978.

Table 36 Recommended Nutrient Content of Rations for Dairy Cattle (Metric System)

Nutrients (Concentration in the Feed Dry matter)	Lactating Cow Rations				Nonlactating Cattle Rations					Maximum Concentrations (All Classes)
Cow Wt (kg): ≤400 / 500 / 600 / ≥700	Daily Milk Yields (kg)				Dry Pregnant Cows	Mature Bulls	Growing Heifers and Bulls	Calf Starter Concentrate Mix	Calf Milk Replacer	
	< 8 / <11 / <14 / <18	8–13 / 11–17 / 14–21 / 18–26	13–18 / 17–23 / 21–29 / 26–35	>18 / >23 / >29 / >35						
Ration No.	I	II	III	IV	V	VI	VII	VIII	IX	Max.
Crude Protein, %	13.0	14.0	15.0	16.0	11.0	8.5	12.0	16.0	22.0	—
Energy										
NE_l, Mcal/kg	1.42	1.52	1.62	1.72	1.35	—	—	—	—	—
NE_m, Mcal/kg	—	—	—	—	—	1.20	1.26	1.90	2.40	—
NE_g, Mcal/kg	—	—	—	—	—	—	0.60	1.20	1.55	—
ME, Mcal/kg	2.36	2.53	2.71	2.89	2.23	2.04	2.23	3.12	3.78	—
DE, Mcal/kg	2.78	2.95	3.13	3.31	2.65	2.47	2.65	3.53	4.19	—
TDN, %	63	67	71	75	60	56	60	80	95	—
Crude Fiber, %	17	17	17	17[a]	17	15	15	—	—	—
Acid Detergent Fiber, %	21	21	21	21	21	19	19	—	—	—
Ether Extract, %	2	2	2	2	2	2	2	2	10	—
Minerals[b]										
Calcium, %	0.43	0.48	0.54	0.60	0.37	0.24	0.40	0.60	0.70	—
Phosphorus, %	0.31	0.34	0.38	0.40	0.26	0.18	0.26	0.42	0.50	—
Magnesium, %[c]	0.20	0.20	0.20	0.20	0.16	0.16	0.16	0.07	0.07	—
Potassium, %	0.80	0.80	0.80	0.80	0.80	0.80	0.80	0.80	0.80	—
Sodium, %	0.18	0.18	0.18	0.18	0.10	0.10	0.10	0.10	0.10	—
Sodium chloride, %[d]	0.46	0.46	0.46	0.46	0.25	0.25	0.25	0.25	0.25	5
Sulfur, %[d]	0.20	0.20	0.20	0.20	0.17	0.11	0.16	0.21	0.29	0.35
Iron, ppm[d,e]	50	50	50	50	50	50	50	100	100	1,000
Cobalt, ppm	0.10	0.10	0.10	0.10	0.10	0.10	0.10	0.10	0.10	10
Copper, ppm[d,f]	10	10	10	10	10	10	10	10	10	80
Manganese, ppm[d]	40	40	40	40	40	40	40	40	40	1,000
Zinc, ppm[d,g]	40	40	40	40	40	40	40	40	40	500
Iodine, ppm[h]	0.50	0.50	0.50	0.50	0.50	0.25	0.25	0.25	0.25	50
Molybdenum, ppm[i,j]	—	—	—	—	—	—	—	—	—	6
Selenium, ppm	0.10	0.10	0.10	0.10	0.10	0.10	0.10	0.10	0.10	5
Fluorine, ppm[j]	—	—	—	—	—	—	—	—	—	30
Vitamins[k]										
Vit A, IU/kg	3,200	3,200	3,200	3,200	3,200	3,200	2,200	2,200	3,800	—
Vit D, IU/kg	300	300	300	300	300	300	300	300	600	—
Vit E, ppm	—	—	—	—	—	—	—	—	300	—

[a] It is difficult to formulate high-energy rations with a minimum of 17 percent crude fiber. However, fat percentage depression may occur when rations with less than 17 percent crude fiber or 21 percent ADF are fed to lactating cows.

[b] The mineral values presented in this table are intended as guidelines for use of professionals in ration formulation. Because of many factors affecting such values, they are not intended and should not be used as a legal or regulatory base.

[c] Under conditions conducive to grass tetany (see text), should be increased to 0.25 or higher.

[d] The maximum safe levels for many of the mineral elements are not well defined; estimates given here, especially for sulfur, sodium chloride, iron, copper, zinc, and manganese, are based on very limited data; safe levels may be substantially affected by specific feeding conditions.

[e] The maximum safe level of supplemental iron in some forms is materially lower than 1,000 ppm. As little as 400 ppm added iron as ferrous sulfate has reduced weight gains (Standish et al., 1969).

[f] High copper may increase the susceptibility of milk to oxidized flavor (see text).

[g] Maximum safe level of zinc for mature dairy cattle is 1,000 ppm.

[h] If diet contains as much as 25 percent strongly goitrogenic feed on dry basis, iodine provided should be increased two times or more.

[i] If diet contains sufficient copper, dairy cattle tolerate substantially more than 6 ppm molybdenum (see text).

[j] Maximum safe level of fluorine for growing heifers and bulls is lower than for other dairy cattle. Somewhat higher levels are tolerated when the fluorine is from less-available sources as phosphates (see text). Minimum requirement for molybdenum and fluorine not yet established.

[k] The following minimum quantities of B-complex vitamins are suggested per unit of milk replacer: niacin, 2.6 ppm; pantothenic acid, 13 ppm; riboflavin, 6.5 ppm; pyridoxine, 6.5 ppm; thiamine, 6.5 ppm; folic acid, 0.5 ppm; biotin, 0.1 ppm; vitamin B_{12}, 0.07 ppm; choline, 0.26 percent. It appears that adequate amounts of these vitamins are furnished when calves have functional rumens (usually at 6 weeks of age) by a combination of rumen synthesis and natural feedstuffs.

Source: *Nutrient Requirements of Dairy Cattle*, Fifth revised edition, National Research Council, National Academy of Sciences, Washington, D.C., 1978.

Table 37 Maximum Dry Matter Intake Guidelines

Body Wt (kg)	400	500	600	700	800
FCM (4% milk)	◄───────────────────		% of Body Wt		───────────────►
10	2.5	2.3	2.2	2.1	2.0
15	2.8	2.5	2.4	2.3	2.2
20	3.1	2.8	2.7	2.6	2.4
25	3.4	3.1	3.0	2.8	2.6
30	3.7	3.4	3.2	3.0	2.8
35	4.0	3.6	3.4	3.2	3.0
40	—	3.8	3.6	3.4	3.2
45	—	4.0	3.8	3.6	3.4

[a] Derived from:

(1) Chandler, P.T., and C.A. Brown. 1975. Adjusting nutrient concentrations according to season of year and make up of group or herd. Unpublished report at Dairy Feed Conference Board, University of Delaware.

(2) Smith, N.E. 1971. Feed efficiency in intensive milk production. Proc. 10th Annual Dairy Cattle Day, p. 40. Dep. Anim. Sci., Univ. Calif., Davis.

(3) Swanson, E.W., S.A. Hinton, and J.T. Miles. 1967. Full lactation response on restricted vs. *ad libitum* roughage diets with liberal concentrate feeding. J. Dairy Sci. 50:1147–1152.

(4) Trimberger, G.W., H.G. Gray, W.L. Johnson, M.J. Wright, D. Van Vleck, and C.R. Henderson. 1963. Forage appetite in dairy cattle. Proc. Cornell Nutr. Conf. Feed Manuf., pp. 33–43.

(5) Trimberger, G.W., H.F. Tyrrell, D.A. Morrow, J.T. Reid, M.J. Wright, W.F. Shipe, W.G. Merrill, J.K. Loosli, C.E. Coppock, L.A. Moore, and C.H. Gordon. 1972. Effects of liberal concentrate feeding on health, reproductive efficiency, economy of milk production, and other related responses of the dairy cow. N.Y. Food Life Serv. Bull. No. 8.

Source: *Nutrient Requirements of Dairy Cattle,* Fifth revised edition, National Research Council, National Academy of Sciences, Washington, D.C., 1978.

Table 38 Daily Nutrient Requirements of Dairy Cattle (Pounds)

Body Weight (lb)	Breed Size, Age (wk)	Daily Gain (lb)	Feed DM (lb)	NE$_m$ (Mcal)	NE$_g$ (Mcal)	ME (Mcal)	DE (Mcal)	TDN (lb)	Total Crude Protein (lb)	Ca (lb)	P (lb)	A (1,000 IU)	D (IU)
						Feed Energy				**Minerals**		**Vitamins**	
Growing Dairy Heifer and Bull Calves Fed Only Milk													
55	S-1[a,b]	0.7	1.00	0.85	0.53	2.14	2.38	1.20	0.25	0.013	0.009	1.0	165
65	S-3	0.8	1.15	0.94	0.63	2.47	2.74	1.38	0.28	0.014	0.010	1.2	200
93	L-1	0.9	1.38	1.25	0.70	2.98	3.31	1.66	0.33	0.018	0.011	1.8	280
106	L-3	1.1	1.65	1.36	0.90	3.51	3.90	1.98	0.40	0.020	0.013	1.9	300
Growing Dairy Heifer (F) and Bull (M) Calves Fed Mixed Diets													
100		0.6	2.7	1.35	0.52	3.65	4.16	2.08	0.31	0.018	0.013	1.9	300
100(F)	S-10	0.8	2.8	1.35	0.69	3.98	4.51	2.25	0.36	0.020	0.013	1.9	300
100(M)	S-10	1.0	2.8	1.35	0.87	4.32	4.84	2.42	0.40	0.022	0.014	1.9	300
100(F)	L-3	1.2	2.8	1.35	1.05	4.58	5.10	2.55	0.44	0.024	0.015	1.9	300
100(M)	L-3	1.4	2.8	1.35	1.22	4.80	5.33	2.67	0.48	0.027	0.015	1.9	300
100		1.6	2.8	1.35	1.40	5.04	5.60	2.80	0.58	0.028	0.016	1.9	300
150		0.8	4.0	1.82	0.70	5.02	5.78	2.89	0.49	0.024	0.015	2.9	450
150(F)	S-19	1.0	4.1	1.82	0.88	5.39	6.17	3.08	0.55	0.026	0.015	2.9	450
150(M)	S-18	1.2	4.1	1.82	1.06	5.76	6.54	3.27	0.59	0.028	0.016	2.9	450
150(F)	L-9	1.4	4.1	1.82	1.24	6.07	6.85	3.42	0.64	0.031	0.017	2.9	450
150(M)	L-8	1.6	4.1	1.82	1.42	6.39	7.16	3.58	0.68	0.033	0.018	2.9	450
150		1.8	4.1	1.82	1.59	6.63	7.40	3.70	0.73	0.034	0.018	2.9	450
200		0.8	5.3	2.26	0.73	6.04	7.05	3.53	0.65	0.031	0.017	3.8	600
200		1.0	5.4	2.26	0.92	6.50	7.53	3.76	0.71	0.032	0.018	3.8	600
200(F)	S-26	1.2	5.4	2.26	1.10	6.86	7.88	3.94	0.75	0.034	0.018	3.8	600
200(M)	S-24	1.4	5.4	2.26	1.29	7.22	8.24	4.12	0.80	0.036	0.019	3.8	600
200(F)	L-14	1.6	5.4	2.26	1.47	7.52	8.53	4.27	0.85	0.038	0.020	3.8	600
200(M)	L-12	1.8	5.4	2.26	1.66	7.89	8.90	4.45	0.90	0.040	0.021	3.8	600
200		2.0	5.4	2.26	1.84	8.27	9.28	4.64	0.94	0.041	0.022	3.8	600
Growing Dairy Heifers													
300		0.8	7.9	3.07	0.83	8.12	9.64	4.82	0.88	0.034	0.021	5.8	900
300	S-38	1.0	7.9	3.07	1.04	8.60	10.11	5.06	0.92	0.036	0.022	5.8	900
300		1.2	7.9	3.07	1.25	8.98	10.51	5.25	0.96	0.037	0.023	5.8	900
300		1.4	7.9	3.07	1.46	9.47	10.97	5.48	1.00	0.039	0.024	5.8	900
300	L-23	1.6	7.9	3.07	1.66	9.89	11.40	5.70	1.04	0.041	0.025	5.8	900
300		1.8	7.9	3.07	1.87	10.26	11.76	5.88	1.09	0.043	0.026	5.8	900
400		0.8	10.5	3.81	0.96	10.16	12.18	6.09	1.15	0.040	0.026	7.7	1200
400	S-52	1.0	10.5	3.81	1.20	10.69	12.70	6.35	1.19	0.042	0.028	7.7	1200
400		1.2	10.5	3.81	1.44	11.24	13.26	6.63	1.22	0.043	0.029	7.7	1200
400		1.4	10.5	3.81	1.68	11.75	13.76	6.88	1.26	0.044	0.030	7.7	1200
400	L-32	1.6	10.5	3.81	1.92	12.17	14.18	7.09	1.30	0.045	0.030	7.7	1200
400		1.8	10.5	3.81	2.16	12.68	14.68	7.34	1.33	0.046	0.031	7.7	1200
500		0.8	12.7	4.50	1.08	11.93	14.38	7.19	1.35	0.046	0.032	9.6	1500
500	S-67	1.0	12.7	4.50	1.35	12.60	15.04	7.52	1.38	0.047	0.033	9.6	1500
500		1.2	12.7	4.50	1.62	13.21	15.65	7.82	1.41	0.048	0.034	9.6	1500
500		1.4	12.7	4.50	1.89	13.77	16.20	8.10	1.44	0.049	0.035	9.6	1500
500	L-41	1.6	12.7	4.50	2.16	14.34	16.76	8.38	1.47	0.050	0.036	9.6	1500
500		1.8	12.7	4.50	2.43	14.90	17.32	8.66	1.50	0.051	0.037	9.6	1500
600		0.8	14.7	5.16	1.18	13.63	16.46	8.23	1.52	0.049	0.034	11.5	1800
600	S-81	1.0	14.7	5.16	1.47	14.37	17.20	8.60	1.53	0.049	0.035	11.5	1800
600		1.2	14.7	5.16	1.76	14.97	17.79	8.89	1.56	0.050	0.036	11.5	1800

Table 38 Daily Nutrient Requirements of Dairy Cattle (Pounds) *continued*

Body Weight (lb)	Breed Size, Age (wk)	Daily Gain (lb)	Feed DM (lb)	NEm (Mcal)	NEg (Mcal)	ME (Mcal)	DE (Mcal)	TDN (lb)	Total Crude Protein (lb)	Ca (lb)	P (lb)	A (1,000 IU)	D (IU)
600		1.4	14.7	5.16	2.06	15.62	18.44	9.22	1.59	0.051	0.037	11.5	1800
600	L-50	1.6	14.7	5.16	2.35	16.31	19.12	9.56	1.61	0.052	0.038	11.5	1800
600		1.8	14.7	5.16	2.65	16.90	19.70	9.85	1.64	0.053	0.039	11.5	1800
700		0.8	16.2	5.79	1.26	15.02	18.14	9.07	1.64	0.050	0.036	13.5	2100
700	S-95	1.0	16.4	5.79	1.57	15.86	19.02	9.51	1.67	0.051	0.037	13.5	2100
700		1.2	16.4	5.79	1.88	16.53	19.68	9.84	1.69	0.052	0.038	13.5	2100
700		1.4	16.4	5.79	2.20	17.22	20.36	10.18	1.72	0.053	0.039	13.5	2100
700	L-59	1.6	16.4	5.79	2.51	17.86	21.00	10.50	1.74	0.054	0.040	13.5	2100
700		1.8	16.4	5.79	2.83	18.55	21.68	10.84	1.76	0.054	0.041	13.5	2100
800	S-109	0.6	16.2	6.40	1.00	15.02	18.14	9.07	1.58	0.048	0.036	15.4	2400
800		0.8	17.6	6.40	1.33	16.33	19.72	9.86	1.75	0.054	0.038	15.4	2400
800		1.0	18.0	6.40	1.66	17.34	20.80	10.40	1.80	0.055	0.039	15.4	2400
800		1.2	18.0	6.40	1.99	18.00	21.46	10.73	1.82	0.056	0.040	15.4	2400
800	L-68	1.4	18.0	6.40	2.32	18.79	22.24	11.12	1.83	0.056	0.041	15.4	2400
800		1.6	18.0	6.40	2.66	19.52	22.96	11.48	1.85	0.057	0.042	15.4	2400
900	S-133	0.4	15.9	6.99	0.70	14.74	17.80	8.90	1.51	0.045	0.035	17.3	2700
900		0.6	17.4	6.99	1.04	16.14	19.49	9.74	1.68	0.050	0.039	17.3	2700
900		1.0	19.2	6.99	1.74	18.58	22.28	11.14	1.88	0.056	0.042	17.3	2700
900		1.2	19.2	6.99	2.09	19.35	23.04	11.52	1.90	0.057	0.043	17.3	2700
900	L-78	1.4	19.2	6.99	2.44	20.20	23.88	11.94	1.90	0.057	0.044	17.3	2700
900		1.6	19.2	6.99	2.78	20.91	24.58	12.29	1.92	0.058	0.045	17.3	2700
1000		0.4	17.1	7.57	0.73	15.86	19.16	9.58	1.60	0.049	0.040	19.2	3000
1000		0.6	18.7	7.57	1.09	17.34	20.94	10.47	1.78	0.054	0.042	19.2	3000
1000		1.0	20.2	7.57	1.82	19.75	23.64	11.82	1.95	0.060	0.045	19.2	3000
1000		1.2	20.2	7.57	2.18	20.56	24.44	12.22	1.96	0.060	0.046	19.2	3000
1000	L-88	1.4	20.2	7.57	2.55	21.39	25.26	12.63	1.97	0.060	0.046	19.2	3000
1000		1.6	20.2	7.57	2.91	22.16	26.02	13.01	1.98	0.060	0.046	19.2	3000
1100		0.4	18.3	8.13	0.76	16.97	20.50	10.25	1.70	0.051	0.042	21.2	3300
1100		0.8	20.9	8.13	1.53	19.79	23.82	11.91	1.98	0.060	0.045	21.2	3300
1100	L-98	1.2	20.9	8.13	2.29	21.69	25.70	12.85	1.99	0.060	0.046	21.2	3300
1100		1.6	20.9	8.13	3.06	23.39	27.38	13.69	2.00	0.060	0.046	21.2	3300
1200		0.4	19.4	8.68	0.79	17.98	21.72	10.86	1.79	0.053	0.042	23.1	3600
1200	L-110	0.8	21.6	8.68	1.58	20.90	25.06	12.53	2.01	0.060	0.044	23.1	3600
1200		1.2	21.6	8.68	2.38	22.64	26.78	13.39	2.02	0.060	0.044	23.1	3600
1200		1.6	21.6	8.68	3.17	24.40	28.52	14.26	2.04	0.061	0.046	23.1	3600
1300		0.4	20.5	9.21	0.82	19.01	22.96	11.48	1.88	0.054	0.040	25.0	3900
1300		0.8	21.9	9.21	1.63	21.63	25.84	12.92	2.01	0.058	0.040	25.0	3900
1300		1.2	21.9	9.21	2.45	23.62	27.82	13.91	2.01	0.058	0.042	25.0	3900
1300		1.6	21.9	9.21	3.26	25.39	29.56	14.78	2.02	0.058	0.042	25.0	3900
Growing Dairy Bulls													
300		1.0	7.9	3.07	1.01	8.53	10.04	5.02	0.93	0.036	0.022	5.8	900
300	S-34	1.4	7.9	3.07	1.41	9.36	10.86	5.43	1.02	0.039	0.024	5.8	900
300		1.8	7.9	3.07	1.82	10.12	11.62	5.81	1.11	0.043	0.026	5.8	900
300	L-20	2.0	7.9	3.07	2.02	10.51	12.00	6.00	1.15	0.045	0.027	5.8	900
300		2.2	7.9	3.07	2.22	10.89	12.38	6.19	1.20	0.047	0.028	5.8	900
400		1.0	10.5	3.81	1.10	10.58	12.60	6.30	1.20	0.041	0.027	7.7	1200
400	S-44	1.4	10.5	3.81	1.54	11.45	13.46	6.73	1.29	0.044	0.029	7.7	1200
400		1.8	10.5	3.81	1.98	12.34	14.34	7.17	1.37	0.047	0.031	7.7	1200

Body Weight (lb)	Breed Size, Age (wk)	Daily Gain (lb)	Feed DM (lb)	Feed Energy					Total Crude Protein (lb)	Minerals		Vitamins	
				NE_m (Mcal)	NE_g (Mcal)	ME (Mcal)	DE (Mcal)	TDN (lb)		Ca (lb)	P (lb)	A (1,000 IU)	D (IU)
400		2.0	10.5	3.81	2.20	12.76	14.76	7.38	1.41	0.048	0.032	7.7	1200
400	L-28	2.2	10.5	3.81	2.42	13.19	15.18	7.59	1.46	0.050	0.033	7.7	1200
500		1.0	12.7	4.50	1.18	12.24	14.68	7.34	1.41	0.048	0.032	9.6	1500
500	S-55	1.4	12.7	4.50	1.65	13.27	15.70	7.85	1.49	0.051	0.033	9.6	1500
500		1.8	12.7	4.50	2.12	14.30	16.72	8.36	1.56	0.053	0.035	9.6	1500
500		2.0	12.7	4.50	2.36	14.72	17.14	8.57	1.60	0.054	0.036	9.6	1500
500	L-34	2.2	12.7	4.50	2.60	15.18	17.60	8.80	1.64	0.056	0.037	9.6	1500
600		1.0	14.8	5.26	1.27	13.97	16.82	8.41	1.59	0.053	0.036	11.5	1800
600	S-65	1.4	14.8	5.26	1.78	15.22	18.06	9.03	1.64	0.055	0.038	11.5	1800
600		1.8	14.8	5.26	2.29	16.27	19.10	9.55	1.71	0.057	0.040	11.5	1800
600		2.0	14.8	5.26	2.54	16.77	19.60	9.80	1.75	0.058	0.041	11.5	1800
600	L-41	2.2	14.8	5.26	2.79	17.24	20.06	10.03	1.79	0.060	0.042	11.5	1800
700		1.0	17.0	6.02	1.38	15.86	19.14	9.57	1.76	0.056	0.040	13.5	2100
700	S-75	1.4	17.0	6.02	1.93	17.13	20.40	10.20	1.82	0.057	0.042	13.5	2100
700		1.8	17.0	6.02	2.48	18.37	21.62	10.81	1.87	0.059	0.043	13.5	2100
700		2.0	17.0	6.02	2.76	18.91	22.16	11.08	1.91	0.060	0.044	13.5	2100
700	S-47	2.2	17.0	6.02	3.04	19.48	22.72	11.36	1.94	0.061	0.045	13.5	2100
800		1.0	18.7	6.78	1.48	17.42	21.02	10.51	1.90	0.058	0.043	15.4	2400
800	S-85	1.4	18.7	6.78	2.07	18.89	22.48	11.24	1.93	0.058	0.044	15.4	2400
800		1.8	18.7	6.78	2.66	20.24	23.82	11.91	1.98	0.060	0.045	15.4	2400
800		2.0	18.7	6.78	2.96	20.85	24.42	12.21	2.01	0.061	0.045	15.4	2400
800	L-54	2.2	18.7	6.78	3.26	21.41	24.98	12.49	2.04	0.062	0.046	15.4	2400
900		1.0	20.0	7.55	1.62	19.15	23.00	11.50	1.96	0.060	0.045	17.3	2700
900	S-95	1.4	20.0	7.55	2.27	20.68	24.52	12.26	1.99	0.061	0.047	17.3	2700
900		1.8	20.0	7.55	2.92	22.18	26.00	13.00	2.02	0.062	0.049	17.3	2700
900		2.0	20.0	7.55	3.24	22.99	26.80	13.40	2.04	0.063	0.050	17.3	2700
900	L-60	2.2	20.0	7.55	3.56	23.39	27.20	13.60	2.08	0.064	0.050	17.3	2700
1000		1.0	21.0	8.33	1.73	20.62	24.66	12.33	2.00	0.061	0.046	19.2	3000
1000	S-106	1.2	21.0	8.33	2.08	21.42	25.45	12.73	2.01	0.062	0.047	19.2	3000
1000		1.6	21.0	8.33	2.77	23.02	27.04	13.52	2.04	0.063	0.048	19.2	3000
1000	L-67	2.0	21.0	8.33	3.46	24.44	28.44	14.22	2.08	0.064	0.050	19.2	3000
1000		2.2	21.0	8.33	3.81	25.10	29.10	14.55	2.10	0.065	0.050	19.2	3000
1100		0.8	22.0	8.94	1.45	20.93	25.17	12.58	2.06	0.062	0.049	21.2	3300
1100	S-118	1.2	22.0	8.94	2.17	22.70	26.92	13.46	2.07	0.062	0.049	21.2	3300
1100		1.6	22.0	8.94	2.90	24.40	28.60	14.30	2.09	0.063	0.050	21.2	3300
1100	L-74	1.8	22.0	8.94	3.26	25.17	29.40	14.70	2.10	0.064	0.050	21.2	3300
1100		2.0	22.0	8.94	3.62	25.87	30.06	15.03	2.12	0.064	0.051	21.2	3300
1200	S-129	0.6	22.6	9.55	1.13	20.96	25.32	12.66	2.08	0.063	0.049	23.1	3600
1200		1.0	23.0	9.55	1.88	22.95	27.37	13.69	2.13	0.064	0.050	23.1	3600
1200	L-82	1.4	23.0	9.55	2.63	24.81	29.22	14.61	2.13	0.064	0.050	23.1	3600
1200		1.8	23.0	9.55	3.38	26.43	30.82	15.41	2.16	0.065	0.051	23.1	3600
1300		0.6	23.7	10.14	1.16	21.97	26.54	13.27	2.16	0.064	0.049	25.0	3900
1300		1.0	23.7	10.14	1.94	24.03	28.58	14.29	2.16	0.064	0.050	25.0	3900
1300	L-29	1.4	23.7	10.14	2.72	25.99	30.52	15.26	2.16	0.065	0.050	25.0	3900
1300		1.8	23.7	10.14	3.49	27.72	32.24	16.12	2.17	0.066	0.051	25.0	3900
1400		0.2	21.5	10.71	0.40	19.94	24.08	12.04	1.89	0.064	0.049	26.9	4200
1400		0.6	24.3	10.71	1.19	22.88	27.60	13.80	2.19	0.065	0.050	26.9	4200
1400	L-102	1.0	24.3	10.71	1.99	25.08	29.74	14.87	2.18	0.065	0.050	26.9	4200
1400		1.4	24.3	10.71	2.79	26.96	31.60	15.80	2.18	0.065	0.051	26.9	4200

Table 38 Daily Nutrient Requirements of Dairy Cattle (Pounds) *continued*

Body Weight (lb)	Breed Size, Age (wk)	Daily Gain (lb)	Feed DM (lb)	NEm (Mcal)	NEg (Mcal)	ME (Mcal)	DE (Mcal)	TDN (lb)	Total Crude Protein (lb)	Ca (lb)	P (lb)	A (1,000 IU)	D (IU)
1500		0.2	22.5	11.28	0.41	20.86	25.20	12.60	1.97	0.064	0.050	28.8	4500
1500	L-116	0.6	24.9	11.28	1.22	23.95	28.74	14.37	2.21	0.065	0.050	28.8	4500
1500		1.0	24.9	11.28	2.03	26.01	30.78	15.39	2.20	0.066	0.051	28.8	4500
1500		1.4	24.9	11.28	2.84	27.86	32.62	16.31	2.20	0.066	0.051	28.8	4500
1600		0.2	23.6	11.84	0.41	21.89	26.44	13.22	2.05	0.064	0.050	30.8	4800
1600	L-140	0.6	25.5	11.84	1.22	24.83	29.74	14.87	2.23	0.066	0.051	30.8	4800
1600		1.0	25.5	11.84	2.04	26.99	31.88	15.94	2.22	0.066	0.051	30.8	4800
1700	L-163	0.2	24.6	12.40	0.41	22.82	27.56	13.78	2.14	0.064	0.051	32.7	5100
1700		0.6	26.1	12.40	1.23	25.62	30.64	15.32	2.27	0.066	0.051	32.7	5100
1700		1.0	26.1	12.40	2.05	27.78	32.78	16.39	2.26	0.066	0.051	32.7	5100
Growing Veal Calves Fed Only Milk													
75	—	1.10	1.5	0.97	0.90	3.16	3.51	1.8	0.38	0.015	0.011	1.4	225
100	L-1.0	1.75	2.3	1.37	1.51	5.03	5.59	2.8	0.57	0.018	0.013	1.9	300
125	L-3.0	2.00	2.7	1.59	1.75	5.83	6.48	3.2	0.65	0.024	0.015	2.4	375
150	L-4.8	2.20	3.0	1.82	1.96	6.57	7.30	3.7	0.72	0.029	0.018	2.9	450
175	L-6.4	2.30	3.3	2.05	2.09	7.12	7.91	4.0	0.74	0.033	0.020	3.4	525
200	L-8.0	2.40	3.6	2.26	2.23	7.68	8.53	4.3	0.77	0.035	0.021	3.8	600
225	L-9.5	2.50	3.8	2.47	2.38	8.25	9.17	4.6	0.81	0.037	0.022	4.3	675
250	L-10.9	2.65	4.1	2.67	2.58	8.94	9.93	5.0	0.86	0.039	0.023	4.8	750
275	L-12.3	2.75	4.4	2.87	2.74	9.53	10.59	5.3	0.89	0.040	0.024	5.3	825
300	L-13.6	2.80	4.6	3.07	2.86	10.02	11.14	5.6	0.92	0.041	0.025	5.8	900
325	L-14.9	2.85	4.9	3.26	2.97	10.48	11.65	5.8	0.94	0.042	0.026	6.3	975
Maintenance of Mature Breeding Bulls													
1200	—	—	18.3	9.98	—	17.01	20.54	10.3	1.58	0.042	0.036	23.1	—
1400	—	—	20.6	11.20	—	19.07	23.04	11.5	1.76	0.049	0.040	26.9	—
1600	—	—	22.7	12.38	—	21.08	25.46	12.7	1.93	0.057	0.045	30.8	—
1800	—	—	24.9	13.53	—	23.05	27.84	13.9	2.09	0.064	0.049	34.6	—
2000	—	—	26.9	14.64	—	24.94	30.12	15.1	2.25	0.071	0.053	38.5	—
2200	—	—	28.9	15.72	—	26.77	32.34	16.2	2.41	0.079	0.057	42.3	—
2400	—	—	30.8	16.78	—	28.58	34.52	17.3	2.56	0.086	0.061	46.2	—
2600	—	—	32.7	17.82	—	30.35	36.66	18.3	2.71	0.093	0.064	50.0	—
2800	—	—	34.6	18.84	—	32.09	38.76	19.4	2.85	0.099	0.068	53.9	—
3000	—	—	36.5	19.84	—	33.79	40.82	20.4	3.00	0.107	0.072	57.7	—

[a] Breed size: S for small breeds (e.g., Jersey); L is for large breeds (e.g., Holstein).

[b] Age in weeks indicates probable age of S or L animals when they reach the weight indicated.

Source: *Nutrient Requirements of Dairy Cattle,* Fifth revised edition, National Research Council, National Academy of Sciences, Washington, D.C., 1978.

Table 39 Daily Nutrient Requirements of Lactating and Pregnant Cows (Pounds)

Body Weight (lb)	Feed Energy				Total Crude Protein (lb)	Calcium (lb)	Phosphorus (lb)	Vitamin A (1,000 IU)
	NE$_l$ (Mcal)	ME (Mcal)	DE (Mcal)	TDN (lb)				
Maintenance of Mature Lactating Cows[a]								
700	6.02	10.00	11.66	5.84	0.71	0.028	0.023	24
800	6.65	11.06	12.89	6.45	0.77	0.032	0.026	28
900	7.27	12.08	14.08	7.05	0.83	0.035	0.028	31
1,000	7.86	13.07	15.23	7.63	0.89	0.038	0.030	35
1,100	8.45	14.04	16.36	8.19	0.95	0.040	0.032	38
1,200	9.02	14.99	17.47	8.75	1.01	0.043	0.034	41
1,300	9.57	15.91	18.55	9.29	1.06	0.046	0.037	45
1,400	10.12	16.82	19.61	9.82	1.12	0.048	0.039	48
1,500	10.66	17.72	20.65	10.34	1.17	0.051	0.041	52
1,600	11.19	18.60	21.67	10.85	1.22	0.053	0.043	55
1,700	11.71	19.46	22.68	11.36	1.27	0.056	0.045	59
1,800	12.22	20.31	23.67	11.86	1.32	0.059	0.047	62
Maintenance Plus Last 2 Months Gestation of Mature Dry Cows								
700	7.82	13.01	15.12	7.60	1.32	0.047	0.033	24
800	8.65	14.38	16.71	8.40	1.45	0.053	0.038	28
900	9.45	15.71	18.25	9.17	1.57	0.059	0.042	31
1,000	10.22	17.00	19.76	9.93	1.69	0.064	0.045	35
1,100	10.98	18.26	21.22	10.66	1.80	0.070	0.050	38
1,200	11.72	19.50	22.65	11.38	1.92	0.075	0.053	41
1,300	12.44	20.70	24.05	12.08	2.03	0.080	0.057	45
1,400	13.16	21.88	25.43	12.78	2.13	0.085	0.060	48
1,500	13.85	23.05	26.78	13.45	2.24	0.090	0.064	52
1,600	14.54	24.19	28.10	14.12	2.34	0.095	0.067	55
1,700	15.22	25.32	29.41	14.78	2.44	0.100	0.071	59
1,800	15.88	26.42	30.70	15.42	2.54	0.105	0.075	62
Milk Production — Nutrients Per Pound of Milk of Different Fat Percentages								
(% Fat)								
2.5	0.27	0.45	0.52	0.260	0.072	0.0024	0.0017	—
3.0	0.29	0.49	0.56	0.282	0.077	0.0025	0.0017	—
3.5	0.31	0.53	0.61	0.304	0.082	0.0026	0.0018	—
4.0	0.34	0.56	0.65	0.326	0.087	0.0027	0.0018	—
4.5	0.36	0.60	0.69	0.344	0.092	0.0028	0.0019	—
5.0	0.38	0.63	0.73	0.365	0.098	0.0029	0.0019	—
5.5	0.40	0.67	0.78	0.387	0.103	0.0030	0.0020	—
6.0	0.42	0.71	0.82	0.410	0.108	0.0031	0.0021	—
Body Weight Change During Lactation — Nutrients Per Pound Weight Change								
Weight loss	−2.23	−3.74	−4.33	−2.17	−0.32			
Weight gain	2.32	3.88	4.52	2.26	0.50			

[a] To allow for growth of young lactating cows, increase the maintenance allowances for all nutrients except vitamin A by 20 percent during the first lactation and 10 percent during the second lactation.

Source: *Nutrient Requirements of Dairy Cattle,* Fifth revised edition, National Research Council, National Academy of Sciences, Washinton, D.C., 1978.

Table 40 Recommended Nutrient Content of Rations for Dairy Cattle (Pounds)

	Lactating Cow Rations				Nonlactating Cattle Rations					Maximum Concentrations (All Classes)
	Cow Wt (lb)	Daily Milk Yields (lb)					Grow-ing	Calf		
Nutrients (Concentration in the Feed Dry matter)	≤ 900 1,100 1,300 ≥1,550	<18 <24 <31 <40	18–29 24–37 31–46 40–57	29–40 37–51 46–64 57–78	>40 >51 >64 >78 Dry Preg-nant Cows	Ma-ture Bulls	Heifers and Bulls	Starter Concen-trate Mix	Calf Milk Re-placer	
Ration No.	I	II	III	IV	V	VI	VII	VIII	IX	Max.
Crude Protein, %	13	14	15	16	11	8.5	12.0	16.0	22.0	—
Energy										
NE$_l$, Mcal/lb	0.64	0.69	0.73	0.78	0.61	—	—	—	—	—
NE$_m$, Mcal/lb	—	—	—	—	—	0.54	0.57	0.86	1.09	—
NE$_g$, Mcal/lb	—	—	—	—	—	—	0.27	0.54	0.70	—
ME, Mcal/lb	1.07	1.15	1.23	1.31	1.01	0.93	1.01	1.42	1.71	—
DE, Mcal/lb	1.26	1.34	1.42	1.50	1.20	1.12	1.20	1.60	1.90	—
TDN, %	63	67	71	75	60	56	60	80	95	—
Crude Fiber, %	17	17	17	17[a]	17	15	15	—	—	—
Acid Detergent Fiber	21	21	21	21	21	19	19	—	—	—
Ether Extract, %	2	2	2	2	2	2	2	2	10	—
Minerals[b]										
Calcium, %	0.43	0.48	0.54	0.60	0.37	0.24	0.40	0.60	0.70	—
Phosphorus, %	0.31	0.34	0.38	0.40	0.26	0.18	0.26	0.42	0.50	—
Magnesium, %[c]	0.20	0.20	0.20	0.20	0.16	0.16	0.16	0.07	0.07	—
Potassium, %	0.80	0.80	0.80	0.80	0.80	0.80	0.80	0.80	0.80	—
Sodium, %	0.18	0.18	0.18	0.18	0.10	0.10	0.10	0.10	0.10	—
Sodium chloride, %[d]	0.46	0.46	0.46	0.46	0.25	0.25	0.25	0.25	0.25	5
Sulfur, %[d]	0.20	0.20	0.20	0.20	0.17	0.11	0.16	0.21	0.29	0.35
Iron, ppm[d,e]	50	50	50	50	50	50	50	100	100	1,000
Cobalt, ppm	0.10	0.10	0.10	0.10	0.10	0.10	0.10	0.10	0.10	10
Copper, ppm[d,f]	10	10	10	10	10	10	10	10	10	80
Manganese, ppm[d]	40	40	40	40	40	40	40	40	40	1,000
Zinc, ppm[d,g]	40	40	40	40	40	40	40	40	40	500
Iodine, ppm[h]	0.50	0.50	0.50	0.50	0.50	0.25	0.25	0.25	0.25	50
Molybdenum, ppm[i,j]	—	—	—	—	—	—	—	—	—	6
Selenium, ppm	0.10	0.10	0.10	0.10	0.10	0.10	0.10	0.10	0.10	5
Fluorine, ppm[j]	—	—	—	—	—	—	—	—	—	30
Vitamins[k]										
Vit A, IU/lb	1,450	1,450	1,450	1,450	1,450	1,450	1,000	1,000	1,720	—
Vit D, IU/lb	140	140	140	140	140	140	140	140	270	—
Vit E, ppm	—	—	—	—	—	—	—	—	300	—

[a] It is difficult to formulate high-energy rations with a minimum of 17 percent crude fiber. However, fat percentage depression may occur when rations with less than 17 percent crude fiber or 21 percent ADF are fed to lactating cows.

[b] The mineral values presented in this table are intended as guidelines for use of professionals in ration formulation. Because of many factors affecting such values, they are not intended and should not be used as a legal or regulatory base.

[c] Under conditions conducive to grass tetany (see text), should be increased to 0.25 or higher.

[d] The maximum safe levels for many of the mineral elements are not well defined; estimates given here, especially for sulfur, sodium chloride, iron, copper, zinc, and manganese, are based on very limited data; safe levels may be substantially affected by specific feeding conditions.

[e] The maximum safe level of supplemental iron in some forms is materially lower than 1,000 ppm. As little as 400 ppm added iron as ferrous sulfate has reduced weight gains (Standish et al., 1969).

[f] High copper may increase the susceptibility of milk to oxidized flavor (see text).

[g] Maximum safe level of zinc for mature dairy cattle is 1,000 ppm.

[h] If diet contains as much as 25 percent strongly goitrogenic feed on dry basis, iodine provided should be increased two times or more.

[i] If diet contains sufficient copper, dairy cattle tolerate substantially more than 6 ppm molybdenum (see text).

[j] Maximum safe level of fluorine for growing heifers and bulls is lower than for other dairy cattle. Somewhat higher levels are tolerated when the fluorine is from less-available sources, such as phosphates (see text). Minimum requirement for molybdenum and fluorine not yet established.

[k] The following minimum quantities of B-complex vitamins are suggested per unit of milk replacer: niacin, 2.6 ppm; pantothenic acid, 13 ppm; riboflavin, 6.5 ppm; pyridoxine, 6.5 ppm; thiamine, 6.5 ppm; folic acid, 0.5 ppm; biotin, 0.1 ppm; vitamin B$_{12}$, 0.07 ppm; choline, 0.26 percent. It appears that adequate amounts of these vitamins are furnished when calves have functional rumens (usually at 6 weeks of age) by a combination of rumen synthesis and natural feedstuffs.

Source: *Nutrient Requirements of Dairy Cattle,* Fifth revised edition, National Research Council, National Academy of Sciences, Washington, D.C., 1978.

Table 41 Feed Composition - Fat and Fatty Acids[a]

Feed Name	International Feed Number	Moisture Basis	Dry Matter (%)	Ether Extract (%)	Saturated Fat[b] (%)	Unsaturated Fat (%)	Linoleic Acid (%)	Arachidonic Acid (%)
ALFALFA								
1 meal dehy 17%	1-00-023	As fed	92.0	2.3	0.30	0.60	0.40	—
protein		Dry	100.0	2.5	0.30	0.70	0.43	—
BARLEY								
2 grain	4-00-549	As fed	89.0	1.8	0.50	1.30	0.24	—
		Dry	100.0	2.1	0.60	1.40	0.27	—
CORN, DENT YELLOW								
3 grain	4-02-935	As fed	89.0	4.0	0.80	3.30	1.82	—
		Dry	100.0	4.5	0.90	3.70	2.05	—
4 gluten, meal	5-02-900	As fed	91.0	7.6	1.40	6.20	3.83	—
		Dry	100.0	8.4	1.50	6.80	4.21	—
FATS AND OILS								
5 oil, fish, menhaden	7-08-049	As fed	100.0	100.0	40.00	60.00	2.70	20.0-25.0
		Dry	100.0	100.0	40.00	60.00	2.70	20.0-25.0
6 oil, flax, common	4-14-502	As fed	100.0	100.0	8.20	91.80	13.90	—
(Linseed oil)		Dry	100.0	100.0	8.20	91.80	13.90	—
7 tallow, animal	4-08-127	As fed	100.0	100.0	47.60	52.40	4.30	0.0-0.2
		Dry	100.0	100.0	47.60	52.40	4.30	0.0-0.2
FISH								
8 solubles, condensed	5-01-969	As fed	51.0	6.5	2.90	3.60	0.20	—
		Dry	100.0	12.8	5.70	7.10	0.39	—
9 menhaden, meal	5-02-009	As fed	92.0	7.7	4.40	3.30	0.11	—
mech extd		Dry	100.0	8.4	4.80	3.60	0.12	—
FLAX								
10 common, meal solv	5-02-048	As fed	91.0	1.7	0.40	1.30	0.37	—
extd (Linseed meal)		Dry	100.0	1.9	0.40	1.50	0.41	—
MEAT Animal								
11 with blood, meal	5-00-386	As fed	92.0	8.1	4.00	4.10	0.28	—
tankage rendered		Dry	100.0	8.8	4.40	4.50	0.30	—
MILK Cattle								
12 skimmed dehy	5-01-175	As fed	94.0	0.9	0.40	0.60	0.01	—
		Dry	100.0	1.0	0.40	0.60	0.01	—
OATS								
13 grain	4-03-309	As fed	89.0	4.5	1.10	3.50	1.49	—
		Dry	100.0	5.1	1.20	3.90	1.67	—
POULTRY								
14 by-products, meal	5-03-798	As fed	93.0	11.6	4.20	7.50	1.72	—
rendered		Dry	100.0	12.5	4.50	8.00	1.98	—
SORGHUM								
15 grain	4-04-383	As fed	90.0	2.9	0.60	2.30	1.08	—
		Dry	100.0	3.2	0.70	2.50	1.20	—
SOYBEAN								
16 seeds, meal solv	5-04-604	As fed	90.0	1.0	0.03	0.07	0.55	—
extd		Dry	100.0	1.1	0.03	0.08	0.61	—
WHEAT								
17 bran	4-05-190	As fed	89.0	4.1	0.80	3.30	2.25	—
		Dry	100.0	4.6	0.90	3.70	2.53	—
18 grain	4-05-211	As fed	89.0	1.7	0.40	1.30	0.58	—
		Dry	100.0	1.9	0.40	1.50	0.65	—
WHEY								
19 dehy (Cattle)	4-01-182	As fed	93.0	0.8	0.50	0.30	0.01	—
		Dry	100.0	0.9	0.60	0.30	0.01	—
YEAST								
20 brewers, dehy	7-05-527	As fed	93.0	1.0	0.20	0.70	0.05	—
		Dry	100.0	1.1	0.20	0.80	0.05	—

[a]Data adapted from Edwards (1964), except arachidonic acids values.

[b]Calculated by assuming that ether extract was all triglyceride (except for alfalfa products). Thus, values were calculated by multiplying percent ether extract by fraction which was saturated or unsaturated. Alfalfa ether extract was presumed to be 40% triglyceride equivalent, and the percentage of ether extract was multiplied by 0.04 and then by the fraction which was saturated or unsaturated.

[c]Data adapted from Hilditch and Williams (1964).

Source: United States - Canadian Tables of Feed Composition, third revision, National Research Council, National Academy of Sciences, Washington, D.C., 1982

Table 42 Feed Composition - Mineral Supplements

Feed Name	International Feed Number	Moisture Basis	Dry Matter (%)	Protein Equivalent N X 6.25 (%)	Calcium (Ca) (%)	Chlorine (Cl) (%)	Magnesium (Mg) (%)	Phosphorus (P) (%)
AMMONIUM								
1 phosphate, monobasic $(NH_4)h_2PO_4$	6-09-338	As fed	97.0	68.8	0.27	—	0.45	24.00
		Dry	100.0	70.9	0.28	—	0.46	24.74
2 phosphate, dibasic $(\tilde{N}H_4)_2HPO_4$	6-00-370	As fed	97.0	112.4	0.50	—	0.45	20.00
		Dry	100.0	115.9	0.52	—	0.46	20.60
BONE								
3 meal (Feeding bone meal)	6-00-397	As fed	95.0	17.8	25.95	—	0.53	12.42
		Dry	100.0	18.8	27.32	—	0.56	13.07
4 meal, steamed	6-00-400	As fed	97.0	12.8	29.82	—	0.32	12.49
		Dry	100.0	13.2	30.71	—	0.33	12.86
CALCIUM								
5 carbonate, $CaCO_3$	6-01-069	As fed	100.0	—	39.39	—	0.05	0.04
		Dry	100.0	—	39.39	—	0.05	0.04
6 phosphate, dibasic, from defluorinated phosphoric acid	6-01-080	As fed	97.0	—	21.30	—	0.57	18.70
		Dry	100.0	—	22.00	—	0.59	19.30
COBALT								
7 carbonate, $CoCO_3$	6-01-566	As fed	99.0	—	—	—	—	—
		Dry	100.0	—	—	—	—	—
COPPER (CUPROUS)								
8 oxide, Cu_2O c p 6-28-224		As fed	99.0_2	—	—	—	—	—
		Dry	100.0	—	—	—	—	—
IRON (FERRIC)								
9 oxide, Fe_2O_3	6-02-431	As fed	92.0^b	—	0.30	—	0.40	—
		Dry	100.0	—	0.32	—	0.43	—
IRON (FERROUS)								
10 oxide, FeO	6-20-728	As fed	97.0	—	—	—	—	—
		Dry	100.0	—	—	—	—	—
11 sulfate, monohydrate $FeSO_4 \cdot H_2O$	6-01-869	As fed	98.0_b	—	—	—	—	—
		Dry	100.0	—	—	—	—	—
LIMESTONE								
12 magnesium (Dolomitic)	6-02-633	As fed	99.0^b	—	22.08	0.12	9.89	0.04
		Dry	100.0	—	22.30	0.12	9.99	0.04
MAGNESIUM								
13 carbonate, $MgCO_3Mg(OH)_2$	6-02-754	As fed	98.0^b	—	0.02	0.00	30.20	—
		Dry	100.0	—	0.02	0.00	30.81	—
oxide, MgO	6-02-756	As fed	98.0	—	3.00	—	54.90	—
		Dry	100.0	—	3.07	—	56.20	—
14 sulfate, hepta-hydrate, $MgSO_4 \cdot H_2O$ (Epsom salts)	6-02-758	As fed	98.0^b	—	0.02	—	9.60	—
		Dry	100.0	—	0.02	—	9.80	—
MANGANESE								
15 dioxide, MnO_2	6-03-042	As fed	98.0^b	—	—	—	—	—
		Dry	100.0	—	—	—	—	—

Potassium (K) (%)	Sodium (Na) (%)	Sulfur (S) (%)	Cobalt (Co) (%)	Copper (Cu) (%)	Fluorine (F) (%)	Iodine (I) (%)	Iron (Fe) (%)	Manganese (Mn) (%)	Selenium (Se) (%)	Zinc (Zn) (%)
0.01	0.06	1.42	0.001	0.001	0.24	—	1.690	0.040	—	0.01
0.01	0.06	1.46	0.001	0.001	0.25	—	1.740	0.040	—	0.01
0.01	0.05	2.10	—	0.001	0.20	—	1.200	0.040	—	0.01
0.01	0.05	2.16	—	0.001	0.21	—	1.240	0.040	—	0.01
—	0.68	—	—	—	0.07	—	0.025	—	—	—
—	0.72	—	—	—	0.07	—	0.026	—	—	—
0.18	5.53	2.44	—	—	—	—	2.600	—	—	0.01
0.19	5.69	2.51	—	—	—	—	2.670	—	—	0.01
0.06	0.06	—	—	—	—	—	0.030	0.030	—	—
0.06	0.06	—	—	—	—	—	0.030	0.030	—	—
0.07	0.05	1.11	0.001	0.001	0.18	—	1.400	0.030	—	0.01
0.07	0.05	1.14	0.001	0.001	0.18	—	1.440	0.030	—	0.01
—	—	0.20	45.54	—	—	—	0.049	—	—	—
—	—	0.20	46.00	—	—	—	0.049	—	—	—
—	—	—	—	87.93	—	—	—	—	—	—
—	—	—	—	88.82	—	—	—	—	—	—
—	—	—	—	—	—	—	57.000	0.300	—	—
—	—	—	—	—	—	—	61.950	0.320	—	—
—	—	—	—	—	—	—	75.370	—	—	—
—	—	—	—	—	—	—	77.700	—	—	—
—	—	18.00	—	—	—	—	32.300	—	—	—
—	—	18.37	—	—	—	—	32.960	—	—	—
0.36	—	—	—	—	—	—	0.076	—	—	—
0.36	—	—	—	—	—	—	0.077	—	—	—
—	—	—	—	—	—	—	0.021	—	—	—
—	—	—	—	—	—	—	0.022	—	—	—
—	—	—	—	—	0.02	—	—	0.01	—	—
—	—	—	—	—	0.02	—	—	0.01	—	—
—	—	12.75	—	—	—	—	—	—	—	—
—	—	13.00	—	—	—	—	—	—	—	—
—	—	—	—	—	—	—	—	61.93	—	—
—	—	—	—	—	—	—	—	63.19	—	—

Table 42 Feed Composition - Mineral Supplements — *continued*

Feed Name	International Feed Number	Moisture Basis	Dry Matter (%)	Protein Equivalent N X 6.25 (%)	Calcium (Ca) (%)	Chlorine (Cl) (%)	Magnesium (Mg) (%)	Phosphorus (P) (%)
MANGANESE (MANGANOUS)								
16 chloride, tetra-hydrate, $MnCl_2 \cdot 4H_2$	6-03-038	As fed	99.0[b]	—	—	35.47	—	—
		Dry	100.0	—	—	35.83	—	—
17 phosphate, tri-hydrate, $MnHPO_43H_2O$	6-29-492	As fed	99.0[b]	—	—	—	—	14.97
		Dry	100.0	—	—	—	—	15.11
18 sulfate, mono-hydrate, $MnSO_4 \cdot H_2O$, c p	6-28-103	As fed	—	—	—	—	—	—
		Dry	100.0	—	—	—	—	—
OYSTERSHELL								
19 ground (Flour)	6-03-481	As fed	99.0	—	37.62	0.01	0.30	0.07
		Dry	100.0	—	38.00	0.01	0.30	0.07
PHOSPHATE								
20 defluorinated	6-01-780	As fed	100.0	—	32.00	—	0.42	18.00
		Dry	100.0	—	32.00	—	0.42	18.00
21 rock	6-03-945	As fed	100.0	—	35.00	—	0.41	13.00
		Dry	100.0	—	35.00	—	0.41	13.00
PHOSPHORIC ACID								
22 H_3PO_4	6-03-707	As fed	75.0	—	0.04	—	0.38	23.70
		Dry	100.0	—	0.05	—	0.51	31.60
POTASSIUM								
23 chloride, KCl	6-03-755	As fed	100.0	—	0.05	47.30	0.34	—
		Dry	100.0	—	0.05	47.30	0.34	—
24 iodate, KIO_3, c p	6-03-758	As fed	100.0[b]	—	—	—	—	—
		Dry	100.0	—	—	—	—	—
25 and magnesium sulfate	6-06-177	As fed	98.0[b]	—	0.06	1.25	11.60	—
		Dry	100.0	—	0.06	1.28	11.84	—
26 sulfate, K_2SO_4, c p	6-06-098	As fed	98.0[b]	—	0.15	1.52	0.60	—
		Dry	100.0	—	0.15	1.55	0.61	—
SODIUM								
27 chloride, NaCl	6-04-152	As fed	100.0	—	—	60.66	—	—
		Dry	100.0	—	—	60.66	—	—
28 phosphate, mono-basic, monohydrate, $NaH_2PO_4 \cdot H_2O$	6-04-288	As fed	97.0	—	—	—	—	21.80
		Dry	100.0	—	—	—	—	22.50
29 tripolyphosphate, $Na_5P_3O_{10}$	6-08-076	As fed	96.0	—	—	—	—	24.00
		Dry	100.0	—	—	—	—	25.00
SULFUR								
30 elemental	6-04-705	As fed	99.0	—	—	—	—	—
		Dry	100.0	—	—	—	—	—
ZINC								
31 carbonate, $ZnCO_3$	6-05-549	As fed	99.0	—	—	—	—	—
		Dry	100.0	—	—	—	—	—
32 chloride, $ZnCl_2$, c p	6-05-552	As fed	98.0[b]	—	—	51.00	—	—
		Dry	100.0	—	—	52.03	—	—
33 oxide, ZnO	6-06-553	As fed	100.0	—	—	—	—	—
		Dry	100.0	—	—	—	—	—
34 oxide, ZnO, c p	6-05-554	As fed	100.0	—	—	—	—	—
		Dry	100.0	—	—	—	—	—
35 sulfate, mono-hydrate, $ZnSO_4 \cdot H_2O$	6-05-555	As fed	99.05[b]	—	0.02	0.015	—	—
		Dry	100.0	—	0.02	0.015	—	—

[a]The composition of mineral ingredients that are hydrated (e.g., $CaSO_4 \cdot 2H_2O$) is shown including the waters of hydration, both on an as-fed and dry matter basis. Mineral composition of feed grade mineral supplements varies by source, mining site, and manufacturer. Use manufacturer's analysis when available.
[b]Dry matter values have been estimated for these minerals.:lm12:ma9

Source: *United States - Canadian Tables of Feed Composition*, third revision, National Research Council, National Academy of Sciences, Washington, D.C., 1982

Potassium (K) (%)	Sodium (Na) (%)	Sulfur (S) (%)	Cobalt (Co) (%)	Copper (Cu) (%)	Fluorine (F) (%)	Iodine (I) (%)	Iron (Fe) (%)	Manganese (Mn) (%)	Selenium (Se) (%)	Zinc (Zn) (%)
—	—	—	—	—	—	—	—	27.48	—	—
—	—	—	—	—	—	—	—	27.76	—	—
—	—	—	—	—	—	—	—	62.17	—	—
—	—	—	—	—	—	—	—	62.80	—	—
—	—	18.97	—	—	—	—	—	32.50	—	—
—	—	18.97	—	—	—	—	—	32.50	—	—
0.10	0.21	—	—	—	—	—	0.284	0.01	—	—
0.10	0.21	—	—	—	—	—	0.287	0.01	—	—
0.08	4.90	—	0.001	0.002	0.18	—	0.670	0.02	—	0.006
0.08	4.90	—	0.001	0.002	0.18	—	0.670	0.02	—	0.006
0.06	0.03	—	0.001	0.001	3.50	—	1.680	0.02	—	0.01
0.06	0.03	—	0.001	0.001	3.50	—	1.680	0.02	—	0.01
0.02	0.03	1.16	0.001	0.001	0.23	—	1.310	0.04	—	0.010
0.02	0.04	1.55	0.001	0.001	0.31	—	1.750	0.05	—	0.013
50.00	1.00	0.45	—	—	—	—	0.060	—	—	—
50.00	1.00	0.45	—	—	—	—	0.060	—	—	—
18.27	—	—	—	—	—	59.30	—	—	—	—
18.27	—	—	—	—	—	59.30	—	—	—	—
18.50	0.76	22.30	—	—	0.001	—	0.010	0.002	—	0.001
18.88	0.78	22.76	—	—	0.001	—	0.010	0.002	—	0.001
41.00	0.09	17.00	—	—	—	—	0.070	0.001	—	—
41.84	0.09	17.35	—	—	—	—	0.071	0.001	—	—
—	39.34	—	—	—	—	—	—	—	—	—
—	39.34	—	—	—	—	—	—	—	—	—
—	16.18	—	—	—	—	—	—	—	—	—
—	16.68	—	—	—	—	—	—	—	—	—
—	29.80	—	—	—	—	—	0.004	—	—	—
—	31.00	—	—	—	—	—	0.004	—	—	—
—	—	99.00	—	—	—	—	—	—	—	—
—	—	99.45	—	—	—	—	—	—	—	—
—	—	—	—	—	—	—	—	—	—	51.63
—	—	—	—	—	—	—	—	—	—	52.15
—	—	—	—	—	—	—	—	—	—	47.00
—	—	—	—	—	—	—	—	—	—	47.97
—	—	—	—	—	—	—	—	—	—	78.00
—	—	—	—	—	—	—	—	—	—	78.00
—	—	—	—	—	—	—	—	—	—	80.33
—	—	—	—	—	—	—	—	—	—	80.33
—	—	17.50	—	—	—	—	0.001	0.001	—	36.00
—	—	17.68	—	—	—	—	0.001	0.001	—	36.36

UNIFORM STATE FEED BILL

Officially Adopted By
ASSOCIATION OF AMERICAN FEED CONTROL OFFICIALS
And Endorsed By
AMERICAN FEED MANUFACTURERS ASSOCIATION
NATIONAL FEED INGREDIENTS ASSOCIATION
PET FOOD INSTITUTE

Although this Bill and Regulations have not been passed into law in all the states the subject matter covered herein does represent the official policy of this Association.

AN ACT

To regulate the manufacture and distribution of commercial feeds in the State of _____ BE IT ENACTED by the Legislature of the State of _____.

Section 1. Title

This Act shall be known as the "_____ Commercial Feed Law of 19____."

Section 2. Enforcing Official

This Act shall be administered by the _____ of the State of _____, hereinafter referred to as the "_____."

Section 3. Definitions of Words and Terms

When used in this Act:
  (a) The term "person" includes individual, partnership, corporation, and association.
  (b) The term "distribute" means to offer for sale, sell, exchange, or barter, commercial feed; or to supply, furnish, or otherwise provide commercial feed to a contract feeder.
  (c) The term "distributor" means any person who distributes.
  (d) The term "commercial feed" means all materials except whole seeds unmixed or physically altered entire unmixed seeds, when not adulterated within the meaning of Sec. 7(a), which are distributed for use as feed or for mixing in feed: Provided, That the _____ by regulation may exempt from this definition, or from specific provisions of this Act, commodities such as hay, straw, stover, silage, cobs, husks, hulls, and individual chemical compounds or substances when such commodities, compounds or

397

substances are not inter-mixed or mixed with other materials, and are not adulterated within the meaning of Section 7(a), of this Act.

(e) The term "feed ingredient" means each of the constituent materials making up a commercial feed.

(f) The term "mineral feed" means a commercial feed intended to supply primarily mineral elements or inorganic nutrients.

(g) The term "drug" means any article intended for use in the diagnosis, cure, mitigation, treatment, or prevention of disease in animals other than man and articles other than feed intended to affect the structure or any function of the animal body.

(h) The term "customer-formula feed" means commercial feed which consists of a mixture of commercial feeds and/or feed ingredients each batch of which is manufactured according to the specific instructions of the final purchaser.

(i) The term "manufacture" means to grind, mix or blend, or further process a commercial feed for distribution.

(j) The term "brand name" means any word, name, symbol, or device, or any combination thereof, identifying the commercial feed of a distributor or registrant and distinguishing it from that of others.

(k) The term "product name" means the name of the commercial feed which identifies it as to kind, class, or specific use.

(l) The term "label" means a display of written, printed, or graphic matter upon or affixed to the container in which a commercial feed is distributed, or on the invoice or delivery slip with which a commercial feed is distributed.

(m) The term "labeling" means all labels and other written, printed, or graphic matter (1) upon a commercial feed or any of its containers or wrapper or (2) accompanying such commercial feed.

(n) The term "ton" means a net weight of two thousand pounds avoirdupois.

(o) The terms "per cent" or "percentages" means percentages by weights.

(p) The term "official sample" means a sample of feed taken by the _____ or his agent in accordance with the provisions of Section 11(c), (e), or (f) of this Act.

(q) The term "contract feeder" means a person who as an independent contractor, feeds commercial feed to animals pursuant to a contract whereby such commercial feed is supplied, furnished, or otherwise provided to such person and whereby such person's remuneration is determined all or in part by feed consumption, mortality, profits, or amount or quality of product.

(r) The term "pet food" means any commercial feed prepared and distributed for consumption by pets.

(s) The term "pet" means any domesticated animal normally maintained in or near the household(s) of the owner(s) thereof.

(t) The term "specialty pet food" means any commercial feed prepared and distributed for consumption by specialty pets.

(u) The term "specialty pet" means any domesticated animal pet normally maintained in a cage or tank, such as, but not limited to, gerbils, hamsters, canaries, psittacine birds, mynahs, finches, tropical fish, goldfish, snakes, and turtles.

Section 4. Registration

(a) No person shall manufacture feed in this State, unless he has filed with the _____ on forms provided by the _____, his name, place of business and location of each manufacturing facility in this State.

(b) No person shall distribute in this State a commercial feed, except a customer-formula feed, which has not been registered pursuant to the provisions of this section. The application for registration shall be submitted in the manner prescribed by the _____. Upon approval by the _____ the registration shall be issued to the applicant. All registrations expire on the 31st day of December of each year. (Option: A registration shall continue in effect unless it is cancelled by the registrant or unless it is cancelled by the _____ pursuant to Subsection (c) of this section.)

(c) The _____ is empowered to refuse registration of any commercial feed not in compliance with the provisions of this Act and to cancel any registration subsequently found not to be in compliance with any provision of this Act: Provided, That no registration shall be refused or canceled unless the registrant shall have been given an opportunity to be heard before the _____ and to amend his application in order to comply with the requirements of this Act.

Section 5. Labeling

A commercial feed shall be labeled as follows:

(a) In case of a commercial feed, except a customer-formula feed, it shall be accompanied by a label bearing the following information:

(1) The net weight.

(2) The product name and the brand name, if any, under which the commercial feed is distributed.

(3) The guaranteed analysis stated in such terms as the _____ by regulation determines is required to advise the user of the composition of the feed or to support claims made in the labeling. In all cases the substances or elements must be determinable by laboratory methods such as the methods published by the Association of Official Analytical Chemists.

(4) The common or usual name of each ingredient used in the manufacture of the commercial feed: Provided, That the _____ by regulation may permit the use of a collective term for a group of ingredients which perform a similar function, or he may exempt such commercial feeds, or any group thereof, from this requirement of an ingredient statement if he finds that such statement is not required in the interest of consumers.

(5) The name and principal mailing address of the manufacturer or the person responsible for distributing the commercial feed.

(6) Adequate directions for use for all commercial feeds containing drugs and for such other feeds as the _____ may require by regulation as necessary for their safe and effective use.

(7) Such precautionary statements as the _____ by regulation determines are necessary for the safe and effecuve use of the commercial feed.

(b) In the case of a customer-formula feed, it shall be accompanied by a label, invoice, delivery slip, or other shipping document, bearing the following information:

(1) Name and address of the manufacturer.

(2) Name and address of the purchaser.

(3) Date of delivery.

(4) The product name and brand name, if any, and the net weight of each registered commercial feed used in the mixture, and the net weight of each other ingredient used.

(5) Adequate directions for use for all customer-formula feeds containing drugs and for

such other feeds as the _____ may require by regulation as necessary for their safe and effective use.

(6) Such precautionary statements as the _____ by regulation determines are necessary for the safe and effective use of the customer-formula feed.

Section 6. Misbranding

A commercial feed shall be deemed to be misbranded:

(a) If its labeling is false or misleading in any particular.

(b) If it is distributed under the name of another commercial feed.

(c) If it is not labeled as required in Section 5 of this Act.

(d) If it purports to be or is represented as a commercial feed, or if it purports to contain or is represented as containing a commercial feed ingredient, unless such commercial feed or feed ingredient conforms to the definition, if any, prescribed by regulation by the _____.

(e) If any word, statement, or other information required by or under authority of this Act to appear on the label or labeling is not prominently placed thereon with such conspicuousness (as compared with other words, statements, designs, or devices in the labeling) and in such terms as to render it likely to be read and understood by the ordinary individual under customary conditions of purchase and use.

Section 7. Adulteration

A commercial feed shall be deemed to be adulterated:

(a) (1) If it bears or contains any poisonous or deleterious substance which may render it injurious to health; but in case the substance is not an added substance, such commercial feed shall not be considered adulterated under this subsection if the quantity of such substance in such commercial feed does not ordinarily render it injurious to health; or

(2) If it bears or contains any added poisonous, added deleterious, or added nonnutritive substance which is unsafe within the meaning of Section 406 of the Federal Food, Drug, and Cosmetic Act (other than one which is (i) a pesticide chemical in or on a raw agricultural commodity; or (ii) a food additive; or

(3) If it is, or it bears or contains any food additive which is unsafe within the meaning of Section 409 of the Federal Food, Drug, and Cosmetic Act; or

(4) If it is a raw agricultural commodity and it bears or contains a pesticide chemical which is unsafe within the meaning of Section 408(a) of the Federal Food, Drug, and Cosmetic Act: Provided, That where a pesticide chemical has been used in or on a raw agricultural commodity in conformity with an exemption granted or a tolerance prescribed under Section 408 of the Federal Food, Drug, and Cosmetic Act and such raw agricultural commodity has been subjected to processing such as canning, cooking, freezing, dehydrating, or milling, the residue of such pesticide chemical remaining in or on such processed feed shall not be deemed unsafe if such residue in or on the raw agricultural commodity has been removed to the extent possible in good manufacturing practice and the concentration of such residue in the processed feed is not greater than the tolerance prescribed for the raw agricultural commodity unless the feeding of such processed feed will result or is likely to result in a pesticide residue in the edible product of the animal, which is unsafe within the meaning of Section 408(a), of the Federal Food, Drug, and Cosmetic Act.

    (5) If it is, or it bears or contains any color additive which is unsafe within the meaning of Section 706 of the Federal Food, Drug, and Cosmetic Act.

(b) If any valuable constituent has been in whole or in part omitted or abstracted therefrom or any less valuable substance substituted therefor.

(c) If its composition or quality falls below or differs from that which it is purported or is represented to possess by its labeling.

(d) If it contains a drug and the methods used in or the facilities or controls used for its manufacture, processing, or packaging do not conform to current good manufacturing practice regulations promulgated by the _____ to assure that the drug meets the requirement of this Act as to safety and has the identity and strength and meets the quality and purity characteristics which it purports or is represented to possess. In promulgating such regulations, the _____ shall adopt the current good manufacturing practice regulations for medicated feed premixes and for medicated feeds established under authority of the Federal Food, Drug, and Cosmetic Act, unless he determines that they are not appropriate to the conditions which exist in this State.

(e) If it contains viable weed seeds in amounts exceeding the limits which the _____ shall establish by rule or regulation.

Section 8. Prohibited Acts

The following acts and the causing thereof within the State of _____ are hereby prohibited:

(a) The manufacture or distribution of any commercial feed that is adulterated or misbranded.

(b) The adulteration or misbranding of any commercial feed.

(c) The distribution of agricultural commodities such as whole seed, hay, straw, stover, silage, cobs, husks, and hulls, which are adulterated within the meaning of Section 7(a), of this Act.

(d) The removal or disposal of a commercial feed in violation of an order under Section 12 of this Act.

(e) The failure or refusal to register in accordance with Section 4 of this Act.

(f) The violation of Section 13(f) of this Act.

(g) Failure to pay inspection fees and file reports as required by Section 9 of this Act.

Section 9. Inspection Fees and Reports

(a) An inspection fee at the rate of _____ cents per ton shall be paid on commercial feeds distributed in this State by the person who distributes the commercial feed to the consumer, subject to the following:

    (1) No fee shall be paid on a commercial feed if the payment has been made by a previous distributor.

    (2) No fee shall be paid on customer-formula feeds if the inspection fee is paid on the commercial feeds which are used as ingredients therein.

    (3) No fee shall be paid on commercial feeds which are used as ingredients for the manufacture of commercial feeds which are registered. If the fee has already been paid, credit shall be given for such payment.

    (4) In the case of a commercial feed which is distributed in the State only in packages of ten pounds or less, an annual fee of _____ shall be paid in lieu of the inspection fee specified above.

(5) The minimum inspection fee shall be _____ per quarter.

(6) In the case of specialty pet food which is distributed in the State in packages of one pound or less, an annual fee of _____ shall be paid in lieu of an inspection fee.

(b) Each person who is liable for the payment of such fee shall:

(1) File, not later than the last day of January, April, July, and October of each year, a quarterly statement, setting forth the number of net tons of commercial feeds distributed in this State during the preceding calendar quarter; and upon filing such statement shall pay the inspection fee at the rate stated in paragraph (a) of this Section. Inspection fees which are due and owing and have not been remitted to the _____ within 15 days following the date due shall have a penalty fee of _____ per cent (minimum _____) added to the amount due when payment is finally made. The assessment of this penalty fee shall not prevent the _____ from taking other actions as provided in this chapter.

(2) Keep such records as may be necessary or required by the _____ to indicate accurately the tonnage of commercial feed distributed in this State, and the _____ shall have the right to examine such records to verify statements of tonnage.

Failure to make an accurate statement of tonnage or to pay the inspection fee or comply as provided herein shall constitute sufficient cause for the cancellation of all registrations on file for the distributor.

(c) Fees collected shall constitute a fund for the payment of the costs of inspection, sampling, and analysis, and other expenses necessary for the administration of this Act.

Section 10. Rules and Regulations

(a) The _____ is authorized to promulgate such rules and regulations for commercial feeds and pet foods as are specifically authorized in this Act and such other reasonable rules and regulations as may be necessary for the efficient enforcement of this Act. In the interest of uniformity the _____ shall by regulation adopt, unless he determines that they are inconsistent with the provisions of this Act or are not appropriate to conditions which exist in this State, the following:

(1) The Official Definitions of Feed Ingredients and Official Feed Terms adopted by the Association of American Feed Control Officials and published in the Official Publication of that organization, and

(2) Any regulation promulgated pursuant to the authority of the Federal Food, Drug, and Cosmetic Act (U.S.C. Sec. 301, et seq.): Provided, That the _____ would have the authority under this Act to promulgate such regulations.

(b) Before the issuance, amendment, or repeal of any rule or regulation authorized by this Act, the _____ shall publish the proposed regulation, amendment, or notice to repeal an existing regulation in a manner reasonably calculated to give interested parties, including all current registrants, adequate notice and shall afford all interested persons an opportunity to present their views thereon, orally or in writing, within a reasonable period of time. After consideration of all views presented by interested persons, the _____ shall take appropriate action to issue the proposed rule or regulation or to amend or repeal an existing rule or regulation. The provisions of this paragraph not withstanding, if the _____, pursuant to the authority of this Act, adopts the Official Definitions of Feed Ingredients or Official Feed Terms as adopted by the Association of American Feed Control Officials, or regulations promulgated pursuant to the authority of the Federal Food, Drug, and Cosmetic Act, any amendment or modification adopted by said Association or by the Secretary of Health, Education and Welfare in the case of regulations promulgated pursuant to the Federal Food, Drug, and Cosmetic Act, shall be adopted automatically under this Act without regard to the publication of the notice required by this paragraph (b), unless the _____, by order specifically determines that said amendment or modification shall not be adopted.

Section 11. Inspection, Sampling, and Analysis

(a) For the purpose of enforcement of this Act, and in order to determine whether its provisions have been complied with, including whether or not any operations may be subject to such provisions, officers or employees duly designated by the _____ , upon presenting appropriate credentials, and a written notice to the owner, operator, or agent in charge, are authorized (1) to enter, during normal business hours, any factory, warehouse, or establishment within the State in which commercial feeds are manufactured, processed, packed, or held for distribution, or to enter any vehicle being used to transport or hold such feeds; and (2) to inspect at reasonable times and within reasonable limits and in a reasonable manner, such factory, warehouse, establishment or vehicle and all pertinent equipment, finished and unfinished materials, containers, and labeling therein. The inspection may include the verification of only such records, and production and control procedures as may be necessary to determine compliance with the Good Manufacturing Practice Regulations established under Section 7(d) of this Act.

(b) A separate notice shall be given for each such inspection, but a notice shall not be required for each entry made during the period covered by the inspection. Each such inspection shall be commenced and completed with reasonable promptness. Upon completion of the inspection, the person in charge of the facility or vehicle shall be so notified.

(c) If the officer or employee making such inspection of a factory, warehouse, or other establishment has obtained a sample in the course of the inspection, upon completion of the inspection and prior to leaving the premises he shall give to the owner, operator, or agent in charge a receipt describing the samples obtained.

(d) If the owner of any factory, warehouse, or establishment described in paragraph (a), or his agent, refuses to admit the _____ or his agent to inspect in accordance with paragraphs (a) and (b), the _____ is authorized to obtain from any State Court a warrant directing such owner or his agent to submit the premises described in such warrant to inspection.

(e) For the purpose of the enforcement of this Act, the _____ or his duly designated agent is authorized to enter upon any public or private premises including any vehicle of transport during regular business hours to have access to, and to obtain samples, and to examine records relating to distribution of commercial feeds.

(f) Sampling and analysis shall be conducted in accordance with methods published by the Association of Official Analytical Chemists, or in accordance with other generally recognized methods.

(g) The results of all analyses of official samples shall be forwarded by the _____ to the person named on the label and to the purchaser. When the inspection and analysis of an official sample indicates a commercial feed has been adulterated or misbranded and upon request within 30 days following receipt of the analysis the _____ shall furnish to the registrant a portion of the sample concerned.

(h) The _____, in determining for administrative purposes whether a commercial feed is deficient in any component, shall be guided by the official sample as defined in paragraph (p) of Section 3 and obtained and analyzed as provided for in paragraphs (c), (e), and (f) of Section 11 of this Act.

Section 12. Detained Commercial Feeds

(a) "Withdrawal from distribution" orders: When the _____ or his authorized agent has reasonable cause to believe any lot of commercial feed is being distributed in violation of any of the provisions of this Act or of any of the prescribed regulations under this Act, he may issue and enforce a written or printed "withdrawal from distribution" order, warning the distributor not to dispose of the lot of commercial feed in any manner until written permission is given by the _____ or the Court. The _____ shall release the lot of commercial feed so withdrawn when said provisions and regulations have been complied with. If compliance is not obtained within 30 days, the _____ may begin, or upon request of the distributor or registrant shall begin, proceedings for condemnation.

(b) "Condemnation and Confiscation": Any lot of commercial feed not in compliance with said provisions and regulations shall be subject to seizure on complaint of the _____ to a court of competent jurisdiction in the area in which said commercial feed is located. In the event the court finds the said commercial feed to be in violation of this Act and orders the condemnation of said commercial feed, it shall be disposed of in any manner consistent with the quality of the commercial feed and the laws of the State: Provided, That in no instance shall the disposition of said commercial feed be ordered by the court without first giving the claimant an opportunity to apply to the court for release of said commercial feed or for permission to process or re-label said commercial feed to bring it into compliance with this Act.

Section 13. Penalties

(a) Any person convicted of violating any of the provisions of this Act or who shall impede, hinder, or otherwise prevent, or attempt to prevent, said _____ or his duly authorized agent in performance of his duty in connection with the provisions of this Act, shall be adjudged guilty of a misdemeanor and shall be fined not less than _____ or more than _____ for the first violation, and not less than _____ or more than _____ for a subsequent violation.

(b) Nothing in this Act shall be construed as requiring the _____ or his representative to: (1) report for prosecution, or (2) institute seizure proceedings, or (3) issue a withdrawal from distribution order, as a result of minor violations of the Act, or when he believes the public interest will best be served by suitable notice of warning in writing.

(c) It shall be the duty of each _____ attorney to whom any violation is reported to cause appropriate proceedings to be instituted and prosecuted in a court of competent jurisdiction without delay. Before the _____ reports a violation for such prosecution, an opportunity shall be given the distributor to present his view to the _____.

(d) The _____ is hereby authorized to apply for and the court to grant a temporary or permanent injunction restraining any person from violating or continuing to violate any of the provisions of this Act or any rule or regulation promulgated under the Act notwithstanding the existence of other remedies at law. Said injunction to be issued without bond.

(e) Any person adversely affected by an act, order, or ruling made pursuant to the provisions of this Act may within 45 days thereafter bring action in the (here name the particular Court in the county where the enforcement official has his office) for judicial review of such actions. The form of the proceeding shall be any which may be provided by statutes of this State to review decisions of administrative agencies, or in the absence or inadequacy thereof, any applicable form of legal action, including actions for declaratory judgments or writs of prohibitory or mandatory injunctions.

(f) Any person who uses to his own advantage, or reveals to other than the _____ , or officers of the _____ (appropriate departments of this State), or to the courts when revelant in any judicial proceeding, any information acquired under the authority of this Act, concerning any method, records, formulations, or processes which as a trade secret is entitled to protection, is guilty of a misdemeanor and shall on conviction thereof be fined not less than $_____ or imprisoned for not less than _____ year(s) or both: Provided, That this prohibition shall not be deemed as prohibiting the _____ , or his duly authorized agent from exchanging information of a regulatory nature with duly appointed officials of the United States Government, or of other States, who are similarly prohibited by law from revealing this information.

Section 14. Cooperation with other entities

The _____ may cooperate with and enter into agreements with governmental agencies of this State, other States, agencies of the Federal Government, and private associations in order to carry out the purpose and provisions of this Act.

Section 15. Publication

The _____ shall publish at least annually, in such forms as he may deem proper, information concerning the sales of commercial feeds, together with such data on their production and use as he may consider advisable, and a report of the results of the analyses of official samples of commercial feeds sold within the State as compared with the analyses guaranteed in the registration and on the label: Provided, That the information concerning production and use of commercial feed shall not disclose the operations of any person.

Section 16. Constitutionality

If any clause, sentence, paragraph, or part of this Act shall for any reason be judged invalid by any court of competent jurisdiction, such judgment shall not affect, impair, or invalidate the remainder thereof but shall be confined in its operation to the clause, sentence, paragraph, or part thereof directly involved in the controversy in which such judgment shall have been rendered.

Section 17. Repeal

All laws and parts of laws in conflict with or inconsistent with the provisions of this Act are hereby repealed. (The specific statute and specific code sections to be repealed may have to be stated.)

Section 18. Effective Date

This Act shall take effect and be in force from and after the first day of _____ .

PROPOSED RULES AND REGULATIONS
under the
UNIFORM STATE FEED BILL

Pursuant to due publication and public hearing required by the provisions of Chapter _____ of the Laws of this State, the _____ has adopted the following Rules and Regulations.

Regulation 1. Definitions and Terms

(a) The names and definitions for commercial feeds shall be the Official Definition of Feed Ingredients adopted by the Association of American Feed Control Officials, except as the _____ designates otherwise in specific cases.

(b) The terms used in reference to commercial feeds shall be the Official Feed Terms adopted by the AAFCO, except as the _____ designates otherwise in specific cases.

(c) The following commodities are hereby declared exempt from the definition of commercial feed, under the provisions of Section 3(d) of the Act: raw meat; and hay, straw, stover, silages, cobs, husks, and hulls when unground and when not mixed or intermixed with other materials: Provided that these commodities are not adulterated within the meaning of Section 7(a), of the Act.

Regulation 2. Label Format

Commercial feeds shall be labeled with the information prescribed in this regulation on the principal display panel of the product and in the following general format:

(a) Net Weight

(b) Product name and brand name if any

(c) If drugs are used:
  (1) The word "medicated" shall appear directly following and below the product name in type size no smaller than one half the type size of the product name.
  (2) The purpose of medication (claim statement).
  (3) The required direction for use and precautionary statements or reference to their location if the detailed feeding directions and precautionary statements required by Regulations 6 and 7 appear elsewhere on the label.
  (4) An active drug ingredient statement listing the active drug ingredients by their established name and amounts in accordance with Regulation 4(d).

(d) The guaranteed analysis of the feed as required under the provisions of Section 5(a)(3) of the Act include the following items, unless exempted in (8) of this subsection, and in the order listed:
  (1) Minimum percentage of crude protein.
  (2) Maximum or minimum percentage of equivalent protein from non-protein nitrogen as required in Regulation 4(e).
  (3) Minimum percentage of crude fat.
  (4) Maximum percentage of crude fiber.
  (5) Minerals, to include, in the following order: (a) minimum and maximum percentages of calcium (Ca), (b) minimum percentages of phosphorus (P), (c) minimum and maximum percentages of salt (NaCl), and (d) other minerals.
  (6) Vitamins in such terms as specified in Regulation 4(c).
  (7) Total Sugars as Invert on dried molasses products or products being sold primarily for their molasses content.

(8) Exemptions.

    (i) Guarantees for minerals are not required when there are no specific label claims and when the commercial feed contains less than 6½% of mineral elements.

    (ii) Guarantees for vitamins are not required when the commercial feed is neither formulated for nor represented in any manner as a vitamin supplement.

    (iii) Guarantees for crude protein, crude fat, and crude fiber are not required when the commercial feed is intended for purposes other than to furnish these substances or they are of minor significance relating to the primary purpose of the product, such as drug premixes, mineral or vitamin supplements, and molasses.

(e) Feed ingredients, collective terms for the grouping of feed ingredients, or appropriate statements as provided under the provisions of Section 5(a)(4) of the Act.

    (1) The name of each ingredient as defined in the Official Definitions of Feed Ingredients published in the Official Publication of the Association of American Feed Control Officials, common or usual name, or one approved by the _____.

    (2) Collective terms for the grouping of feed ingredients as defined in the Official Definitions of Feed Ingredients published in the Official Publication of the Association of American Feed Control Officials in lieu of the individual ingredients; Provided that:

        (i) When a collective term for a group of ingredients is used on the label, individual ingredients within that group shall not be listed on the label.

        (ii) The manufacturer shall provide the feed control official, upon request, with a listing of individual ingredients, within a defined group, that are or have been used at manufacturing facilities distributing in or into the State.

    (3) The registrant may affix the statement, "Ingredients as registered with the State" in lieu of the ingredient list on the label. The list of ingredients must be on file with the _____. This list shall be made available to the feed purchaser upon request.

(f) Name and principal mailing address of the manufacturer or person responsible for distributing the feed. The principal mailing address shall include the street address, city, state and zip code; however, the street address may be omitted if it is shown in the current city directory or telephone directory.

(g) The information required in Section 5(a)(1)-(5) of the Act must appear in its entirety on one side of the label or on one side of the container. The information required by Section 5(a)(6)-(7) of the Act shall be displayed in a prominent place on the label or container but not necessarily on the same side as the above information. When the information required by Section 5(a)(6)-(7) is placed on a different side of the label or container, it must be referenced on the front side with a statement such as "see back of label for directions for use." None of the information required by Section 5 of the Act shall be subordinated or obscured by other statements or designs.

Regulation 3. Brand and Product Names

(a) The brand or product name must be appropriate for the intended use of the feed and must not be misleading. If the name indicates the feed is made for a specific use, the character of the feed must conform therewith. A mixture labeled "Dairy Feed," for example, must be suitable for that purpose.

(b) Commercial, registered brand or trade names are not permitted in guarantees or ingredient listings.

(c) The name of a commercial feed shall not be derived from one or more ingredients of a mixture to the exclusion of other ingredients and shall not be one representing any components of a mixture unless all components are included in the name: Provided, That if any ingredient or combination of ingredients is intended to impart a distinctive characteristic to the product which is of significance to the purchaser, the name of that ingredient or combination of ingredients may be used as a part of the brand name or product name if the ingredient or combination of ingredients is quantitatively guaranteed in the guaranteed analysis, and the brand or product name is not otherwise false or misleading.

(d) The word "protein" shall not be permitted in the product name of a feed that contains added non-protein nitrogen.

(e) When the name carries a percentage value, it shall be understood to signify protein and/ or equivalent protein content only, even though it may not explicitly modify the percentage with the word "protein": Provided, That other percentage values may be permitted if they are followed by the proper description and conform to good labeling practice. When a figure is used in the brand name (except in mineral, vitamin, or other products where protein guarantee is nil or unimportant), it shall be preceded by the word "number" or some other suitable designation.

(f) Single ingredient feeds shall have a product name in accordance with the designated definition of feed ingredients as recognized by the Association of American Feed Control Officials unless the _____ designates otherwise.

(g) The word "vitamin", or a contraction thereof, or any word suggesting vitamin can be used only in the name of a feed which is represented to be a vitamin supplement, and which is labeled with the minimum content of each vitamin declared, as specified in Regulation 4(c).

(h) The term "mineralized" shall not be used in the name of a feed, except for "TRACE MINERALIZED SALT". When so used, the product must contain significant amounts of trace minerals which are recognized as essential for animal nutrition.

(i) The term "meat" and "meat by-products" shall be qualified to designate the animal from which the meat and meat by-products is derived unless the meat and meat by-products are from cattle, swine, sheep and goats.

Regulation 4. Expression of Guarantees

(a) The guarantees for crude protein, equivalent protein from non-protein nitrogen, crude fat, crude fiber and mineral guarantees (when required) will be in terms of percentage by weight.

(b) Commercial feeds containing 6½% or more mineral elements shall include in the guaranteed analysis the minimum and maximum percentages of calcium (Ca), the minimum percentage of phosphorus (P), and if salt is added, the minimum and maximum percentage of salt (NaCl). Minerals, except salt (NaCl), shall be guaranteed in terms of percentage of the element. When calcium and/or salt guarantees are given in the guaranteed analysis such shall be stated and conform to the following:

   (1) When the minimum is 5.0% or less, the maximum shall not exceed the minimum by more than one percentage point.

   (2) When the minimum is above 5.0%, the maximum shall not exceed the minimum by more than 20% and in no case shall the maximum exceed the minimum by more than 5 percentage points.

(c) Guarantees for minimum vitamin content of commercial feeds and feed supplements, when made, shall be stated on the label in milligrams per pound of feed except that:

    (1) Vitamin A, other than precursors of vitamin A, shall be stated in International or USP units per pound.

    (2) Vitamin D, in products offered for poultry feeding, shall be stated in International Chick Units per pound.

    (3) Vitamin D for other uses shall be stated in International or USP units per pound.

    (4) Vitamin E shall be stated in International or USP units per pound.

    (5) Guarantees for vitamin content on the label of a commercial feed shall state the guarantee as true vitamins, not compounds, with exception of the compounds, Pyridoxine Hydrochloride, Choline Chloride, Thiamine, and d-Pantothenic Acid.

    (6) Oils and premixes containing vitamin A or vitamin D or both may be labeled to show vitamin content in terms of units per gram.

(d) Guarantees for drugs shall be stated in terms of percent by weight, except:

    (1) Antibiotics present at less than 2,000 grams per ton (total) of commercial feed shall be stated in grams per ton of commercial feed.

    (2) Antibiotics present at 2,000 or more grams per ton (total) of commercial feed shall be stated in grams per pound of commercial feed.

    (3) Labels for commercial feeds containing growth promotion and/or feed efficiency levels of antibiotics, which are to be fed continuously as the sole ration, are not required to make quantitative guarantees except as specifically noted in the Federal Food Additive Regulations for certain antibiotics, wherein, quantitative guarantees are required regardless of the level or purpose of the antibiotic.

    (4) The term "milligrams per pound" may be used for drugs or antibiotics in those cases where a dosage is given in "milligrams" in the feeding directions.

(e) Commercial feeds containing any added non-protein nitrogen shall be labeled as follows:

    (1) Complete feeds, supplements, and concentrates containing added non-protein nitrogen and containing more than 5% protein from natural sources shall be guaranteed as follows:

        Crude Protein, minimum, _____%

        (This includes not more than _____% equivalent protein from non-protein nitrogen).

    (2) Mixed feed concentrates and supplements containing less than 5% protein from natural sources may be guaranteed as follows:

        Equivalent Crude Protein from Non-Protein Nitrogen, minimum _____%

    (3) Ingredient sources of non-protein nitrogen such as Urea, Di-Ammonium Phosphate, Ammonium Polyphosphate Solution, Ammoniated Rice Hulls, or other basic non-protein nitrogen ingredients defined by the Association of American Feed Control Officials shall be guaranteed as follows:

        Nitrogen, minimum, _____ %

        Equivalent Crude Protein from Non-Protein Nitrogen, minimum, _____%

(f) Mineral phosphatic materials for feeding purposes shall be labeled with the guarantee for minimum and maximum percentage of calcium (when present), the minimum percentage of phosphorus, and the maximum percentage of fluorine.

Regulation 5. Ingredients

(a) The name of each ingredient or collective term for the grouping of ingredients, when required to be listed, shall be the name as defined in the Official Definitions of Feed Ingredients as published in the Official Publication of American Feed Control Officials, the common or usual name, or one approved by the _____.

(b) The name of each ingedient must be shown in letters or type of the same size.

(c) No reference to quality or grade of an ingredient shall appear in the ingredient statement of a feed.

(d) The term "dehydrated" may precede the name of any product that has been artificially dried.

(e) A single ingredient product defined by the Association of American Feed Control Officials is not required to have an ingredient statement.

(f) Tentative definitions for ingredients shall not be used until adopted as official, unless no official definition exists or the ingredient has a common accepted name that requires no definition, (i.e. sugar).

(g) When the word "iodized" is used in connection with a feed ingredient, the feed ingredient shall contain not less than 0.007% iodine, uniformly distributed.

Regulation 6. Directions for Use and Precautionary Statements

(a) Directions for use and precautionary statements on the labeling of all commercial feeds and customer-formula feeds containing additives (including drugs, special purpose additives, or non-nutritive additives) shall:

(1) Be adequate to enable safe and effective use for the intended purposes by users with no special knowledge of the purpose and use of such articles; and,

(2) Include, but not be limited to, all information prescribed by all applicable regulations under the Federal Food, Drug and Cosmetic Act.

(b) Adequate directions for use and precautionary statements are required for feeds containing non-protein nitrogen as specified in Regulation 7.

(c) Adequate directions for use and precautionary statements necessary for safe and effective use are required on commercial feeds distributed to supply particular dietary needs or for supplementing or fortifying the usual diet or ration with any vitamin, mineral, or other dietary nutrient or compound.

Regulation 7. Non-Protein Nitrogen

(a) Urea and other non-protein nitrogen products defined in the Official Publication of the Association of American Feed Control Officials are acceptable ingredients only in commercial feeds for ruminant animals as a source of equivalent crude protein and are not to be used in commercial feeds for other animals and birds.

(b) If the commercial feed contains more than 8.75% of equivalent crude protein from all forms of non-protein nitrogen, added as such, or the equivalent crude protein from all forms of non-protein nitrogen, added as such, exceeds one-third of the total crude protein, the label shall bear adequate directions for the safe use of feeds and a precautionary statement:

CAUTION: "USE AS DIRECTED"

The directions for use and the caution statement shall be in type of such size so placed on the label that they will be read and understood by ordinary persons under customary conditions of purchase and use.

(c) On labels such as those for medicated feeds which bear adequate feeding directions and/ or warning statements, the presence of added non-protein nitrogen shall not require a duplication of the feeding directions or the precautionary statements as long as those statements include sufficient information to ensure the safe and effective use of this product due to the presence of non-protein nitrogen.

Regulation 8. Drug and Feed Additives

(a) Prior to approval of a registration application and/or approval of a label for commercial feed which contain additives (including drugs, other special purpose additives, or non-nutritive additives) the distributor may be required to submit evidence to prove the safety and efficacy of the commercial feed when used according to the directions furnished on the label.

(b) Satisfactory evidence of safety and efficacy of a commercial feed may be:
   - (i) When the commercial feed contains such additives, the use of which conforms to the requirements of the applicable regulation in the Code of Federal Regulations, Title 21, or which are "prior sanctioned" or "generally recognized as safe" for such use, or
   - (ii) When the commercial feed is itself a drug as defined in Section 3(g) of the Act and is generally recognized as safe and effective for the labeled use or is marketed subject to an application approved by the Food and Drug Administration under Title 21 U.S.C. 360(b).

Regulation 9. Adulterants

(a) For the purpose of Section 7(a)(1) of the Act, the terms "poisonous or deleterious substances" include but are not limited to the following:
   - (1) Fluorine and any mineral or mineral mixture which is to be used directly for the feeding of domestic animals and in which the fluorine exceeds 0.30% for cattle; 0.35% for sheep; 0.45% for swine; and 0.60% for poultry.
   - (2) Fluorine bearing ingredients when used in such amounts that they raise the fluorine content of the total ration above the following amounts: 0.009% for cattle; 0.01% for sheep; 0.014% for swine; and 0.035% for poultry.
   - (3) Soybean meal, flakes or pellets or other vegetable meals, flakes or pellets which have been extracted with trichlorethylene or other chlorinated solvents.
   - (4) Sulfur dioxide, Sulfurous acid, and salts of Sulfurous acid when used in or on feeds or feed ingredients which are considered or reported to be a significant source of vitamin B_1 (Thiamine).

(b) All screenings or by-products of grains and seeds containing weed seeds, when used in commercial feed or sold as such to the ultimate consumer, shall be ground fine enough or otherwise treated to destroy the viability of such weed seeds so that the finished product contains no more than _____ viable prohibited weed seeds per pound and not more than _____ viable restricted weed seeds per pound.

Regulation 10. Good Manufacturing Practices

(a) For the purposes of enforcement of Section 7(d) of the Act the _____ adopts the following as current good manufacturing practices:
   - (1) The regulations prescribing good manufacturing practices for medicated feeds as published in the Code of Federal Regulations, Title 21, Part 133, Sections 133.100-133.110.
   - (2) The regulations prescribing good manufacturing practices for medicated premixes as published in the Code of Federal Regulations, Title 21, Part 133, Sections 133.200-133.210.

INDEX

Abomasum, 18; functions of, 20
Absorption of nutrients, 21; in avian digestion, 22; in nonruminant digestion, 14; in ruminant digestion, 14
Accumulator plants, 67
Acetic acid, 92, 149
Acid detergent fiber (ADF), 127
Acid detergent lignin, 128
Acid insoluble lignin, 128
Acidic amino acid, 40
Active transport, 22
Additives, 75; antioxidants, 93; for swine, 211; health concerns, 90; regulations, 90
ADF (acid detergent fiber), 127
Aflatoxin, 278, 280, 284, 287; B_1, 285, 288; B_2, 285; carcinogenic properties, 285, 286; effects of, 285; effects on dairy cattle, 286; effects on ruminants, 285; effects on swine, 285; G_1, 285; G_2, 285; in corn, 287; M_1, 285, 286; M_2, 285
Agricultural industry, 1
Agriculture, 1; assets, 1; labor, 2; livestock production, expenses, and receipts, 1; research, 2
Air-dry, 126
Air-dry feeds, 100
Alfalfa grass, 246
Alfalfa hay, 210, 247
Alfalfa meal, 210
Alfalfa-brome grass, 247
Alfalfa-timothy mixtures, 247
Algae, 112
Alimentary tract, 14
Aluminum, 70
Amides, 39
Amino acids, 4, 16, 22, 40; acidic, 40; basic, 40; essential, 43, 135; limiting, 43, 135, 140; neutral, 40; nonessential, 43; structural formulas of, 42; structure of, 40
Aminopeptidase, 17
Ammonia, 46; toxic levels, 47; treating crop residues, 112
Ammoniated citrus pulp, 49
Ammoniated cottonseed meal, 49
Ammoniated molasses, 49
Ammoniated rice hulls, 49
Ammonium polyphosphate, 49
Amylase: pancreatic, 16; salivary, 17
Angora goats, 231; range feeding, 236
Animal byproducts, 9, 10
Animal functions, 8
Animal origin protein supplements, 46
Animal production, rations for, 107
Animal protein: feeding to swine, 210; for poultry, 266
Animal wastes, 111
Animals: as power sources, 10, 11; finishing for market, 106; fitting for show, 106; functions in society, 8; normal temperatures of, 106; use in recreation, 10

Annual pastures, 246
Anthelmintics, 93
Antibacterials, 91
Antibiotics, 91; for horses, 251; for poultry, 264; for sheep, 222; regulations, 7
Antimicrobials, 6
Antitrypsin factor, 48
Anus, 17
Aquatic plants: algae, 112; as feed source, 112; kelp, 112; water hyacinth, 112
Arsenic, 70
Arsenicals, 91
Artificial drying, 149
As-fed data calculations, 126
Ash, 125
Aspergillus flavus, 284
Aspergillus parasiticus, 284
Association of the American Feed Control Officials, 130
Auger wagons, 147
Aureomycin, 91, 222
Automatic concentrate feeders, 190
Avian digestive systems: absorption of minerals, 22; absorption of nutrients, 21; ceca, 21; cloaca, 21; crop, 21; gizzard, 21; grit in, 21; gullet, 21

B-complex vitamins, 83; deficiency symptoms, 83, 84; need in horses, 83; need in poultry, 83; need in ruminants, 83; need in swine, 83
Bacitracin, 91
Backgrounding, 173
Bacteria, 19; cellulolytic, 19
Bacteriodes succinogenes, 19
Bakery wastes, 209
Balancing rations, 135; with electronic spreadsheets, 143; with mainframes, 142; with microcomputers, 142
Balansia epichloe, 281
Baling hay, 117
Bambermycins, 91
Barley, 33, 208; for horses, 249
Basal feeds, 32
Basal metabolism, 36
Basic amino acid, 40
Batch feed mixers, 146; portable, 147; stationery, 147
Batch mixing feed, 146
Beef cattle, 159; calculating feed requirements, 181; calves, backgrounding, 173; calves, creep-feeding, 171; effects of cold weather, 160; finishing, 178; optimum temperature range, 159; pasturing, 170
Beef cattle, rations: balancing rations, 165; bulls, 176; minerals in rations, 181; nursing mothers, 167; pregnant cows, 165; urea in rations, 169, 182
Beet pulp, 49, 50
Beriberi, 75

Beta-carotene, 77; conversion to vitamin A, 78
Bile, 16; functions of, 17
Biotin, 87; deficiency symptoms, 87; sources of, 87
Birdsfoot trefoil, 246
Biuret, 49
Black persimmon, 236
Bleaching, 116
Bleeding disease, 82
Blending feeds with mycotoxins, 287, 288
Bloating, 244
Body heat in animals, 153
Bone scraps, 210
Boron, 70
Bottle feeding lambs, 227
Bovatec, 92
Breeding heifers, rations for, 175
Briar, 236
Bright greenish-yellow (BGY) fluorescence, 286
Broiler litter, 111
Brome grass, 248
Browse, 233; for Angora goats, 236
Bulk, 130; adding to swine rations, 214
Bulk feed, sampling, 122
Bulky feeds, 12
Bulls: mature, rations for, 177; replacement, rations for, 176; yearling, rations for, 177
Bunker silos, 119
Burr mill, 144
Buttermilk, 210
Butyric acid, 92
Byproduct feeds, 103
Byproducts, 9, 10, 197; as protein sources, 49; gland extracts, 10

Cadmium, 71; toxicity, 71
Calf starter, 194
Calcium, 53, 55, 136; deficiency symptoms, 55; functions of, 55; interrelationships, 57; poultry, 263; ratio to phosphorus, 136; sources of, 57, 58; trace-mineralized, 55
Calcium hydroxide, 112
Calcium/Phosphorus ratio, 59
Calorie (cal), 26
Calorimetry: direct, 129; indirect, 129
Calves: backgrounding, 173; dry matter, 193; growth stimulants, 174; urea, 174; water intake, 174; winter rations, 174
Cane Molasses, 249
Carbadox, 91
Carbohydrates, 3, 125; compounds of, 27; fiber, 28; in fitting for show, 106; nitrogen-free extract (NFE), 28; sugars, 27
Carotene, 76; structural formula, 77
Cashmere goats, 231
Catclaw, 236

Cattle: See Beef cattle, Dairy cattle, Heifers.
Ceca, 21
Cecum, functions of, 17
Cedar, 236
Cellulolytic bacteria, 19
Cellulose, 19, 27, 124, 125, 126
Cereal grain screenings, 50
Challenge feeding, 187
Changes in temperature, 153
Chemobiotic compounds, 91
Chlorine, 53, 59. See also Salt.
Chlorophylls, 124
Chlorotetracycline, 92; Aureomycin, 91
Choline, 86; deficiency symptoms, 86; functions of, 86; sources of, 86; structural formula, 86
Chromium, 53, 70
Chyme, 16
Classification of feeds, 100
Claviceps paspali, 280, 281
Claviceps purpurea, 280
Cloaca, 21
Clothing, animal sources, 9
Clovers, 247
Cob meal, 32
Cobalt, 53, 67; deficiency symptoms, 68; deficient areas in US, 68; functions of, 67; in rations, 68; toxicity, 68
Coccidiosis, 93
Coccidiostats, 93
Code of Federal Regulations (CFR), 90
Cold stress, 155
Collagen, 89
Colon, 17
Colostrum, 225
Colostrum milk, 192, 193, 214
Complete ration feeding, 187
Complex sugars, 28
Compound lipids, 30
Computers, 142. See also Balancing rations.
Concentrate feeders, 190
Concentrates, 103, 134; as a percentage of feed, 2; bulky feeds, 12; corn, 3; for sheep, 222; protein feeds, 2
Conditioning, 116
Consumption of water, 95
Contamination of water, 95
Continuous flow feed mixers, 146
Conventional silos, 118
Cooking, 146
Cooking feeds, garbage, 146
Copper, 53, 64, 210; deficiency symptoms, 65; functions of, 64; interrelationships, 64; sources of, 65; toxicity levels, 65
Coral bean, 236
Core sampling, 122
Corn, 3; contamination, 288; for horses, 248; harvesting for silage, 120
Corn meal, 32
Corn oil, 32
Corn silage, 34, 173
Corn starch, 32
Corn stover, 34
Cottonseed meal, 209; for horses, 249
Cracking, 144
Creep-feeding, 171, 214, 225; disadvantages, 172; foals, 255; guidelines, 172;

lambs, 225; limitations, 172; when to use, 172
Crimping, 144
Crimson clover, 170
Critical temperatures, 154
Crop, 21
Crop residues: treating with ammonia, treating with calcium hydroxide, 112
Crop residues as feed source, 111, 112
Crops, harvesting for silage, 120
Crude fat, 124; measuring content in feed, 124
Crude fiber, 100, 124; measuring content in feed, 124
Crude protein (CP), 40, 135; calculating in feed, 40; digestible amount, 40; measuring content in feed, 123
Crumbles, 145
Cubing, 145
Cystine, 220
Cystitis, 246

Dairy calves, 158; feeding yearling calves, 194; frequency of feeding, 194; newborn calves, 192; replacement calves, rations for, 192; tolerance to temperature variations, 158; weaning calves, 194
Dairy cattle, 185; effects of aflatoxin, 286; effects of heat stress, 159; optimum temperature range, 158; protein deficiency, 44
Dairy cattle, feeding: challenge feeding, 187; feed analysis, 188; feeds for, 185; lead feeding, 187; traditional feeding, 186
Dairy cattle, rations: complete rations, 187, 188; dry rations for, 192; grains in rations, 196; grouping for complete rations, 188; herd replacements, rations for, 192; rations for reproduction, 201
Dairy cows: lactating, rations for, 190, 191; lactating, water intake, 158; milk production, rations for, 185; tolerance to temperature variations, 158
Dairy goats: feeding bucks, 234; feeding kids, 236; gestation rations, 234; lactation feeding, 234; rations for, 233, 234
Dallis grass poisoning, 281
Damage, insects, 149
DDD, 96
DDT, 96
Defluorinated phosphate, 165
Defluorinated rock phosphate, 210
Dent corn, 206
Deoxynivenol, 278, 282
Depeptidase, 17
Deprivation of water, 96
Derived lipids, 30
Dermatoxins, 278
Diammonium phosphate, 49
Dicalcium phosphate, 165, 210
Diet, 133, 135
Diethylstilbestrol (DES), 92
Diffusion, 22
Digestible energy (DE), 26, 135
Digestable protein (DP), 40, 135
Digestion, 14
Digestive trials, 128
Digestive disturbances, 129
Digestive systems, 4, 14; avian, 14, 21 (see also Avian digestive systems); 4, 14

(see also Nonruminant digestive systems); ruminant, 4, 14, 17 (see also Ruminant digestive systems)
Digestive tract, 14
Dilan, 96
Dipeptides, 16
Disaccharides, 28
Dissolved minerals, absorption of in avian digestion, 22
Dried bakery product, 35
Dried beet pulp, 34
Dried citrus pulp, 34
Dried whey, 35
Dry matter, 121
Drying feed, artificial methods, 149
Dryland pastures, 245
Drylot, 167, 168
Ducks, 275
Duodenum, 16

EAT (effective ambient temperature), 153
Economics of rations, 134, 135
Effective ambient temperature (EAT), 153
Electronic spreadsheets, 143
Elm, 236
Energy, 4; calorimetry, 129; conversion efficiency for animals, 10; measures of, 135; measuring losses in feed, 128; requirements for lactating dairy cows, 191; requirements for poultry, 261; requirements for swine, 204; retained in feed, 129; terminology, 26
Energy concentrates, 32
Energy conversion efficiency, 10
Energy feeds for poultry, 265
Energy nutrients, 26; byproducts, 34; carbohydrates, 27; concentrated sources of, 33; effects of deficiencies, 36, 37; fats, 27, 35; forages, 34; functions of, 35, 36; molasses, 35; protein, 45; roughages, 34
Environment, 157; adjusting nutrition to, 158; effects of, 153; effects on nutritional efficiency, 157
Enzymes, 14
Epichloe typhina, 281
Epithelium, 21
Equipment for force feeding, 194
Equipment for complete rations, 188
Ergot, 281, 287
Ergotism, symptoms of, 281
Esophageal groove, 18
Esophagus, cardia, 16
Essential amino acids, 41, 135
Essential fatty acids, 31
Essential functions of minerals, 53
Ether, use in feed analysis, 31
Ether extract, 31, 124
Ewes, rations for, 223, 224
Exploding, 146
Extruding, 146

F-2 toxin, 282
Farm economy, stabilization of, 12
Farrowing, 214
Fats, 30, 35, 209; extraction methods, 48; heat increment in swine, 161; in finishing for market, 106; marbling in tissues, 106; rations, 31; synthesis of, 30; unsaturated, 32

Fattening animals, 106
Fatty acids, 16, 17; essential (EFA), 31; saturated, 30; unsaturated, 30
Feces, 17
Federal Register, 91
Federal Regulations, Code of (CFR), 90
Feed: air-drying, 100; effects on animal reproduction,104; effects on growth, 104; effects on fleece production in goats, 236; forms of, 101; labeling, 130; moisture content, 122; moisture content of stored feed, 148; nutritive value, 130; palatability, 129; partially digested, 16; physical form, effects on water consumption, 156; preparation for poultry, 266; proximate analysis of, 123; sampling, 122; silage, analysis of, 121; storage, infestation, 284; tagging, 130; uses of by animals, 104; water content, 75, 123. *See also* Feeds.
Feed Additive Compendium, 90
Feed additives, 6, 75, 90, 134, 147; antimicrobials, 6; hormones, 92; premixes, 6
Feed analysis, 121, 188
Feed components, 3; cellulose, 19; crude protein, 123; hemicellulose, 19; vitamins, 75
Feed composition basis, 126; converting from one to another, 126
Feed concentrates, 2, 12, 100; bulky, 12; protein feeds, 2
Feed Control Officials, Association of, 130
Feed conversion: after condemnations, 272; before condemnations, 272; for poultry, 272; to human food, 8
Feed costs, 1; cattle, 1, 2; dairy, 1; efficiencies, 1; poultry, 1; research, 2; swine, 1
Feed energy, measurement of, 4
Feed intake, 155; effects of stress on, 160; poultry, factors affecting, 262
Feed mixers: auger wagons, 147; batch, 146; continuous flow, 146
Feed mixes, 141
Feed mixing, 146; area maintenance, 147; precautions, 147
Feed names, 101
Feed processing, mechanically, 143; burr mill, 144; cracking, 144; crimping, 144; fineness of grind, 144; flaking, 144; grinding, 143; grinding grains, 144; grinding hay, 144; hammer mill, 143; rolling, 144
Feed quality, 115; analyzing quality, 115; effects of amino acid content, 115; effects of vitamin content, 115; palatability, 49, 115
Feed regulations, 130
Feed storage, 148; damage by rodents, 149; effects of insects, 149
Feeding standards, 136
Feeding trials, 130
Feedlot cattle, 159, 178; optimum temperature range, 159
Feeds, 2; antimicrobials in, 6; basal, 32; classification of, 100; digestibility, 128; effects of improper handling, 45; energy, 4; forms of, 101; guidelines, 136; high-moisture, 100; international classes of, 101; international numbers, 101; mois-

ture content, 126; mycotoxins in, 287; names of, 101; net energy values, 128; niacin in, 86; preparation of, 143; prices, 135; roughages, 3, 100; underutilized sources of, 107
Feedstuffs, 100
Fermentation wastes, 110
Fertility, 224
Fescue foot, 281; symptoms in cattle, 282
Fiber, 28; in rations for mature breeding animals, 30
Finishing animals for market, 106
Finishing rations: for cattle, 179; for swine, 215
Fish meal, 210
Fitting animals for show, 106
Fixed ingredients, 141
Flaking, 144
Fleece production, 236
Fluorine, 53, 69; deficiency symptoms, 69; functions of, 69; removal of excess, 69; toxicity, 69
Flushing, 221
Folic acid (Folacin), 87; deficiency symptoms, 87; sources of, 87; structural formula, 88
Food, oxidation, 105
Food and Drug Administration (FDA), 90; limits on mycotoxins, 287
Food-processing wastes, 108
Forages, 34, 155; evaluation, Van Soest method, 127; extracts of, 124; grass, 50; intake, 155; legumes, 50; moisture content, 149
Force feeding, 194
Forest residues, 110
Formic acid, 149
Formulating rations, 133
Fruit-processing wastes, 109
Fumigation of feed, 149
Fungi, 148, 278; *Aspergillus*, 280; *Penicillium*, 280
Fusarium, 282
Fusarium graminearum, 282
Fusarium roseum, 282
Fusarium tricinctum, 283

Galactose, 27
Garbage, feeding to swine, 146
Gas-tight silos, 118
Gastric juice, 16; enzymes in, 16
Gastric lipase, 16
Gastrointestinal tract, 14
Geese, 275
Genitotoxins, 278
Gibberella zeae, 282, 283
Gizzard, 21
Glandular stomach, 21
Glucose, 22, 27
Glycerol, 16, 17
Glycolipids, 30
Goats: Angora, 231, 236; browse feeding, 233; Cashmere, 231; dairy, rations for, 233; energy requirements, 231; fat requirements, 231; fleece production, 236; herbage feeding, 233; mineral requirements, 232; nutrient requirements, 231; protein deficiency, 44; protein requirements, 231; range-fed, 232; stable-fed, 232; urea in rations, 232;

vitamin requirements, 232; water requirements, 232, 233
Grain consumption: by animals, 8; by humans, 8
Grain sorghum, 33
Grains, 50, 103; fumigation, 149; grinding, 144; high-moisture storage, 149; palatability for horses, 248; processing with heat, 146; sampling, 122; small, 120
Granules, 145
Grass hays, 247
Grass staggers, 61
Grass tetany, 61
Grasses: for pastures, 246; harvesting for silage, 120; mycotoxins in, 281; sorghum-sudan, 120; sudan, 120
Grazing, horses, 244
Greases, 209
Grinding, 143; feeds for horses, 252; fineness, 144; grains, 144; hay, 144
Grit, 21
Gross Energy (GE), 26
Ground ear corn, 208
Ground limestone, 210
Ground snapped corn, 32
Growth, 104
Growth implants, 172
Gullet, 21

Hammer mill, 143
Hand sampling, 122
Hardware disease, 20
Hay, 115; baling, 117; conditioning, 116; core sampling, 122; cubing, 145; for dairy goats, 234; grinding, 144; harvesting, 115; maintaining quality of, 115; quality loss, 116; shattering, 115; types of, 247; windrowing, 116
Haylage, 165
Heat: in animals, 153; in ruminants, 155; sources for farm animals, 154
Heat gain in animals, 154
Heat Increment (HI), 26
Heat loss in animals, 153; convection, 154; precipitation, 154
Heat stress, 155, 161
Heifers, 175, 195; breeding, rations for, 175; dairy cattle, feeding, 195; replacement, rations for, 175
Hemicellulose, 19, 28, 124, 125, 126
Hemoglobin, 63
Hepatotoxins, 278
Herbage, 233
Herbicides, avoiding problems with, 112
Hexoses, 28
High-lysine corn (HLC), 206, 216
High-moisture corn, 206
High-moisture feeds, 100
HLC (high-lysine corn), 206
Hominy feed, 33
Horizontal batch mixers, 147
Horizontal silos, 118
Hormones, 92
Horses: antibiotics, 251; large intestine, 17 protein deficiency, 45; protein supplements, 249; rations, grinding, 252; stomach, 16; weaning foals, 255
Horses, energy requirements for, 239; lactation, 240; maintenance, 239; pregnancy, 239; work, 239

Horses, feeding: feed guidelines, 251, 252; forages for, 242, 246; grains, 248, 249; grazing, 244
Horses, pastures, 242, 243; bloating, 244; rotating, 222
Horses, rations: dry mares, 257; foals, 254, 255; geldings, 257; growing, 254; lactating mares, 254; pregnant mares, 254; stallions, 253; two-year-olds, 256; yearlings, 256
Horses, requirements: mineral requirements, 241; nutrient requirements, 239; protein requirements, 240; vitamin requirements, 241, 242; water requirements, 242
Hyperestrogenism, 282
Hyperthermia, 153

Ileum, 16
Ill-scented sumac, 236
Imino acids, 40
Industrial food-processing wastes, 107
Inorganic phosphorus, for poultry, 263
Inositol, 89; structural formula, 89
Insecticides, avoiding problems with, 112
Insects, 149
Intake, 155; effects of temperature on, 155; water, 94
Internal parasite cycle, 222
Internal parasite eggs, 244
International feed nomenclature, 100
International Units (IU), 76
Intestinal juice, enzymes in, 17
Iodine, 53, 68; deficiency symptoms, 68; deficient areas in USA, 68; in rations, 69
Iron, 53, 210; deficiency symptoms, 63; functions of, 63; in rations, 63; interrelationships, 63; sources of, 63
Irrigated pastures, 245
Isobutyric acid, 149

Jejunum, 16

Kelp, 112
Kids, 236
Kilcalorie (kcal), 26
Kjeldahl process, 123

Labeling, 130
Lactase, 17
Lactobacilli, 19
Lactose, 27, 193
Lambing mortality rate, 222
Large intestine: cecum, 17; colon, 17; rectum, 17
Lasolacid (Bovatec), 92
Layer waste, 111
Laying ratios, 269
Leaching, 116
Lead, 71
Lead feeding, 187
Legume hays, 247
Legumes, harvesting for silage, 120
Lignin, 27, 28, 124, 125, 126; acid detergent, 128; acid insoluble, 128
Limestone, for poultry, 263
Limiting amino acid, 135, 140
Lincomycin, 91
Linseed meal, 209; for horses, 249
Lipase, pancreatic, 17

Lipids, 3, 30; compounds of, 30; derived, 30; simple, 30
Lipoproteins, 30
Liquid protein supplements, 169
List of CFR Sections Affected, 91
Liver: bile, 17; conversion of vitamin D, 79; secretions of, 16
Livestock: feed costs, 1; human consumption, 8; labor costs, 2; use in conservation, 11
Livestock production: labor force, 7; occupations, 7
Losses of water, 95
Low-lysine rations for poultry, 273
Lower critical temperature, 154
Lysine, 140

Magnesium, 61; deficiency symptoms, 61; functions of, 61; interrelationships, 62; sources of, 62
Maintenance rations, 105
Maintenance requirements, 36, 106
Major minerals, 6, 53
Malt sprouts, 50
Maltase, 17
Maltose, 16
Manganese, 53, 63; deficiency symptoms, 64; functions of, 63; in rations, 64; interrelationships, 64
Manure, 11; as fertilizer, 11; as fuel source, 11; methane gas digesters, 11
Marbling, 106
Market lambs, 226
Mashes, 270
ME (metabolizable energy), 135
Measuring energy losses, 128
Meat scraps, 210
Mechanical processing of feeds, 143
Medicating Ingredient Brochure, 90
Medications in water, 95
Megacalorie (Mcal), 26
Melengestrol acetate (MGA), 92, 93
Mercury, 71
Merino sheep, 157
Mesquite, 236
Metabolites, 278
Metabolizable Energy (ME), 26, 135, 261
Metabolizable protein, 43
Methionine, 220
Methoxychlor, 96
Methylene blue, 49
Micro minerals, 53
Micronizing, 146
Micronutrients, 134
Microorganisms, absorption in the bloodstream, 182
Milk production, 107; effects of temperature variations, 158; rations for, 185
Milk yield, 158
Milo, 144, 208. See also sorghum grains.
Minerals, 3, 6, 53, 136, 197; deficiency symptoms, 54; functions of, 53; in bones, 54; in egg production, 54; in rations, 54; in water, 94; major, 6, 53; maximum levels in animals, 56; measuring content in feed, 125; micro, 53; minor, 53; sources of, 54; toxic, 70; trace, 6, 53
Mineral requirements: for cattle, 181; for goats, 232; for horses, 241; for poultry,

263, 266; for sheep, 220; for swine, 205, 210
Minor minerals, 53
Mixes, single-mineral, 55
Moisture basis, 136
Moisture content of feed, 122
Moisture content of stored feed, 148
Molasses, 35, 209
Molybdenum, 53, 64, 66; functions of, 66; in rations, 66; toxicity, 66
Monensin (Rumensin), 91, 92
Monoammonium phosphate, 49
Monoglycerides, 17
Monosaccharide, 28
Mouth, 17
Muncipal solid wastes, 110
Mycelia, 110
Mycotoxicoses: commonly found in USA, 279; incidence in USA, 280
Mycotoxins, 278; acute effects of, 279; carcinogenic effects of, 279; detecting in feed, 286; residues, 286; teratogenic effects of, 279

Native ranges, 245
NDF (neutral detergent fiber), 127
NDS (neutral detergent solubles), 127
NE (net energy), 135
Neomycin, 91
Nephrotoxins, 278
Net Energy (NE), 26, 135, 261
Net energy values, 128
Neurotoxins, 278
Neutral amino acids, 40
Neutral detergent fiber (NDF), 127
Neutral detergent solubles (NDS), 127
Newborn calves: fat intake levels, 193; monitoring intake, 192; protein sources, 193
NFE (Nitrogen-Free Extract), 125
Niacin, 85; deficiency symptoms, 85, 86; sources of, 85; structural formula, 86
Nickel, 70
Nipple pail, 192
Nirites, 47
Nitrates, 47
Nitrate poisoning, 49; symptoms of, 49; treatment of, 49
Nitrofurans, 91
Nitrogen free extract (NFE), 28, 125
Nonfood industrial wastes, 109
Nonprotein nitrogen (NPN), 19, 39, 220, 240; sources, 39
Nonprotein products, 103
Nonruminants, 4; protein requirements, 4; vitamin requirements, 6
Nonruminant digestive systems: components of, 14; esophagus, 16; large intestine, 17; mouth, 14; small intestine, 16; stomach, 16
Nursing mothers, rations for: beef cattle, 167; dairy cattle, 190; dairy goats, 234; horses, 254; ewes, 223; swine, 214
Nutrients: absorption of, 21; cost of, 136; digestion coefficients, 128; energy producing, 26
Nutrition, 153; adjusting to the environment, 158; effects on reproduction, 104
Nutritive value, 130

Oat groats, 32
Oats, 32, 170, 208; for horses, 248
Occupations, 7
Ochratoxin: A, 280, 286, 287; B, 280; in poultry, 280; in ruminants, 280; in swine, 280
Oils, 30
Oleandomycin, 91
Oligopeptides, 16
Omasum, 18; functions of, 20
Orchard grass hay, 248
Organic acids, 125
Organic chemical wastes, 109
Organophosphorus insecticides, 96
Osteomalacia, 55
Osteoporosis, 55
Overgrazing, 244
Oxidation of food, 105
Oxytetracycline (Terramycin), 91
Oyster shell, as feed for poultry, 275

Palatability, 129, 130, 135; effects of fungi, 148
Pancreatic amylase, 16
Pancreatic juice, 16; enzymes in, 16
Pancreatic lipase, 17
Pantothenic acid, 84; deficiency symptoms, 85; structural formula, 85
Papillae, 20
Para-aminobenzoic acid, 89; structural formula, 89
Parakeratosis, 65
Paspalum staggers, 281
Pastures, 34; for geese, 275; for horses, 242, 243; for swine, 210; fungi in, 281; grasses for, 246; mycotoxins in, 280, 281; overgrazing, 244; types of, 245
Peanut meal, 209
Pearson Square, 137; completing, 137
Pelleting, 144, 145; complete rations, 145; crumbles, 145; granules, 145; storage space, 145
Pellets, for horses, 250
Penicillin, 7, 91
Pentosans, 30
Pepsin, 16
Peristaltic waves, 16
Permanent pastures, 245
Perthane, 96
Pesticides, 96
Phase feeding, 270
Phospholipids, 30
Phosphorus, 53, 136; deficiency symptoms, 58; functions of, 58; interrelationships, 58; trace mineralized, 55
Pica, 58
Plant proteins: for poultry, 265; for swine, 209
Plant protein supplements, 46
Plants: protein in, 39; stages of maturity, 102
Poison, 278
Polypeptide chain, formation of, 41
Polypeptides, 16
Polysaccharides, 19, 28
Popping, 146
Portable batch mixers, 147
Potassium, 60; deficiency symptoms, 61; functions of, 60; sources of, 61; toxicity levels, 61

Potato meal, 35
Potatoes, 208
Poultry: antibiotics, 264; digestive system, 21; grit, 267; optimum temperature range, 162; protein deficiency, 45; protein supplements, 265; slowing down maturity, 273; unidentified nutrient factors, 264; water intake, 162
Poultry, feeding: energy feeds, 265; feed costs, 1; feed preparation, 266; feeding broilers, 270; feeding guidelines, 267; feeds for, 265; free choice feeding, 261; intake, effects of temperature on, 162, 262; laying hens, systems for feeding, 270; low-lysine rations, 273; mashes, 270
Poultry, requirements: amino acid requirements, 262; calcium requirements, 263; energy requirements, 261; nutrient requirements, 261; protein requirements, 262; water requirements, 162, 265
Poultry, turkeys, 273; grit, 273; poults, 274
Preganancy, effects of Vitamin D, 81
Premixes, 6, 147
Preparation of feeds, 143
Preservatives, 121
Pressure flaking, 146
Prices of feeds, 135
Propionic acid, 92, 149
Protein, 3, 4, 135; as an energy source, 45; biological value of, 46; conversion efficiency for animals, 10; crude, 40; cubed, 169; effects of fermentation, 241; effects of heat damage, 148; effects of microbial synthesis, 41; functions of, 44; metabolizable, 43; need for, 4; pelleted, 169; quality of, 41; solubility, 46; unavailable feed, 45
Protein conversion efficiency, 10
Protein deficiency: in beef cattle, 44; in dairy cattle, 44; in goats, 44; in horses, 45; in poultry, 45; in sheep, 44; in swine, 45
Protein in feeds: in byproduct feeds, 49; in grains, 50; in grass, 50; in legumes, 50; in soybeans, 48
Protein in human diet: animal sources, 9; vegetable sources, 9
Protein requirements: for cattle, 44, 165, 196; for horses, 240; for poultry, 262; for sheep, 220; for swine, 205
Protein supplements, 134, 168, 197; animal origin, 46; for horses, 249, 250; for poultry, 265; plant, 46
Proteins, amino acids in, 16
Protozoa, 19
Proventriculus, 21
Provitamin A, 76
Proximate analysis, 126; limitations of, 125
Pyridoxine: B6, 85; deficiency symptoms, 85; sources of, 85; structural formula, 85

Ralgro, 92
Range feed, Angora goats, 236
Ratio, calcium/phosphorus, 59
Rations, 133; cobalt in, 68; cubing, 145; economics of, 134; energy requirements, 37; essential amino acid, 43; for animal production, 197; formulating, 133; functions of, 134; iodine in, 69; iron in, 63; maintenance, 105; manga-

nese in, 64; minerals in, 54; molybdenum in, 66; palatability, 130; pelleting, 145; salt, 60; sulphur in, 62; using computers to formulate, 142; Vitamin A in, 79; vitamins in, 76
Rations, balancing: one nutrient and two feeds, 137; Pearson Square, 137; principles of balancing, 135; steps involved, 136, 137; two grains with a supplement, 138; using algebraic equations, 139; using simultaneous algebraic equations, 140; with fixed ingredients, 141; with computers, 142; with electronic spreadsheets, 143
Rations, beef cattle: balancing, 165; bulls, 176, 177; calves, 174; finishing, 179; heifers, 175; nursing mothers, 167; pregnant cows, 165; urea in, 181, 182
Rations, dairy cattle: balancing, 196; calves, 193; complete, 187, 188; dry, 192; for milk production, 185; heifers, 195; herd replacements, 192; lactating, 191; minerals, 197; replacement calves, 192; reproducing, 201; vitamins, 198; water, 198
Rations, dairy goats, 233: bucks, 234; gestation, 234; kids, 236; lactating, 234
Rations, fat in: nonruminants, 31; fat in ruminants, 31
Rations, goats, urea in, 232
Rations, horses: dry mares, 257; foals, 254, 255; geldings, 257; general, 251, 252; grains, 248, 249; growing, 254; lactating mares, 254; pellets, 250; pregnant mares, 254; protein supplements, 249; stallions, 253; two-year-olds, 256; yearlings, 256
Rations, poultry: broilers, 270, 271; ducks, 275; for meat production, 273; geese, 275; grit, 267; laying hens, 269, 270; mineral requirements, 263; preparation, 266; protein requirements, 262, 263; protein supplements, 265; pullets, 267; turkeys, 274
Rations, sheep, 219; ewes, 223; lactating, 223, 224; lambs, 225; market lambs, 226; orphan lambs, 227, 228; pregnant, 221; rams, 224
Rations, swine, 205; baby pigs, 214; complete, 216; for finishing swine, 215; hog, 161; lactating, 214; market pigs, 214, 215; for production, 107
Rations, turkeys, 273, 274; breeding, 274; poults, 274; vitamin requirements, 264
Regurgitation, 19
Relative palability, 135
Rennin, 16
Resins, 124
Respiratory quotient (RQ), 129
Reticulum, 18; functions of, 20
Rhizoctonia leguminicola, 284
Riboflavin: B2, 84; deficiency symptoms, 84; sources of, 84; structural formula, 84
Rice bran, 34
Roasting, 146
Rodents, 149
Rolling, 144
Roughages, 3, 34, 103, 134; for dairy cattle, 196; for dairy goats, 233; for horses, 242; for pregnant sheep, 221;

Roughages, *continued*: for swine, 210; grasses, 103; forest residues, 110; legumes, 103

RQ (respiratory quotient), 129

Rumen, 18; functions of, 19; gas production in, 20; microorganisms in, 19, 182; papillae, 20; protein in, 19

Rumensin, 92

Ruminants, 4; effects of aflatoxin, 285; feeding treated crop residues, 112; metabolic heat in, 155; protein requirements, 4; vitamin requirements, 6

Ruminant digestive systems, 4, 14, 17; mouth, 17, stomach, 18

Rumination, 18; cattle, 18

Ruminococcus flavefaciens, 19

Rye, 33, 170, 208; for horses, 249

Ryegrass, 170

Saliva: enzymes in, 14; functions of, 15

Salmonella, 7; resistance to antibiotics, 7

Salt, 55, 59; deficiency symptoms, 59; functions of, 59; concentrations in water, 94; ration requirements, 60; sources of, 60; toxicity, 60; trace-mineralized, 55

Sampling, grain, 122

Saturated fats, 30

Scab, 282

Scours: in swine, 214; preventing, 192, 194

Scurvy, 75

Selenium, 53, 66; deficiency symptoms, 66; FDA regulations, 67; functions of, 66; poisoning, 67; toxicity, 67

Shattering, 115, 116

Sheep: antibiotics for, 222; breeding, flushing, 221; grazing, 222; intake, effects of temperature variations, 160; optimum temperature range, 160; protein deficiency, 44; self-feeding ewes, 223; water intake, 156

Sheep, lambs: colostrum, 225; mortality rate, 222; weaning, 225, 226

Sheep, rations for: ewes, 223; lactating ewes, 223; lambs, 225; market lambs, 226; orphan lambs, 227, 228; pregnant ewes, 221; rams, 224

Sheep, requirements: energy requirements, 219; mineral requirements, 220; nutritional requirements, 219; protein requirements, 220; vitamin requirements, 220; water requirements, 221

Shelled corn, 32

Shin oak, 236

Shorts, 33

Silage, 117, 165, 210; adding dry matter, 121; adding preservatives, 121; analyzing samples, 121; corn, 173; for horses, 248; quality of, 117; sample collection, 121; surplus storage, 119

Silicon, 53, 70

Silos, 118; bunker, 119; conventional, 118; gas-tight, 118; harvesting crops for, 120; horizontal, 118; loading, 119; physiological changes in, 120; stack, 119; trench, 118; unloading, 119; vertical, 118

Simple lipids, 30

Simple sugars, 28

Single-cell protein feed, 109

Single-mineral mixes, 55

Skim milk, 210

Slobber, 284. *See also* Saliva.

Small grains, 120

Small intestine, 16; duodenum, 16; ileum, 16; jejunum, 16

Sodium, 53, 59. *See also* salt.

Solar energy, use by plants, 8

Solubility, protein, 46

Sorghum grains, 144

Sorghum-sudan hybrid grass, 120

Sources of minerals, 54

Soybean meal, for horses, 249

Soybeans, 48; antitrypsin factor, 48; urease, 48

Spanish oak, 236

Stack silos, 119

Starch, 16, 30

Starches, 125

Starter, 194

Starter rations, 214

Stationery batch mixers, 147

Steam flaking, 146

Steam rolling, 146

Steamed bone meal, 165, 210

Stiffs, 55

Stilbestrol, 92

Stomach, 16, 18; abomasum, 18; gastric juice, 16; in horses, 16; omasum, 18; reticulum, 18; rumen, 18

Storage space, for pellets, 145

Straws, 34

Streptococci, 19

Streptomycin, 91

Strontium, 70

Sucrase, 17

Sudan grass, 120, 170, 246

Sugars, 27, 125; complex, 28; simple, 28

Sulfa drugs, 91

Sulfamethazine, 92

Sulphur, 53, 62; deficiency symptoms, 62; functions of, 62; in rations, 62; sources of, 62

Supplements: amounts of fluorine, 69; balancing two grains with a supplement, 138; calcium, 58; mineral, 103; protein, 134; protein, for beef cattle, 168; protein, for horses, 249; protein, for poultry, 265; vitamin, 103

Sweet potatoes, 35

Sweetclover poisoning, 82

Swine: additives, 211; effects of aflatoxin, 285; effects of heat stress, 161; farrowing, 214; flushing, 212; intake, effects of temperature variations, 161; lysine in diet, 140; optimum temperature range, 161; weaning rations, 214

Swine digestive systems, effect of starch in young pigs, 204

Swine, feeding: baby pigs, 214; boars, 211; breeding herd, 211; creep-feeding, 214; energy feeds, 206; gilts, 211; market pigs, 214, 215; sows, 212

Swine, feeds: bakery wastes, 209; barley, 208; corn, 207; energy feeds, 206; fats, 209; garbage, 146; grain sorghum, 208; greases, 209; milo, 208; molasses, 209; oats, 208; potatoes, 208; roughages, 210; rye, 208; bakery wastes, 209; barley, 208; corn, 207; fats, 209; garbage, 146;

grain sorghum, 208; greases, 209; milo, 208; molasses, 209; oats, 208; potatoes, 208; roughages, 210; rye, 208; tallow, 209; triticale, 208; wheat, 208

Swine, protein: deficiency, 45; plant protein, 209

Swine, rations: breeding rations, 212; complete rations for, 216; gestation rations, 213; lactating, rations for, 214; starter rations, 214

Swine, requirements: energy requirements, 204; mineral requirements, 205, 210; protein requirements, 205; vitamin requirements, 205, 211; water requirements, 205

Swine, water intake, effects of temperature on, 157

Synovex H, 93

Synovex S, 93

Synthetic proteins, 4

T-2 toxins, 283

Tagging, 130

Tall fescue grass, 246

Tallow, 209

TDN (Total digestible nutrients), 134, 135

Temperature: critical, 154; extremes, 157; thermoneutral zone, 154

Temperature, effects of changes: on milk yield, 158; on animal production, 157; on beef cattle, 159; on dairy calves, 158; on dairy cows, 158; on dairy heifers, 158; on feedlot cattle, 159; on forage intake, 155; on intake, 155

Temperatures of animals, 106

Temporary pasture crops, 170

Terrain, effects on nutrient requirements, 107

Terramycin, 91

Tetracycline, 7, 91; uses with cattle, 92

The Federal Register Index, 91

Thermoneutral zone, 154

Thiamin: B_1, 83; deficiency symptoms, 84; sources of, 84; structural formula, 84

Thryoxin, 93

Timothy grass, 246

Tin, 70

Titration, 124

Tocopherols, 81

Total digestible nutrients (TDN), 100, 134, 135

Toxic elements in water, 96

Toxic minerals, 70; arsenic, 70; cadmium, 71; lead, 71; mercury, 71

Toxin, 278

Trace minerals, 6, 53, 210

Trace-mineralized salt, 165

Transmethylation, 86

Trench silo, 118

Trials, 130

Trichothecenes (T-2 toxins), 283

Triticale, 208

True stomach, 20

Tryptophan, 206

Turkeys, rations: for breeding, 274; for finishing, 274; for growing, 274

Tylosin (Tylan), 91 92

Unavailable feed protein, 45

Uniform State Feed Bill, 130, 397

Unsaturated fats, 30; in swine, 32
Upper critical temperature, 154
Urea, 4, 47, 103, 169; components of, 48; need for limitation, 47; toxicity, 47
Urea, feeding: amounts to feed, 182; beef cattle, 182; goats, 232; in finishing rations, 181; palatability in feeds, 47; sheep, 220
Urea fermentation potential, 182
Urease, 48

Van Soest method, 127
Vanadium, 70
Vegetable proteins, 103
Vegetable-processing wastes, 108
Ventilation, 147
Vertical batch mixers, 147
Vertical silos, 118
Vesticular Exanthema (VE), 146
Villi, 22
Virginiamycin, 91
Vital amine, 75
Vitamin A, 76, 165; body storage, 79; deficiency symptoms, 78; excess, 79; functions of, 77; in rations, 79; requirements, 136; sources of, 78; structural formula, 77
Vitamin B$_{12}$, 87; cobalt in, 87; deficiency symptoms, 87; structural formula, 88
Vitamin C, 89; structural formula, 89
Vitamin D, 79; deficiency symptoms, 80; forms of, 79; solubility, 79; sources of, 80; toxicity, 80
Vitamin E, 81; deficiency symptoms, 81; forms of, 81; functions of, 81; isomeric tocopherols, 82; sources of, 81; structural formula of, 82; with excessive levels of nitrite, 81
Vitamin K, 82; deficiency symptoms, 82; functions of, 82; sources of, 82
Vitamin requirements: goats, 232; horses, 241, 242; poultry, 264; sheep, 220; swine, 205, 211
Vitamin supplements, 103; poultry, 266; swine, 205
Vitamins, 3, 6, 75, 136; composition of, 76; compounds in, 75; history of, 75; in rations, 76; naming of, 76; solubility, 76; water soluble, 76
Vitamins B-complex, 83. See also B-complex vitamins.
Volatile oils, 124
Vomitoxin, 278

Wastes as feed source: animal, 111; fermentation, 110; food-processing, 108; fruit-processing, 109; industrial food-processing, 107; municipal solid, 110; nonfood industrial, 109; organic chemical, 109; vegetable-processing, 108
Water, 3, 6, 75, 93, 155, 198; contamination, 95; deprivation, 96; functions of, 93; intake, 94; loss of, 95, 155, 156; measuring content in feed, 123; percent of body weight, 94; pH level, 95; sources of, 155; temperature, 94; tolerance for lack of, 157; toxicity, 96
Water, components: medications in, 95; minerals in, 94; nitrate level in, 49; pesticides in, 96; salt concentrations in, 94

Water consumption, 95; effects of air temperature, 156; effects of availability, 156
Water hyacinth, 112
Water requirements, 94; ducks, 275; geese, 275; goats, 232, 233; horses, 242; poultry, 265; sheep, 221; swine, 205
Weaning: foals, 255; lambs, 225
Weende Experiment Station, 123
Wheat, 33, 208; for horses, 249
Wheat bran, 33, 50; for horses, 249
Wheat middlings, 33
Wheat poisoning, 61
Whey, 50, 210
White brush, 236
Whole cooked soybeans, 209
Wild plum, 236
Wind chill index, 154
Windrowing, 116
Work production, 107

Xanthophylls, 264

Yaupon, 236

Zearalenone (F-2 toxin), 282
Zeranol, 93
Zinc, 53, 65, 211; deficiency symptoms, 65; functions of, 65; interrelationships, 65; sources of, 66